T0137806

Communications
in Computer and Information Science   1451

More information about this series at http://www.springer.com/series/7899

Jianchao Zeng · Pinle Qin ·
Weipeng Jing · Xianhua Song ·
Zeguang Lu (Eds.)

Data Science

7th International Conference
of Pioneering Computer Scientists,
Engineers and Educators, ICPCSEE 2021
Taiyuan, China, September 17–20, 2021
Proceedings, Part I

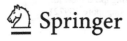 Springer

Editors
Jianchao Zeng
North University of China
Taiyuan, China

Pinle Qin
North University of China
Taiyuan, China

Weipeng Jing
Northeast Forestry University
Harbin, China

Xianhua Song
Harbin University of Science
and Technology
Harbin, China

Zeguang Lu
National Academy of Guo Ding
Institute of Data Science
Beijing, China

ISSN 1865-0929   ISSN 1865-0937 (electronic)
Communications in Computer and Information Science
ISBN 978-981-16-5939-3   ISBN 978-981-16-5940-9 (eBook)
https://doi.org/10.1007/978-981-16-5940-9

This Springer imprint is published by the registered company Springer Nature Singapore Pte Ltd.
The registered company address is: 152 Beach Road, #21-01/04 Gateway East, Singapore 189721, Singapore

Preface

As the program chairs of the 7th International Conference of Pioneer Computer Scientists, Engineers and Educators 2021 (ICPCSEE 2021, originally ICYCSEE), it is our great pleasure to welcome you to the conference proceedings. ICPCSEE 2021 was held in Taiyuan, China, during September 17–20, 2021, and hosted by the North University of China and the National Academy of Guo Ding Institute of Data Science, China. The goal of this conference series is to provide a forum for computer scientists, engineers, and educators.

This year's conference attracted 256 paper submissions. After the hard work of the Program Committee, 81 papers were accepted to appear in the conference proceedings, with an acceptance rate of 31%. The major topic of this conference series is data science. The accepted papers cover a wide range of areas related to basic theory and techniques for data science including mathematical issues in data science, computational theory for data science, big data management and applications, data quality and data preparation, evaluation and measurement in data science, data visualization, big data mining and knowledge management, infrastructure for data science, machine learning for data science, data security and privacy, applications of data science, case studies, multimedia data management and analysis, data-driven scientific research, data-driven bioinformatics, data-driven healthcare, data-driven management, data-driven eGovernment, data-driven smart city/planet, data marketing and economics, social media and recommendation systems, data-driven security, data-driven business model innovation, and social and/or organizational impacts of data science.

We would like to thank all the Program Committee members, a total of 215 people from 102 different institutes or companies, for their hard work in completing the review tasks. Their collective efforts made it possible to attain quality reviews for all the submissions within a few weeks. Their diverse expertise in each research area helped us to create an exciting program for the conference. Their comments and advice helped the authors to improve the quality of their papers and gain deeper insights.

We thank Lanlan Chang and Jane Li from Springer, whose professional assistance was invaluable in the production of the proceedings. A big thanks also to the authors and participants for their tremendous support in making the conference a success.

Besides the technical program, this year ICPCSEE offered different experiences to the participants. We hope you enjoyed the conference.

July 2021   Pinle Qin
                                                                                           Weipeng Jing

Organization

The 7th International Conference of Pioneering Computer Scientists, Engineers and Educators (http://2021.icpcsee.org) was held in Taiyuan, China, during September 17–20, 2021, and hosted by the North University of China and the National Academy of Guo Ding Institute of Data Science, China.

General Chair

Jianchao Zeng   North University of China, China

Program Chairs

Pinle Qin   North University of China, China
Weipeng Jing   Northeast Forestry University, China

Program Co-chairs

Yan Qiang   Taiyuan University of Technology, China
Yuhua Qian   Shanxi University, China
Peng Zhao   Taiyuan Normal University, China
Lihu Pan   Taiyuan University of Science and Technology, China
Alex kou   University of Victoria, Canada
Hongzhi Wang   Harbin Institute of Technology, China

Organization Chairs

Juanjuan Zhao   Taiyuan University of Technology, China
Fuyuan Cao   Shanxi University, China
Donglai Fu   North University of China, China
Xiaofang Mu   Taiyuan Normal University, China
Chang Song   Institute of Coal Chemistry, CAS, China

Organization Co-chairs

Rui Chai   North University of China, China
Yanbo Wang   North University of China, China
Haibo Yu   North University of China, China
Yi Yu   North University of China, China
Lifang Wang   North University of China, China
Hu Zhang   Shanxi University, China
Wei Wei   Shanxi University, China
Rui Zhang   Taiyuan University of Science and Technology, China

Publication Chair

Guanglu Sun   Harbin University of Science and Technology, China

Publication Co-chairs

Xianhua Song   Harbin University of Science and Technology, China
Xie Wei   Harbin University of Science and Technology, China

Forum Chairs

Haiwei Pan   Harbin Engineering University, China
Qiguang Miao   Xidian University, China
Fudong Liu   Information Engineering University, China
Feng Wang   RoarPanda Network Technology Co., Ltd., China

Oral Session and Post Chair

Xia Liu   Sanya Aviation and Tourism College, China

Competition Committee Chairs

Peng Zhao   Taiyuan Normal University, China
Xiangfei Cai   Huiying Medical Technology (Beijing) Co., Ltd.,
                               China

Registration and Financial Chairs

Chunyan Hu   National Academy of Guo Ding Institute of Data
                               Science, China
Yuanping Wang   Shanxi Jinyahui Culture Spreads Co., Ltd., China

Steering Committee Chair

Hongzhi Wang   Harbin Institute of Technology, China

Steering Committee Vice Chair

Qilong Han   Harbin Engineering University, China

Steering Committee Secretary General

Zeguang Lu   National Academy of Guo Ding Institute
                               of Data Science, China

Steering Committee Vice Secretary General

Xiaoou Ding   Harbin Institute of Technology, China

Steering Committee Secretaries

Dan Lu   Harbin Engineering University, China
Zhongchan Sun   National Academy of Guo Ding Institute
                               of Data Science, China

Steering Committee

Xiaoju Dong   Shanghai Jiao Tong University, China
Lan Huang   Jilin University, China
Ying Jiang   Kunming University of Science and Technology, China
Weipeng Jing   Northeast Forestry University, China
Min Li   Central South University, China
Junyu Lin   Institute of Information Engineering, CAS, China
Xia Liu   Hainan Province Computer Federation, China
Rui Mao   Shenzhen University, China
Qiguang Miao   Xidian University, China
Haiwei Pan   Harbin Engineering University, China
Pinle Qin   North University of China, China
Xianhua Song   Harbin University of Science and Technology, China
Guanglu Sun   Harbin University of Science and Technology, China
Jin Tang   Anhui University, China
Ning Wang   Xiamen Huaxia University, China
Xin Wang   Tianjin University, China
Yan Wang   Zhengzhou University of Technology, China
Yang Wang   Southwest Petroleum University, China
Shengke Wang   Ocean University of China, China
Yun Wu   Guizhou University, China
Liang Xiao   Nanjing University of Science and Technology, China
Junchang Xin   Northeastern University, China
Zichen Xu   Nanchang University, China
Xiaohui Yang   Hebei University, China
Chen Ye   Hangzhou Dianzi University, China
Canlong Zhang   Guangxi Normal University, China
Zhichang Zhang   Northwest Normal University, China
Yuanyuan Zhu   Wuhan University, China

Program Committee

Witold Abramowicz   Poznan University of Economics and Business, Poland
Chunyu Ai   University of South Carolina Upstate, USA
Jiyao An   Hunan University, China

Yutai Hou	Harbin Institute of Technology, China
Wei Hu	Nanjing University, China
Xu Hu	Xidian University, China
Lan Huang	Jilin University, China
Hao Huang	Wuhan University, China
Kuan Huang	Utah State University, USA
Hekai Huang	Harbin Institute of Technology, China
Cun Ji	Shandong Normal University, China
Feng Jiang	Harbin Institute of Technology, China
Bin Jiang	Hunan University, China
Xiaoyan Jiang	Shanghai University of Engineering Science, China
Wanchun Jiang	Central South University, China
Cheqing Jin	East China Normal University, China
Xin Jin	Beijing Electronic Science and Technology Institute, China
Chao Jing	Guilin University of Technology, China
Hanjiang Lai	Sun Yat-sen University, China
Shiyong Lan	Sichuan University, China
Wei Lan	Guangxi University, China
Hui Li	Xidian University, China
Zhixu Li	Soochow University, China
Mingzhao Li	RMIT University, Australia
Peng Li	Shaanxi Normal University, China
Jianjun Li	Huazhong University of Science and Technology, China
Xiaofeng Li	Sichuan University, China
Zheng Li	Sichuan University, China
Mohan Li	Jinan University, China
Min Li	South University, China
Zhixun Li	Nanchang University, China
Hua Li	Changchun University of Science and Technology, China
Rong-Hua Li	Shenzhen University, China
Cuiping Li	Renmin University of China, China
Qiong Li	Harbin Institute of Technology, China
Qingliang Li	Changchun University of Science and Technology, China
Wei Li	Georgia State University, USA
Yunan Li	Xidian University, China
Hongdong Li	Central South University, China
Xiangtao Li	Northeast Normal University, China
Xuwei Li	Sichuan University, China
Yanli Liu	Sichuan University, China
Hailong Liu	Northwestern Polytechnical University, China
Guanfeng Liu	Macquarie University, Australia
Yan Liu	Harbin Institute of Technology, China

Xia Liu	Sanya Aviation Tourism College, China
Yarong Liu	Guilin University of Technology, China
Shuaiqi Liu	Tianjin Normal University, China
Jin Liu	Central South University, China
Yijia Liu	Harbin Institute of Technology, China
Zeming Liu	Harbin Institute of Technology China
Zeguang Lu	National Academy of Guo Ding Institute of Data Sciences, China
Binbin Lu	Sichuan University, China
Junling Lu	Shaanxi Normal University, China
Mingming Lu	Central South University, China
Jizhou Luo	Harbin Institute of Technology, China
Junwei Luo	Henan Polytechnic University, China
Zhiqiang Ma	Inner Mongolia University of Technology, China
Chenggang Mi	Northwestern Polytechnical University, China
Tiezheng Nie	Northeastern University, China
Haiwei Pan	Harbin Engineering University, China
Jialiang Peng	Norwegian University of Science and Technology, Norway
Fei Peng	Hunan University, China
Yuwei Peng	Wuhan University, China
Jianzhong Qi	The University of Melbourne, Australia
Xiangda Qi	Xidian University, China
Shaojie Qiao	Southwest Jiaotong University, China
Libo Qin	Research Center for Social Computing and Information Retrieval, China
Zhe Quan	Hunan University, China
Chang Ruan	Central South University of Sciences, China
Yingxia Shao	Peking University, China
Yingshan Shen	South China Normal University, China
Meng Shen	Xidian University, China
Feng Shi	Central South University, China
Yuanyuan Shi	Xi'an University of Electronic Science and Technology, China
Xiaoming Shi	Harbin Institute of Technology, China
Wei Song	North China University of Technology, China
Shoubao Su	Jinling Institute of Technology, China
Yanan Sun	Oklahoma State University, USA
Minghui Sun	Jilin University, China
Guanghua Tan	Hunan University, China
Dechuan Teng	Harbin Institute of Technology, China
Yongxin Tong	Beihang University, China
Xifeng Tong	Northeast Petroleum University, China
Vicenc Torra	University of Skövde, Sweden
Hongzhi Wang	Harbin Institute of Technology, China
Yingjie Wang	Yantai University, China

Dong Wang	Hunan University, China
Yongheng Wang	Hunan University, China
Chunnan Wang	Harbin Institute of Technology, China
Jinbao Wang	Harbin Institute of Technology, China
Xin Wang	Tianjin University, China
Peng Wang	Fudan University, China
Chaokun Wang	Tsinghua University, China
Xiaoling Wang	East China Normal University, China
Jiapeng Wang	Harbin Huade University, China
Qingshan Wang	Hefei University of Technology, China
Wenfeng Wang	CAS, China
Shaolei Wang	Harbin Institute of Technology, China
Yaqing Wang	Xidian University, China
Yuxuan Wang	Harbin Institute of Technology, China
Wei Wei	Xi'an Jiaotong University, China
Haoyang Wen	Harbin Institute of Technology, China
Huayu Wu	Institute for Infocomm Research, Singapore
Yan Wu	Changchun University of Science and Technology, China
Huaming Wu	Tianjin University, China
Bin Wu	Institute of Information Engineering, CAS, China
Yue Wu	Xidian University, China
Min Xian	Utah State University, USA
Sheng Xiao	Hunan University, China
Wentian Xin	Xidian University, China
Ying Xu	Hunan University, China
Jing Xu	Changchun University of Science and Technology, China
Jianqiu Xu	Nanjing University of Aeronautics and Astronautics, China
Qingzheng Xu	National University of Defense Technology, China
Yang Xu	Harbin Institute of Technology, China
Yaohong Xue	Changchun University of Science and Technology, China
Mingyuan Yan	University of North Georgia, USA
Yu Yan	Harbin Institute of Technology, China
Cheng Yan	Central South University, China
Yajun Yang	Tianjin University, China
Gaobo Yang	Hunan University, China
Lei Yang	Heilongjiang University, China
Ning Yang	Sichuan University, China
Xiaochun Yang	Northeastern University, China
Shiqin Yang	Xidian University, China
Bin Yao	Shanghai Jiao Tong University, China
Yuxin Ye	Jilin University, China
Xiufen Ye	Harbin Engineering University, China

Minghao Yin	Northeast Normal University, China
Dan Yin	Harbin Engineering University, China
Zhou Yong	China University of Mining and Technology, China
Jinguo You	Kunming University of Science and Technology, China
Xiaoyi Yu	Peking University, China
Ye Yuan	Northeastern University, China
Kun Yue	Yunnan University, China
Yue Yue	SUTD, Singapore
Xiaowang Zhang	Tianjin University, China
Lichen Zhang	Shaanxi Normal University, China
Yingtao Zhang	Harbin Institute of Technology, China
Yu Zhang	Harbin Institute of Technology, China
Wenjie Zhang	University of New South Wales, Australia
Dongxiang Zhang	University of Electronic Science and Technology of China, China
Xiao Zhang	Renmin University of China, China
Kejia Zhang	Harbin Engineering University, China
Yonggang Zhang	Jilin University, China
Huijie Zhang	Northeast Normal University, China
Boyu Zhang	Utah State University, USA
Jin Zhang	Beijing Normal University, China
Dejun Zhang	China University of Geosciences, China
Zhifei Zhang	Tongji University, China
Shigeng Zhang	Central South University, China
Mengyi Zhang	Harbin Institute of Technology, China
Yongqing Zhang	Chengdu University of Information Technology, China
Xiangxi Zhang	Harbin Institute of Technology, China
Meiyang Zhang	Southwest University, China
Zhen Zhang	Xidian University, China
Jian Zhao	Changchun University, China
Qijun Zhao	Sichuan University, China
Bihai Zhao	Changsha University, China
Xiaohui Zhao	University of Canberra, Australia
Peipei Zhao	Xidian University, China
Bo Zheng	Harbin Institute of Technology, China
Jiancheng Zhong	Hunan Normal University, China
Jiancheng Zhong	Central South University, China
Fucai Zhou	Northeastern University, China
Changjian Zhou	Northeast Agricultural University, China
Min Zhu	Sichuan University, China
Yuanyuan Zhu	Wuhan University, China
Yungang Zhu	Jilin University, China
Bing Zhu	Central South University, China
Wangmeng Zuo	Harbin Institute of Technology, China

Contents – Part I

Basic Theory and Techniques for Data Science

Machine Learning for Data Science

Multimedia Data Management and Analysis

Contents – Part II

Applications of Data Science

**Education Research, Methods and Materials for Data Science
and Engineering**

Research Demo

Big Data Management and Applications

A Blockchain-Based Scheme of Data Sharing for Housing Provident Fund

Yang Song[1], Jiawen Wang[2], Shili Yang[1], Xiaojun Zhu[3], and Keting Yin[3(✉)]

[1] Chongqing Municipal Housing Provident Fund Administration Center, Chongqing, China
[2] Hangzhou Echaincity Technology Co., Ltd., Hangzhou, China
[3] School of Software Technology, Zhejiang University, Hangzhou, China
yinkt@zju.edu.cn

Abstract. This paper proposes a secure, reliable and collaborative data-sharing system for China's housing provident fund based on blockchain. Firstly, federal computing was introduced to realize "available but invisible" sharing of data about housing provident fund, which reduces the data leakage risk and improves the data availability. Secondly, four data sharing modes were proposed to deal with different situations with different amount of data provider and data. Lastly, to realize individual data deep sharing on the premise of security, an enterprise and personal information query authorization mechanism was established to provide solutions to personal and institutional authorization. This system helps to realize both the internal and external data sharing of the housing provident fund system under the premise of security and privacy protection. This system improves the efficiency of housing provident fund issue, and fully taps the value of data comprehensively.

Keywords: Blockchain · Housing provident fund · Data sharing · Literature study · Case study

1 Introduction

As a kind of social, mutual aid and policy-oriented housing social security system, HPF (housing provident fund) is attached to great importance by governments at all levels and relevant functional departments. Therefore, HPF policy is rapidly implemented at an inspiring speed. In 2019, the number of employees benefited from the HPF system continued to grow, with 3.224 million paid in units and 148.8138 million paid in employees, showing an increase of 10.57% and 3.08% respectively over the previous year. Xiaoqing Zhou [1] analyses that HPF policy is expected to raise the rate of homeownership by 8.7 percentage points and to increase the average home size by 20%. With the development of HPF, its data can be used in all aspects of people's daily lives such as applying for a loan. However, due to the existing data management regulations of vertical government systems at all levels, it is still difficult to break through the data barriers of various vertical systems and business fields. It is also difficult to carry out deeper sharing and collaboration of HPF data, which causes much inconvenience to both residents and government workers. By introducing a data-sharing system based on blockchain, this paper

© Springer Nature Singapore Pte Ltd. 2021
J. Zeng et al. (Eds.): ICPCSEE 2021, CCIS 1451, pp. 3–14, 2021.
https://doi.org/10.1007/978-981-16-5940-9_1

aims to solve the problems encountered in current HPF data system. Overall, the main contribution of this paper can be summarized as follows:

- First, we apply blockchain into HPF system, which simplifies the data sharing, also makes it more convenient and secure.
- Second, we introduce four data sharing modes fitting to different usage scenarios, which meets the data sharing need between various departments.
- Third, federal computing is used in this system to ensure the security and collaboration of data sharing.

The rest of the paper is organized as follows. In Sect. 2, we discuss the development and existing problems in China's current HPF system. Meanwhile, we illustrate the advantages of adopting blockchain. In Sect. 3, related works on data sharing with blockchain and traditional sharing strategy of HPF data are introduced. In Sect. 4, we give an overview of our system design, from business design to system architecture. In Sect. 5, the specific application of this system is given. We discuss functional models including data sharing, storing, exchange, authentication and federal computing. In Sect. 6, the performance of this system is illustrated. Finally, we conclude with a discussion in Sect. 7.

2   Motivation

With the development of economy, the mobility of people's work has greatly increased, and there are still information barriers in the existing provident fund center systems in different cities, leading to difficulties in cross regional provident fund loans and other business. Residents need to run back and forth in different cities to handle simple affairs, causing a waste of human and material resources.

HPF is such complicating business that the data of public security, civil affairs and other external departments are also involved in the system. However, the existing administrative framework cannot even carry out direct data interface call in the process of external investigation in different places, which is normal in the HPF business. It leads to the limitation that HPF centers in different places can only exchange their own business information at present, and the deeper information exchange is difficult to carry out.

The data is not only difficult to circulate in HPF centers in different places, but also isolated among different functional organizations such as State Administration for Industry and Commerce. The HPF [2] requires that both employee and employer contribute a certain part of the employee's monthly wages to the program; the savings belong to the employee and are deposited into his/her HPF account for the individual's future housing consumption. The contribution from the employer can be treated as an additional subsidy for the employee, which employers of enterprises choose whether to pay or not. By 2003, almost all sectors have been mandated to contribute to HPFs. However, even today it has not been taken up by the less formal sectors of the Chinese economy. [3] In order to obtain high profits, some enterprises refuse to pay the provident fund for their employees on the grounds of poor income, which needs to be supervised and managed.

However, the provident fund management center can't get access to the data from the Administration for Industry and Commerce and the tax bureau to judge the business situation of the enterprise leading to the situation that it can't take management measures for the enterprise's behavior of evading to pay the provident fund.

In order to solve the problems listed above, we build a credible data sharing platform based on blockchain technology, realizing cross-department, cross-industry and cross-regional credible data sharing. It also helps with the integration of government data.

3 Related Works

In recent years, many scholars have studied data-sharing based on blockchain. J. Yang [4] observes that blockchain is expected to become the new engine for building a future data sharing platform because of its information-sharing characteristics and decentralized management principles. Zhang, Jing [5] finds the technical features of blockchain, including decentralization, data transparency, tamper-proofing, traceability, privacy protection and open-sourcing, make it a suitable technology for solving the information asymmetry problem. Kim, Keonhyeong [6] demonstrates that a blockchain-based information-sharing that verifies the shared data and the sharing process using a public blockchain can protect the privacy of the shared information. In the case of traceability, Hui HUANG [7] suggests consortium blockchain technique being used so that any user from different groups in the system can easily verify the validity of the shared data without interacting with a third-party auditor. In the medical field, Sihua Wu [8] brings up a safe and efficient electronic medical record sharing model based on blockchain with data masking technology and Inter Planetary File System (IPFS), which can not only guarantee the security of medical data, but also save resources in blockchain. In the field of knowledge, Shuang Hu [9] proposes a Reputation Based Knowledge Sharing system in blockchain to exploit the copyright protection of the knowledge owner using proposed fine-grained access control system. Mu Yang [10] proposes a blockchain-based approach that enables data owners to control the anonymization process and that enhances the security of the services.

As an emerging development mode, "blockchain + Digital Government" has strong vitality [11]. And due to the growing demand of applying blockchain to the field of people's livelihood [12] and the excellent performance of blockchain in the field of data sharing, some policies and scholars are optimistic about its application in the field of HPF. Li Lizhen [13] believes that blockchain can help complete data collection, transmission, storage and unified backup of each HPF center to ensure data security. From the perspective of data sharing, blockchain is considered to be the most widely used in HPF data [14]. With the help of blockchain, the data sharing platform of HPF is expected to solve the problem of scattered data islands in various regions, realize the sharing of national HPF data, and make the organization, supervision and protection of HPF data more standardized [15].

Regional governments are also taking actions. Some of them have already been working on data-sharing in HPF system. Chengdu and Chongqing have reached a consensus on establishing working mechanism to promote information sharing and deepen mutual loan. It will also help in linkage governance of HPF arrears, arbitrage, fraud and dishonesty, as well as financing between the two places. Cross regional transfer and mutual recognition and loan mechanism have been initially established, so that employees can use HPF loans more conveniently. Thanks to this collaboration, the transfer of HPF between Chengdu and Chongqing can achieve a complete service for the same location instead of back and forth in the past. The processing time has been reduced from one month to 2–3 working days, and the processing requirements have been simplified to one table.

The system above mainly focuses on the help of data-sharing inside provident fund system for the convenience of business. But provident fund data is also valuable for some external departments such as tax and social security. From the perspective of external service of provident fund data, personal credit analysis with provident fund business data has high credibility and convenience. It can also form the overall credit evaluation of enterprises by integrating the historical payment of provident fund. These credit models are of great significance to financial institutions, personal loan business and enterprise credit. But in the premise of unable to ensure personal privacy protection, the relevant provident fund business data cannot be fully used.

Based on our works, to make a better provident fund data sharing, we concentrate on solving the existing inconvenience and inefficiency by proposing a blockchain based solution.

4   System Design Overview

The system architecture of this system is divided into five layers (see Fig. 1. System structure diagram).

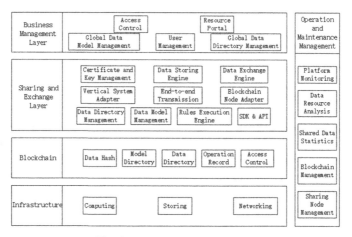

Fig. 1. System structure diagram

1. *Infrastructure layer.* The Infrastructure Layer is the foundation providing the underlying physical deployment environment support. The first layer contains basic computing, storing and networking devices.

2. *Blockchain layer.* Based on the underlying platform, nodes of blockchain are deployed locally or remotely in each business participant. The second layer contains data directory, model directory, business rules, access control rules, etc. defined through smart contract.
3. *Sharing and exchange layer.* Based on the trusted data grid, the data sharing node is deployed locally in each business participant, and is also used to connect the internal vertical pipe system and the blockchain layer. The main functions of the data sharing node in this layer include: managing the local data directory and data calculation model of the vertical pipe system; managing the local certificate and key of the data sharing node. It also works as an adapter between data sharing node and vertical pipe system local data source, technical components for data sharing and exchange logic, including data exchange engine, data storage engine, point-to-point transmission component, data calculation rule engine. In addition, in order to be compatible with different blockchain underlying technology platforms and vertical pipe systems, blockchain node adapter, SDK and API interfaces for interaction are also included in this layer.
4. *Business management layer.* This layer implements some business management functions of the main data sharing platform, such as global data directory management, data model management, resource portal, user and permission control, etc.
5. *Operation and maintenance management.* This layer provides platform monitoring, data resource analysis and shared data statistics to platform operation and maintenance manager. For the underlying data platform, it provides the management function of blockchain management and data sharing node.

5 The Specific Program

5.1 Data Sharing

Sharing Mode. The amount of data provider and data vary in realistic scenarios. To deal with different situations, this system proposes four data sharing modes.

- *single data provider and low privacy requirement*

In the case of single data provider and low privacy requirement, a simple visible sharing mode is adopted. The data requester applies for the authorization of the data provider through the blockchain. After obtaining the data access certification, the data provider encrypts the original data and transmits it to the requester point-to-point through the data sharing network. For instance, the latest list of dishonest persons containing personal details such as the ID number can be passed from the court to the Chongqing provident fund center in this model.

- *single data provider with high privacy requirement*

In the case of single data provider and high privacy requirement, an invisible sharing mode is adopted. The data requester is informed of the information of data used in model calculation through the blockchain. After obtaining the authorization, the data requester transmits the calculation model and parameters to the data provider for the calculation. Then the data provider transmits the result of calculation after it verifies that the model and calculation results meet the requirements. In this mode, there is no need for heterogeneous data from different participants to be delivered, migrated, integrated and cleaned. It can be used in normal practical scenarios in the field of HPF because of its high collaboration and security. For instance, if a bank requests Chongqing provident fund center to obtain the floating rate of the provident fund deposits amount of a loans enterprise in the latest period to help judge its business growth, Chongqing provident fund center can use this mode to transmit the result instead of the original data.

- *multi data providers with high privacy requirement and small amount of data*

In the case of multiple data providers with high privacy requirements and small amount of data, the sharing mode with federal computing based on TEE (Trusted Execution Environment) is adopted. The data and model of each node are encrypted externally and transmitted to the TEE secure isolated computing intermediary which can be trusted by multiple parties through the encrypted channel. Only in TEE can the data be decrypted and calculated while the external world cannot get to know the exchanged data. When the return value is returned, it is encrypted again to prevent the leakage of data and model algorithm. For example, this mode can be applied in the verification of some suspected forged unit income certificate or wage bank income flow for loans apply if Chongqing provident fund center or a bank need.

- *multi data providers with high privacy requirement and large amount of data*

In the case of multiple data providers with high privacy requirements and large amount of data, the sharing mode with federal computing based on secure multi-party computing (MPC) is adopted. In the environment of multiple participants, secure multi-party computing can realize that each participant has its own private information while it can also use other people's information to complete the calculation process together securely. Taking the privacy secure aggregation (PSA) implemented by MPC as an example, the real data of multiple data providers are obfuscated based on secret sharing, and the aggregation calculation is carried out on the premise of data security. After the calculation, the obfuscation value is offset to get the accurate result, and the data requester can only get the final aggregation result. For example, the situation that Chongqing provident fund center needs data fusion with tax and social security units for extended analysis can be applied to this mode.

Data Type. In addition to structured data such as employee basic information, provident fund data also includes unstructured data such as pictures. This system designs different data sharing methods for these two types of data.

Structured Data. Enterprise information, employee information and government administrative approval in provident fund are structured data that can be logically expressed and realized by two-dimensional table structure.

- *data model*

Data model sharing is based on the data model that can be broadcasted by data sharing node. Data sharing node undertakes the work of data opening, access control, privacy protection, etc., so as to avoid the complicating data exchange burden to the upper application system. In the process of data sharing, the data provider directly provides the data model to the data user. The data user defines the data model according to the data source of the data provider and obtains the approval of the data provider. After the user initiates a data query request through the data model, the data provider verifies the authorization information on the blockchain and calls the corresponding model to calculate the original data. Then the calculation results are transmitted through the point-to-point network. The call and calculation results are both hashed and stored on the chain as the certification.

- *interface service*

Data can also be shared with data user through data service interface. Data provider needs to define data service interface information, including service interface address, input parameter, output parameter, interface description and so on. After the interface definition is completed, the data provider can publish the service interface on the blockchain, and the interface in and out reference description information will be synchronized to the blockchain as index information. The data provider can set the visibility of the service interface to make the service interface visible to all data sharing nodes in the chain or only to the data sharing nodes in the specified range. In the process of data sharing, the data user initiates the interface call. After verifying the authorization information on the blockchain, the data provider completes the interface call, passes the call result to the user. The call and calculation results are both hashed and stored on the chain as the certification.

Unstructured Data. In the data of provident fund, there are some data that are inconvenient to be represented by two-dimensional logical tables of database, namely unstructured data, including office documents, pictures, images and audio/video information, etc.

- *file sharing*

Unstructured data is mainly exchanged by file sharing. For those small files with frequent updating and sharing, they can be uploaded to the data sharing node. File modification records, historical versions and updating file data fingerprints in real time can be better saved with the help of distributed storage. And it can also carry out file disaster recovery through distributed storage. For large files, there is no need to upload the files to the data sharing network, instead, you can directly connect to the big data file storage center with quick access and system storage cost reduction.

The file data source is stored on the blockchain with the index information provided by the data provider such as data hash, data source name, data source description and data source display permission, which can be viewed by the data user with permission. The data display permission can be modified by the data provider in real time, and the granularity can be a single file. If the file in the directory is not set separately, the display permission will be based on the file directory permission. After the data user obtains the file data. The unique hash generated according to the content of the file data will be compared with the file hash recorded on the blockchain to determine whether the file data has been tampered.

5.2 Data Storing

On-Chain. The blockchain mainly stores metadata of management and hash of shared data, instead of specific shared data. The information stored on the blockchain includes the identity and authentication information of participants, file directory, model directory and data authorization information.

Off-Chain. The main purpose of off-chain storage is to expand the storage capacity of blockchain and protect data privacy. The off-chain storage stores large files in the structure of the Merkel tree which can ensure the correspondence between the data on and off chain by the way of Merkel tree root and original information on the chain. At the same time, a secure and reliable storage proof mechanism is constructed through cryptography algorithm.

This storage structure and proof mechanism realizes the modification detection, transmission traceability, and cross verification of data on and off the chain, improves the storage and management ability of the whole blockchain, also achieves seamless collaboration of data on and off the chain.

5.3 Data Exchange

After the data requester successfully obtains authorization from the data provider, the data provider will return a token to the data requester. Every time the authorization is successful, a new token will be generated. Each token can specify the validity period during which the token can be used all the time. The data requester initiates a data request with the successfully authorized token. After the data provider checks that the token is correct, it will return the data requester according to the data storage situation. Structured data will be returned directly through point-to-point connection, while unstructured data may be returned as file or file module download information.

P2P. In the process of data exchange, end-to-end encryption is carried out. After the node establishes the connection, asymmetric encryption is used to verify the identity of the node, and then symmetric encryption is used in communication to ensure the security and efficiency of communication.

First, the node of the data requester generates the symmetric key for the subsequent communication, and then encrypts the symmetric key with the public key of the data provider, and sends the encrypted information together with the hash of the symmetric key to the data provider. After the data provider obtains the symmetric key decrypted with its own private key and checks whether the hash of the symmetric key is the same as that of the sent one. If the hash is the same, it indicates that the decryption is successful and subsequent communication can be carried out; if not, the communication stops; the data provider then encrypts the symmetric key with the public key of the data requester and returns it to the data requester. After receiving the data, the requester decrypts with the private key to verify whether the symmetric key is the same. If the symmetric key is the same, it means that a secure connection has been established, and the subsequent communication can be carried out based on the symmetric key. Through the point-to-point communication, the data leakage between nodes is prevented, and the data is protected by directional encryption to prevent the third party from parsing.

Access Control. Data exchange requires accurate authorization by the data owner, and the authorization records of data management should be recorded and traceable. Through the smart contract of the blockchain, we can flexibly formulate the business-related data sharing access control strategy to achieve fine-grained access control and traceback of call records.

There are two types of authorization management: directory authorization and request authorization. Directory authorization refers to the operation permission setting of data according to the directory structure. Through the smart contract and access control policy, the participants and the shared application accounts can be authorized separately or in batches. Only the users within the directory can obtain the corresponding data. Request authorization means that users or participants who have not been authorized send authorization requests to the data provider when they need to access the data. If they pass the audit, they can obtain the access token. Authorization application, audit results, authority update and other records need to be recorded on the blockchain to provide reliable basis for later data traceability and audit.

Data List. To facilitate the data demander to locate the required data quickly and accurately, it is necessary to establish a shared data list display service, which will display the meta information of the shared data within the blockchain and provide the following functions:

1. Resource classification. The published data can be classified into interface, model and data source, and be searched respectively;
2. Resource retrieval. Fuzzy search can be carried out for the title, type and description of the published data, and the search results can be filtered and sorted by time, data type and other dimensions;
3. Resource Recommendation. According to the release time, request volume, browsing records and other information, the relevant resources are reasonably recommended for the data demander.

Analysis. In order to help data providers and data demanders to make statistical summary and detailed analysis on the usage of published data and acquired data, this system establishes a statistical analysis display service. For data providers, it can count the number of access applications and request calls of all data published by participants, and further analyze on this basis. For the data demander, the obtained data information and model calculation results can be counted.

5.4 Authorization

On the business data sharing that is not established by the superior department or covered by the data exchange platform, the information management of some vertical pipe system may not be open to the third party because of the protection of the privacy right of enterprise and personal information in the process of data sharing, which is an obstacle in the way of individual data deep sharing. To solve this problem, an enterprise and personal information query authorization mechanism is established in the data collaboration alliance chain. That is to say, an independent third-party query authorization registration platform is established on the blockchain, on which enterprises or individuals can register information, query the effective authorization certificate information, data requester and data provider, and store the authorization certificate on the chain for future reference and pursue the responsibility for possible forgery.

5.5 Federal Computing

Federal computing is a kind of distributed computing that combines multi-party data through specific rules to achieve privacy protection of data aggregation. In the scenario where there are multiple data providers and data privacy needs to be protected, the data "available but invisible" can be realized based on federal computing. The original data of data providers is not sent out of the organization, but only parameters with encryption mechanism are exchanged among the providers. A virtual computing model is built without violating the data privacy protection regulations, in which the data itself does not move, only the data calculation results are shared and transferred among providers.

6 Case Study

In this section, two use cases are introduced to illustrate the practical application of the system.

6.1 Sichuan & Chongqing HPF System Integration

Due to factors like geography, culture and central government policies, there are frequent economic, trade and population exchanges between Sichuan and Chongqing. At present, in terms of provident fund, Chengdu and Chongqing are both the most frequent places for the other party to handle business. There used to be data barriers in business systems of provident fund centers, which lead to many difficulties in provident fund loans and other businesses in these two well-developing cities. Employees needed to run back and forth, wasting a lot of human and material resources. Relying on the data sharing system of Sichuan & Chongqing HPF, Chongqing and Sichuan realize paperless deposit certificates in different places, simplifying the procedures for handling loans in different places. Employees applying for provident fund loans in different places have changed from "running in two places" to "running in one place". Without the data obstacles between two places, employees can fully enjoy more convenient intelligent processing services of provident fund. However, this system is lack of data circulation between different institutions due to traditional technologies used. This deficiency urges this system to try to integrate the system promoted in this paper.

6.2 Extension of HPF Policy and Management in Chongqing

In order to effectively expand the influence of the provident fund system and improve the coverage of the HPF system, Chongqing spent a lot of manpower and material resources to carry out thorough investigation work. However, due to the lack of available data resources, the investigation efficiency is low. Based on this system, the HPF center can get access to the data from industrial and commercial bureau and tax bureau. Those data can reflect the current business situation and subsequent development trend of enterprises helping to predict, analyze and evaluate the expansion feasibility of enterprises that have not deposited provident fund. A normal analysis and evaluation mechanism can be built to quickly find and identify benign business enterprises, establishing accurate image and evaluation of the business situation of enterprises that have not deposited provident fund Evaluation. This system helps to achieve the purpose of efficient and accurate expansion.

7   Conclusion and Future Work

In this paper, we propose a data-sharing system for HPF based on blockchain taking advantages of blockchain to solve two main problems: the difficulties of data sharing between HPF centers in different places, data isolation among different functional organizations and HPF centers. At the meantime of sharing available data, we put up federal computing and access control to protect the security and privacy.

Nevertheless, there are still some problems. There is so much data in HPF system that may cause the low efficiency of storage and query. Besides, the departments involved in the data sharing of provident fund are complex, which also involve the interests of different parties. Policies are still needed to support the transformation of data sharing.

In the future research, we may consider optimizing data storage strategy and the performance of blockchain in the case of large amount of data. And we will follow up relevant policies and adjust some details to meet different needs.

Acknowledgment. This research was supported by the National Key R&D Program of China No. 2019YFB1404903.

References

1. Zhou, X.: A quantitative evaluation of the housing provident fund program in China. China Econ. Rev. **61**(3), 101436 (2020)
2. Chen, M., Wu, Y., Liu, G., Wang, X.: The effect of the housing provident fund on income redistribution: the case of China. Hous. Policy Debate **30**(6), 879–899 (2020)
3. Tang, M., Coulson, N.: The impact of China's housing provident fund on homeownership, housing consumption and housing investment. Region. Sci. Urban Econ. **63**, 25–37 (2016)
4. Yang, J., Wen, J., Jiang, B., Wang, H.: Blockchain-based sharing and tamper-proof framework of big data networking. IEEE Netw. **34**(4), 62–67 (2020)
5. Zhang, J., Tan, R., Li, Y.-D.: Design of personal credit information sharing platform based on consortium blockchain. In: Xu, G., Liang, K., Su, C. (eds.) FCS 2020. CCIS, vol. 1286, pp. 166–177. Springer, Singapore (2020). https://doi.org/10.1007/978-981-15-9739-8_14
6. Kim, K., Kim, T., Jung, I.: Blockchain-based Information sharing between smart vehicles for safe driving. In: 2020 IEEE 91st Vehicular Technology Conference (VTC2020-Spring). IEEE (2020)
7. Huang, H., Chen, X., Wang, J.: Blockchain-based multiple groups data sharing with anonymity and traceability. Sci. China Inf. Sci. **63**(3), 1–13 (2019)
8. Wu, S., Du, J.: Electronic medical record security sharing model based on blockchain. In: Proceedings of the 3rd International Conference on Cryptography, Security and Privacy (ICCSP 2019), pp. 13–17. Association for Computing Machinery, New York, NY, USA (2019)
9. Hu, S., Hou, L., Chen, G., Weng, J., Li, J.: Reputation-based Distributed Knowledge Sharing System in Blockchain. In: Proceedings of the 15th EAI International Conference on Mobile and Ubiquitous Systems: Computing, Networking and Services (MobiQuitous 2018), pp. 476–481. Association for Computing Machinery, New York, NY, USA (2018)
10. Yang, M., Margheri, A., Hu, R., Sassone, V.: Differentially private data sharing in a cloud federation with blockchain. IEEE Cloud Comput. **5**(6), 69–79 (2018)
11. Ning, S.: Exploration and application of blockchain technology in Qingdao Digital Government. Party School of Qingdao municipal Party committee. J. Qingdao Admin. College **03**, 115–119 (2020)
12. Chen, G.: Blockchain should serve the people better. China Inf. Ind. **01**, 26–29 (2020)
13. Li, L., Jiang, Y., Wang, W., Dai, P.: Research on information sharing construction in standardized management of housing provident fund. Coast. Enterp. Technol. **01**, 63–67 (2020)
14. Huang, X.: "Blockchain + government service" enables local governance. Decis. Making **04**, 32–34 (2020)
15. Liu, G., Xing, L., Qiu, W.: Analysis on the application of blockchain technology in banking industry. Banker **09**, 110–113 (2020)

Social Media and Recommendation Systems

Item Recommendation Based on Monotonous Behavior Chains

Tiepai Yang$^{(\boxtimes)}$

Heilongjiang University, Harbin 150080, Heilongjiang, China

Abstract. In the field of e-commerce, recommendation systems can accurately provide users with products and services of potential interest, thereby enhancing users' online shopping experience. "Explicit" feedback and "implicit" feedback are mostly studied in two relatively independent research fields. In the actual interaction process between users and commodities, there is a kind of signal between the two Monotonic dependence, that is, sparse and reliable explicit signals must imply dense and noisy implicit signals. In this paper, a special "monotonic behavior chain" structure is proposed to constrain the two signals, and a series of user-commodity interaction behaviors is mapped into a user-commodity multi-stage binary interaction diagram. The two feedback signals were combined and the complete interaction was simulated between the user and the product. Then a depth model framework GAERE was proposed based on the graph auto-encoder, which converts the matrix completion problem of the traditional recommendation system into the problem of graph link prediction. Four realistic data sets were applied to evaluate the effectiveness of the proposed method. The model shows competitiveness on standard collaborative filtering benchmarks. In addition, the application of graph convolutional network was further explored to process graph structure data in recommendation system from the perspective of user behavior intention.

Keywords: Recommendation system · Graph convolution model · Matrix completion

1 Introduction

Many e-commerce companies hope to use the recommendation system to analyze the actual needs or potential needs of users based on the historical interaction records of users and products, and provide users with accurate and personalized recommendation results. Collaborative filtering (CF) has become the most successful case among many recommendation algorithms due to its high efficiency and refinement characteristics [1]. Collaborative filtering models user preferences by tracking different types of historical interaction records of users, such as user click records, viewing time, and purchase records, and predicts the probability or rating of possible interactions between users and commodities in the future. The user's feedback in historical interaction records usually includes: 1) "Explicit feedback" that can show the user's clear preference information, for example: in the movie scoring system, the user's star rating for the movie. But

© Springer Nature Singapore Pte Ltd. 2021
J. Zeng et al. (Eds.): ICPCSEE 2021, CCIS 1451, pp. 17–31, 2021.
https://doi.org/10.1007/978-981-16-5940-9_2

in real life, most applications cannot provide a large amount of effective rating data. 2) Compared with "explicit feedback", "implicit feedback" is easier to collect and has more information sources. These implicit feedbacks refer to the indirect interest preferences expressed behind observable user actions, such as purchase history, browsing history, etc. At present, due to the differences in the representation methods and data distribution of the two interactive signals in the model, "explicit" and "implicit" feedback are mostly studied in two independent topics in the recommendation system.

1.1 User-Commodity Interaction Monotonous Behavior Chain

In the complete interaction between users and commodities, most of the two kinds of signals exist at the same time and can reflect the user's interest and preference for commodities from different perspectives. Some studies [2–4] tried to find the connection between the two signals by using other feedback signals as auxiliary information, thereby optimizing the rating prediction. In this paper, we propose a "behavior chain" structure, which unifies the "explicit" and "implicit" feedback signals of different interaction stages in the e-commerce system on the interaction chain between users and commodities by using binary coding. The meaning of the expression is there any interaction between the user and the product at a certain stage. As shown in Fig. 1, the marked vector code (1, 1, 0, 0) represents the interaction behavior of the user who clicks on the product, purchases, but does not score and recommend the product. At the same time, the "activation" of users in the "behavior chain" stage by stage means that users are more and more interested in commodities. And each element in the user behavior coding vector presents a monotonically increasing structure from left to right.

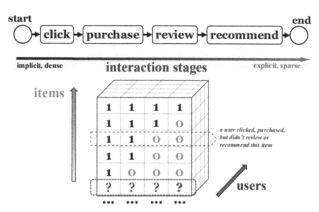

Fig. 1. The monotonous behavior chain between users and products (click-purchase-comment-forward) and an explanation of related item recommendation issues.

1.2 Bipartite Link Prediction Considers Matrix Completion

This coding method allows each interaction stage in the behavior chain to be defined as a binary interaction matrix between the user and the commodity. When we regard both users and commodities in the recommendation system as nodes in the interaction graph, and express the observable interactions as links on the label, the recommendation goal is simplified from predicting user-commodity interaction to predicting users on the bipartite graph-Commodity link mark. Based on the existing graph neural network research on non-graphical data [5, 6], we propose a recommendation framework (GAERE) for graph convolutional auto-encoders based on monotonic behavior chains.

First, the graph convolutional auto-encoder maps the user and the commodity node to the same latent space, and generates the behavior intention feature vector of the user and the commodity node by using the method of transmitting information on the user-commodity interactive graph. Then, the potential feature vector can be used to reconstruct the interactive link of the next interactive stage through the nonlinear decoder. We try to use the implicit monotony of the behavior chain to simulate the user's behavioral decision-making process on the product, and to predict the unobserved behavioral interaction between the user and the product. The later the interaction stage between the user and the product is, the more interested the user is in the product and the clearer the recommendation performance is. Therefore, we regard the last stage of the monotonous behavior chain as the main evaluation criterion of GAERE. Our contributions are as follows:

1. We have observed the recommendation problem of the fusion of multiple interactive signals, and proposed a method of uniformly representing multiple types of user and product interaction signals-monotonic behavior chain, and proposed a deep learning of product recommendation based on monotonic behavior chain frame.
2. We propose a new model framework, which effectively explores the characteristics of users' potential behavior intentions through graph neural networks, and solves the problem of product recommendation from the perspective of users' behavior intentions on the monotonous behavior chain.
3. The model is evaluated on four different real-world data sets, and the experimental results show that the recommendation performance of the algorithm is better than the existing recommendation methods.

2   Related Work

Matrix decomposition [7] is the first collaborative filtering algorithm proposed in recommender systems. This type of method attempts to find two low-rank and representative user and product feature vectors, and approximate the observed scores by inner product. In order to solve the problem of data sparseness, Probabilistic Matrix Factorization (PMF) [8] introduces Gaussian distribution into the data and optimizes the traditional MF using a probability model. In recent years, with the shift of research focus, the MF method has been extended to build models of implicit data. Unlike display feedback, implicit feedback recommendation is to provide users with a personalized product recommendation list based on their richer historical behaviors (clicks, purchases), and was later defined as a Top-K item recommendation task [9, 10].

Some studies have tried to use both signals for modeling at the same time, breaking the long-term independent research status of explicit feedback and implicit feedback

[11–14]. Among them [13] found that there is a positive correlation between the two signals, using implicit feedback as auxiliary information to predict the user's rating level. [15, 16] uses neural networks to explore the interaction of various signals in the user's "micro-behavior" session to predict the user's next action, but their research does not start from the overall perspective of the user's behavioral interaction. Consider the recommendation issue.

In recent years, GCN graph convolutional neural network, as an extension of CNN convolutional neural network application on non-European data, has achieved outstanding performance and rapid development [5, 17–19]. In the recommendation system, a lot of data can also be expressed as a multi-form graph structure [19–21], for example, the historical behavior of users and commodities is regarded as the edge of the bipartite graph of users and commodities. In fact, graph neural network is a method of network embedding, such as the algorithm based on graph auto-encoder. User or item-based auto-encoder [22, 23] is the most advanced collaborative filtering model of the latest category. In the research of Auto-Rec [24], the edge information in the graph is integrated into the matrix as auxiliary information. Completing. Specifically, the user and product network topology and node content information are simultaneously handed over to the convolutional neural network for processing, and the recurrent neural network is used to reconstruct and model the final evaluation results. The CF-NADE algorithm [25] assigns the unobserved score to level 3, and performs data segmentation, retention, and batch training after random arrangement of nodes, so CF-NADE can also be defined as a denoising auto-encoder.

3   Problem Definition

3.1   User-Item Bipartite Graph Representation of Each Stage of Monotone Behavior Chain

N_U and N_V represent the number of users and items, respectively. According to the user and items a series of interactive type definition into a $N_U \times N_V$ user interaction matrix M_l, which l is the number of interaction stage, $\forall l = \{1 \cdots L\}$.It is assumed that the element $M_{uv,l}$ in this matrix represents the observable interaction between user u and item v in the L stage (the interaction between user u and item v in this stage is 1), or the invisible interaction (the fact that the interaction between user and item can not be observed in this stage, the element value is 0). Suppose that for user u and item v, we have a set of labels $M_{uv} = [M_{uv,1}, M_{uv,2}, \cdots , M_{uv,L}]^T$ for user-item interactions.

$$M_{uv,l} = \begin{cases} 1, & \text{if the interaction } (u, v) \text{ at stage } l \text{ is observed} \\ 0, & \text{otherwise} \end{cases} \qquad (1)$$

Where $M_{uv,l}$ is 1 indicates that there is an interaction between user u and item v in stage l, that is, the user performs an action on the item (for example, clicking on it), and that action is considered an instance of "positive" feedback. However, a value of $M_{uv,l}$ of 0 is not simply a "negation" of the click operation. There may be a lack of cognitive information or noisy signals about the user's preferences.We want the product that the user interacts with to rank higher in the final recommendation than the other items. If there is $M_{uv,1} \geq M_{uv,2} \geq \cdots \geq M_{uv,L}$, we believe that the behavior chain M_{uv} is monotonous. The matrix completion or recommendation task can be thought of as predicting the value of unobserved entries in $M_{uv,l}$, which are used to rank the items.

3.2 Graph Link Prediction Solves the Matrix Completion Problem

Since the size of the multiple models is $N_U \times N_V$, the adjacency matrix data format describing the user-item correlation chain is regarded as the binary graph structure of node interaction, so the matrix completion problem in the recommendation can be considered from the perspective of link prediction on the user-item binary interaction graph. Before [26, 27] to study, using the undirected graph $G = (W, E, R)$ to represent the transformed structure interaction data, W represents the set of all nodes, the nodes entity is divided into user node $u_i \in U$ and i $\in \{1, \cdots N_U\}$ and item node $v_j \in V$ and $j \in \{1, \cdots N_V\}$, W = U $\cup$ V, then between nodes in the graph edges $(u_i, l, v_j) \in E$ value between users and items in the original data of phase interactive information, specifically, the l value value is only 0 (unobserved edge) or 1 (observed edge), r $\in \{0, 1\} \in R$.

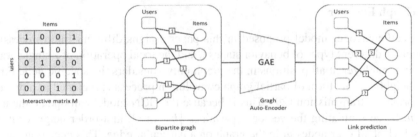

Fig. 2. Phase user-commodity interaction matrix M_l, where entries correspond to user-item interactions. Right: user-commodity interaction diagram with bipartite structure. The number on the edge represents the user's observable interaction with a particular product.

4   Method

We propose a graph auto-encoder variant based on monotone behavior chain recommendation task, referring to end-to-end unsupervised learning model [28] and implementation link prediction model on undirected graphs. The model not only realizes weight sharing, but also effectively uses user nodes and item nodes as auxiliary information to obtain the embedded feature vectors of users and items at specific stages, so as to achieve the main goal of enhancing the performance of recommendation prediction.

4.1 Graph Auto-encoder

Graph model is composed of encoder model $Z = f(X, A)$ decoder model of user interaction behavior and reconstruct $\tilde{A} = g(Z)$ in two parts, figure structure used for extracting feature information. graph structure data encoding model receive size of $N \times D$ feature vector matrix and size for $N_u \times N_V$ adjacency matrix A, the feature vector matrix X is used to store each entity node i feature vector x_i the output $N \times E$ node embedded matrix $Z = \left[z_1^T, z_2^T, \cdots, z_N^T \right]^T$, where N number of nodes, D represent the number of input elements, E is the feature vector dimension size. Decoding model, the use of the encoder output come to embedded matrix of node embedded (z_u, z_v), predict goal matrix corresponding entries in the $\tilde{M}_{uv}$. As show in Fig. 2.

In the equivalent multi-interaction bipartite recommendation graph G = (W, E, R), on the monotone behavior chain, the graph auto-encoder is redefined as $[U, V] = f(X, M_1, \cdots, M_L)$, where $M_l \in \{0, 1\}_{N_u \times N_V}$ is the binary adjacency matrix in the interaction stage between user and item l. According to the original data, when there is interaction between users and items in the interaction l stage, the corresponding position value of $M_{uv,l} = 1$. After processing by the graph autoencoder, the user feature vector matrices U and item feature vector matrices V with the shapes of $N_u \times E$.

U and $N_v \times E$ are obtained. Likewise, the decoder is restructured as $\tilde{M} = g(U, V)$, it will get the user and item embedded coding through the linear function reconstruction of size $N_u \times N_V$ goods user interaction matrix $\tilde{M}_{l+1}$ the behavior of the stage.

4.2 Graph Encoder

The specific encoder model proposed in this paper assigns different filter processing channels to different types of behavior stages l. and the local operation of a single graph convolutional layer at all positions in the graph only considers the specific node First-order neighbor. This form of partial graph convolution process is regarded as a process of information transmission [29]. This is because the GCN model can update the user node u_i by accumulating the vector eigenvalues $\{h_j\}$ of all first-order neighbor items nodes $\{v_j\}$ of all user nodes u_i in the graph on a particular edge. This encoding can be used as an expression of the similarity of the local graph structure. As show in Fig. 3.

The specific message $h_{j \to i, l}$ about the edge type l from the product j to the user i can be obtained as follows:

$$h_{j \to i, l} = \sigma \left(\frac{1}{c_{ij}} W_l x_j \right) \tag{2}$$

J is the index of the neighbor node of the node u_i, c_{ij} is the normalization constant of edge (u_i, v_j). Usually its value can be selected as $|N_{IJ}|$ (left normalization), N_I represents the set of neighbors of node u_i and x_j is the initial feature vector of node j. $\sigma(x)$ represents the element activation function, for example, ReLU(x) = max(0,x). The message $h_{i \to j,l}$ from the user to the project can be processed in a similar way through message passing:

$$h_{i \to j,l} = \sigma \left(\frac{1}{c_{ji}} W_l x_i \right) \tag{3}$$

The definition of each parameter in the above formula is similar to the definition in the items-user message. To obtain user node i and item node j in the final embedded coding, the need for the intermediate model output h_i, h_j converted as follows:

$$\text{For user : the calculation formula is } u_i = \sigma(W_u h_i) \tag{4}$$

$$\text{For item : the calculation formula is } v_j = \sigma(W_v h_j) \tag{5}$$

We use separate parameter matrices W_u and W_v to calculate user code u_i and item code v_j respectively. The calculation process of h_i and h_j is through a layer of graph convolutional neural network model, which is called graph convolutional layer. The calculation process of user code u_i and item code v_j corresponds to a fully connected neural network, which is called a dense layer.

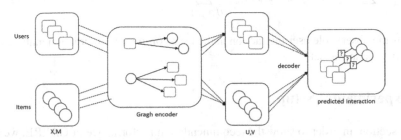

Fig. 3. Schematic diagram of forward pass through graph auto-encoder model

4.3 Monotonic Behavior Decoder

Modeling the probability $p\left(\tilde{M}_{u_i v_j,l+1}\right)$ that there is an edge between the user and the product, the purpose is to reconstruct the user-item interaction diagram in the next behavior stage. The decoder is similar to the preference score of user u and item v by multiplying the target user feature vector and the product feature vector. $\tilde{M}_{u_i v_j}$ indicates the refactoring interaction between user u_i and project v_j. Then apply the softmax function to obtain the probability distribution of interaction that may occur in the next stage of the monotonic behavior chain:

$$P_{u_i v_j,l+1} = p\left(\tilde{M}_{u_i v_j,l+1} = 1\right) = \sigma\left(s_{u_i v_j,l+1}\right) = \frac{1}{1 + \exp\left(-s_{u_i v_j,l+1}\right)} \tag{6}$$

Where $s_{u_i v_j,l+1}$ is the preference score, we use tensor decomposition to directly decompose the preference score.

$$s_{u_i v_j,l+1} = \langle u_{i,l}, v_{j,l} \rangle \tag{7}$$

Specifically, the user conducts the next stage of interaction on whether the product exists, and we model this logic in the final observation result.

4.4 Model Training

When training the model, the model is trained by minimizing the difference between the reconstructed mid-interaction value $\tilde{M}_{u_i v_j,l+1}$ and the real value observed by $M_{u_i v_j,l+1}$.

Each interaction stage in the monotonic behavior chain is defined as a separate recommendation problem. Therefore, in actual operation, different behavior stages are regarded as different interaction types, and cross entropy is used as a metric for reconstruction errors.

$$L = - \sum_{(U,V)\in Y} \log \tilde{M}_{uv,l+1} - \sum_{(U,V)\in Y^-} \log\left(1 - \tilde{M}_{uv,l+1}\right)$$

$$= - \sum_{(U,V)\in Y} M_{uv,l+1} \log \tilde{M}_{uv,l+1} - \sum_{(U,V)\in Y^-} \left(1 - M_{uv,l+1}\right) \log\left(1 - \tilde{M}_{uv,l+1}\right) \quad (8)$$

Y is the observable user interaction in the user-commodity interaction matrix, Y^- is a negative sampling set.

5 Experimental Setup

In this section, in order to show the recommendation performance of GAERE, we have conducted an experimental comparison between GAERE and popular recommendation algorithms in recent years on four real data sets. Specifically, we tried to answer the following three questions through experimental comparison: (1) Compared with the single feedback signal collaborative filtering model, does the GAERE model improve the performance of the recommendation system by fusing multiple feedback signals on the monotonic behavior chain recommendation. (2) Can the GAERE model use the information transmission on the graph structure or the graph convolutional neural network for information expansion to better obtain the user's potential preference for the behavioral intention of the product, thereby improving the recommendation for the most sparse feedback (explicit feedback) Performance; (3) Whether the recommendation based on user behavior intention is better than the recommendation model designed to obtain user and product characteristics.

5.1 Datasets

In this paper, we choose four real data sets of Steam, YooChoose, Yelp, and Goodreads to evaluate the effect of the experiment. These data can all reflect the user's multi-stage interaction process (click, purchase, comment, and forward) in the structure of multiple bipartite graphs.

In the process of data processing, for these four data sets Steam, YooChoose, Yelp, Goodreads, the same preprocessing standards are applied: (1) Filter out less than 5 interactive chains and none of the interactive chains that reach the final stage User; (2) Delete products with less than 10 interactions with the user. At the same time, using each data set with subgraphs with 3000 users and 3000 products, we randomly select 80%, 10%, and 10% of the interactive data as the training set, validation set, and test set of the model. Among these three sets, each interaction chain corresponding to the user is different. After preprocessing, the statistical information and distribution of the above data set are included in Table 1.

Table 1. Basic dataset statistics

Dataset	#Item	#User	#Interaction
Steam	3000	3000	27176
YooChoose	3000	3000	126894
Yelp	3000	3000	5379
Goodreads	3000	3000	394730

- Steam: This data set is a set of user interaction data disclosed by online game video sites. The data set includes user purchase game records, viewing time, user reviews, and user's recommendation intention ("recommended" or "not recommended"). Based on the original data, we construct a user behavior chain of "purchase-game-comment-forward" for each user.
- YooChoose: This is a data set composed of multiple sets of interactive sessions of a series of products being clicked and purchased. Because the entity user ID is not indicated in this data set, when constructing the interactive chain, we regard each session ID as a user ID. This set of data can be constructed into a user's "click-buy" behavior chain.
- Yelp: It is a set of explicit feedback data sets that users express to their friends their preferences for things through product reviews or ratings. In this data set, there are only [0–5] user ratings. We stipulate that when the user's rating is greater than or equal to 3, a recommendation operation will be performed. In this way, we construct a user "rating-forwarding" behavior chain.
- GoodRead: This is a set of data sets disclosed on the online reading platform that includes a rich and detailed interactive record. The data set has a total of 876,145 users, 2,360,655 books, and 229,154,523 observable user interaction records. We choose the "shelf-reading-rating-recommendation" that appears the most frequently among many feedback signals as the basis for constructing the behavior chain.

5.2 Comparison Methods and Evaluation Methods

In this section, we compare the GAERE model with other latest classic collaborative filtering algorithms BPRMF, logMF, and the machine learning algorithm ChainRec, which integrates multiple types of interactive feedback, for the recommendation problem proposed in (1) and (2) above. The first is to consider the impact of the independent interaction phase on the recommendation results. For the Bayesian personalized ranking BPRMF model of pairwise ranking, the independent latent factor model that ranks known items better than unknown items is obtained at different interaction stages. The WRMF model optimizes the model by minimizing the root mean square error between the predicted user preference score and the actual interaction label, and at the same time introduces other weights to adjust the observed interaction. Similarly, logMF also uses an independent latent factor model, and uses binary cross entropy to optimize the model. The above models ignore the potential relationship between different entity interaction stages, and combine different perspectives to reflect the user's preferences. Next, we consider the impact of the relationship between different interaction stages on the recommendation results. Using conditional optimization criterion independent latent factor model condMF, using conditional probability to predict user behavior interaction, that is, when there is interaction in the previous stage, the probability of interaction in the next stage. In SliceTF, the optimization standard combines the estimation and tensor decomposition in the vertical "slice" of the interaction matrix at each stage of the interaction. This model also models different types of signals together on the behavior chain. Recently, [30] proposed chainRec through the application of a monotonic scoring function, so that the recommendation performance of the model has been significantly improved compared with the above models.

5.3 Parameter Settings

Since we focus on recommending top-K products for each user, we use AUC@K and NDCG@K as the measurement indicators of the model, just like the evaluation indicators of other ranking models, where K is the personalized recommendation list we provide to users. The number of goods, we set it to 5, 10, 15. Specifically, the AUC measurement model is based on the relative position of user preference data to dislike data in the known user Top-K preference list. NDCG@K will consider the popular position of the product. If the popular position is higher, the score will be higher. We repeat this operation 10 times and take the average as the data in the report.

5.4 Experimental Results

Table 2 shows the results of AUC@10 and NDCG@10. In all comparative experiments, the first group of classic collaborative filtering bprMF and logMF only considers a single interactive signal as the basis for recommendation. chainRec is a way to improve the range of user preferences of the first group by combining the monotonicity of the user's multi-behavior interactions, showing a better prediction of user preferences. Similarly, our proposed GAERE is also superior to the benchmark method on these data sets. On Yelp, GAERE is 5% better than the latest collaborative filtering logMF and 3% on YooChoose. Proof problem (1) Compared with the current popular single feedback signal collaborative filtering model, the GAERE model can improve the performance of the recommendation system by fusing multiple types of feedback signals on the monotonic behavior chain recommendation. In terms of model parameters, compared with the latest chainRec machine learning method, because the node attributes are different, and each user and product separately applies for storage space, the space complexity of the model will increase as the number of nodes increases. Our model graph convolution model realizes parameter sharing between all users and commodities, making the dimension of each parameter far smaller than the sum of the number of users and commodities, which also greatly saves storage and computing costs.

The GAERE model has a 0.014 (1.5%) increase in the AUC value of the last interaction stage on the Steam data set compared to the chainRec model, and the AUG value has increased by 0.01 (1%) compared to the chainRec in the last interaction stage on the YooChoose data set, and the AUG value The last interaction phase on the Yelp data set increased by 0.009 (1%) relative to chainRec. However, the performance on the Goodreads data set is slightly worse than sliceTF by 1.4%. The reason may be that Goodreads is a data set with detailed records. The observable data in the defined interaction diagrams at each stage is relatively dense, and users can comment on public social networks, which increases traffic to popular books. Therefore, there is no need to define feedback for modeling through model design in the learning process. But in the remaining three data sets, the GAERE model can mine the effective information in the sparse data well and achieve higher prediction accuracy. This also shows that the problem (2) is established.

Table 2. The results of the model on AUC@10 and NDCG@10 on datasets

Dataset	Metric	bprMF	WRMF	LogMF	condTF	sliceTF	chainRec	GNMRE
Steam	AUC	0.826	0.829	0.830	0.832	0.833	0.832	**0.836**
	NDCG	0.124	0.127	0.137	0.148	0.147	0.145	**0.149**
YooChoose	AUC	0.781	0.783	0.784	0.791	0.794	0.802	**0.807**
	NDCG	0.126	0.127	0.133	0.137	0.140	0.151	**0.152**
Yelp	AUC	0.821	0.824	0.826	0.832	0.836	0.838	**0.840**
	NDCG	0.086	0.087	0.088	0.097	0.113	0.114	**0.116**
Goodreads	AUC	0.808	0.831	0.830	0.831	**0.841**	0.840	0.840
	NDCG	0.097	0.107	0.114	0.125	**0.131**	0.127	0.130

In terms of node feature learning, the ultimate goal is to obtain the traditional matrix decomposition method of two low-rank vector matrices representing users and commodities. In the GAERE model, users and commodities use GCN to make the feature vectors of each node in the form of information transfer in the graph or recursively. Aggregate and update node characteristics, which makes the characteristics of the target node (users and commodities) model independent learning, which effectively improves time efficiency and reflects better performance. Therefore, the problem (3) GAERE model can use the graph structure to better obtain the user's potential preference for product behavior intentions, and improve the recommendation performance for the sparsest feedback (explicit feedback).

Parameter setting: In the model, for each user and product, both the convolutional layer with a hidden unit of 500 and the dense layer with output dimensions {8,16,32,64,128} use the RELU function as the activation function; from Figs. 4, 5 and 6. It can be seen in Fig. 7 that the change of d affects the performance of the model. The recommended performance on each data set increases with the increase of d, but it also shows that the sensitivity of each data to d is different during the rising process. It is optimal at the time, so we choose the model parameters.

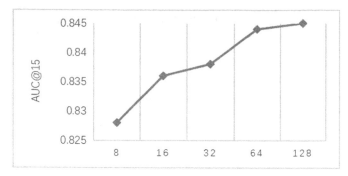

Fig. 4. Influence of different d to GNMRE model on Yelp data

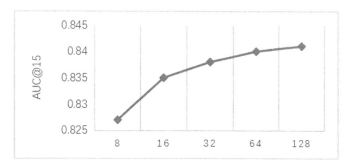

Fig. 5. Influence of different d to GNMRE model on Steam data

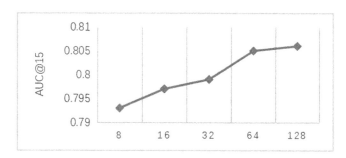

Fig. 6. Influence of different d to GNMRE model on YooChoose data

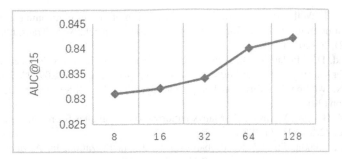

Fig. 7. Influence of different d to GNMRE model on Goodread data

6 Conclusions and Future Work

In this paper, by defining the monotonic chain graph structure between the user and the item interaction, multiple types of interaction signals generated during the interaction between the user and the item can be integrated into the recommendation system at the same time. On this basis, we propose a recommendation model GAERE based on a graph structure, which uses graph convolution to obtain the user's behavioral intentions and infer whether the user will establish a link with a specific item in the next stage. The GAERE model effectively simulates the user's monotonous and complete interaction behavior, and completes the matrix completion problem in the recommendation system from the perspective of link prediction in the two-part interaction diagram of the user's item. On four real-world data sets, the experimental results clearly show that our model shows better recommendation performance compared to the benchmark single-signal algorithm and the machine learning method that uses multiple signals with limited use of the GAERE model. For example, using NDCG@10 on the Yelp dataset, GAERE increased 1% than the optimum reference, than the traditional algorithm logMF improved 5%. In the future, we hope to explore other multi-type graph structures hidden among users or items to extend graph-based methods to other recommendation applications. Such as user-user social network and project-project relationship network. At the same time, we also use the NLP depth model to analyze the relevant comment text in the user's historical interaction records to reflect the user's preferences, thereby contributing to the recommendation performance. This also reflects the flexibility and effectiveness of the GAERE model.

References

1. Qiang, B., Lu, Y., Yang, M., et al.: sDeepFM: multi-scale stacking feature interactions for click-through rate prediction. Electronics **9**(2), 350 (2020)
2. Qin, J., Ren, K., Fang, Y., et al. Sequential recommendation with dual side neighbor-based collaborative relation modeling. In: WSDM 2020: The Thirteenth ACM International Conference on Web Search and Data Mining. ACM (2020)
3. Parra, D., Karatzoglou, A., Amatriain, X., Yavuz, I.: Implicit feedback recommendation via implicit-to-explicit ordinal logistic regression mapping. In: CARS (2011)

4. Jawaheer, G., Weller, P., Kostkova, P.: Modeling user preferences in recommender systems: a classification framework for explicit and implicit user feedback. ACM Trans. Interact. Intell. Syst. **4**(2), 26 (2014)
5. Duvenaud, D., Maclaurin, D., Aguilera-Iparraguirre, J., et al. Convolutional networks on graphs for learning molecular fingerprints. arXiv preprint arXiv:1509.09292 (2015)
6. Kipf, T.N., Welling, M.: Variational graph auto-encoders. In: NIPS Bayesian Deep Learning Work shop (2016)
7. Koren, Y., Bell, R., Volinsky, C.: Matrix factorization techniques for recommender systems. Computer **42**(8), 30–37 (2009)
8. Mnih, A., Salakhutdinov, R.R.: Probabilistic matrix factorization. In: Advances in Neural Information Processing Systems, pp. 1257–1264 (2008)
9. Hu, Y., Koren, Y., Volinsky, C.: Collaborative filtering for implicit feedback datasets. In: ICDM (2008)
10. Strub, F., Gaudel, R., Mary, J.: Hybrid recommender system based on autoencoders. In: Proceedings of the 1st Workshop on Deep Learning for Recommender Systems (2016)
11. Liu, N.N., Xiang, E.W., Zhao, M., Yang, Q.: Unifying explicit and implicit feedback for collaborative filtering. In: CIKM (2010)
12. Li, X., Chen, H.: Recommendation as link prediction in bipartite graphs: a graph kernel-based machine learning approach. Decis. Support Syst. **54**(2), 880–890 (2013)
13. Li, Y., Tarlow, D., Brockschmidt, M., Zemel, R.: Gated graph sequence neural networks. In: ICLR (2016)
14. Jawaheer, G., Weller, P., Kostkova, P.: Modeling user preferences in recommender systems: a classification framework for explicit and implicit user feedback. ACM Trans. Interact. Intell. Syst. **2**(4), 26 (2014)
15. Gurbanov, T., Ricci, F.: Action prediction models for recommender systems based on collaborative filtering and sequence mining hybridization. In: Proceedings of the Symposium on Applied Computing (2017)
16. Zhou, M., Ding, Z., Tang, J., Yin, D.: Micro behaviors: a new perspective in e-commerce recommender systems. In: WSDM. ACM (2018)
17. Wu, S., Tang, Y., Zhu, Y., et al.: Session-based recommendation with graph neural network on Artificial Intelligence. Proc. AAAI Conf. Artif. Intell. **33**(01), 346–353 (2019)
18. Wang, H., et al. Knowledge-aware graph neural networks with label smoothness regularization for recommender systems. In: Proceedings of the 25th ACM SIGKDD International Conference on Knowledge Discovery & Data Mining (2019)
19. Zhao, L., et al.: T-gcn: a temporal graph convolutional network for traffic prediction. IEEE Trans. Intell. Transp. Syst. **21**(9), 3848–3858 (2019)
20. Xu, F., et al.: Relation-aware graph convolutional networks for agent-initiated social e-commerce recommendation. Proceedings of the 28th ACM International Conference on Information and Knowledge Management (2019)
21. Kipf, T.N., Welling, M.: Semi-supervised classification with graph convolutional networks. arXiv preprint arXiv:1609.02907 (2017)
22. Hamilton, W.L., Ying, R., Leskovec, J.: Representation learning on graphs: methods and applications. IEEE Data Eng. Bull. **40**(3), 52–74 (2017)
23. Kipf, T.N., Welling, M.: Variational graph auto-encoders. arXiv preprint arXiv:1611.07308 (2016)
24. Monti, F., et al. Geometric deep learning on graphs and manifolds using mixture model CNNs. In: 2017 IEEE Conference on Computer Vision and Pattern Recognition (CVPR). IEEE (2017)
25. Zheng, Y., et al. A neural autoregressive approach to collaborative filtering. In: International Conference on Machine Learning. PMLR (2016)

26. Johnson, C.C.: Logistic matrix factorization for implicit feedback data. In: NIPS (2014)
27. Ying, R., et al. Graph convolutional neural networks for web-scale recommender systems. In: Proceedings of the 24th ACM SIGKDD International Conference on Knowledge Discovery & Data Mining (2018)
28. Zhou, F., Wu, H., Trajcevski, G., et al.: Semi-supervised trajectory understks. In: Proceedings of the AAAI Conference and with POI Attention for End-to-End Trip Recommendation. ACM Transactions on Spatial Algorithms and Systems (TSAS) (2020)
29. Sedhain, S., Menon, A.K., Sanner, S., et al. Autorec: autoencoders meet collaborative filtering. In: Proceedings of the 24th International Conference on World Wide Web, pp. 111–112 (2015)
30. Wan, M., McAuley, J.: Item recommendation on monotonic behavior chains. In: Proceedings of the 12th ACM Conference on Recommender Systems (2018)

Design and Implementation of Collaborative Filtering Recommendation Algorithm for Multi-layer Networks

Ling Gou[1,2,3], Lin Zhou[1,2,3], and Yuzhi Xiao[1,2,3(✉)]

[1] Computer Department, Qinghai Normal University, Xining 810016, Qinghai, China
[2] Tibetan Information Processing and Machine Translation Key Laboratory
of Qinghai Province, Xining 810008, Qinghai, China
[3] Key Laboratory of Tibetan Information Processing, Ministry of Education, Xining 810008, Qinghai, China
qh_xiaoyuzhi@139.com

Abstract. With the continuous development of mobile communications and Internet technologies, the marketing model of the communications industry has shifted from calling-based to social APP-based personalized recommendations. In order to improve the accuracy of recommendation, this paper proposes a recommendation algorithm for social analysis. Empirical data was firstly used to construct a "user-APP" two-layer communication network model, and then the traditional collaborative filtering recommendation technology was integrated to reconstruct similar users and similar APP network model. The bipartite graph weight distribution method was taken to recommend targets in the obtained network model. The experimental simulation shows that, in view of the characteristics of the two-layer communication network, compared with the traditional recommendation algorithm, the algorithm effectively improves the accuracy of the score prediction.

Keywords: Two-layer communication network · Social network analysis · Recommendation algorithm · Collaborative filtering algorithm

1 Introduction

With the advent of Web 2.0 and the increase of Internet penetration, social networks have become one of the important research areas that have received much attention [1]. Social networks is becoming a major vehicle for information sharing, experience exchange and social opinion, social individuals express their personal emotions through ratings, retweets and comments, and amplify these personal behaviors into social behaviors [2]. Therefore, many scholars have used this characteristic to analyze user behavior of communication networks and to perform personalized recommendation [3], information dissemination [4], and network destructiveness analysis of such social networks based on network dynamics theory and related knowledge [5]. At the same time, the empirical network has the characteristics of multilevel and intersectionality, and there

© Springer Nature Singapore Pte Ltd. 2021
J. Zeng et al. (Eds.): ICPCSEE 2021, CCIS 1451, pp. 32–50, 2021.
https://doi.org/10.1007/978-981-16-5940-9_3

is heterogeneity in both the objects represented by the nodes and the connections represented by the connected edges [6]. Buldyrev S. V. et al. [7] proposed that a dependent network in the multi-layer network is a network consisting of two networks and more, in which the nodes within a single network are homogeneous, while the nodes between networks are heterogeneous. Gao J. [8] considered multi-layer networks as a set of networks consisting of multiple single-layer networks, each of which constitutes a network layer. Nicosia V et al. [9] extended the degree centrality, betweenness centrality and proximity centrality in a single network to multi-layer networks in the form of vectors, but there is a problem of how to rank the importance of nodes.

Information dissemination on social networks is one of the hotspots of current research. Information dissemination refers to the act of people exchanging information with each other, which is an important part of people's daily life and can promote people's understanding of the world and each other. With the popularity of mobile terminals, users are not only consumers of information, but also creators and disseminators of information. It is especially important to find the user group that can most influences other users to use a certain product and to use corresponding marketing strategies for this group [10]. However, the traditional collaborative filtering techniques only uses the user's rating matrix for items, which does not take full advantage of the more social relationships between users and items, resulting in sparse data. Balabanov proposed a solution to build a multi-layer network for collaborative filtering recommendation algorithms in which the sparse rating matrix leads to a serious decrease in the recommendation accuracy of the recommendation system [11]. Thus, improving the accuracy of recommendation is the most important problem to be solved at present, so various personalized recommendation algorithms with different ideas have gradually become one of the research hotspots in academia, among which the content-based recommendation algorithm considers the user's liking of the item, Its accuracy is relatively high and it is also easy to understand, but the algorithm is more difficult to deal with non-text information [12]. The recommendation algorithm based on bipartite graph has higher accuracy, but the time as well as space overhead increases as a result [13]. Collaborative filtering-based recommendation algorithms has the advantage of being independent of data format and it can also handle complex data for effective and fast recommendations, but they also has problems such as sparse data and cold start [14]. Each of these algorithms has its own advantages and disadvantages, and how to analyze the user's interest preferences more accurately and efficiently through recommendation algorithms is the direction of all personalized recommendation algorithms.

In this paper, based on the actual communication data set, it will study the idea of the traditional base collaborative filtering recommendation algorithm and the bipartite graph resource redistribution [15] and combine with the social network relationship to establish the multi-layer network social relationship graph, it will research on the basis of the traditional collaborative filtering recommendation algorithm combining with social relations to alleviate the data sparsity problem, and in combination with the social communication network, it will add the now prevalent user multi-layer social relationship and user social behavior analysis filtering to recommend resources, comparing the three evaluation indicators of RMSE, MSE and Precision with several classical recommendation algorithms in order to improve the accuracy of the recommendation system.

2 Pre-requisite Knowledge

2.1 User-Based Collaborative Filtering Algorithm

The idea of the user-based collaborative filtering algorithm is that the items that the user likes are similar to the preferences of his similar friends, and recommendations are made for the target users based on the size of similar users. Therefore, the algorithm construction can be divided into two steps:

Step 1: Build a set of user similarities. By calculating the similarity between users, the users with higher similarity are regarded as the similar users of the target users;

Step 2: The similarity user set of the target user is used to predict the user's preference degree to the item, after obtaining the user's similarity matrix and the user's neighbor user set, the predicted score of user u on item i is calculated. As shown in formula (1) [16]:

$$p(u, i) = \sum_{v \in S(u,K) \cap N(i)} w_{uv} r_{vi} \tag{1}$$

Where $p(u, i)$ denotes the predicted score of user u for item i, $S(u, K)$ denotes the set of k neighbors of user u, $N(i)$ denotes the set of all users for which item i has been rated, w_{uv} denotes the similarity value between user u and user v, and r_{vi} denotes user v's rating of item i.

2.2 Project-Based Collaborative Filtering Algorithm

The idea of item-based recommendation algorithms is that items that are most similar to items that users have liked in the past are more likely to be liked by users, and that users' preferences can be predicted by their level of preference for historical items [17]. Therefore, the algorithm process can be divided into two steps:

Step 1: Build a collection of neighborhood items for the item. The neighbor set of an item refers to the top k items most similar to the item. The similarity between items can be calculated by calculating the correlation coefficient between the vectors of the items being scored;

Step 2: Predict the score based on the set of neighbor items. After the similarity matrix and neighbor items of the item are obtained, the predicted score of user u for item i can be calculated by formula (2):

$$p(u, j) = \sum_{v \in N(u) \cap S(j,K)} w_{ji} r_{ui} \tag{2}$$

where $p(u, j)$ denotes the predicted score of user u by item j, $N(u)$ denotes the set of items rated by the user historically, $S(j, K)$ denotes the set of neighbors of item j, w_{ji} denotes the similarity between item i and item j, and r_{ui} denotes the rating of item u by user i.

2.3 Bipartite Graph Recommendation Algorithm

The bipartite graph is composed of two classes of nodes, one class of nodes is the user that generates the behavior; the other class of nodes is the user's behavioral activity. The bipartite graph recommendation algorithm specifies that nodes of the same class are not connected to edges initially, and the relationship between them is obtained through the connections between nodes of different classes and the topology of the graph [18]. The resource weights assigned by item j to item i are shown in Eq. (3).

where $k(y_i)$ denotes the sum of the weights of the edges between all the items connected by user y_l; a_{il} denotes the rating matrix of the items by the user of $n \times m$, $x_i y_l$ denotes the rating between user y_l and item x_i of the bipartite graph, i.e.rating a_{il}, w_{ij} denotes the weight obtained by node x_i from the nodes.

$$f(x_i) = \sum_{i=1}^{m} \frac{a_{ij}}{k(y_l)} f(y_l) = \sum_{j=1}^{n} w_{ij} f(x_j) \tag{3}$$

3   Recommendation Algorithms for Two-Layer Networks

3.1   Problem Formulation

Currently, major operators have adopted various recommendation methods to further improve their marketing, among which increasing the usage of app to improve traffic usage is a common means [19]. Using the user's traffic usage in APP to measure the user's preference degree, so the targeted traffic package can be designed to achieve its own purpose based on the user's traffic usage in a certain type of app. However, the number of users and items in the actual application is increasing, but users generally have only used a small portion of apps, i.e., the number of users using the same app together is very small, and the same is true between apps. Thus causing serious data sparsity problems [20]. The traditional collaborative filtering algorithm is based on the preference model established by the scoring matrix to help users discover new types of items that may be of interest. However, the traditional collaborative filtering algorithm has less characterization of attribute relationships, and it is difficult to describe the potential dependence and relevance of complex attributes.

In this case, the two-layer network is built according to the multi-attribute and intersecting characteristics of social networks, the two-layer network built by using "user-APP" rating data contains two types of nodes, one is the user node and the other is the app node. The relationship between users and users, app and app as well as the relationship between users and app cause the intersection of the network. Therefore, the recommendation algorithm based on the two-layer network with collaborative filtering is used in order to improve the accuracy of the algorithm.

3.2   Dataset Pre-processing

For app recommendations, users do not have clear ratings or clear verbal text comments on apps, so users' preferences can be evaluated based on the number of views or traffic usage of a mobile app. Analyze the real data of app usage of students in a school. The popular applications and their traffic usage data were obtained by setting the threshold value, and finally 861 users were obtained, and there were 207 apps in total.

As the obtained empirical data has data sparsity and data untidiness, the data is firstly pre-processed, and the traffic is divided into 5 parts according to the size and the number of users, and the data is reasonably scored as 1, 2, 3, 4, 5, according to the smallest to the largest to reflect the user's favorite degree. The minimum app traffic usage is 32 MB, and the maximum is 470,500 MB. By calculating the number of users in each traffic segment several times, after comparison and analysis, it can be divided according to the flow segment as shown in Table 1, which is the most reasonable after application verification.

Table 1. Traffic usage of each segment

Scope	0–0.5	0.5–2	2–10	10–100	100–500
Number	107	122	106	512	14
Rating	1	2	3	4	5

When the data is divided into training set and test set according to the ratio of 8:2, data cleaning is required when performing algorithm validation because of the randomness of the segmentation, and the cleaning steps are shown below.

Step 1: Removal of users: After the user-APP rating matrix obtained by pre-processing, users who have only generated less than 3 ratings are removed, and the abnormal values are also recorded as 0.

Step 2: Filling of missing values: Similar user matrix is obtained by collaborative filtering recommendation algorithm based on users, and similar mobile application matrix is obtained by collaborative filtering algorithm based on items, but since the target users have chosen colder mobile applications causing their similarity with other users to be all 0 values, the corresponding target mobile applications are similar to other similar mobile applications with 0 values, so new values are obtained by properly adjusting unpopular mobile applications, separately presenting and reconstructing the similarity matrix, or filling these 0 values appropriately based on subjective feelings, or directly deleting these unpopular mobile applications.

Step 3: Abnormal processing: After the two-part graph resource redistribution recommendation algorithm to obtain the user rating matrix, or cause abnormal data, the *inf* value that appears is due to the denominator is zero, resulting in the value obtained is infinity, at this time, set $A(A==inf) = 1$, and the *nan* that appears is a meaningless value obtained by *0/0, Inf/Inf, Inf-Inf, Inf*0*, etc., in this case $A(A==nan) = 0$.

3.3 Two-Layer Network Characterization Analysis

The "user-APP" rating matrix is established by the user's traffic usage of mobile applications, and each mobile application can be classified according to the category it belongs to. As shown in Table 2, Every mobile application has its corresponding category. When designing a package in a communications company, directional traffic packages can be designed according to the categories of mobile applications, if a user spends a lot of data on the video, then you can a traffic package that combines video and call duration. Through the algorithm idea of resource redistribution of the bipartite graph, the "user-APP" matrix is formed from user to app, and the "APP-category" matrix is formed from mobile application to category, which can be mapped into "user-category" matrix, according to the category of the situation recommended the corresponding targeted traffic packages, Such as game oriented traffic packages, video oriented traffic packages and other directional traffic packages. The third layer of the network can just reflect the network of directed traffic packages, and the category relationship of this network is not connected, and it can be treated as a bipartite graph. Figure 1 shows the three-layer network model of communication users, i.e. "user-APP-package". Combine the UBP recommendation algorithm and the classification of mobile applications to recommend corresponding targeted traffic packages to users.

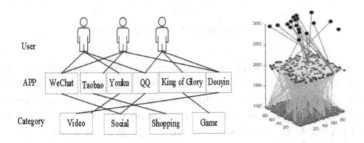

(a)User-APP-Package bipartite graph   (b) Network topology diagram

Fig. 1. User-mobile application-set network diagram

Table 2. Classification of some apps

APP	Dou Yin	Cool Dog	Baidu	WeChat	AikiYi	Jing Dong	Tao Bao	QQ	King of Glory	Jedi Quest
Category	Video	Music	Browser	Communication	Video	Shop	Shop	Commu-nication	Game	Game

Based on the modeling of the communication network, the degree distribution, clustering coefficients and average paths of the "user-APP" network were analyzed, Fig. 2 shows the degree distribution of the network obtained by formula (5).

$$k_i = \sum_{j=1}^{N} a_{ij} \tag{4}$$

$$< k >= \frac{1}{N} \sum_{i=1}^{N} k_i = \frac{1}{N} \sum_{i,j=1}^{N} a_{ij} \tag{5}$$

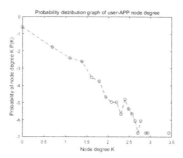

Fig. 2. User-APP network degree distribution

It can be seen that the network as a whole is scale-free distribution, which is in line with the self-growth characteristics of social networks. As the network continues to grow, newly added nodes are more inclined to older nodes with greater connectivity. For the "user-APP" network, new users are more inclined to choose popular apps. For those originally suitable for their own app will be ignored, so a comprehensive and accurate recommendation algorithm is needed to meet the needs of users to improve their experience. Table 3 shows the clustering coefficients and average paths of the user layer network, mobile application layer network and the whole network calculated by Eq. 6 and Eq. 7 respectively.

According to the table, it can be seen that the clustering coefficient of the network in the user layer is the smallest, the clustering coefficient of the whole network is the largest, and the clustering coefficient of the network in the mobile application layer tends to be in the middle, it can be concluded that the choice of mobile applications by the middle-layer users increases the clustering coefficient of the network, users are more closely connected with mobile applications, so the algorithm construction according to this model can improve the accuracy of the recommendation.

And the shorter the average path and the more path entries of the network, it shows that the closer the relationship between nodes is, the easier it is to connect, and the accuracy of the recommendation can be improved. According to the data in this table, it is seen that the average path of the user layer is the largest, the average path of the whole network is the smallest, and the average path of the mobile application layer tends to the middle, it shows that through the user's choice of mobile applications, the compactness of the network is improved, and the establishment of this model can further improve the accuracy of the recommendation.

$$C_i = \frac{E_i}{(k_i(k_i-1))/2} = \frac{2E_i}{(k_i(k_i-1))} \qquad (6)$$

$$L = \frac{1}{1/2N(N-1)} \sum_{i \geq j} d_{ij} \qquad (7)$$

Table 3. Clustering coefficient values and mean path values of the network

Category	User layer network	APP layer network	Whole network
Clustering coefficient	0.1548	0.2365	0.3120
Average path	8.7320	6.1843	5.2915

3.4 Recommendation Algorithm Implementation

Proposal of UBP Algorithm. Based on the analysis of the characteristics of the two-layer network and the problems faced by the recommendation algorithm, we propose the information recommendation algorithm in the two-layer communication network, namely UBP algorithm. Figure 3 is a flowchart of UBP algorithm. The "user-APP" rating matrix is obtained through the user's rating of the app. The similar user matrix is obtained by using the user-based collaborative filtering algorithm, the similar matrix is obtained by using the users with high similarity as the target users, and the similar app matrix is obtained by using the item-based collaborative filtering recommendation algorithm. The app with high similarity is used as the similarity matrix of the target app, and then the new relationship matrix is mapped into a "similar user-similar app" bipartite graph, with the user's rating of the app as the weight of the edge. Thus, a weighted bipartite graph is constructed. Using the idea of resource reallocation algorithm of bipartite graph, the recommendation result of the target user is obtained.

Constructing the Similarity Matrix. By constructing a 207 × 207 app similarity matrix through collaborative filtering algorithm based on items, the number of ratings of some apps are listed in Table 4, the total rated scores of some apps are listed in Table 5, as well as the similarity values of apps according to Table 10, it can be observed that the apps similar to the target apps are the popular mobile applications with more scoring times. The reason for this phenomenon is that the popular app is more likely to be chosen by the same user as other mobile apps, so it is more similar to other apps. However, the number of times a mobile app is rated is not absolutely proportional to the similarity ranking of the mobile app, which indicates that the popular mobile app is not the only criterion for the similarity measure of the mobile app, so the recommendation model can avoid the recommendation of the popular mobile app to a certain extent and can bring diversity of recommendations.

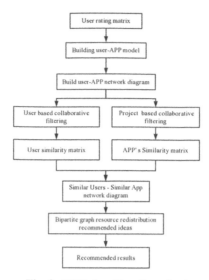

Fig. 3. UBP algorithm flow chart

A user similarity matrix of 861 × 861 can be constructed by the user-based collaborative filtering algorithm. The number of ratings of some users is listed in Table 6, the total rated scores of some users are listed in Table 7, and the similarity values of some users are listed in Table 9. It can be observed that the users who are similar to the target users are the users who have selected more mobile applications, which indicates that the users who have selected more mobile applications users are more similar to other users. However, the number of user ratings is not absolutely proportional to the user similarity ranking, which indicates that the users who are enthusiastic about mobile applications are not the only criteria for user similarity measurement. So the recommendation model can avoid the recommendation of popular items to a certain extent and can bring diversity of recommendations.

Table 4. Number of times some apps were rated

APP	DouYin	CoolDog	Baidu	WeChat	AikiYi	Jing Dong	TaoBao	QQ	King of Glory	Jedi Quest
Number of times	458	312	325	861	56	653	883	860	256	123

Algorithm Description. Therefore, the UBP algorithm, a recommendation algorithm based on a two-layer social network that fits the characteristics of realistic social networks, is proposed to address the shortcomings of collaborative filtering and the rationality of the similarity matrix. The algorithm works by inferring the user's interest

Table 5. Total scores of some apps being rated

APP	Dou Yin	Cool Dog	Baidu	WeChat	AikiYi	Jing Dong	Tao Bao	QQ	King of Glory	Jedi Quest
Score	580	458	532	2031	193	983	1058	2243	543	322

Table 6. Total number of partial user ratings statistics

User	User1	User2	User3	User4	User5	User6
Number of times	13	16	2	14	6	7

Table 7. Total scores of some user ratings

User	User1	User2	User3	User4	User5	User6
Number of times	38	18	6	29	19	15

preference for items that do not produce behavior through the intuitive differences in the user's ratings on the items. The algorithm considers two main aspects, the ratings of different apps by the same user and information on the use of the same app by different users. In order to illustrate the idea of algorithm implementation more intuitively, the score of 10 apps from 6 users is used for specific explanation. The main steps of the algorithm are as follows.

Step 1: Construct the user-APP matrix G. Suppose the user node is $U = \{u_1, u_2, u_3, ..., u_n\}$, the item set is $V = \{u_1, u_2, u_3, ..., u_n\}$, and Element a_{ij} in user project matrix G represents the score of user i on the project; as shown in Table 8.

Table 8. User-APP rating matrix

User/APP	Dou Yin	Cool Dog	...	QQ	King of Glory	Jedi Seeker
User1	5	3	...	1	5	0
User2	4	0	...	0	4	2
...	...	...	...	...	...	...
User6	4	0	...	4	4	0

Step 2: Calculate the user similarity. Firstly, find out the apps used by user u_i and user u_j together, calculate the number of apps used by users, and find the user similarity matrix by the cosine similarity formula, as shown in Eq. (8) [19]; the obtained results are shown in Table 9.

$$sim_{cc}(u, v) = \frac{\sum (r_{u,j} - \overline{r_u})(r_{v,j} - \overline{r_v})}{\sqrt{\sum (r_{u,j} - \overline{r_u})^2} \times \sqrt{\sum (r_{v,j} - \overline{r_v})^2}} \tag{8}$$

Table 9. Matrix of similar users

User	User1	User2	...	User6
User1	0	0.852322729	...	−0.029339267
User2	0.852322729	0	...	0
...	...	...	...	...
User6	−0.029339267	0	...	0

Step 3: Calculate app similarity: firstly, find all app v_i and the set of users of v_j. The number of apps evaluated by the user and the sum of the ratings are calculated. The modified cosine similarity formula is used to find the similar APP matrix, and the calculation formula is shown in Eq. 8; the results obtained from the calculation are shown in Table 10.

Table 10. Similar APP matrix

APP	Dou Yin	Cool Dog	...	QQ	King of Glory	Jedi Seeker
Dou Yin	0	0.049493453	...	−0.822719171	0	−1
Cool Dog	0.049493453	0	...	0.252719435	−1	0
...	...	...	...	...	...	...
QQ	−0.822719171	0.252719435	...	0	−0.452059977	0

Step 4: According to the similarity matrix obtained in step 4 and step 3, the users whose similarity to the target user is greater than the threshold are extracted, as shown in Fig. 4, which is the network graph of the APP selected by user 1 obtained according to Table 4, which is known according to the user similarity matrix in Table 5, and the similar users of the target user when the similarity is greater than the threshold. From this, it can be seen that the users similar to user 1 are, user 2, user 5 and user 6. Figure 5 shows the network diagram composed after extracting the similar users of user 1.

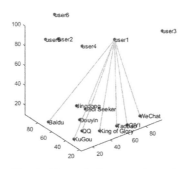

Fig. 4. Network diagram of user 1

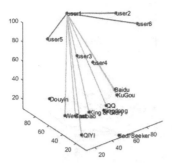

Fig. 5. Similar users of user 1

Step 5: According to the user1-APP network diagram obtained in step 4, the APP ratings performed by user1 are composed into a new "user-APP" rating matrix. According to the similar APP matrix, we extract these apps which are similar to apps with similarity greater than a certain threshold, and use the APP similarity matrix in Table 10 and the apps with similarity greater than the threshold as the similar apps of the target apps, and the apps which are similar to the target apps are all apps by calculation. As shown in Fig. 6, the user-APP two-layer network diagram shown in Fig. 7 is obtained by using these similar apps and similar users as well as the initial rating matrix.

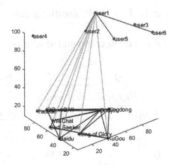

Fig. 6. User 1 target APP similar to APP

Step 6: According to the result obtained in step 5, the connection within the layer is undone, and the standard bipartite graph of "User-APP" shown in Fig. 8 is obtained to form the final "User-APP" rating matrix, according to which the resource reallocation bipartite graph recommendation is used. According to this matrix, we use the resource reallocation bipartite recommendation formula (see formula 3 in Sect. 2.3) to get the user 1 recommendation result. In this way, the final recommendation results for all users are obtained; as shown in Table 11.

Step 7: Based on the two-layer social network obtained from the above steps, the similar users connected by the target users and all the similar items corresponding to these users are extracted in turn to form the final network graph. As shown in Fig. 9.

Step 8: The accuracy of the algorithm is obtained based on the predicted scoring matrix as well as the actual scoring matrix.

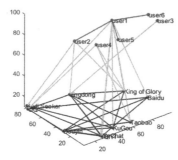

Fig. 7. New user-APP network

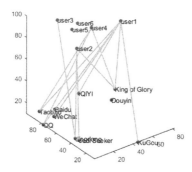

Fig. 8. User-APP bipartite graph

Table 11. Recommended results

	Jitterbug Cool Dog	Cool Dog	...	QQ	King of Glory	Jedi Quest
User1	3.9286	3.4676	...	3.9286	3.9603	4
User2	3.6728	3.4615	...	0	2.5967	3.4615
...	...	...	...	...	...	...
User6	3.5463	0	...	3.5882	3.5882	0

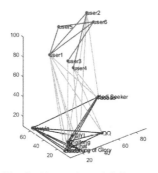

Fig. 9. Network model diagram

3.5   Algorithm Implementation Process

Input: target user U_a, user-APP score matrix

Output: the recommended item set of target user U_a

---

for each U_a in user-APP data U_b matrix
App label evaluated by users U_a and U_b
Row vector stores apps jointly evaluated by users U_a and U_b
If p_ab=NULL
  sim_u_ab=0

  Else
Use the corrected cosine similarity to obtain the user similarity matrix
Similarity u_ab
    End
for each M_a in User-APP Matrix
App evaluated jointly by user M_a and M_b, row vector stores APP evaluated jointly by user M_a and M_b
If p_ab=NULL
  sim_u_ab =0
  Else
Use the corrected cosine similarity to obtain the user similarity matrix
Similarity m_ab
    End
For each U_a matrix in Similarity u_ab
  Find (Similarity u_ab>0.1)
End
For each M_a matrix in Similarity m_ab
        Find (Similarity m_ab>0.1)
          Get cell matrix
End
For each U_a-M_a matrix
Use bipartite graph to calculate the secondary allocation matrix of resources
  score(P==inf) =1;
    score (is nan(P)==1) = 0;
End
Recommended result of adjustment
Calculate recommended error and accuracy rate
RMSE=sqrt(Eval(:,3)'*Eval(:,3)/x);
MSE=sum(Eval(:,3))/x;
Precision=sum((yyh(:,1)-per(:,1)=0(:)))/x

---

4 Experimental Results

4.1 Evaluation Criteria

Root Mean Square Error (RMSE). Usually the smaller the value of RMSE, the more accurate the recommendation result is. Where p_i denotes the actual rating of item i and q_i denotes the predicted rating of item i [20], the calculation is shown in Eq. (9).

$$RMSE = \sqrt{\frac{\sum_{i=1}^{N} |p_i - q_i|^2}{N}} \tag{9}$$

Mean Square Error (MSE). The mean squared error MSE is the sum of the absolute values of the user's rating errors for all items, and it describes the total deviation of the predicted ratings. Denoted by p_i for the actual rating of item q_i and i for the predicted rating of item X [20], the MSE is shown in Eq. (10).

$$RMSE = \frac{\sum_{i=1}^{N} |p_i - q_i|^2}{N} \tag{10}$$

Precision (Precision). Precision refers to the ratio of the number of correct predictions to the number of all predictions, and measures the search accuracy of the retrieval system, calculated as follows, where TP indicates the number of correct predictions and FP indicates the number of all predictions.

$$\text{Precision} = \frac{\text{TP}}{\text{TP} + \text{FP}} \tag{11}$$

4.2 Selection of Parameters

In order to compare the effectiveness and the advantages and disadvantages of the recommendation results, the data from the communication dataset were applied to the improved UBP algorithm with the user-based collaborative filtering recommendation algorithm (U) and the item-based collaborative filtering recommendation algorithm (P), and their root mean square error (RMSE), mean square error (MSE) and precision (Precision) were compared, and the results showed that the proposed UBP algorithm has a higher precision and lower RMSE and MSE.

From Figs. 10, 11, and Fig. 12, we can see that the accuracy rate of the traditional collaborative filtering recommendation algorithm is low, and it can hardly bring effective recommendations. This indicates that when the sparsity is very high, direct collaborative filtering on the original data cannot produce accurate recommendation results. By comparison, it can be seen that the accuracy rate of the improved recommendation algorithm is much higher than that of the traditional collaborative filtering recommendation algorithm. In the comparison between the user-based collaborative filtering recommendation algorithm and the item-based collaborative filtering recommendation algorithm, the error rate and accuracy of the user-based collaborative filtering recommendation algorithm are all due to the item-based collaborative filtering recommendation algorithm. It shows that the accuracy of the algorithm that incorporates the ideas of collaborative filtering as well as bipartite graph is higher.

In addition, the trend of the curves of root mean square error and mean square error shows that the RMSE and MSE values of UBP algorithm as well as U algorithm have a tendency to increase with the increase of neighbor user similarity until they tend to be stable. The P algorithm is gradually decreasing until it stabilizes. It indicates that the higher the similarity of neighboring users, the easier it is to ignore the user's overall favorite items, which will affect the recommendation results. But overall, the UBP algorithm has the lowest RMSE and MSE values. According to the trend of the accuracy rate change curve, the accuracy rate of U algorithm tends to decrease as the similarity of neighboring users increases, the accuracy rate of the P algorithm has an upward trend until it stabilizes. The UBP algorithm fluctuates between about 0.9. When it reaches the range of 0.15–0.2, UBP is the least accurate algorithm. Overall, the UBP algorithm has the highest precision among the three algorithms.

In summary, when the neighbor user similarity is 0.01, the root mean square error and mean square error of the UBP algorithm are the smallest, and the accuracy rate of the UBP algorithm is the highest.

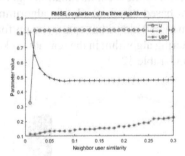

Fig. 10. RMSE metrics

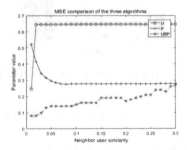

Fig. 11. MSE metrics

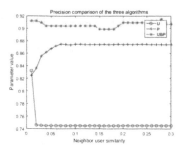

Fig. 12. Precision metrics

4.3 Experimental Results

To verify the accuracy of the UBP algo1est value of Precision, indicating that the algorithm has the best rating prediction on mobile applications; the accuracy rate of user-based collaborative filtering recommendation algorithm is second only to UBP algorithm; the accuracy of item-based recommendation algorithms is relatively poor; the accuracy and mean square error of the recommendation algorithm based on the bipartite graph are between the item-based recommendation algorithm and the user-based recommendation algorithm; the UBP algorithm can make up for the shortcomings of the traditional collaborative filtering algorithm in the real network and improve the accuracy of the prediction (Fig. 13 and Table 12).

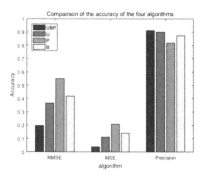

Fig. 13. Graph comparing the metrics of the four algorithms

Table 12. Experimental comparison results of algorithms

Category	RMSE	MSE	Precision
UBP	0.1966	0.0386	0.9101
U	0.3672	0.1124	0.8989
P	0.5510	0.2089	0.8189
B	0.4189	0.1421	0.8719

5 Summary and Prospect

In this paper, we combine the advantages of collaborative filtering algorithm and bipartite graph resource reallocation recommendation algorithm, and build a multi-layer social network to study the recommendation algorithm with higher accuracy. Firstly, the "User-App" matrix is constructed, and the "User-App" two-layer network graph is constructed based on the matrix. Then obtain the similarity matrix between users and items through collaborative filtering. Then, the similarity matrix of items is constructed according to the theory of resource secondary allocation, and the collaborative filtering recommendation system of multi-layer network is proposed. In order to measure the effectiveness of the recommender system, two control groups were set up: They are based on user collaborative filtering recommendation algorithm and item-based collaborative filtering recommendation algorithm. The performance of the recommendation is evaluated by comparing the root mean square error, mean square error and precision index of the experimental group and the two control groups. According to the empirical results, the accuracy and error rate of the experimental group is much higher than that of the control group. Therefore, the recommendation system built by UBP, a collaborative filtering recommendation algorithm based on multi-layer network, is reasonable and can greatly optimize the traditional collaborative filtering recommendation system. The results show that the proposed UBP algorithm is more accurate. However, with the advancement of recommendation technology, users' requirements for recommendation systems are getting higher and higher, in the future research, it is necessary to further optimize the complexity of the algorithm, and use more social relations to further improve the rationality of the algorithm.

Acknowledgements. This work was supported by the National Science Foundation of China (No.61763041) and the Science Found of Qinghai Province (No.2020-GX-112).

References

1. Liu, Z.: Research on maximizing the influence of social networks. Anhui University (2018)
2. Liu, Q.: Research and application of information dissemination in single-layer social network and its traction control recommendation model. Chongqing University (2015)
3. Zhang, S., Wang, C., Li, Q., Li, H.: Recommendations based on the circle of friends in the online social network environment. J. Nanjing Normal Univ. (Natural Science Edition) **41**(04), 72–78 (2018)
4. Lv, W., Ke, Q., Li, K.: Dynamical analysis and control strategies of an SIVS epidemic model with imperfect vaccination on scale-free networks. Nonlinear Dyn. **99**(2), 1507–1523 (2020)
5. Jia, C., Han, H., Wan, Y., Lv, Y.: Research on network invulnerability based on network model feature attack. Complex Syst. Complexity Sci. **14**(04), 43–50 (2017)
6. Zhang, X.: Research progress of multi-layer complex network theory: concepts, theories and data. Complex Syst. Complexity Sci. **12**(02), 103–107 (2015)
7. Buldyrev, S.V., Parshani, R., Paul, G., et al.: catastrophic cascade of failures in interdependent networks. Nature **464**(7291), 1025–1028 (2010)
8. Gao, J., Buldyrev, S.V., Stanley, H., et al.: Networks formed from interdependent networks. Nat. Phys. **8**(1), 40–48 (2011)

9. Battiston, F., Nicosia, V., Latora, V., et al.: Structural measures for multiplex networks. Phys. Rep. **544**(1), 1–122 (2014)

10. Gao, W.: Research on marketing strategy of communication company in mobile Internet Era. Inf. Commun. **2019**(07), 271+273 (2019)

11. Felipe, T., Monteiro, C., José, L.C, et al.: Experimental evaluation of pulse shaping based 5G multicarrier modulation formats in visible light communication systems. Opt. Commun. **457** (2020)

12. Balabanov, M., Shoham, Y., et al.: Fab: content-based, collaborative recommendation. Commun. ACM, **40**(3), 66–72 (1997)

13. Ma, J.: Research on weighted hybrid recommendation technology based on improved collaborative filtering and bipartite graph network. East China Normal University (2014)

14. Yang, K., Wang, L., Zhou, Z., et al.: Personalized recommendation of scientific literature based on content and collaborative filtering. Inf. Technol. **43**(12), 11–14 (2019)

15. Zhang, X., Jiang, S.: Personalized recommendation algorithm based on weighted bipartite graph. Comput. Appl. **32**(03), 654–657+678 (2012)

16. Jiang, S., Zhang, L., Zhou, N.: Collaborative filtering recommendation based on dynamic changes of user preferences. Comput. Modern. 2020(01), 75–80 (2020)

17. Li, X.: Research on collaborative filtering recommendation algorithm. Comput. Digital Eng. **47**(09), 2118–2122+2136 (2019)

18. Ying, Y.: Research on Recommendation System Based on Bipartite Graph Resource Allocation. University of Jinan (2019)

19. Zhang, H.: Research on personalized recommendation algorithm for social network. Beijing University of Posts and Telecommunications (2018)

20. Nong, Y., Tang, Z.: A collaborative filtering recommendation algorithm based on user attributes and similarity. Microcomput. Appl. **35**(11), 27–29 (2019)

Integrating Local Closure Coefficient into Weighted Networks for Link Prediction

JieHua Wu$^{(\boxtimes)}$ [iD]

Guangdong Polytechnic of Industry Commerce, Guangzhou 510510, GD, China

Abstract. Triadic closure is a simple and fundamental kind of link for-mulation mechanism in network. Local closure coefficient (LCC), a new network property, is to measure the triadic closure with respect to the fraction of length-2 paths for link prediction. In this paper, a weighted format of LCC (WLCC) is introduced to measure the weighted strength of local triadic structure, and a statistic similari-ty-based link prediction metric is proposed to incorporate the definition of WLCC. To prove the metrics effectiveness and scalability, the WLCC formula-tion was further investigated under weighted local Naive Bayes (WLNB) link prediction framework. Finally, extensive experimental studies was conducted with weighted baseline metrics on various public network datasets. The results demonstrate the merits of the proposed metrics in comparison with the weighted baselines.

Keywords: Link prediction · Local closure coefficient · Clustering coefficient · Complex network · Weighted network

1 Introduction

In real world, the complex relationships between various types of entities can be represented as network [1]. In network, links between nodes are the basic and core component. How to utilize link information as an effective computational tool for network analysis has received a fair amount of attention. One of an impor-tant research orient is called link prediction [2]. Link prediction is to predict the existence of links that do not have link relationship yet, or to predict future links by mining the historical network structure information [3]. This technol-ogy is of great significance in various fields. In social network, service providers recommend potential friendships to users with link prediction techniques [4]. In bioinformatics, ideas of many protein-protein iteration function prediction meth-ods origin from some novel ideas of link prediction algorithms [5]. In knowledge graph mining, link prediction is a powerful tool for learning and inferring the potential relationships between entities [6,7]. It have been demonstrated to be helpful to various tasks, such as correlation analysis of businessman, identifica-tion of potential risks of anti-fraud and management of lost customers.

One popular link prediction [8] task can be described as follows: given a known network structure, each node pair (also commonly referred to as potential node

© Springer Nature Singapore Pte Ltd. 2021
J. Zeng et al. (Eds.): ICPCSEE 2021, CCIS 1451, pp. 51–64, 2021.
https://doi.org/10.1007/978-981-16-5940-9_4

pair) that has not yet generated a link relationship is given a similarity score based on the hidden information of network. The higher the score, the more likely two nodes will formulate a link relationship. Therefore, similarity-based link prediction metrics (also called method) is gaining increasing and renewed attention from multiple domains. Most of existing metrics have focused on the topological structure of network, such as common neighbors [8,9], path information [10,11], random walks [12,14], global adjacency matrix [15,16] and so on.

Very recently, cluster [17] or motif [13,18] information such as local closure coefficient has shown great potential for fast and accurate link prediction. The recently proposed Local Closure Coefficient (LCC) is derived form a standard property for measuring the frequency of triadic closure on networks - Clustering Coefficient (CC). Different from CC, LCC measures triadic closure from the center of a wedge implicitly accredits the closure to the center node. It has been proved that LCC-based link prediction metrics [17] are more effective than some classic similarity-based metrics [8,9]. However, the definition of LCC are only defined on unweighted networks. In weight networks, weight information appears between nodes aims to indicate the strength of relationship between nodes or local triadic structure. For example, in social networks mentioned above, the weight indicates the number of comments among users, or the number of common friends between users; for a scientific collaborative network, nodes are researchers and links are their collaborative relationship, weight information indicates the number of conferences or co-authored papers they cooperate together. Obviously, it is very important for the introduction of weight information to judge whether two non-friends will have a friend relationship and to identify the future cooperation possibility between two researchers who have not been work together. Most of the existing similarity-based metrics are based on unweighted networks, that is, when performing link prediction, only attributes of common neighbors are paid attention to, and the strength-weight information of links between potential nodes and common neighbors is ignored.

Although much effort has been made to improve the performance of weighted format of similarity-based metrics, such as weighted proximity measures [19], weighted clustering coefficient [20], weighted local naive Bayes [21] metrics and so on, they usually only focus on improving the efficiency of neighborhood similarity metrics by the definition of weighted clustering coefficient, while neglecting other kinds of local triadic closure mechanism. To achieve this goal, we introduce a new statistic clustering-based measure, referred to as Weighed Local Closure Coefficient (WLCC) [17] and propose a WLCC-based similarity link prediction metric. The innovation is that WLCC concern a more natural way to consider the weight value of neighbors and structural formulation pattern in local view. To further highlight the effectiveness and scalability of proposed metric, we plug WLCC into a well-known weighted network link prediction framework - WLNB (Weighted Local Naive Bayes). In addition, we present three variant of WLNB-LCC based on classic CN, AA, RA metrics. Finally, We conduct experimental evaluations using real-world weighted networks. The results indicate that our method is able to improve the weighted network link prediction accuracy.

Organization. Section 2 reviews the related works. Section 3 formulates the problem and details the proposed metrics and its extend format. Section 4 shows experimental results that prove the effectiveness of our metrics. Section 5 concludes the paper.

2   Related Works

Link prediction [1] predicts the existence of links between nodes based on the observed network structure properties. This technology is generally divided into two categories: one portion are based on the similarity between nodes and the other are learning methods based on network features.

The basic idea of the first type of methods [8] is to combine the concepts of statistics, sociology, graph theory, and topology to mine the network structural information to calculate the link probability (similarity score) between two potential node pairs. Generally, the greater similarity between two potential node pairs, the more likely they formulate a link. Because network structure information can be expressed as many types, the link prediction algorithms can be divided into similarity based on neighbor nodes, path information, random walks, community information and so on. More detail, the representative metrics based on similarity of neighbor nodes are: Common Neighbors (CN) [8], Adamic Adar (AA) [9], Adaptive degree penalization (ADP) [22], representative metrics based on similarity of path information are: Katz [10], Local Path (LP) [11], the representative algorithms based on the similarity of random walk are: Local Random Walk (LRW) [12], Return Random Walks (RRW) [14]). These metrics are simple and have the advantages of fast speed and high efficiency. Thus, research in recent years has made some progress in other fields. MPD (Matrix Perturbation and Decomposition) [15] and Low Rank (LR) [16] try to capture the network global structural property and obtain a better performance. With deeper research on network analysis, new metrics are constantly being proposed. How to explore the latent influence of network structure on link formation worth further research. Nevertheless, the focus of our article is weighted link prediction problem.

Weighted network represents a network that contains weight information. Directly using the unweighted similarity algorithm will ignore the existence of weight information between nodes, affecting the performance and authenticity of link prediction. Tsuyoshi Murata et al. [23] noticed the weighted link prediction problem early, and they proposed some weighted proximity measures for social link prediction. Lin Lv et al. [19] incorporated the Weak-Ties Theory to local similarity metrics, and prove that weak ties play a important role in the link prediction task. Boyao Zhu et al. [24] presented a weighted model based on the mutual information of local network structures. JieHua Wu et al. [21] adapt the definition of clustering coefficient into a weighted scenario and proposed a weighted local naive Bayes (WLNB) probabilistic link prediction framework. On the other hand, HR De S et al. [25] investigated weight information with network structure to construct the classification feature and improve supervised link prediction task. Behnaz Moradabadi et al. [26] take advantage of using learning automata to learn the true weight of the corresponding link according to the current networks links weight information. However, the learning mechanism of these methods are totally different from that of similarity-based focused in this article.

3 Preliminaries

3.1 Notation

Given a weighted network $G = (V, E, W)$, where V is the set of all nodes in the network, $E \subseteq V \times V$ is the set of links between all nodes in the network, A and W are the non-weighted and weight matrix of G respectively. There is a functional relationship: $\varphi : e(u, v) \rightarrow w_{uv}$, which represents the link between nodes u and v, and w_{uv} represents the weight of the corresponding link; if there is no link between two nodes, then the corresponding $w_{uv} = 0$. The degree of node u is expressed as $d_u = |N(u)|$, where $N(u)$ is the set of adjacent (neighbor) nodes.

3.2 Problem Definition

The task of traditional weighted network link prediction can be described as the following process:

1. Divide G into training network G_T and prediction network G_P according to ratio r.
2. Calculate the similarity index of each node pair (or potential node pairs, node pairs that need to be predicted) in G_T.
3. Sort the similarity scores of all node pairs, take the largest N as candidate sequences, and then compare with the set of links that actually exist in the G_P, thus hit rate is the prediction accuracy.

3.3 General Weighted Similarity Metric

Link prediction metrics generally rely on network topological features, which usually can be induced as a similarity function. Most features are based on common neighbors $CN(u, v)$, i.e., forming closing triangles structure with the potential node pair. Since it is directly connected to a potential node pair, many weighted network link similarity metrics are defined according to this structure:

$$sim_w(u, v) = \sum_{\omega \in CN(u,v)} f_w(\omega) \tag{1}$$

Such traditional metrics that rely on common neighbors aims to define $f_w(\omega)$, such as the Jaccard Coefficient metric [2] and Adamic/Adar metric [9], etc.

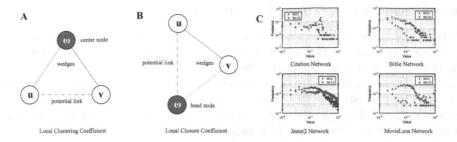

Fig. 1. Illustration of local clustering coefficient (A), local closure coefficient (B) [17] and their distribution of four weighted networks (C).

3.4 The Local Closure Coefficient

For a better understanding of our approach, we first overview the novel definition of Local Closure Coefficient (LCC) [17,27] which describes the mechanism formation of triangles in complex networks. The above mentioned traditional similarity metrics are effective and efficient as they derive from the local triadic structure. Fortunately, recent works have proved that weighted local clustering coefficient is more capable of capturing weighted triadic information for link prediction [21]. This metric is the sum of the weighted clustering coefficients of nodes u and v. The definition is:

$$WCC_w(u, v) = \sum_{\omega \in CN(u,v)} cc_w(\omega) \tag{2}$$

where $cc_w(\omega)$ denotes the weighted format of clustering coefficient containing ω.

In the explanation of local clustering coefficient, it is defined as the fraction of wedges (i.e., length-2 paths) with center node (this is node ω in Fig. 1A) that are closed. This means that there is an edge connecting the two ends of the wedge, inducing a triangle. However, some real complex systems, there exits some other phenomena. For example in social networks with local clustering nature, if a user follows an artificial intelligence user, then he may also concern about this Weibo user's follower in the same field. Or in a knowledge graph, if one entity A infers another entity B, and entity B can lead to entity C, then entity A is likely to have a causal relationship with entity C. This statistic measures clustering from the head node (this is node ω in Fig. 1B) of a triangle (instead of from the common node, which is the definition of clustering coefficient). More detail, the explanation of LCC is derived from the structure of length-2 path (i.e., wedge). Thus, local closure coefficient can be calculated as:

$$lcc(u) = \frac{2T(u)}{\sum_{v \in N(u)} (d(v) - d(u))} \tag{3}$$

3.5 Weighted Local Closure Coefficient

Furthermore, it is well known that many real-world networks involve relationships with rich information - link/node weighted. In weight networks, weight is crucial and it plays an important role in link prediction task. It provides additional information for describing the network properties. Obviously, Eq. (3) do not take into account the weight information. To overcome this issue, we try to generalize local closure coefficient to weighted networks by considering different ways to weight the neighbours of a node.

In weighted network, weights are given to links, so weight is usually called edge weight. Here, when an end-node u of a link is specified, we denote the weight of a node called node weight. Node weight describes the closeness of the node to the surrounding network structure. It is formed by summing the edge weights of this node and its adjacent nodes and is defined as:

$$s_u = \sum_{v \in N(u)} w_{uv} \tag{4}$$

For any node u and its neighbor node $v \in N(u)$, each edge containing v (apart from the edge (u, v)) contributes one wedge headed at node u and use $d(u)$ and $d(v)$ to denote. As regard to this issue in weighted network, the unweighted definition of $d(u)$ and $d(v)$ should be extended to weighted format, which is naturally be considered as s_u and s_v respectively, the formal definition of WLCC can be provided as follows:

$$lcc_w(u) = \frac{2T(u)}{\sum_{v \in N(u)}(s_v - s_u)} \tag{5}$$

Here we highlight again the fundamental difference between these two network statistic measures and the advantage of our proposed one. On one hand, Eq. (5) takes into account the wedges that a head node form, which preserve the idea of the local closure coefficient. On the other hand, the weights of these triangles structure also have been incorporated. Moreover, the distribution of A and B are basically the same (Fig. 1C), so the WLCC definition can reflect the structural characteristics of network. It's worth noting that the weighted definition of in $T(u)$ WLCC is slightly different from that of WCC. In WCC, $T(u)$ is extended to $\sum_{(u,v)} \frac{w_{uw}+w_{wv}}{2} a_{uv} a_{uw} a_{wv}$, which is the triangle number in such local structure formulated by the potential links and its common neighbors. a_{uv} is binary entry in adjacent matrix A, where 1 means there exists a link between node v_i and v_j and 0 represents the opposite. For reasons that are not easy to normalize and poor experimental results, the number of unweighted triangles retained in the definition of numerator in Eq. (5).

3.6 Metric Extension

In our previous work [21], we have proved that WCC can be effectively embedded in the Local Naive Bayes (LNB) model [28] and form a new weighted link prediction framework: WLNBs. WLNBs contains WLNCN, WLNBAA, WLNBRA based on 3 classic link prediction metrics. For simplicity, we use WLNCN an example to illustrate and its similarity-based equation is:

$$WLNBCN(u,v) = |CN(u,v)|\log r + \sum_{\omega \in N(u,v)} \log \frac{cc_w(\omega)}{1 - cc_w(\omega)} \qquad (6)$$

The ratio r equals between the number of connection links $|E|$ and disconnection links scale $|V|(|V|-1)/2 - |E|$. In order to the highlight the effectiveness of WLCC, we also attempt to extend such definition to WLNBs. In this context, how to bridge the different between these frameworks is the key issue. Since the left half of the formula $|CN(u,v)|\log r$ is a constant value, the crucial part which affects the similarity score is in the right. Thus, we can naturally replace $cc_w(\omega)$ with $lcc_w(\omega)$ and rewrite Eq. (6) as follows:

$$WLNBCN - LCC(u,v) = |CN(u,v)|\log r + \sum_{\omega \in N(u,v)} \log \frac{lcc_w(\omega)}{1 - lcc_w(\omega)} \qquad (7)$$

Similarity, the corresponding form of WLCC to the Weighted Local Naive Bayes (WLNB) form of AA and RA metrics respectively are:

$$WLNBAA - LCC(u,v) = \sum_{\omega \in N(u,v)} \frac{\log r}{\log |N(\omega)|} + \sum_{\omega \in N(u,v)} \frac{\log \frac{lcc_w(\omega)}{1 - lcc_w(\omega)}}{\log |N(\omega)|} \qquad (8)$$

$$WLNBRA - LCC(u,v) = \sum_{\omega \in N(u,v)} \frac{\log r}{|N(\omega)|} + \sum_{\omega \in N(u,v)} \frac{\log \frac{lcc_w(\omega)}{1 - lcc_w(\omega)}}{|N(\omega)|} \qquad (9)$$

4 Experiments

4.1 Datatset

We consider the following 6 weighted networks drawn from disparate fields:

- **Citation Network** [29]. It is a country level paper citation network. Each node represents a city and edge represents that the author from one place cites the author from the other place.
- **Bible Network** [30]. This undirected network contains nouns (places and names) of the King James Version of the Bible and information about their co-occurrences. A node represents one of the above noun types and an edge indicates that two nouns appeared together in the same Bible verse. The weighted information denotes how often two nouns occurred together.
- **MovieLens Network** [31]. This network contains ten million movie ratings from http://movielens.umn.edu/. Left nodes are users and right nodes are movies. An edge between a user and a movie indicates that the user has rated the movie with the rating value attached to the edge.

– **Jester2 Network** [30]. This network is in the collection of Miscellaneous Networks. Edges weighted represent ratings (of items, movies, music, etc.).
– **Bitcoinalpha Network** [32]. This is who-trusts-whom social network based on a Bitcoin trading platform called Bitcoinalpha. The weight information is the source's rating for the target.
– **Bitcoinotc Network.** [32] This is who-trusts-whom This is who-trusts-whom social network based on a Bitcoin trading platform called Bitcoinotc. The weight information is the source's rating for the target.

The statistics of the datasets are shown in Table 1, including the number of nodes and links, the average degree $\langle k \rangle$, average cluster coefficient $\langle c \rangle$, the triad structure number $\langle t \rangle$ and the average common neighbors number $\langle O \rangle$.

Table 1. Statistics of networks.

Dataset	#Nodes	#Links	$\langle c \rangle$	$\langle k \rangle$	$\langle t \rangle$	$\langle O \rangle$
Citation network	226	28869	0.8597	142.389	699783	103.643
Bible network	1773	16401	0.7208	18.501	19966	1.902
MovieLens network	1682	100000	0.3636	112.763	2408182	22.494
Jester2 network	4338	178352	0.7683	81.64	4003188	18.359
Bitcoinalpha network	3783	24186	0.1766	3.714	22153	1.531
Bitcoinotc network	5881	35592	0.1775	7.158	33493	1.406

4.2   Comparison Methods and Evaluation Metric

Comparison Methods. We consider several similarity-based traditional and state-of-the-art link prediction baselines. We first use three aforementioned weighted local metrics: WCN [23], WAA [23] and WCC [20]. Then we also adopt a path based method Weighted Path (WLP) [11]. The detail formula of baselines are:

– **Weighted CN metric (Weighted CN, WCN).** [23] In an unweighted network, this formula calculates the number of common neighbors, which is generalized to the weighted network. The weight is added to the calculation formula, that is, each common neighbor is multiplied by the mean of the edge weights of it and u and v. The definition is: $f_w(\omega) = \frac{w_{u\omega} + w_{\omega v}}{2}$.
– **Weighted AA metric (Weighted AA, WAA).** [23] The degree of nodes in WAA should be added with the node weight accordingly. The definition is: $f_w(\omega) = \frac{w_{u\omega} + w_{\omega v}}{2 \cdot log(1 + s_\omega)}$
– **Weighted Clustering Coefficient metric (Weighted CC, WCC).** [20] This metric is the sum of the weighted clustering coefficients of nodes u and v. The definition is: $f_w(\omega) = cc_w(\omega)$.

- **Weighted Local Path (Weighted LP, WLP).** [11] It considers two-step information and the influence of three-step path. The formula is as follows: $WLP(u, v) = W^2 + \alpha W^3$. W^2 and W^3 denote weighted adjacency matrix where the nodes having 2-length and 3-length path respectively.

To show the superiority of the weighted algorithm over the unweighted format, we also use non weighted metric LCC. Our proposed metric is referred as WLCC. Furthermore, WLNB-LCC denotes the summarization of our extensive metrics.

Evaluation Metric. We adopt a general matric called Precision to evaluate these baselines. The 10-folder cross-validation method is used. Each folder is performed 10 times. Each time a subset is selected as the test network. The average cross-validation precision of 10 times is used as the evaluation index. The hit ratio Precision@N reflects the accuracy of prediction and measures the proportion of correct hit links p_N versus candidate Top N. The definition is:

$$Precision@N = \frac{p_N}{N} \tag{10}$$

Table 2. Prediction performance on six weighted networks with all baselines.

Network	TopN	WCN	WAA	WLP	LCC	WCC	WLCC
Citation network	600	0.7883	0.7911	0.7816	0.2883	0.9833	**0.9858**
	800	0.7273	0.7267	0.7221	0.2662	0.9461	**0.9521**
	1000	0.6649	0.6656	0.6756	0.2261	0.8892	**0.8953**
Bible network	600	0.1366	0.1551	0.1366	0.0483	0.1133	**0.3002**
	800	0.1200	0.1412	0.1187	0.0462	0.0975	**0.2575**
	1000	0.1009	0.1224	0.1000	0.0562	0.0833	**0.2328**
MovieLens network	600	0.2983	0.2983	0.3053	0.0783	**0.3183**	0.3116
	800	0.2887	0.2854	0.2962	0.0757	0.2837	**0.3062**
	1000	0.2751	0.2740	0.2787	0.0682	0.2846	**0.2869**
Jester2 network	600	0.1466	0.0016	0.1550	0.0633	0.3033	**0.4016**
	800	0.1425	0.0012	0.1487	0.0625	0.2622	**0.3527**
	1000	0.1337	0.0010	0.1411	0.0610	0.2366	**0.3188**
Bitcoinalpha network	600	0.1628	0.1643	0.1628	0.1121	0.0226	**0.1863**
	800	0.1337	0.1378	0.1337	0.1071	0.0214	**0.1487**
	1000	0.1131	0.1226	0.1131	0.0965	0.0188	**0.1272**
Bitcoinotc network	600	0.1283	0.0833	0.1283	0.0151	0.0233	**0.1633**
	800	0.1054	0.0875	0.1062	0.0137	0.0187	**0.1475**
	1000	0.0941	0.0870	0.0950	0.0116	0.0116	**0.1327**

4.3 Experiment Results

In the rest of this section, we first assess how well different methods perform over various datasets and compare their performance based on different measures. Next, we discuss the efficiency of our proposed method by measuring and comparing its running time against the other baselines. Finally, we analyze the effect of different parameters and problem configurations on the performance of competitive methods.

Comparative Performance. In the first part of experiment, we evaluate the prediction performance of our proposed WLCC with other baselines and report the results with different N value (600, 800 and 1000) in Table 2. The best results in each case are marked in bold. We have the following observations: (1) The overall performance drops as the N value increases in each network. This indicates that the experimental setting and results are effective and conform to the general rules of link prediction task. (2) We can see that in all network, WLCC achieve better performance than other baselines under all different N value. For example in Citation Network, WLCC obtain an average (25.054%, 24.611%, 26.126%, 241.935%, 0.254%) improvements compare with WCN, WAA, WLP, LCC, WCC when N value is 600, and (30.905%, 31.017%, 31.852%, 257.663%, 0.634%), (34.652%, 34.510%, 32.519%, 295.976%, 0.686%) improvements when N value is 800 and 1000 respectively. This results shows the extension of LCC into a weighted format can effectively deliver local triadic weighted information into the similarity framework and make link prediction more accurately. And also deduce that WLCC can be adopted to the prediction task in different type of weight networks. (3) The metric with weighted local coefficient WCC performs slightly worse than WLCC in most cases and outperform three local similarity weighted metrics in half cases. It not merely demonstrates WLCC exhibit a better triad closure mechanism in weighted network evolution process, but also show that local triadic density are more telling than the degree and path information in similarity-based link prediction metrics. Meanwhile, it is evident that the non-weighted metric LCC dose not work as well as WLCC, even showing very poor results in some cases. Especially in MovieLens Network, the results of LCC have -297.957%, -304.491%, -320.674% weaker predicability in terms of N value 600, 800, 1000 comparing with WLCC. The result clearly yields that WLCC perform well on capturing the weighted information of network and can extend to the weighted link prediction scenario. (4) WCN, WAA and WLP perform worse than WLCC in all cases, which demonstrates the superiority of local closure coefficient on weighted network link prediction. However, We also notice that they outperform LCC in most cases, which indicates that the incorporation of link weighted or node weighted should be considered.

In this part, we analyze the effect of different ratio $r \subseteq \{0.6, 0.65, 0.7, 0.75, 0.8, 0.85, 0.9\}$ of training network on the prediction performance. The results on all networks are shown in Fig. 1. First, we can see that results in different cases show similar trends by increasing the number of training samples. That means as training ratio increase, the overall performance gradually decrease. This

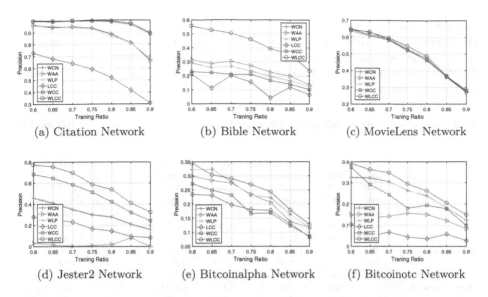

(a) Citation Network   (b) Bible Network   (c) MovieLens Network

(d) Jester2 Network   (e) Bitcoinalpha Network   (f) Bitcoinotc Network

Fig. 2. Performance of link prediction with different percentage of training network.

phenomenon indicates a negative effect of sample of training data for weighted network link prediction. But the most important is, WLCC (red line with circle point curve) noticeably lie below against all other baselines and its decline is the smallest. For example in Citation Network, when predicting weighted links, the precision of WLCC drops down from 0.999 at ratio $r = 0.6$ to 0.902 at top $r = 0.9$, which the decrease is 10.753%, while this number of WCN, WAA, WLP, LCC, WCC are 42.456%, 42.878%, 45.359%, 12.415%, 11.87%. It demonstrates that WLCC obtain a better, stable performance and the advantage of connect local closure coefficient and weighted information (Fig. 2).

Performance Analysis. To explore the contributions of weighted local closure coefficient definition to the prediction task, we further embed it to the WLNB model for extensive experiments. The results are shown in Table 3. We can have the following observations: (1) In most cases, three versions of WLNB exploiting LCC (including WLNBCN-LCC, WLNBAA-LCC, WLNBRA-LCC, called WLNB-LCC) are better than corresponding WLNB metrics. More detail, WLNBCN-LCC averagely outperform WLNBCN with a (0.1638%, 4.812%, 3.660%, 10.763%, 106.318%, 94.662%) increase in each network. Similarly, we can also find that WLNBAA-LCC and WLNBRA-LCC outperform the corresponding metric WLNBAA and WLNBRA with a (0.179%, 10.995%, 2.048%, 14.760%, 85.985%, 122.311%) and (−0.633%, −7.346%, 21.172%, 7.642%, 110.632%, −60.309%) increase respectively. Notice that performance of WLNBAA-LCC is slightly better than that of WLNBCN-LCC. Such phenomenon is consistent with the unweighted WLNB link prediction metrics. Some exception cases is the WLNBCC-RA metric in Citation and Bible Network. The reason may be related to the structural properties of the network - the number of triangles in the network is relatively small. And the RA metric use node degree as denominator of the formula, thus obtain poor results. Generally, the results show that WLNB-LCC can take advantage of definition of LCC and archive the best and stable performance.

Table 3. Performance of link prediction with different WLNB-LCC and WLNB model.

Network	TopN	WLNBCN	WLNBCN-LCC	WLNBAA	WLNBAA-LCC	RA	WLNBRA-LCC
Citation network	500	0.9988	1.0000	0.9988	0.9992	0.9209	0.9082
	1000	0.8914	0.8911	0.8886	0.8919	0.8466	0.8431
	1500	0.7348	0.7382	0.7382	0.7392	0.7094	0.7099
Bible network	500	0.2282	0.2258	0.3882	0.3963	0.5837	0.5280
	1000	0.1752	0.1756	0.2676	0.2882	0.4003	0.3569
	1500	0.1496	0.1568	0.2110	0.2342	0.2981	0.2762
MovieLens network	500	0.3424	0.3439	0.3413	0.3445	0.2628	0.3484
	1000	0.2707	0.2956	0.2764	0.2937	0.2299	0.2809
	1500	0.2541	0.2634	0.2538	0.2590	0.2064	0.2501
Jester2 network	500	0.3650	0.4552	0.3477	0.4132	0.2001	0.2403
	1000	0.2532	0.3075	0.2384	0.2933	0.1650	0.1874
	1500	0.2109	0.2336	0.2019	0.2317	0.1518	0.1634
Bitcoinalpha network	500	0.0895	0.1693	0.0604	0.1786	0.0242	0.0761
	1000	0.0539	0.1146	0.0571	0.1214	0.0365	0.0807
	1500	0.0459	0.0947	0.0528	0.0982	0.0348	0.0733
Bitcoinotc network	500	0.0532	0.1481	0.0697	0.1521	0.0318	0.0318
	1000	0.0575	0.1249	0.0539	0.1265	0.0375	0.0108
	1500	0.0562	0.1094	0.0502	0.1116	0.0388	0.0154

5   Conclusion

In this article, we studied the problem of link prediction in weighted network. Specifically, we formulate the problem as a ranking task and proposed a simple but novel link prediction method. We first introduce a recently proposed connectivity patterns of network: local closure coefficient, then embed it into weighted similarity-based metric WLCC. To showcase the effectiveness and expansibility of WLCC, we extend such definition to the WLNB framework. The experimental results show that WLCC can substantially enhance link prediction accuracy in weighted network. More importantly, we find that each variant of WLNB-LCC can be further configured with LCC and outperforms corresponding comparison methods.

For future works, there are several directions need to be considered. First, we will investigate how to use the notion of another local clustering structure: motifs to extend the link prediction techniques. Second, it is also necessary to examine the property of LCC for better link prediction performance in other types of networks, such as multiplex network, signed network, and so on.

Acknowledgements. This work is supported by Basic and Applied Basic Research Foundation of Guangdong Province (No. 2020A1515011495) and Guangzhou Science and Technology Foundation Project (No. 202002030266).

References

1. Getoor, L., Diehl, C.P.: ACM SIGKDD Explor. Newslett. **7**, 3–12 (2005)
2. Liben-Nowell, D., Kleinberg, J.: The link prediction problem for social networks. In: Proceedings of the Twelfth International Conference on Information and Knowledge Management, New Orleans, LA, USA, vol. 9999, pp. 556–559. ACM (2003)
3. Martinezctor, V., Berzal, F., Cubero, J.-C.: A survey of link prediction in complex networks. ACM Comput. Surv. **49**, 1–33 (2016)
4. Li, S., Song, X., Lu, H., Zeng, L., Shi, M., Liu, F.: Friend recommendation for cross marketing in online brand community based on intelligent attention allocation link prediction algorithm. Expert Syst. Appl. **139**, 112839 (2020)
5. Benson, A.R., Abebe, R., Schaub, M.T., Jadbabaie, A., Kleinberg, J.: Proc. Nat. Acad. Sci. 115, E11221–E11230 (2018)
6. Kazemi, S.M., Poole, D.: Simple embedding for link prediction in knowledge graphs. In: Proceedings of Advances in Neural Information Processing Systems, Montreal, Canada, pp. 4284–4295. ACM (2018)
7. Tay, Y., Luu, A.T., Hui, S.C., Brauer, F.: Random semantic tensor ensemble for scalable knowledge graph link prediction. In: Proceedings of the Tenth ACM International Conference on Web Search and Data Mining, Cambridge, UK, p. 751760. ACM (2017)
8. Lv, L., Zhou, T.: Link prediction in complex networks: a survey. Physica A Stat. Mech. Appl. **390**, 1150–1170 (2011)
9. Adamic, L.A., Adar, E.: Friends and neighbors on the web. Soc. Netw. **25**, 211–230 (2003)
10. Katz, L.: A new status index derived from sociometric analysis. Psychometrika **18**, 39–43 (1953)
11. Lv, L., Jin, C.-H., Zhou, T.: Similarity index based on local paths for link prediction of complex networks. Phys. Rev. E **80**, 046122 (2009)
12. Liu, W., Lu, L.: Link prediction based on local random walk. EPL (Europhys. Lett.) **89**, 58007 (2010)
13. Wang, L., Ren, J., Xu, B., et al.: MODEL: motif-based deep feature learning for link prediction. IEEE Trans. Comput. Soc. Syst. **7**(2), 503–516 (2020)
14. Manuel, C.: Return random walks for link prediction. Inf. Sci. **51**, 99–107 (2020)
15. Wang, W., Cai, F., Jiao, P., Pan, L.: A perturbation-based framework for link prediction via non-negative matrix factorization. Sci. Rep. **6**, 38938 (2016)
16. Pech, R., Hao, D., Pan, L., Cheng, H., Zhou, T.: Link prediction via matrix completion. EPL (Europhys. Lett.) **117**, 38002 (2017)
17. Yin, H., Benson, A.R., Leskovec, J.: The local closure coefficient: a new perspective on network clustering. In: Proceedings of the Twelfth ACM International Conference on Web Search and Data Mining, Melbourne, Australia, pp. 303–311. ACM (2019)
18. Liu, Y., Li, T., Xu, X.: Link prediction by multiple motifs in directed networks. IEEE Access **8**, 174–183 (2019)
19. Lv, L., Zhou, T.: Link prediction in weighted networks: the role of weak ties. EPL (Europhys. Lett.) **89**, 18001 (2010)
20. Wu, J., Zhou, B., Shen, J.: Algorithm of integrating local weighted clustering coefficients for link prediction. Appl. Res. Comput. **12**, 15 (2018)
21. Wu, J., Zhang, G., Ren, Y., Zhang, X., Yang, Q.: Weighted local naive bayes link prediction. J. Inf. Process. Syst. **13**, 914–927 (2017)

22. Rafiee, S., Salavati, C., Abdollahpouri, A.: CNDP: link prediction based on common neighbors degree penalization. Physica A Stat. Mech. Appl. **539**, 122950 (2020)
23. Murata, T., Moriyasu, S.: Link prediction of social networks based on weighted proximity measures. In: Proceedings of International Conference on Web Intelligence, California, USA, pp. 85–88. IEEE/WIC/ACM (2007)
24. Zhu, B., Xia, Y.: Link prediction in weighted networks: a weighted mutual information model. PloS ONE **11**, e0148265 (2016)
25. De Sa, H.R., Prudencio, R.B.C.: Supervised link prediction in weighted networks. In: Proceedings of the 2011 International Joint Conference on Neural Networks, San Jose, California, USA, pp. 2281–2288. IEEE (2011)
26. Moradabadi, B., Meybodi, M.R.: Link prediction in weighted social networks using learning automata. Eng. Appl. Artif. Intell. **70**, 16–24 (2018)
27. Yin, H., Benson, A.R., Ugander, J.: Measuring directed triadic closure with closure coefficients. arXiv preprint arXiv:1905.10683 (2019)
28. Liu, Z., Zhang, Q.-M., Lu, L., Zhou, T.: Link prediction in complex networks: a local naive bayes model. EPL (Europhys. Lett.) **96**, 48007 (2011)
29. Pan, R.K., Kaski, K., Fortunato, S.: World citation and collaboration networks: uncovering the role of geography in science. Sci. Rep. **2**, 902 (2012)
30. Kunegis, J.: KONECT: the Koblenz network collection. In: Proceedings of international Conference on World Wide Web, Seoul, Korea, pp. 1343–1350. ACM (2011)
31. Harper, F.M., Konstan, J.A.: The movielens datasets: history and context. ACM Trans. Interact. Intell. Syst. (TIIS) **5**, 1–19 (2015)
32. Kumar, S., Hooi, B., Makhija, D., Kumar, M., Faloutsos, C., Subrahmanian, V.S.: Rev2: fraudulent user prediction in rating platforms. In: Proceedings of the Eleventh ACM International Conference on Web Search and Data Mining, California, USA, pp. 333–341. ACM (2018)

A Transformer Model-Based Approach to Bearing Fault Diagnosis

Zhenshan Bao[1], Jialei Du[1], Wenbo Zhang[1(✉)], Jiajing Wang[2], Tao Qiu[3], and Yan Cao[3]

[1] Faculty of Information Technology, Beijing University of Technology, Beijing 100124, China
zhangwenbo@bjut.edu.cn
[2] China North Vehicle Research Institute, Beijing 100072, China
[3] Faculty of Environment and Life, Beijing University of Technology, Beijing 100124, China

Abstract. Bearings are an important component in rotating machinery and their failure can lead to serious injuries and economic losses, therefore the diagnosis of bearing faults and the guarantee of their smooth operation are essential steps in maintaining the safe and stable operation of modern machinery and equipment. Traditional bearing fault diagnosis methods focus on manually designing complex noise reduction, filtering, and feature extraction processes, however, these processes are too cumbersome and lack intelligence, making it increasingly difficult to rely on manual diagnosis with large amounts of data. With the development of information technology, convolutional neural networks have been proposed for bearing fault detection and identification. However, these convolutional models have the disadvantage of having difficulty handling fault-time information, leading to a lack of classification accuracy. So this paper proposes a transformer-based fault diagnosis method, using the short-time Fourier transform to convert the one-dimensional fault signal into a two-dimensional image, and then input the two-dimensional image into the transformer model for classification. Experimental results show that the fault classification can reach an accuracy of 98.45%.

Keywords: Fault diagnosis · Transformer · Convolutional neural network · Deep learning

1 Introduction

Fault diagnosis is an essential step in maintaining the normal operation of modern machinery and equipment, and bearings are an essential part of modern machinery, especially in rotating machinery. Bearing failures account for a large proportion of mechanical equipment failures [1]. It is therefore essential to ensure the proper functioning of the bearings. Early bearing fault diagnosis is the use of artificial listening stick contact bearings, and then the use of hearing to determine the health of the bearing, this method is of subjective influence and very vulnerable to environmental interference. With the development of fault diagnosis methods, a variety of bearing fault detection means began to emerge, and the fault diagnosis technology of bearings is also constantly developing.

© Springer Nature Singapore Pte Ltd. 2021
J. Zeng et al. (Eds.): ICPCSEE 2021, CCIS 1451, pp. 65–79, 2021.
https://doi.org/10.1007/978-981-16-5940-9_5

Traditional algorithms require complex signal filtering and noise reduction processes and feature design and selection before performing bearing fault diagnosis, however, with the advent of Industry 4.0 and Big Data, purely manual methods of extracting fault features will become increasingly unfeasible in the face of massive amounts of data. With the progress of information technology in recent years, data-driven bearing fault diagnosis methods have gradually become a hot spot for research [2]. However, most data-driven methods usually deal with relatively simple scenarios of bearing fault diagnosis and should focus more on the identification of the type of bearing fault.

In recent years, artificial intelligence techniques have been used in various fault detection and diagnosis [3], but some problems are difficult to solve. With the development of computer information technology, deep learning has been proposed as an emerging neural network that enables the approximation of complex functions and distributed representation of input data by learning deep nonlinear network structures. Its robustness, learning ability, powerful characterization, and non-linear modeling capabilities have led to success in signal detection [4], image recognition [5], natural language processing [6], audio recognition [7], and feature extraction [8].

Among the deep network model structures, there are several common types of model structures, such as deep confidence networks [9], auto-coding networks, convolutional neural networks [10], and recurrent neural networks [11]. Convolutional neural networks have a comparative advantage in extracting the underlying features and visual structure, and also in speed due to traditional methods. Based on the advantages of convolutional neural networks, there have been researchers who have applied them to bearing fault diagnosis [12]. However, at the same time, convolutional neural networks also have some disadvantages, such as the possible loss of temporal information when dealing with time-dependent fault sequences, resulting in poor classification accuracy.

Therefore, this paper proposes a transformer-based fault diagnosis method to address the shortcomings of traditional convolutional neural networks in fault diagnosis applications, to obtain better classification accuracy. The paper is structured as follows: Sect. 2 presents the reasons for using the transformer to solve fault diagnosis; Sect. 3 presents the specific methodology; Sect. 4 presents the experimental results and discussion; Sect. 5 summarizes the work of the paper and future research directions.

2   Related Work

2.1   Status of Fault Diagnosis

Methods for solving fault diagnosis using traditional artificial intelligence techniques generally include expert systems [13], Bayesian networks [14] rough sets [15], Petri nets [16], neural networks [17], and other methods. Expert systems slow down when the model becomes complex and the number of rules is large because their inference engines need to search for all the rules at each inference; Bayesian networks are complex to train and cannot handle the changing results based on the combination of features. This means that the variables are required to be independent of each other; Rough sets become poorly diagnosed when key information is missing and are prone to combinatorial explosion problems; Petri nets also suffer from the same low fault tolerance and combinatorial explosion. For artificial neural networks, on the other hand, there are sample acquisition problems, the problem of falling into local minima and the problem of gradient disappearance explosion, etc. Moreover, most of the above methods are aimed at fault detection or diagnosis of switching information, which has a lag compared to real-time data monitoring and is less capable of handling large data.

Therefore, to solve the problems of traditional artificial intelligence techniques, deep learning-based fault diagnosis methods began to emerge. The self-encoder approach allows a high-level feature representation of mechanical data to be obtained in an automatic learning manner [18]; Deep confidence networks can extract features from vibration signals containing three morphologies: temporal, frequency and time-frequency, then construct a Bernoulli Gaussian deep Boltzmann machine (GDBMS) to apply each of the three morphological features, and finally classify the three GDBMSs through a support vector framework and output the final results [19]; Convolutional neural networks were initially mainly used to process two-dimensional images. Due to their powerful cognitive computing capabilities, scholars began to introduce them into the field of fault diagnosis, which can well characterize the complex mapping relationships between signals and health states and improve the diagnostic analysis of diverse, non-linear, and high-dimensional health monitoring data in the context of big data [20–23]. Although convolutional neural networks have many advantages, such as their use of convolutional kernels or filters to continuously extract abstract high-level features, many studies have shown that their actual field of perception is much smaller than their theoretical field of perception, which does not allow us to fully exploit contextual information for feature capture. Although we could stack deeper and deeper convolutional layers, this would obviously cause the model to become too bloated and the computation would increase dramatically, defeating the original purpose. The advantage of the transformer is that it captures global contextual information in an attentional way to create a long-range dependency on the target, allowing for more powerful features to be extracted in fault analysis.

2.2 Status of the Transformer

The transformer is a landmark model proposed by Google in 2017, and key technology in the language AI revolution. Previous SOTA models were based on recurrent neural networks (RNN, LSTM, etc.). Essentially, recurrent neural networks serially process data, which corresponds to the NLP task of processing one word at each time step in the order of the words in the sentence. The great innovation of Transformer is the parallelization of language processing compared to this serial model: All words in the text can be analyzed at the same time, rather than in sequential order. To support this parallelized processing, the Transformer relies on attention mechanisms. The attention mechanism allows the model to consider the interrelationship between any two words, independent of their position in the text sequence. Decide which words or phrases should be given more attention by analyzing the two-by-two relationships between words. The Transformer uses the Encoder-Decoder architecture, Fig. 1 shows the structure of the Transformer. The left half of the middle is Encoder and the right half is Decoder [24].

Most of the existing Transformer-based models are only relevant to NLP tasks. However, some recent papers have pioneered the cross-domain application of the Transformer model to computer vision tasks, with good results [25]. This is also considered by many AI scholars to be ushering in a new era in CV, and may even replace traditional convolutional operations altogether.

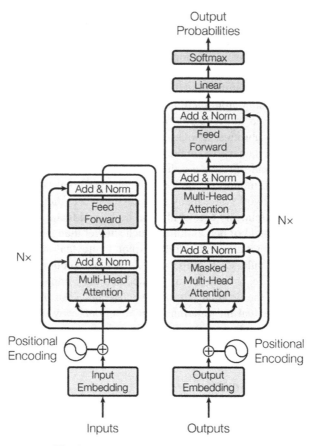

Fig. 1. Encoder-Decoder Architecture [24]

This paper is also inspired by existing transformer models working in the CV field and proposes a transformer-based approach to fault diagnosis image classification. The method takes advantage of the transformer's ability to capture contextual information and can make better use of the temporal information in the fault sequence, thus solving the problem of poor classification accuracy of convolutional neural networks due to the lack of temporal information.

3   Fault Diagnosis Method Based on Transformer Model

The fault diagnosis method proposed in this paper is divided into three main steps:

(1)   Input of raw data.
(2)   Conversion of raw data to 2D time-frequency images using the short-time Fourier transform.
(3)   Classification of 2D time-frequency images as input to the transformer model for fault diagnosis.

Figure 2 depicts the main flow of the method.

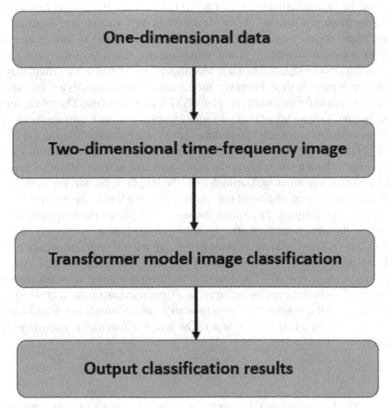

Fig. 2. The main process of fault diagnosis method based on transformer

3.1   Data Pre-processing Using the Short-Time Fourier Transform

In this paper, the Short Time Fourier Transform (STFT) is used directly to convert the input 1D signal directly into a 2D time-frequency image for the following reasons:

(1) The input part of the transformer model used in this paper is two-dimensional data, so converting the one-dimensional input to two-dimensional data is an essential step. (2) Most fault diagnosis methods usually use one-dimensional data directly as input to deep learning methods [26], due to the faster training time and lower time complexity of one-dimensional data. However, in practical industrial fault diagnosis, noise interference is very common when acquiring vibration signals due to sensor noise and other environmental factors. These noises lead to vibration signals that are often non-stationary, and using these non-stationary signals directly into the model for fault diagnosis is often not effective. To overcome this limitation, it is necessary to transform the original 1D time-domain data into a 2D time-frequency image. Such transformation methods are usually the Fourier Transform (FT), Fast Fourier Transform (FFT), Wavelet Transform (WT), and Short Time Fourier Transform (STFT). One serious disadvantage of FT is that the time information is lost when the signal is transformed from the time domain to the frequency domain. When we are using the FT to analyze a specific signal, we do not know which frequency corresponds to which point in time it appears and which point in time it disappears. If the frequency of a signal does not vary with time, then we call it a smooth signal. Then it is less important to know which frequency signal appears at which point in time. However, the actual industrial signals are non-stationary and FT is not suitable for this analysis, where STFT performs best. Therefore, the use of STFT for data pre-processing of bearing faults can improve the robustness of the overall approach and reduce the need for large data sets to satisfactorily train the model.

STFT [27] is one of the most commonly used methods of time-frequency analysis, which represents the signal characteristics at a moment in time through a segment of the signal within a time window. During STFT, the length of the window determines the temporal and frequency resolution of the spectrogram; the longer the window length, the longer the intercepted signal. The longer the signal, the higher the frequency resolution and the worse the time resolution after Fourier transformation; conversely, the shorter the window length, the shorter the intercepted signal, the worse the frequency resolution and the better the time resolution. This means that there is no trade-off between temporal and frequency resolution in STFT and that the trade-off should be made according to the specific needs. This is achieved by multiplying a time window function g(t − u) with the source signal f(t) in the Fourier transform to achieve an additive window and translation around u, followed by a Fourier transform. The window function is shown in Eq. 1; as shown in Fig. 3.

$$G_f(\varepsilon, u) = \int f(t)g(t - u)e^{jwt}dt \tag{1}$$

In the method proposed in this paper, the duration of a set of samples sampled is typically between 60 ms and 63 ms due to different data set loadings. The length of the selected window function, g(t − u), is around one-third of the sampling duration, typically between 20 ms and 21 ms. This is taken to have higher accuracy in frequency and also to make the signal relatively smooth during this time window length. The original one-dimensional signal data is subjected to a short-time Fourier transform and the results are fed into the transformer classification model. A representative example of different fault conditions after using the short-time Fourier transform on the original one-dimensional signal data is shown in Fig. 4.

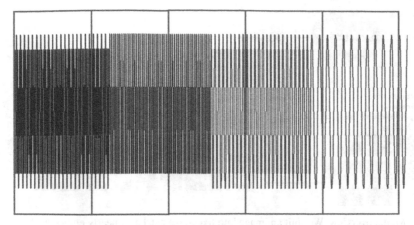

Fig. 3. Window function example

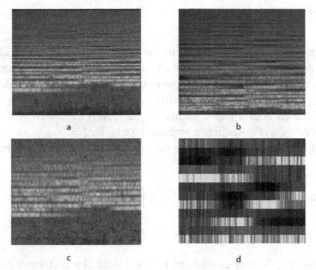

Fig. 4. Image conversion for four different conditions (a) no fault; (b) inner race fault; (c) ball fault; (d) outer race fault

3.2 Transformer-Based Approach to Image Classification

To classify the short-time Fourier transformed images effectively, this paper proposes a transformer-based approach to image classification (Fig. 5).

To process the 2D data at the input, this paper transforms the 2D image $x \in R^{H*W*C}$ into a set of 2D image block sequences $x_p \in R^{N*(P^2*C)}$. Where H, W is the resolution of the original image, C is the number of image channels, and (P, P) is the resolution of each image block. $N = HW/P^2$ is the number of image blocks calculated from this formula and N is also the length of the transformer's effective input sequence. All layers of the transformer use a constant potential vector size D. Therefore, a trainable linear mapping

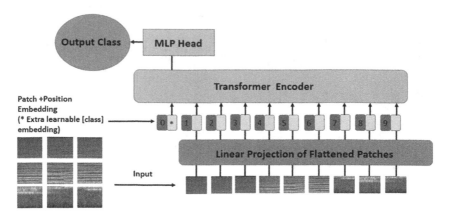

Fig. 5. Model overview. We split an image into fixed-size patches, linearly embed each of them, add position embedding, and feed the resulting sequence of vectors to a standard Transformer encoder. To perform classification, we use the standard approach of adding an extra learnable "classification token" to the sequence.

is used to flatten each block and map it to the D dimension, as shown in Eq. (2). We refer to the output of this mapping as a patch embedding. Besides, position information is added to each image block, and this paper uses the standard learnable 1D position embedding $E_{pos} \in R^{(N+1)*D}$, and the resulting sequence of embedding vectors is added to the input of the encoder. Finally, similar to the [class] flag of BERT, we prepend a learnable embedding to the sequence embedding block $z_0^0 = x_{class}$, and the state z_l^0 at the transformer encoder output is used as the image output, represented by y, as shown in Eq. (3).

$$z_0 = \left[x_{class}; x_p^1 E; x_p^2 E; \cdots; x_p^N E \right] + E_{pos}, \quad E \in R^{(P^2*C)*D} \tag{2}$$

$$y = LN(z_L^0) \tag{3}$$

Furthermore, the transformer consists of L = 6 identical transformer encoders, each containing two layers. The first layer is the Multiple Self-Attention Mechanism (MSA), which is related as shown in Eq. (4).

$$z_l' = MSA(LN(z_{l-1})) + z_{l-1} \quad l = 1 \ldots L \tag{4}$$

The second layer is the multilayer perceptron (MLP), as shown in Eq. (5)

$$z_l = MLP(LN(z_l')) + z_l' \tag{5}$$

The Layer norm (LN) is used before each block and the residuals are connected after each block. This MLP contains two layers with GELU nonlinearity, as shown in Fig. 6.

The explanation of the Multiple Self-Attention Mechanism (MSA) is as follows: The standard qkv attention mechanism (SA) [24] is a module that is currently very common in neural networks. For each element $z \in R^{N*D}$ in the input sequence, we compute the weighted sum of all values in the sequence. The attention weight A_{ij} is determined by the pairwise similarity between two elements and their respective q(query) and k(key), as shown in Eqs. (6) to (8).

$$[q, k, v] = zU_{qkv} \tag{6}$$

$$A = softmax\left(qk^T / \sqrt{D_h}\right) \tag{7}$$

$$SA(z) = AV \tag{8}$$

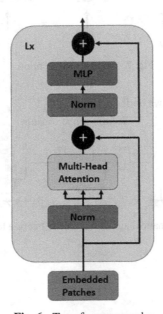

Fig. 6. Transformer encoder

The Multi-head Self-Attention Mechanism (MSA) is an extension of SA where we run k self-attentive operations in the MSA and call it a "head". These "heads" are calculated in parallel and output in series. To ensure that the amount of calculation and the number of parameters remain the same when changing the value of k, D_k is usually set to D/K. This is shown in Eq. (9) and Fig. 7.

$$MSA(z) = [SA_1(z); SA_2(z); \cdots; SA_k(z)]U_{msa} \tag{9}$$

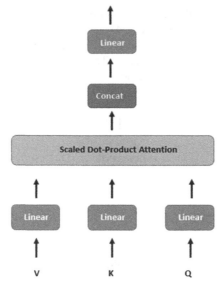

Fig. 7. Multi-head attention

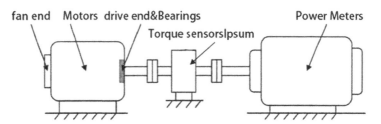

Fig. 8. Schematic diagram of CWRU bearing test platform

4 Experiment

4.1 Data Set Description

The proposed algorithm was experimentally validated using a publicly available database, the one used is the Case Western Reserve University bearing vibration database [28]. This database is widely used in bearing fault diagnosis and is therefore suitable as a benchmark database for comparing the performance of algorithms [29].

The experimental setup used for this database is shown in Fig. 8. The experimental set-up is divided into three main parts from left to right: motor, torque sensor, and power meter (motor load). The fan end and the drive end of the motor are equipped with a bearing and an acceleration sensor respectively, where the acceleration sensor is used to measure the vibration signal of the bearing. The signals used in this experiment were collected from the drive end bearing at a sample rate of 48000 Hz, with motor loads ranging from 0 to 3 HP and motor speeds from 1730 rpm to 1797 rpm. The drive end bearing is a deep groove ball bearing, type 6205-2RSJEMSKF. Bearing failure types can be classified according to their fault location: outer ring failure, roller failure, inner ring failure, where the outer ring failure is selected as the outer ring failure data in the central direction. Bearings are also available in three failure sizes, 0.007″, 0.014″ and 0.021″. Each type of failure will have different failure data at different loads and different bearing failure sizes, representing different levels of severity. This is shown in Table 1.

Table 1. Analytical data description of the CWRU dataset

Bearing Condition Status	Bearing Defect Diameter (Inches)	Load (Hp)			
		0	1	2	3
		Speed (rpm)			
		1797	1772	1750	1730
no fault	X	O	O	O	O
Inner race fault	0.007	O	O	O	O
	0.014	O	O	O	O
	0.021	O	O	O	O
Ball fault	0.007	O	O	O	O
	0.014	O	O	O	O
	0.021	O	O	O	O
Outer race fault	0.007	O	O	O	O
	0.014	O	O	O	O
	0.021	O	O	O	O

Each bearing fault diagnosis sample is a time series of 1600 sampling points, corresponding roughly to the number of data points collected in one rotation of the point machine. For each of the different health conditions of the bearings, 60 samples were taken. Therefore, as can be seen from Table 1, a total of 2400 samples were extracted for classification. The training and test data were randomly selected in a ratio of 8:2.

4.2 Experimental Setup Details and Benchmarking Algorithms

Details of the Experimental Setup. Once the training and test data are prepared, the next step is the process of training the network. First, the structure of the network is defined and the network parameters are initialized. Then take a specified batch size of data and feed it into the network to forward calculate the output value of the network, then compare the error between the output value of the network and the real label. As the network parameters are initially initialized randomly, the error will be large and the aim of training the network is to make this error smaller by continuously adjusting the parameters of the network. The backpropagation algorithm is then used to calculate the error layer by layer and the optimization algorithm is used to adjust the parameters of the network. The process of cycling through a fixed batch size of data is fed into the network until all the training data is iterated through, thus completing a cycle of training. Usually, a total training period is set during the training process, and the process of training the network is to continuously use the training data to update the network parameters until the total number of iterations reaches a set value. When training is complete, the weights of the current network are saved. The testing phase of the network is to read in test samples using the trained model and then output the final diagnostic results. Both the proposed method and the benchmark network were trained using the Adam algorithm

[30] as the optimization function and cross-entropy as the loss function. To accelerate the training speed of the network and reduce the volatility in the training process, a power decay learning rate planning algorithm was used in this paper with an initial value of 0.001 for the learning rate and a decay coefficient of 0.9. Based on the proportional classification of the dataset in the previous section, 1920 samples were selected as the training set and 480 samples as the test set.

Effect of Batch Size on Experimental Results. The experiments were conducted using batch sizes of 64, 128, and 256, controlling for other conditions, and the classification accuracy of the model was 98.75%, 98.45%, and 95.55% respectively. When the batch size is 256, the network ends up with the lowest classification accuracy of the three, while the difference in classification accuracy is not very large for batch sizes of 128 and 64. As batch normalization is used in the network, the overall statistics of the current batch of samples are used in the batch normalization calculation to represent the characteristics of the batch, so when the batch size is large each sample's characteristics will be masked by information from other samples, thus affecting the final accuracy. Furthermore, it has been shown that using larger batch sizes may cause the network to converge more easily to local minima during training, thus affecting the performance of the network [31]. However, choosing a relatively large batch size during training can speed up the training process, as the more batches used at a time when the total number of samples is fixed, the fewer iterations per cycle, thus making full use of the graphics memory capacity of the card. Therefore, considering both accuracy and training speed, a batch size of 128 was finally chosen as the parameter for subsequent experiments.

Effect of Training Period on Experimental Results.
To determine the best-fitparameters for this method, while controlling for other conditions, the training data was again divided into a training and validation set in a ratio of 8:2. The results observed after several training sessions are shown in Fig. 9. The vertical axis is the accuracy of training β and the horizontal axis is the training period epoch.

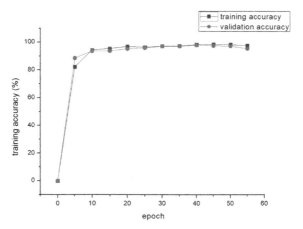

Fig. 9. Accuracy of training and validation during the training process

The graph shows that the model accuracy is at its best when the training period is kept at around 45, after which it decreases, probably due to over-fitting of the training data.

4.3 Experimental Results and Comparison Experiments

To demonstrate the effectiveness of the model classification method proposed in this paper, we compared the method in this paper with traditional machine learning and deep learning methods, such as Conv-LSTM [32], Bi-LSTM [33], CNN [34], DBN [35] and DNN [36]. Table 2 clearly shows that the results obtained by the method proposed in this paper are superior to those obtained by the method described above. Therefore, the authors believe that the fault diagnosis classification method proposed in this paper can be used as a highly accurate system for bearing fault diagnosis.

Table 2. Comparison results with other machine learning competitive methods

methods	accuracy
Dense Neural Network (DNN)	80.00%
Conv-LSTM	91.20%
CNN	92.60%
Bi-LSTM	85.80%
DBN with ensemble learning	96.95%
Transformer	98.45%

5 Conclusion and Future Works

In this paper, a transformer-based fault diagnosis method is proposed, whereby a one-dimensional fault signal with a time series is transformed into a two-dimensional image using a short-time Fourier transform and then fed into a transformer model for the classification of faults. The experimental results show that the classification results perform better compared to traditional deep learning methods. Future works will not only be limited to the classification of bearing fault data, but hopefully, this method can be applied to more fault datasets with temporal information, and the generalisability of this method can also be checked.

References

1. Rai, A., Upadhyay, S.H.: A review on signal processing techniques utilized in the fault diagnosis of rolling element bearings. Tribol. Int. **96**, 289–306 (2016)
2. Cerrada, M., et al.: A review on data-driven fault severity assessment in rolling bearings. Mech. Syst. Signal Process. **99**, 169–196 (2018)
3. Chen, K., Huang, C., He, J.: Fault detection, classification and location for transmission lines and distribution systems: a review on the methods. High Voltage **1**(1), 25–33 (2016)
4. Khan, F.N., et al.: Modulation format identification in coherent receivers using deep machine learning. IEEE Photon. Technol. Lett. **28**(17), 1886–1889 (2016)

5. Zhang, H., et al.: Object-level video advertising: an optimization framework. IEEE Trans. Ind. Inform. **13**(2), 520–531 (2016)
6. Zhang, H., et al.: Understanding subtitles by character-level sequence-to-sequence learning. IEEE Trans. Ind. Inform. **13**(2), 616–624 (2016)
7. Ali, H., et al.: Speaker recognition with hybrid features from a deep belief network. Neural Comput. Appl. **29**(6), 13–19 (2018)
8. Muhammad, U.R., et al.: Goal-driven sequential data abstraction. In: Proceedings of the IEEE/CVF International Conference on Computer Vision, pp. 71–80 (2019)
9. Liu, F.Y., Wang, S.H., Zhang, Y.D.: Survey on deep belief network model and its applications. Comput. Eng. Appl. **54**(1), 11–18 (2018)
10. Khan, A., et al.: A survey of the recent architectures of deep convolutional neural networks. Artif. Intell. Rev. **53**(8), 5455–5516 (2020)
11. Karpathy, A.: The unreasonable effectiveness of recurrent neural networks. Andrej Karpathy Blog **21**, 23 (2015)
12. Ma, L., et al.: Bearing fault diagnosis based on convolutional neural network learning of time-domain vibration signal imaging. In: 2019 Chinese Control and Decision Conference (CCDC). IEEE, pp. 659–664 (2019)
13. Liu, Y., et al.: Hierarchical independence thresholding for learning Bayesian network classifiers. Knowl.-Based Syst. **212**, 106627 (2021)
14. Sun, W., Paiva, A.R.C., Xu, P., Sundaram, A., Braatz, R.D.: Fault detection and identification using Bayesian recurrent neural networks. Comput. Chem. Eng. 141 (2020)
15. Chen, X., et al.: Rolling bearings fault diagnosis based on tree heuristic feature selection and the dependent feature vector combined with rough sets. Appl. Sci. **9**(6), 1161 (2019)
16. Tong, T., Xu, X.: Improvement of power system fault diagnosis algorithm based on Petri Net. Int. Core J. Eng. **6**(9), 319–334 (2020)
17. Long, J., Shelhamer, E., Darrell, T.: Fully convolutional networks for semantic segmentation. In: Proceedings of the IEEE Conference on Computer Vision and Pattern Recognition, pp. 3431–3440 (2015)
18. Sun, W., et al.: A sparse auto-encoder-based deep neural network approach for induction motor faults classification. Measurement **89**, 171–178 (2016)
19. Li, C., et al.: Multimodal deep support vector classification with homologous features and its application to gearbox fault diagnosis. Neurocomputing **168**, 119–127 (2015)
20. Kiranyaz, S., et al.: Real-time fault detection and identification for MMC using 1-D convolutional neural networks. IEEE Trans. Industr. Electron. **66**(11), 8760–8771 (2018)
21. Eren, L., Ince, T., Kiranyaz, S.: A generic intelligent bearing fault diagnosis system using compact adaptive 1D CNN classifier. J. Signal Process. Syst. **91**(2), 179–189 (2019)
22. Abdeljaber, O., et al.: 1-D CNNs for structural damage detection: verification on a structural health monitoring benchmark data. Neurocomputing **275**, 1308–1317 (2018)
23. Eren, L.: Bearing fault detection by one-dimensional convolutional neural networks. Math. Probl. Eng., 34–37 (2017)
24. Vaswani, A., et al.: Attention is all you need. In: Proceedings of the 31st International Conference on Neural Information Processing Systems, pp. 6000–6010 (2017)
25. Zheng, S., et al.: Rethinking semantic segmentation from a sequence-to-sequence perspective with transformers. arXiv Preprint, arXiv:2012-15840 (2020)
26. Zhao, R., et al.: Machine health monitoring using local feature-based gated recurrent unit networks. IEEE Trans. Ind. Electron. **65**(2), 1539–1548 (2017)
27. Gabor, D.: Theory of communication. Part 1: The analysis of information. J. Instit. Electr. Eng. Pt. III Radio Commun. Eng. **93**(26), 429–441 (1946)
28. Lou, X., Loparo, K.A.: Bearing fault diagnosis based on wavelet transform and fuzzy inference. Mech. Syst. Signal Process. **18**(5), 1077–1095 (2004)

29. Smith, W.A., Randall, R.B.: Rolling element bearing diagnostics using the Case Western Reserve University data: A benchmark study. Mech. Syst. Signal Process. **64**, 100–131 (2015)
30. Kingma, D.P., Ba, J.: Adam: a method for stochastic optimization. arXiv Preprint, arXiv: 1412.6980 (2014)
31. Keskar, N.S., et al.: On large-batch training for deep learning: generalization gap and sharp minima. In: 5th International Conference on Learning Representations, ICLR 2017 (2019)
32. Hinchi, A.Z., Tkiouat, M.: Rolling element bearing remaining useful life estimation based on a convolutional long-short-term memory network. Procedia Comput. Sci. **127**, 123–132 (2018)
33. Fan, Y., et al.: Study on a small sample rolling bearing fault diagnosis method based on BI-LSTM. Noise Vib. Control **40**(4), 103–108 (2020)
34. Lu, C., Wang, Z., Zhou, B.: Intelligent fault diagnosis of rolling bearing using hierarchical convolutional network based health state classification. Adv. Eng. Inform. **32**, 139–151 (2017)
35. Liang, T., et al.: Bearing fault diagnosis based on improved ensemble learning and deep belief network. In: Journal of Physics: Conference Series, vol. 1074, no. 1, p. 012154. IOP Publishing (2018)
36. Jia, F., et al.: Deep neural networks: a promising tool for fault characteristic mining and intelligent diagnosis of rotating machinery with massive data. Mech. Syst. Signal Process. **72**, 303–315 (2016)

Research and Simulation of Mass Random Data Association Rules Based on Fuzzy Cluster Analysis

Huaisheng Wu[1]([✉]), Qin Li[2], and Xiuming Li[1]

[1] Qinghai Minzu University, Xining, China
wuhuaisheng@sina.com
[2] Beijing University of Technology, Beijing, China

Abstract. Because the traditional method is difficult to obtain the internal relationship and association rules of data when dealing with massive data, a fuzzy clustering method is proposed to analyze massive data. Firstly, the sample matrix was normalized through the normalization of sample data. Secondly, a fuzzy equivalence matrix was constructed by using fuzzy clustering method based on the normalization matrix, and then the fuzzy equivalence matrix was applied as the basis for dynamic clustering. Finally, a series of classifications were carried out on the mass data at the cut-set level successively and a dynamic cluster diagram was generated. The experimental results show that using data fuzzy clustering method can effectively identify association rules of data sets by multiple iterations of massive data, and the clustering process has short running time and good robustness. Therefore, it can be widely applied to the identification and classification of association rules of massive data such as sound, image and natural resources.

Keywords: Fuzzy clustering · Massive random data · Management rules · Cut-set levels

1 Introduction

With the rapid development of cloud computing and database technology, people can easily obtain and store more data. However, in the face of massive amounts of data, traditional data analysis tools can only do some surface-level queries and processing, but cannot obtain the internal relationships and association rules between the data [1]. There is a need for a technology and tool that can intelligently and automatically convert data into useful information and knowledge. This urgent need for powerful data analysis tools has led to the emergence of data mining technology [2].

The traditional cluster analysis is Crisp Partition, which strictly divides each object to be identified into a certain category, which has the nature of "either-or". Therefore, the boundary of this type of classification is clear. However, in fact, most objects do not have strict attributes. They are intermediary in nature and category, and have the nature of "this and other", so they are suitable for soft partition [3]. The fuzzy set theory proposed by

The original version of this chapter was revised: The name of the third author has been corrected as "Xiuming Li" and the acknowledgement section has been added. The correction to this chapter is available at https://doi.org/10.1007/978-981-16-5940-9_42

J. Zeng et al. (Eds.): ICPCSEE 2021, CCIS 1451, pp. 80–89, 2021.
https://doi.org/10.1007/978-981-16-5940-9_6

Zadeh provides a powerful analysis tool for this soft partition. People begin to use fuzzy methods to deal with clustering problems, and call it fuzzy clustering analysis. Fuzzy clustering obtains the degree of uncertainty of the sample belonging to each category, expresses the intermediary of the sample category, and establishes a description of the sample's uncertainty of the category, which can more objectively reflect the real world, thus becoming a cluster analysis the mainstream of research [4].

Fuzzy cluster analysis is a mathematical method to classify things according to certain requirements when it involves fuzzy boundaries between things. Cluster analysis is a multivariate analysis method in mathematical statistics. It uses mathematical methods to quantitatively determine the closeness of the sample, so as to objectively classify the types [5]. The boundaries between things, some are exact, some are fuzzy. For example, the boundary between the degrees of similarity in the crowd is fuzzy, and the boundary between cloudy and sunny weather is also fuzzy. When clustering involves fuzzy boundaries between things, the fuzzy clustering analysis method can provide information on the "similarity" of different categories and overcome the shortcomings of the crisp partition methods [6].

2 Construction of Association Rule Pattern for Mass Random Data Based on Fuzzy Cluster Analysis

2.1 The Principle of Fuzzy Cluster Analysis

Fuzzy cluster analysis is a type of cluster analysis. It classifies samples with greater similarity or smaller distance into one category. Unlike general cluster analysis, fuzzy cluster analysis compares the similarity of two fuzzy sets [7]. According to the fuzzy equivalence relation, the degree is divided by different λ cut-set levels, so that the classification result is more discernible and reasonable [8]. First, establish the fuzzy similarity relationship between samples, and calculate the statistics $r_{ij}(i, j = 1, 2, 3, 4, .., n)$ of the degree of similarity between samples, so that all objects in the fuzzy matrix $R = (r_{ij})_{n \times n}$ The range value of is in the closed interval of [0,1]; then the fuzzy matrix is transformed into a fuzzy equivalent matrix $R = (r_{ij})_{nxn}$ using the transitive closure method, and the fuzzy relationship must satisfy the following two items at the same time: (1) Reflexivity, $r_{ij} = 1(i = j)$; (2) Symmetry, $r_{ij} = r_{ij}(i, j = 1, 2, 3, 4.., n)$; Fuzzy similarity matrix only has reflexivity and symmetry, not has is transitive, and the premise of finding the cut matrix is that R is a fuzzy equivalence relation on X [9]. Therefore, we must first obtain the R transitive closure, and transform the fuzzy similar matrix into a fuzzy equivalent matrix. Finally, the cut-set classification can be performed according to different cut-set levels λ [10], as shown in Fig. 1.

The Principle of Fuzzy Cluster Analysis
The method based on fuzzy cluster analysis can be divided into three steps: (1) data normalization, (2) construction of fuzzy similarity matrix, (3) fuzzy classification. Fuzzy classification can be implemented with different algorithms [11].

This paper uses the fuzzy transitive closure method for fuzzy classification. Suppose the set of classified objects is $X = \{x_1, x_2, \ldots, x_n\}$ [12], each object x_i has m characteristic indexes, that is, x_i can be expressed by the following formula (1) as an m dimensional characteristic index vector:

$$x_i = (x_{i1}, x_{i2}, \ldots, x_{in}), i = 1, 2, \ldots, n \tag{1}$$

Where x_{ij} represents the jth characteristic index of the ith object. Then all the characteristic indexes of n objects form a matrix, denoted as $X^* = (x_{ij})_{n \times m}$, call X^* the characteristic index matrix of X, which is expressed by formula (2).

$$X^* = \begin{pmatrix} x_{11} & x_{12} & \cdots & x_{1m} \\ x_{21} & x_{22} & \cdots & x_{2m} \\ \vdots & \vdots & \ddots & \vdots \\ x_{m1} & x_{n2} & \cdots & x_{nm} \end{pmatrix} (2) \tag{2}$$

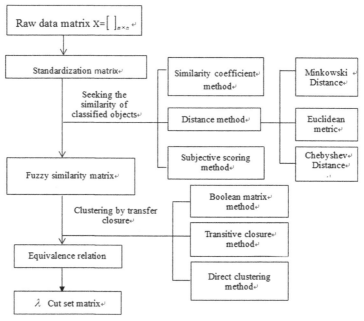

Fig. 1. Schematic diagram of association classification of random massive data Standardization of Sample Data.

There are various dimensions to describe the characteristics of things, so matrices should be standardized for the convenience of analysis and comparison. Since the dimension and order of magnitude of the m feature indexes are not necessarily the same, the function of the feature indexes of a very large order of magnitude may be highlighted in the process of calculation [12], while the function of some feature indexes of a very small order of magnitude may be reduced or even excluded. Data normalization enables each index value to be unified in some common range of numerical characteristics.there are many kinds of methods to standardize the matrix, For example, the translation standard deviation transformation can transform the matrix into a normalized matrix as much as possible [13]. Data normalization makes each index value unified in a common range of numerical characteristics [6]. In this paper, the method of translation standard deviation is used to calculate the mean value and standard deviation of the j-th column of the characteristic index matrix X*, which are expressed by formula (3) and formula (4) respectively, Then the j-th normalized data value of the i-th object is obtained by transformation, which is expressed by formula (5).

$$\overline{x_j} = \frac{1}{n}\sum_{i=1}^{n} x_{ij}i = 1, 2, \ldots, n \tag{3}$$

$$\sigma_j = \sqrt{\frac{1}{n}\sum_{i=1}^{n}(x_{ij} - \overline{x_j})^2} \tag{4}$$

$$x_{ij}' = \frac{x_{ij} - \overline{x_j}}{\sigma_j}, i = 1, 2, \ldots, n; j = 1, 2, \ldots, m \tag{5}$$

In formula (5), x_{ij} represents the j-th characteristic index of the i-th object, $\overline{x_j}$ represents the mean value of all objects of the j-th characteristic index, and σ_j represents the standard deviation of the j-th characteristic index.

Constructing fuzzy similarity matrix
Clustering is to identify the closeness of objects in matrix x according to a certain standard and classify the objects close to each other[14]. The similarity coefficient in classical clustering analysis and the closeness between fuzzy sets can be used as the similarity degree. r_{ij} is usually used to express the closeness or similarity between x_i and x_j in matrix x. Let the data $x_{ij}(i = 1, 2., \ldots, n; j = 1, 2, \ldots, m)$ have been normalized, The degree of similarity between $x_i = (x_{i1}, x_{i2}, \ldots, x_{im})$ and $x_j = (x_{j1}, x_{j2}, \ldots, x_{jm})$ is recorded as $r_{ij} \in [0, 1]$, Then the fuzzy similarity matrix $R = (r_{ij})_{n \times n}$ between objects is obtained. For the determination of the similarity coefficient, this paper adopts the quantitative product method, which is expressed by formula (6):

$$rij = \begin{cases} 1 & i = j \\ \frac{1}{M}x_i \times x_j & i \neq j \end{cases} \quad x_i \times x_j = \sum_{k=1}^{m} x_{ik}x_{jk} \tag{6}$$

In formula (6), M > 0 for the appropriate choice of parameters and meet $M \geq max\{x_i \times x_j | i \neq j\}$. In this case, $x_i \times x_j$ is the dot product of x_i and x_j.

Fuzzy Classification
As the fuzzy relation matrix $R = (r_{ij})_{n \times n}$ between objects constructed by the above methods, it is generally only a fuzzy similarity matrix, but not necessarily transitive. Therefore, a new fuzzy equivalence matrix should be constructed from R, and then dynamic clustering can be carried out based on the fuzzy equivalence matrix. As mentioned above, the transitive closure $t(R)$ of the fuzzy similarity matrix R is a fuzzy equivalence matrix. The clustering method based on t(R) classification is called fuzzy transitive closure method. The specific steps are as follows:

(1) The transitive closure $t(R)$ of the fuzzy similarity matrix R is obtained by using the square self-synthesis method [15].
(2) Properly select the level value of the cut-set $\lambda \in [0, 1]$, and obtain the λ cut-set matrix $t(R)_\lambda$ of $t(R)$, which is an equivalent Boole matrix on X. Then, $t(R)_\lambda$ is classified, and the obtained classification is equivalent classification at the level of λ, which is expressed by formula (7).

$$\text{Set } t(R) = (r'_{ij})_{n \times n}, t(R)_\lambda = (r'_{ij}(\lambda))_{n \times n}, \text{ then } r'_{ij}(\lambda) = \begin{cases} 1, r'_{ij} \geq \lambda \\ 0, r'_{ij} < \lambda \end{cases} \quad (7)$$

In the formula $x_i, x_j X$, if $r'_{ij}(\lambda) = 1$, objects x_i and x_j are classified in the same class at the λ level.

Draw dynamic cluster diagram: in order to intuitively see the correlation degree among the classified objects, all the different objects in $t(R)$ are usually arranged in order from large to small: $1 = \lambda_1 > \lambda_2 > \dots$ to obtain a series of classifications according to $t(R)_\lambda$. This series of classification is plotted on the same graph to obtain the dynamic cluster graph.

3   Simulation Experiment Analysis

3.1   Raw Data Matrix Normalization

Feature extraction and original data matrix were established. The original data matrix X was transformed into X_{ij} and the specification was changed into X, as shown in Fig. 2 and Fig. 3.

Standardized method is adopted to construct fuzzy similarity matrix R^8, as shown in Fig. 4.

Calculate the transitive closure t(R) by using the square self-synthesis method [16], and calculate R^2, R^4, and R^8 in turn. Since $R^8 = R^4$. then $t(R) = R^4$.

Select the appropriate cut-set level $\lambda \in [0, 1]$ and conduct dynamic clustering according to λ cut-set matrix $t(R)_\lambda$. The order of the objects in t(R) from the largest to the smallest is as follows: $1 > 0.8499 > 0.7329 > 0.5902 > 0.4944 > 0.4608 > 0.4014 > 0.3679$ Take $\lambda = 1, 0.8499, 0.7329, 0.5902, 0.4944, 0.4608, 0.4014, 0.3679$ in order to get Series cut matrixs.

3.2   Dynamic Clustering

When $\lambda = 1$, X is classified into eight classes: {x1}, {x2}, {x3}, {x4}, {x5}, {x6}, {x7}, {x8}; When $\lambda = 0.8499$, X is classified into seven categories: {x1}, {x2}, {x3}, {x4}, {x5, x6}, {x7}, {x8}; When $\lambda = 0.7329$, X is classified into six types: {x1}, {x2}, {x3}, {x4}, {x5, x6}, {x7, x8}; When $\lambda = 0.5902$, X is classified into five types: {x2}, {x3}, {x4}, {x1, x5, x6}, {x7, x8}; When $\lambda = 0.4944$, X is classified into four classes: {x2}, {x3}, {x4}, {x1, x5, x6, x7, x8}; When $\lambda = 0.4608$, X is classified into three

$$X = \begin{bmatrix} 56 & 55 & 15 & 42 & 45 & 34; \\ 59 & 73 & 71 & 22 & 44 & 18; \\ 73 & 86 & 39 & 27 & 68 & 39; \\ 37 & 38 & 32 & 16 & 24 & 33; \\ 65 & 55 & 45 & 53 & 34 & 41; \\ 65 & 54 & 54 & 55 & 29 & 47; \\ 57 & 53 & 64 & 70 & 53 & 27; \\ 38 & 57 & 67 & 75 & 62 & 27]; \end{bmatrix}$$

Fig. 2. Raw data matrix

0.5278	0.3542	0	0.4407	0.4773	0.5517
0.6111	0.7292	1.0000	0.1017	0.4545	0
1.0000	1.0000	0.4286	0.1864	1.0000	0.7241
0	0	0.3036	0	0	0.5172
0.7778	0.3542	0.5357	0.6271	0.2273	0.7931
0.7778	0.3333	0.6964	0.6610	0.1136	1.0000
0.5556	0.3125	0.8750	0.9153	0.6591	0.3103
0.0278	0.3958	0.9286	1.0000	0.8636	0.3103

Fig. 3. Standardization matrix

1.0000	0.3359	0.3011	0.4014	0.5902	0.4401	0.4842	0.2560
0.3359	1.0000	0.2761	0.1291	0.2855	0.2254	0.4608	0.2704
0.3011	0.2761	1.0000	0.0149	0.3679	0.2178	0.1427	0.0368
0.4014	0.1291	0.0149	1.0000	0.3016	0.2268	0.0982	0.1266
0.5902	0.2855	0.3679	0.3016	1.0000	0.8499	0.4944	0.2506
0.4401	0.2254	0.2178	0.2268	0.8499	1.0000	0.4649	0.2095
0.4842	0.4608	0.1427	0.0982	0.4944	0.4649	1.0000	0.7329
0.2560	0.2704	0.0368	0.1266	0.2506	0.2095	0.7329	1.0000

Fig. 4. Fuzzy similarity matrix

classes: $\{x3\}, \{x4\}, \{x1, x2, x5, x6, x7, x8\}$; When $\lambda = 0.4014$, X is classified into two classes: $\{x4\}, \{x1, x2, x3, x5, x6, x7, x8\}$; When $\lambda = 0.3679$, X is classified into one categories: $\{x1, x2, x3, x4, x5, x6, x7, x8\}$. In this paper, we select 8λ in order to show the change process of the generation matrix, as shown in Fig. 5 and Fig. 6.

λ=1

```
1 0 0 0 0 0 0 0
0 1 0 0 0 0 0 0
0 0 1 0 0 0 0 0
0 0 0 1 0 0 0 0
0 0 0 0 1 0 0 0
0 0 0 0 0 1 0 0
0 0 0 0 0 0 1 0
0 0 0 0 0 0 0 1
```

λ=0.8499

```
1 0 0 0 0 0 0 0
0 1 0 0 0 0 0 0
0 0 1 0 0 0 0 0
0 0 0 1 0 0 0 0
0 0 0 0 1 1 0 0
0 0 0 0 1 1 0 0
0 0 0 0 0 0 1 0
0 0 0 0 0 0 0 1
```

λ=0.7329

```
1 0 0 0 0 0 0 0
0 1 0 0 0 0 0 0
0 0 1 0 0 0 0 0
0 0 0 1 0 0 0 0
0 0 0 0 1 1 0 0
0 0 0 0 1 1 0 0
0 0 0 0 0 0 1 1
0 0 0 0 0 0 1 1
```

λ=0.5902

```
1 0 0 0 1 1 0 0
0 1 0 0 0 0 0 0
0 0 1 0 0 0 0 0
0 0 0 1 0 0 0 0
1 0 0 0 1 1 0 0
1 0 0 0 1 1 0 0
0 0 0 0 0 0 1 1
0 0 0 0 0 0 1 1
```

Fig. 5. Matrix change Part 1

λ=0.4944

```
1 0 0 0 1 1 1 1
0 1 0 0 0 0 0 0
0 0 1 0 0 0 0 0
0 0 0 1 0 0 0 0
1 0 0 0 1 1 1 1
1 0 0 0 1 1 1 1
1 0 0 0 1 1 1 1
1 0 0 0 1 1 1 1
```

λ=0.4608

```
1 1 0 0 1 1 1 1
1 1 0 0 1 1 1 1
0 0 1 0 0 0 0 0
0 0 0 1 0 0 0 0
1 1 0 0 1 1 1 1
1 1 0 0 1 1 1 1
1 1 0 0 1 1 1 1
1 1 0 0 1 1 1 1
```

λ= 0.4014

```
1 1 0 1 1 1 1 1
1 1 0 1 1 1 1 1
0 0 1 0 0 0 0 0
1 1 0 1 1 1 1 1
1 1 0 1 1 1 1 1
1 1 0 1 1 1 1 1
1 1 0 1 1 1 1 1
1 1 0 1 1 1 1 1
```

λ=0.3679

```
1 1 1 1 1 1 1 1
1 1 1 1 1 1 1 1
1 1 1 1 1 1 1 1
1 1 1 1 1 1 1 1
1 1 1 1 1 1 1 1
1 1 1 1 1 1 1 1
1 1 1 1 1 1 1 1
1 1 1 1 1 1 1 1
```

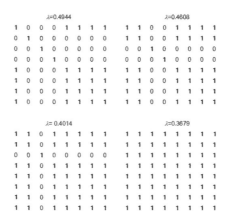

Fig. 6. Matrix change Part 2

3.3 Dynamic Clustering Diagram

The cut-set levels $\lambda \in [0, 1]$ is successively taken. The cut-set relation R_λ, R_λ is a classical equivalence relation, which induces a partition X/R on X and divides X into some equivalence classes [17]. It can be classified by determining the corresponding λ cut-set matrix. As λ changes from large to small and the classification changes from fine to coarse, a dynamic classification graph is formed. This is shown in Fig. 7.

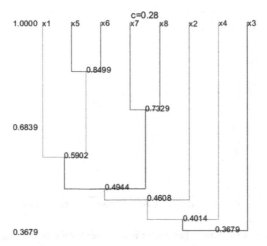

Fig. 7. Dynamic classification of pedigree

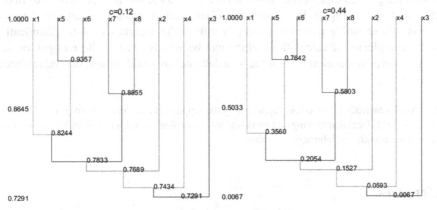

Fig. 8. Comparison diagram of dynamic classification of pedigree

4 Conclusion

In the classification of fuzzy clustering analysis of association rules in practice, for the cut-set of different level $\lambda \in [0, 1]$, can get a different classification, thus forming a kind of dynamic clustering figure, the comprehensive understanding of sample classification is more image and intuitive [18]. In constructing a fuzzy similar matrix, if used Euclidean distance to solve the compatible parameter c values will affect the assembly level values, but does not affect the assembly level of the total number, and the classification of the basic will not affect the dynamic clustering results [19]. When $c = 0.12$, effective assembly level $\lambda \in [0.7292, 0.9357]$, when $c = 0.44$, effective assembly level $\lambda \in [0.0067, 0.7642]$, as shown in Fig. 8. Compatible parameter c can be inferred by the experimental results and the effective range of values of the cut-set levels λ negative correlation relationship. Moreover, with the increase of the compatible parameters c in

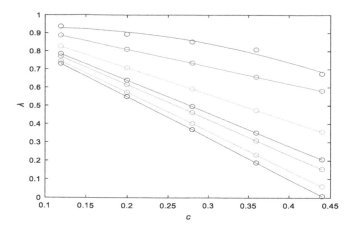

Fig. 9. Compatible parameter and cut set level relation graph

a certain range, the range of the cut-set level λ decreases, and also presents a diffusion trend, as shown in Fig. 9. For associative classification of massive random data, the selecton of compatible parameters directly affects the accuracy of data classification and the complexity of calculation. which can be widely used in the recognition and classification of association rules of massive data such as sound, image, natural resources and so on.

Acknowledgements. This work is supported by the Applied Basic Research Program of Qinghai (2019–ZJ–7017) Decision making and early warning of ecological animal husbandry development based on multimodal collaborative learning.

References

1. Kang, Y., Feng, L., Zhang, Z.: Simulation of cloud sea big data fuzzy clustering method based on grid index. Comput. Simulat. **36**(12), 341–344 (2019)
2. Qiu, D.: Four problems in fuzzy cluster analysis. China Statist. **21**(3), 70–72 (2021)
3. Yangl, H.: Precipitation regionalization based on fuzzy clustering algorithm. Meteorol. Sci. Technol. **39**(5), 582–586 (2011)
4. Han, B.H., et al.: Fuzzy Clustering Method Based on Improved Weighted Distance. Mathematical Problems in Engineering (2021)
5. Tian, F., Yang, Y.: Automatic classification method of intelligent electronic archives based on fuzzy clustering algorithm. Appl. Microcomput. **37**(02), 87–90 (2021)
6. Kuifeng, Y., Duan, G., Shi, X.: College entrance examination volunteer recommendation algorithm based on multi feature weight fuzzy clustering. J. Central South Univ. (NATURAL SCIENCE EDITION) **51**(12), 3418–3429 (2020)
7. Chen, X., Fan, B., Shiqi, W.: Research on inventory classification of auto parts based on fuzzy clustering analysis. Manuf. Autom. **42**(03), 110–116 (2020)
8. Jie, H.Y., Pan, T., et al.: TW-Co-MFC: two-level weighted collaborative fuzzy clustering based on maximum entropy for multi-view data. Tsinghua Sci. Technol. **26**(2), 53–66 (2021)

9. Vvan, T., Phamtoan, D.: Interval forecasting model for time series based on the fuzzy clustering technique. IOP Conf. Ser. Materials Sci. Eng. **1109**(1), 12–30 (2021)

10. Huang, R., Chen, L., Yuan, X.: A visual uncertainty analytics approach for weather forecast similarity measurement based on fuzzy clustering. J. Visual. **24**(2), 317–330 (2021). https://doi.org/10.1007/s12650-020-00709-z

11. Jie, H.Y., Pan, T., et al.: TW-Co-MFC: two-level weighted collaborative fuzzy clustering based on maximum entropy for multi-view data. Tsinghua Sci. Technol. **26**(02), 53–66 (2021)

12. Zhenwei, L., Liu, K.: Method and application of fuzzy cluster analysis. Pract. Understand. Math. **49**(6), 288–291 (2019)

13. Karlekar, A., Seal, A., Krejcar, O., et al.: Fuzzy K-means using non-linear s-distance. IEEE Access **7**, 55121–55131 (2019)

14. Ying, Z., Feng, L., Chen, M., et al.: Evaluating Multi-Dimensional Visualizations for Understanding Fuzzy Clusters (2019)

15. Tai, V., Lethithu, T.: A fuzzy time series model based on improved fuzzy function and cluster analysis problem. Commun. Math. Statist. (2020). https://doi.org/10.1007/s40304-019-00203-5

16. Vovan, T., Ledai, N.: A new fuzzy time series model based on cluster analysis problem. Int. J. Fuzzy Syst. **21**(3), 852–864 (2019). https://doi.org/10.1007/s40815-018-0589-x

17. Kuo, R.J., Lin, J.Y., Nguyen, T.: An application of sine cosine algorithm-based fuzzy possibilistic c-ordered means algorithm to cluster analysis. Soft. Comput. **25**(11), 1–16 (2021)

18. Zhang, T., Li, Z., Ma, F., et al.: Rough fuzzy k-means clustering algorithm based on unbalanced measure of cluster size. Inf. Control **34**(3), 281–288 (2020)

19. Liu, Z.: A fuzzy c-means clustering algorithm and its implementation. Mod. Navigat. **11**(2), 122–125 (2020)

Predicting Stock Price Movement with Multiple Data Sources and Machine Learning Models

Yang Xia and Yue Wang$^{(\boxtimes)}$

School of Information, Central University of Finance and Economics, Beijing, China

Abstract. Stock price trend prediction is a challenging issue in the financial field. To get improvements in predictive performance, both data and technique are essential. The purpose of this paper is to compare deep learning model (LSTM) with two ensemble models (RF and XGboost) using multiple data. Data is gathered from four stocks of financial sector in China A-share market, and the accuracy and F1-measure are used as performance measure. The data of the past three days is applied to classify the rise and fall trend of price on the next day. The models' performance are tested under different market styles (bull or bear market) and different market activities. The results indicate that under the same conditions, LSTM is the top algorithm followed by RF and XGBoost. For all models applied in this study, prediction performance in bull markets is much better than in bear markets, and the result in active period is better than inactive period by average. It is also found that adding data sources is not always effective in improving forecasting performance, and valuable data sources and proper processing may be more essential than providing a large quantity of data source.

Keywords: Stock market prediction · Multiple data sources · Deep learning · Machine learning · LSTM · Random forest · XGBoost

1 Introduction and Literature

Forecasting Stock market is a hot problem in the field of both finance and information science. Fama proposed the efficient market hypothesis (EMH) which asserts that financial market is efficient taking that the stock price is always correct as a prerequisite. Therefore, the market direction is random, and no one can predict the future direction of the market [1]. The theory states that fundamental or technical analysis would not yield any consistent profit to investors. However, there are scholars who do not approve of EMH [2]. They set researches in the purpose of establish effective prediction models. Fundamental and technical analysis are to be the beginning of the endeavor. The development of machine learning also cast its light on stock market prediction, e.g., support vector machine (SVM) and logistic regression [3], random forests (RF) [4], and

Y. Xia and Y. Wang---Contributed equally to this work.

This work is supported by: Engineering Research Center of State Financial Security, Ministry of Education, Central University of Finance and Economics, Beijing, 102206, China; Program for Innovation Research in Central University of Finance and Economics.

© Springer Nature Singapore Pte Ltd. 2021
J. Zeng et al. (Eds.): ICPCSEE 2021, CCIS 1451, pp. 90–105, 2021.
https://doi.org/10.1007/978-981-16-5940-9_7

XGBoost [5]. After 2012, with the significant improvement of hardware, data and algorithms during years, deep learning began to be widely employed in various fields. White was the forerunner to predict the stock price of IBM using neural network. Although it encountered failure of a simple network to find evidence against the efficient markets hypothesis, it yielded practical benefits for subsequent studies [6]. Yoon and Swales examined the ability of a neural network technique and compares its predictive power with multiple analysis methods. They demonstrated the efficiency of neutral network by concluding that the neural network could predict the stock prices more precisely than the traditional multivariate analysis technology [7].

The long short-term memory (LSTM) neural network was developed by Hochreiter and Schmidhuber in 1997 [8], which was an important result of the two researchers studying the problem of disappearing gradient. As a kind of recursive neural network (RNN) which is most suitable for stock forecasting among the forms of neural networks, LSTM is good at stock-market-like time series prediction for its ability to capture the dependencies between data related to time effectively. Nelson et al. studied the LSTM networks to predict future movements of stock prices based on the past price and technical indicators. It obtained the promising result that an average of 55.9% of accuracy when predicting price movement in the near future [9]. Althelaya et al. evaluated a number of RNN based on LSTM and GRU using multiple directional stacked structures with multiple inputs. The study found that for both short and long term, the stacked LSTM architecture has reached the highest prediction performance [10].

The type and quantity of data sources are essential for stock price forecasting. They are roughly divided into traditional data sources, also referenced as technical analysis such as historical data and technical indicator, and non-traditional data sources. Deng et al. used the frequency and overall sentiment analysis of news and comments as well as technical analysis to predict famous Japan companies' stocks and found that features other than mining from stock prices themselves improved the performance [11]. Michał et al. conducted sentiment classification of large datasets available from Twitter and stock market records using machine learning to estimate future stock prices and discussed the results for different time intervals and input datasets [12]. Roy et al. proposed a deep learning approach to predict if the stock price would increase by 25% for the following period, with the comparison against random forest and gradient boosted machines, and the Korea composite stock price index (KOSPI) of companies for the period 2007 to 2015 was used as input. They showed that all the methods performed satisfactorily [13]. Teng et al. devised an effective time-sensitive data augmentation method for stock trend prediction by corrupting high-frequency patterns of original stock price data and preserving low-frequency ones to augment limited stock price data. They also proposed a transformation technique to recognize the importance of the patterns at varied time points. A series of experiments were carried out on a 50 corporation stocks' dataset [14]. Li and Pan proposed to use sentiment analysis to extract useful information from multiple textual data sources and use a blending ensemble deep models to predict stock movement to further improve prediction accuracy. The blending models contained two RNNs, one LSTM and one GRU, followed by a fully connected neural network as the second level. The paper showed that ensemble deep learning methods can predict stock price movements more effectively than traditional methods [15].

Our research aims to predict the stock price directions in ten years period of four stocks in same sector of China A-share market using traditional data sources but more machine learning methods in varies periods.

Contributions. The main contribution of this research is a benchmark comparing 3 machine learning models using multiple data sources in predicting the stock price direction. In different market style: bull and bear market and different activity time periods in the market, we also test the models and hypothesize that there will be significant differences in forecasting effects.

The rest of this study is organized as follows: Sect. 2 gives an overview of the papers we cover; Sect. 3 describes the major results and discussion; We conclude this research in Sect. 4 and point out our study's limitation and some possible future research directions.

2 Methodology

2.1 Data and Variables

We select four stocks in financial sector with top high ranking of market value and turnover volume, which make the sample considered to be an appropriate representation of the financial industry sector in Chinese A-share stock market. Information including codes, names and industry sectors about these stocks are presented in Table 1. All the data, including stock price, technical indicators, historical transaction data and Baidu Index are obtained from uqer.datayes.com, or downloaded and calculated later. These data forms our entire data set.

Table 1. Description of the stocks to be tested

Stock code	Stock name	Industry sector
600030.SH	CITIC Securities Co., Ltd	Capital market services, Financial sector
600016.SH	China Minsheng Banking Co., Ltd	Monetary and financial services, Financial sector
601628.SH	China Life Insurance Co., Ltd	Insurance, Financial sector
300033.SZ	Zhejiang Hexin Flush Network Information Co., Ltd	Other financial industry, Financial sector

The sample period has a range of 10 years, from January 24, 2011 to January 22, 2021. We set 3-day forecasting horizons, with always using the data of the past 3 days to predict the closing price of today. As shown in Fig. 1, all data from the opening of the $t - 3$ day to the closing of the $t - 1$ day are put into our models to predict the closing price's movement of the t day, which is calculated to become the target (or label, the true value to predict) of our study in the following way.

In this study, the previous re-rights (or re-weighting) closing price is used as the basis to calculate the daily rise and fall labels, as is defined in Eq. (1). Re-rights is to repair the stock price with stock dividend.

$$Label = \begin{cases} 1, & SMA_{t-3} < SMA_{t+3} \\ 0, & SMA_{t-3} \geq SMA_{t+3} \end{cases} \tag{1}$$

When the label is 1, it means that the previous re-rights closing price showed an upward trend on day t, and 0 means that the price is a downward trend. Here, we use the simple moving average (SMA) to denote stock price to eliminate noise, and SMA_t means the moving average on time t. This makes the prediction be recognized as a binary classification problem. We use this calculation result as the target for subsequent model training, validation and test.

(1) Historical trading data.
Stocks' historical transaction data is gathered in each trading day including highest prices, lowest prices, previous re-rights closing prices, opening prices, and trading volume.

(2) Technical indicators transformed into discrete representation.
We select twelve technical indicators, and they are: 5 days' moving average (MA5), 10 days' moving average (MA10), 5 days' exponential moving average (EMA5), 10 days' Exponential moving average (EMA10), 10 days' moving average of Momentum Index (MTMMA), stochastic oscillators (STCK%, STCD% and Williams R%), Relative Strength Index (RSI), Moving Average Convergence Divergence (MACD), 10 days' Commodity Channel Index (CCI10), Accumulation/Distribution Line (A/D). Each data's update time is 17:00 for each day. All of them are downloaded from the uqer.datayes.com. In addition, we calculate the value of Williams R% based on historical transaction data by using the Talib library in Python. We refer to the technical indicators' discretization method from Patel et al. [16], which certify that the discretization technical indicators perform better in the forecast of stock price movements.

We employ a data processing technique which converts continuous valued technical indicators to discrete value, representing the future trend of stock prices. The purpose of this method is to convert these continuous values to '+1' or '−1' by considering indicator's intrinsic principle during the process, where '+1' shows up movements and '−1' implies down trend. Details about how each of the technical indicators is discretized is mentioned in Table 2.

(3) Baidu index.
The Baidu Index is based on the search behavior data of many Baidu Internet users. Through the Baidu Index, people can study keyword search trends and gain insights into the interests and needs of netizens. The Baidu index depends on the user's active search behavior record. Each user's search behavior in the Baidu search engine is a display of active willingness to see and may become an expression of the consumer's willingness to consume. The popularity of a stock is considered to be reflected in the search volume which is the number of times a keyword is retrieved on the network every day. We attain the daily Baidu Index for each stock in ten years period using the stock name as the search term.

Table 2. Selected technical indicators and their discretization methods

Technical indicators	Ups and downs signals	Judgement condition
SMA5	+1	*Closing Price$_t$ > SMA5$_t$*
	−1	*Closing Price$_{t_t}$ ≤ SMA5$_t$*
SMA10	+1	*Closing Price$_{t_t}$ > SMA10$_t$*
	−1	*Closing Price$_{t_t}$ ≤ SMA10$_t$*
EMA5	+1	*Closing Price$_t$ > EMA5$_t$*
	−1	*Closing Price$_t$ ≤ EMA5$_t$*
EMA10	+1	*Closing Price$_t$ > EMA10$_t$*
	−1	*Closing Price$_t$ ≤ EMA10$_t$*
MTMMA	+1	*MTMMA > 0*
	−1	*MTMMA$_t$ ≤ 0*
STCK %	+1	*STCK < 10; 10 ≤ STCK ≤ 90 and STCK$_t$ > STCK$_{t-1}$*
	−1	*STCK > 90; 10 ≤ STCK ≤ 90 and STCK$_t$ ≤ STCK$_{t-1}$*
STCD %	+1	*STCD < 20; 20 ≤ STCD ≤ 80 and STCD$_t$ > STCD$_{t-1}$*
	−1	*STCD > 80; 20 ≤ STCD ≤ 80 and STCD$_t$ ≤ STCD$_{t-1}$*
RSI	+1	*RSI < 30; 30 ≤ RSI ≤ 70 and RSI$_t$ > RSI$_{t-1}$*
	−1	*RSI > 70; 30 ≤ RSI ≤ 70 and RSI$_t$ ≤ RSI$_{t-1}$*
MACD	+1	*MACD$_t$ > MACD$_{t-1}$*
	−1	*MACD$_t$ ≤ MACD$_{t-1}$*
WILL_R%	+1	*WILL_R < 20; 20 ≤ WILL_R ≤ 80 and WILLR$_t$ > WILLR$_{t-1}$*
	−1	*WILL_R > 80; 20 ≤ WILL_R ≤ 80 and WILL_R$_t$ ≤ WILL_R$_{t-1}$*
A/D	+1	*A/D$_t$ > A/D$_{t-1}$*
	−1	*A/D$_t$ ≤ A/D$_{t-1}$*
CCI	+1	*CCI < −200; −200 ≤ CCI ≤ 200 and CCI$_t$ > CCI$_{t-1}$*
	−1	*CCI > 200; −200 < CCI ≤ 200 and CCI$_t$ ≤ CCI$_{t-1}$*

(4) Dataset and combination of data sources.

The sample size and dataset time interval of stocks and different training or testing period are summarized in Table 3. The collected datasets are arranged in chronological order. Especially we set the proportion of training period and test period to 0.7:0.3, and inside training period we use last one third data to validate when using neutral network to predict. This is the standard setting in the initial experiment. The dataset can change in subsequent experiments, and the validation dataset of ensemble model is different, both of which are illustrated in the next section and Sect. 3.4, respectively.

Table 3. The dataset time period and sample size of stocks

Stock code	Sample size	Training period	Test period
600030.SH	2433	2011/1/24–2018/6/19	2018/6/19–2021/1/22
600016.SH	2433	2011/1/24–2018/6/19	2018/6/19–2021/1/22
601628.SH	2433	2011/1/24–2018/6/19	2018/6/19–2021/1/22
300033.SZ	2433	2011/1/24–2018/6/19	2018/6/19–2021/1/22

We aim to observe and test whether the accumulation of data sources can improve the prediction performance. We set previous re-rights closing price and discretized technical indicators together as baseline, and the other sources: historical transaction data, Baidu Index are added one by one, as presented in Table 4.

Table 4. Combinations of different data sources

Combination no	Description
1	Previous re-rights closing price, technical indicators
2	Previous re-rights closing price, technical indicators, historical transaction data
3	Previous re-rights closing price, technical indicators, Baidu index
4	Previous re-rights closing price, technical indicators, historical transaction data, Baidu index

The average daily Baidu Index of each stock are summarized in Table 5. We find that the public's attention and interest reflected by the index values to these four stocks is quite different. Each of the stocks may have its own active and inactive period, and the performance between those periods of different models' prediction may vary.

Table 5. Average daily Baidu Index of each stock

Stock code	600030.SH	600016.SH	601628.SH	300033.SZ
Baidu search volume	8120.76	17376.52	6620.25	13129.74

2.2 Machine Learning Models

In this section we describe the three techniques used in this study: Long short-term memory algorithm (LSTM), Random Forest (RF) and XGBoost (XGB). The first is a deep learning algorithm, which is also a single classifier here implemented under the support of tensorflow 1.9.0 and keras 2.1.2. The last two are ensemble methods, implemented using sklearn. All analyses are performed using Python version 3.6.5.

(1) LSTM

LSTM is a variant of recurrent neural network (RNN), known as learning the long-term dependence between information due to the vanishing gradient problem. LSTM solves the problem by adding a method to save information for later use, thereby preventing earlier signals from fading away during processing [17]. A one-or-two-layer LSTM neural network is employed in our research. On the final layer an activation function sigmoid is set. This produces a continuous output between 0 and 1 with a threshold of 0.5 to classify the upward or downward of stock price. With the output value not less than 0.5, stock price is considered to going up and the predicted label is 1, and vice versa. Each of the layers employs relu as the activation function. The parameters and their levels are summarized in Table 6.

Table 6. LSTM parameters and levels to be tested in parameter tuning

Parameters	Level(s)
Learning rate (lr)	0.001
Number of hidden layer neurons (n)	1024, 512, 256, 128, 64, 32, 16
Number of layers (nl)	1, 2
Number of epochs (ep)	Use best epoch in validation set
L1–L2 regularization	0.001 or none
Drop out and recurrent drop out	0.1 or none

(2) Random Forest (RF)

Decision tree learning is an efficient technique for classification. Random forest learns representation trough ensembles of decision tree [16]. The idea of reducing the degree of overfitting by combining multiple overfitting evaluators is actually an integrated learning method called a bagging algorithm. The bagging algorithm uses parallel evaluators to integrate the data with replacement (or a hodgepodge). Each evaluator overfits the data, and better classification results can be obtained by averaging. The ensemble algorithm of random decision trees is called random forest [18].

We employed the sklearn library's RandomForestClassifier and adjust 4 parameters: the number of trees in the forest (n_estimators) varied from 200 to 1000, the maximum depth of the tree (max_depth) varied from 5 to 20, the minimum number of samples required to split an internal node (min_samples_split) varied from 3 to 9, and the minimum number of samples required to be at a leaf node (min_samples_leaf) varied from 3 to 9.

(3) XGBoost (XGB)

XGBoost, also known as the gradient booster algotithm, is an open source machine learning project which is expert in handling structured data. It efficiently implements the Gradient Boosting Decision Tree (GBDT) algorithm. XGboost library is one of the dominators in Kaggle competitions and has achieved excellent results [19].

XGBoost is essentially a GBDT, but strives to maximize speed and efficiency by explicitly add a regularization term to control the complexity of the model, and perform a second-order Taylor expansion with the loss price function, etc. Both are boosting methods compared to Random Forests' bagging method. The Boosting method uses a serial method to train the base classifiers, and there are dependencies between each base classifier. Its basic idea is to superimpose the weak classifiers layer by layer, and each layer gives higher weight to the samples that are incorrectly classified by the previous layer of base classifiers during training. GBDT needs to accumulate the scores of multiple trees to get the final prediction score, and in each iteration, on the basis of the existing tree, add a tree to fit the residual between the prediction result of the previous tree and the true value [20]. The XGboost parameters and their levels which are tested are summarized in Table 7.

Table 7. XGBoost parameters and their levels tested in parameter setting

Parameters	Level(s)
Learning rate (*eta*)	0.01, 0.1, 0.3
Maximum depth of tree (*max_depth*)	From 3 to 20
Gamma	0, 0.01, 0.05, 0.1, 1
The number of trees(*n_estimators*)	From 200 to 30000
Subsample	0.1, 0.3, 0.5, 0.7, 0.9
Colsample_bytree	0.1, 0.3, 0.5, 0.7, 0.9
Alpha	0, 0.01, 0.05, 0.1, 0.5
Lambda	0, 0.01, 0.05, 0.1, 1

2.3 Model Evaluation Criteria

Several metrics are used for models' evaluation. First, the precision and recall in the traditional sense are computed from True Negative (TN), False Negative (FN), True Positive (TP), and False Positive (FP). These formulas are shown in Eq. (2)-(5).

$$\text{Precision}_{positive} = \frac{TP}{TP + FP} \tag{2}$$

$$\text{Precision}_{negative} = \frac{TN}{TN + FN} \tag{3}$$

$$\text{Recall}_{positive} = \frac{TP}{TP + FN} \tag{4}$$

$$\text{Recall}_{negative} = \frac{TN}{TN + FP} \tag{5}$$

Accuracy and F1-measure which is used as our models' metrics are shown in Eqs. (6) and (7).

$$\text{Accuracy} = \frac{TP + TN}{TP + FP + TN + FN} \tag{6}$$

$$\text{F1} = \frac{2 \times \text{Precision} \times \text{Recall}}{\text{Precision} + \text{Recall}} \tag{7}$$

2.4 Model Validation

In applied machine learning, the k-fold cross validation is often applied by dividing the data into K partitions of the same size. For each partition i, train the model on the remaining partitions, and then evaluate the model on partition i. The final score is the average of K scores. This method cannot be directly used with time series data because it assumes that there is no time relation between the samples, which is not the fact. Data should be split up with the respect of the temporal order [21]. We use different validation methods between deep learning model and the ensemble models.

In LSTM we adopt holdout set, as have mentioned above. We always use training set's last one third data as validation set. In Random Forest and XGBoost, we use TimeSeriesSplit in sklearn which is specially designed for time series data. The split parameter n_splits is set to 2, 3, or 5 based on needs, illustrated in Fig. 1.

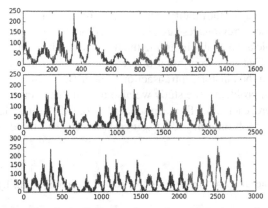

Fig. 1. Illustration using TimeSeriesSplit (n_splits = 3; x axis is the number of observations and y axis is data value; data in blue is for training and data in green is for validation)

3   Results and Discussion

3.1   Performance of Three Models Using Multiple Data Sources

By using the LSTM, RF and XGBoost classification algorithms, the datasets of four data source combinations of four stocks are trained and tested. The forecast accuracy of four stocks are collected as shown in Table 8. It is obvious that all three models have beat the dumb baseline with the accuracy of 0.50 in the two-label classification problem.

From Table 8, we first conclude that the deep learning model LSTM achieves the highest average predictive accuracy of four stocks comparing to the other two ensemble models. Though, the lead of performance is rather slight, while considering it is the average based on both four stocks and four data combinations, the divergence is not supposed to be ignored. Meanwhile, according to the detailed results of the experiment we find that ensembled models show more instability while dealing with more input data. To specify it, among all the results, we find 5 outcomes: in model RF, stock 601628.SH with combination 3 and 4; in model XGBoost, stock 601628.SH with combination 3 and 4 and stock 300033.SZ with combination 4. Not only are the accuracy and F1-measure of those significantly lower than others, but it seems that model can't manage to learn effective representation from data viewing from confusion matrix. We did not find results like this in LSTM. Those seems to be convincing evidences on LSTM's stronger ability and relatively high stability dealing with time series data.

Another conclusion we arrived at is adding more sources of data not always lead to the improvement on performance of prediction in classification problems. More than that, combining more data resources and put them into models can resulting in more unstable and even worse results. The number of our data sources goes up as the combination No. (Table 4) increase, and LSTM achieve its best score in combination 1 with least data source. Also, different models reach their own best scores through totally different combinations. LSTM with combination one, RF with two and XGboost with three. None model reaches best score in four, although it consists all data sources.

One explanation may be the quality of the data source matters, especially what kind of information it includes. Previous research has pointed out about the discretized technical indicators that people are actually inputting the trend when the data is given as the input to the model [16]. That may imply that discretized technical indicator contains more valuable information than historical transaction data or Baidu Index.

In the regression problems, the study we also did on this dataset with three models but not presented here, each model's mean squared error (MSE) and Mean absolute error (MAE) do decrease each time new data source is added. But the degree of reduction varies greatly.

Table 8. Prediction accuracy of four stocks under each data combinations by different models

Model/stock codes	No. of data source combination (Table 4)				Average
	1	2	3	4	
LSTM					
600030.SH	0.6620	0.6390	0.6563	0.6344	0.6479
600016.SH	0.6641	0.6604	0.6531	0.6531	0.6577
601628.SH	0.6410	0.6297	0.6438	0.6313	0.6365
300033.SZ	0.6453	0.6422	0.6547	0.6641	0.6516
Average	**0.6531**	0.6428	0.6520	0.6457	**0.6484**
RF					
600030.SH	0.6396	0.6575	0.6437	0.6637	0.6511
600016.SH	0.6534	0.6561	0.6451	0.6369	0.6479
601628.SH	0.6121	0.6135	0.6162	0.5887	0.6076
300033.SZ	0.6217	0.6080	0.6547	0.6355	0.6300
Average	0.6317	0.6338	**0.6399**	0.6312	0.6342
XGB					
600030.SH	0.6093	0.6685	0.6396	0.6437	0.6403
600016.SH	0.6217	0.6396	0.6341	0.6479	0.6358
601628.SH	0.6355	0.6272	0.6052	0.6052	0.6183
300033.SZ	0.6176	0.5928	0.6024	0.5997	0.6031
Average	0.6210	**0.6320**	0.6203	0.6241	0.6244

3.2   Comparison of Bull and Bear Markets

In order to compare models' performance on different market style, all stocks are put together and divided by market style during ten years. The time period of bull market or bear market in China A-share market during ten years are shown in Table 9. We connect different time periods of bull market together, the same with bear market, considering single period of time is too short for the models to learn fully.

According to the results of the first experiment, the neural network is trained with data source combination 1 and the ensemble models are trained with data source combination 2 to seek the best performance and avoid instability.

In second experimentation, other settings are identical to the first one except that the training, validation and test dataset is repartitioned according to the proportion mentioned in the previous section. Table 10 show each stock's number of days in bull market and bear market after connection.

Results on two market styles for all the proposed models is reported in Table 11. Comparison shows that all the models performed better and the result is very close after putting the four stocks into the model at one time. The difference in forecast performance of the bull and bear market is quite clear, with 3 metrics: accuracy, F1 and precision of bull market all better than bear market for each model. Thus, we can conclude that a model may predict more accurately in uptrend markets than downtrend markets totally.

Table 9. The time period of bull market and bear market during ten years

Bull market	Bear market
2012/12/4–2013/2/18	2011/1/24–2012/12/4
2013/6/25–2013/9/12	2013/2/18–2013/6/25
2014/3/12–2015/6/12	2013/9/12–2014/3/12
2016/1/27–2018/1/29	2015/6/12–2016/1/27
2019/1/4–2019/4/19	2018/1/29–2019/1/4

Table 10. Each stock's number of days in bear or bull market during ten years

Stock codes	Bull market	Bear market
600030.SH	970	1033
600016.SH	970	1033
601628.SH	970	1033
300033.SZ	970	1033

3.3 Comparison of Inactive and Active Periods

In order to compare models' performance on market with different activity, each stock is divided by activity during ten years. If a trading day's Baidu Index value is higher than ten years' average, it is considered active and vice versa. The time period of active or inactive period during ten years are shown in Table 12.

Table 11. Performance of prediction models on bull and bear market data set

Models	Market style					
	Bull market			Bear market		
	Accuracy	F1	Precision	Accuracy	F1	Precision
LSTM	0.7905	0.8107	0.8157	0.7331	0.6756	0.7152
RF	0.7915	0.8124	0.81	0.7361	0.6518	0.74
XGB	0.7919	0.8144	0.82	0.7522	0.6975	0.72

Similar as experiment 2, we connect different time period of active periods together, and the same with inactive period with correctly dealing with intermittent point between different time periods to ensure the 3-day forecast horizon.

Experiment 3's other settings are identical to the second one except that the training, validation and test dataset is repartitioned according to the proportion mentioned in the previous section. Tables 13 show each stock's number of days in two kinds of periods.

Table 12. The active and inactive period of four stocks in ten years

Stock codes	Activity	
	Active period	Inactive period
600030.SH	2014/10/29–2016/7/15, 2019/2/18–2019/4/19	2011/1/24–2014/10/28, 2016/7/16–2019/2/17, 2019/4/20–2021/1/22
600016.SH	2012/11/26–2017/9/29	2011/1/24–2012/11/25, 2017/9/30–2021/1/22
601628.SH	2014/11/24–2019-4/19	2011/1/24–2014/11/23, 2019/4/20–2021/1/22
300033.SZ	2014/11/24–2016/12/23, 2017/2/6–2017/4/23, 2019/2/25–2019/4/21	2011/1/24–2014/11/23, 2016/12/24–2017/2/5, 2017/4/24–2019/2/24, 2019/4/22–2021/1/22

Results on two types of periods for all the proposed models is reported in Table 14. We can conclude that the average and best performance of the neural network is still stronger than the two ensemble models, both in active periods and inactive periods, which verifies the result of experiment one.

We notice that after dividing the stocks into periods, some stocks performed significantly better in a certain period than in the full data set, which has not appeared before. For example, there are results of accuracy reaching or even exceeding 0.7.

Table 13. Each stock's number of days in active and inactive periods in ten years

Stock codes	Activity		Total
	Active	Inactive	
600030.SH	465	1968	2433
600016.SH	1181	1251	2432
601628.SH	1075	1357	2432
300033.SZ	604	1828	2432

Table 14. Forecast performance of models in active and inactive period of four stocks

Models	Activate			Inactivate		
	Accuracy	F1	Precision	Accuracy	F1	Precision
LSTM						
600030.SH	0.6341	0.6354	0.6930	0.5238	0.1954	0.7391
600016.SH	0.6431	0.5227	0.6216	0.6257	0.6714	0.5983
601628.SH	0.6522	0.6889	0.5990	0.4975	0.0286	0.9999
300033.SZ	**0.7207**	0.6914	0.6914	**0.7039**	0.6864	0.6591
Average	0.6625	0.6346	0.6513	0.5877	0.3955	0.7491
RF						
600030.SH	0.6087	0.5	0.82	0.5850	0.4455	0.71
600016.SH	0.6374	0.4128	0.69	0.5856	0.6192	0.57
601628.SH	0.6242	0.5926	0.61	0.4901	0.0000	0.00
300033.SZ	**0.6536**	0.6125	0.62	**0.6514**	0.6388	0.66
Average	0.6310	0.5295	0.6850	0.5780	0.4259	0.4850
XGB						
600030.SH	0.5507	0.5571	0.60	**0.6259**	0.5089	0.78
600016.SH	0.6289	0.4903	0.61	**0.6260**	0.4962	0.60
601628.SH	**0.7019**	0.6800	0.69	0.4951	0.0191	1.00
300033.SZ	0.5698	0.6010	0.52	0.5982	0.4222	0.75
Average	0.6128	0.5821	0.6050	0.5863	0.3616	0.7825

For each model, the average result during the active period is better than the average result during the inactive period. It seems that the predicting performance of stocks' active period is quite likely to be stronger than that of the inactive periods. However, XGB for 600030.SH and XGB for 300033.SZ show results contrary to the hypotheses at an individual case, which are worthy of attention. In spite of that, the consistency of the outcomes of the active and inactive periods between different models is convincing as well.

4   Conclusion

Our study adds to the field of stock price direction forecasting by making predictions about the price trends of four stocks of the financial sector in China A-share market. Evaluating 3 machine learning models and 4 kinds of traditional data sources, we test the forecast ability of our technique on every stock on the total dataset, different market styles and different active periods respectively. Our work's conclusions are as follows. (1) No matter on whole dataset or split dataset, under the same conditions, LSTM gain better performance than two other ensemble models, both on predicting accuracy and stability. Adding data sources does not always improve forecasting performance in classification problems, and choose valuable data sources and process properly may be more essential than quantity of data source. (2) For all models applied in this study, prediction performance in bull markets is better than in bear markets. (3) For each model, the average result during the active period is often better than the average result during the inactive period, but should still notice that the results of some stocks do not conform to this principle.

There are some future works. (1) Previous study has shown the importance of non-traditional sources in predicting stock price's trend [22]. That includes investor sentiments, news and posts' polarity scores, etc. Better data source can tremendously improve the forecast performance. As we mainly based our research on data from the traditional sources, further research might incorporate more types of data. (2) It is recommended to use a model that is good at handling time series to predict stock price movement. It will be more beneficial to predict during the rising and active periods of the stock market. (3) The models in our study are limited to deep learning model and ensemble models and the prediction scores are fairly close. Including other machine learning techniques, especially single classifier may be an option to consider in further study. (4) Changing the forecast horizon may contribute to differentiate model performance as well.

References

1. Fama, E.F.: The behavior of stock-market prices. J. Bus. **38**, 34–105 (1965)
2. Malkiel, B.G.: The efficient market hypothesis and its critics. J. Econ. Perspect. **17**(1), 59–82 (2013)
3. Alpaydin, E.: Introduction to machine learning. MIT Press (2020)
4. Khaidem, L., Saha, S., Dey, S.R.: Predicting the direction of stock market prices using random forest. arXiv Preprint, arXiv:1605.00003 (2013)
5. Dey, S., Kumar, Y., Saha, S., Basak, S.: Forecasting to Classification: Predicting the Direction of Stock Market Price Using Xtreme Gradient Boosting. PESIT South Campus (2016)
6. White, H.: Economic prediction using neural networks: the case of IBM daily stock returns. ICNN **2**, 451–458 (1988)
7. Yoon, Y., Swales, G.: Predicting stock price performance: a neural network approach. In: Proceedings of the Twenty-Fourth Annual Hawaii International Conference on System Sciences, vol. 4, pp. 156–162 (1991)

8. Bengio, Y., Simard, P., Frasconi, P.: Learning long-term dependencies with gradient descent is difficult. IEEE Trans. Neural Netw. **5**(2), 157–166 (1994)
9. Nelson, D.M., Pereira, A.C., de Oliveira, R.A.: Stock market's price movement prediction with LSTM neural networks. In: 2017 International Joint Conference on Neural Networks (IJCNN), pp. 1419–426 (2017)
10. Althelaya, K.A., El-Alfy, E.S.M., Mohammed, S.: Stock market forecast using multivariate analysis with bidirectional and stacked (LSTM, GRU). In: 21st Saudi Computer Society National Computer Conference (NCC), pp. 1–7 (2018)
11. Deng, S., et al.: Combining technical analysis with sentiment analysis for stock price prediction. In: 2011 IEEE Ninth International Conference on Dependable, Autonomic and Secure Computing, pp. 800–807 (2011)
12. Skuza, M., Romanowski, A.: Sentiment analysis of Twitter data within big data distributed environment for stock prediction. In: 2015 Federated Conference on Computer Science and Information Systems (FedCSIS), pp. 1349–1354 (2015)
13. Roy, S.S., Chopra, R., Lee, K.C., Spampinato, C., Mohammadi-ivatlood, B.: Random forest, gradient boosted machines and deep neural network for stock price forecasting: a comparative analysis on South Korean companies. Int. J. Ad Hoc Ubiquitous Comput. **33**(1), 62–71 (2020)
14. Teng, X., et al.: Enhancing stock price trend prediction via a time-sensitive data augmentation method. Complexity (2020)
15. Li, Y., Pan, Y.: A novel ensemble deep learning model for stock prediction based on stock prices and news. arXiv Preprint, arXiv:2007.12620 (2020)
16. Patel, J., Shah, S., Thakkar, P., Kotecha, K.: Predicting stock and stock price index movement using trend deterministic data preparation and machine learning techniques. Expert Syst. Appl. **42**(1), 259–268 (2015)
17. Chollet, F.: Deep Learning with Python. Manning Publications (2018)
18. VanderPlas, J.: Python Data Science Handbook. People Post Press (2018)
19. XGBoost: https://github.com/NLP-LOVE/ML-NLP/blob/master/Machine%20Learning/3.3%20XGBoost/3.3%20XGBoost.md (2019)
20. GBDT: https://github.com/NLP-LOVE/ML-NLP/blob/master/Machine%20Learning/3.2%20GBDT/3.2%20GBDT.md (2019)
21. Brownlee, J.: How to Backtest Machine Learning Models for Time Series Forecasting. https://machinelearningmastery.com/backtest-machine-learning-models-time-series-forecasting/ (2016)
22. Zhou, Z., Gao, M., Liu, Q., Xiao, H.: Forecasting stock price movements with multiple data sources: evidence from stock market in China. Physica A: Statis. Mech. Appl. **542**(3), 123389 (2020)

Data-Driven Prediction of Foodborne Disease Pathogens

Xiang Chen[ID] and Hongzhi Wang[(✉)][ID]

Faculty of Computing, Harbin Institute of Technology, Harbin, China
20s003052@stu.hit.edu.cn, wangzh@hit.edu.cn

Abstract. In recent years, foodborne diseases have become one of the most Data analysis technology has been widely used in the field of public health, and greatly facilitates the preliminary judgment of medical staff. Foodborne pathogens, as the main factor of foodborne diseases, play an important role in the treatment and prevention of foodborne diseases. However, foodborne diseases caused by different pathogens lack specificity in clinical features, and the actual clinical pathogen detection ratio is very low in reality. This paper proposes a data-driven foodborne disease pathogen prediction model, which paves the way for early and effective patient identification and treatment. Data analysis was implemented to model the foodborne disease case data. The best model achieves good classification accuracy for *Salmonella, Norovirus, Vibrio parahaemolyticus, Staphylococcus aureus, Shigella* and *Escherichia coli*. With the patient data input, the model can conduct rapid risk assessment. The experimental results show that the data-driven approach reduces manual intervention and the difficulty of testing.

Keywords: Foodborne disease · Pathogens prediction · Data-driven healthcare

1 Introduction

The World Health Organization defines foodborne diseases as: "all infectious or toxic diseases caused by pathogenic agents (pathogens) that enter the human body through ingestion." Including common food poisoning, intestinal infectious diseases, Diseases caused by zoonotic infectious diseases, parasitic diseases, and chemical toxic and harmful substances [1]. Foodborne diseases include three basic elements: the carrier of the disease (food), the causative factors of foodborne diseases (toxic and harmful substances in food), and clinical features (toxic or infectious manifestations). Due to the globalization of food production, the application of new food processing technologies, the integration of Eastern and Western eating habits, the sudden changes in the natural environment (such as floods, earthquakes, etc.) or the social environment (such as wars, leakage of toxic and hazardous substances, etc.), And the emergence of various new pathogens and transmission vectors, diseases caused by food contamination have

© Springer Nature Singapore Pte Ltd. 2021
J. Zeng et al. (Eds.): ICPCSEE 2021, CCIS 1451, pp. 106–116, 2021.
https://doi.org/10.1007/978-981-16-5940-9_8

become the most widespread health problems in the world today [2]. In 2015, the WHO released the "Estimation Report on the Global Burden of Foodborne Diseases", which pointed out that as many as 600 million people or nearly one in ten people get sick from eating contaminated food every year, causing 420,000 deaths, including 125,000 children under the age of five [3]. At present, the incidence of foodborne diseases is the highest among all diseases. Epidemiological surveillance data show that the incidence of foodborne diseases is still rising. This is an important public health issue in both developed and developing countries. However, many public health departments have not fully realized the importance of food safety, and the research and prevention and control work related to foodborne diseases is imminent.

Foodborne disease pathogens are the most direct cause of foodborne diseases, and they enter the patient's body after the patient eaten the food that has been infected by them. However, there are a wide range of foodborne disease pathogens and the clinical characteristics of foodborne diseases caused by different pathogens are still unclear. For clinicians, it is difficult to judge whether this is caused by food and accurate food information after only 3 to 5 min of consultation time [4]. Most primary hospitals often do not have the ability to detect the causative factors of common foodborne diseases, and this is also a factor that affects the accurate diagnosis of foodborne diseases [5].

Accurately predict the foodborne diseases pathogens so that foodborne diseases can be targeted and effectively treated as soon as possible. Therefore, the main research goal of this article is to propose a reliable data-driven computer-aided diagnosis model, which can calculate the individual's risk through various index data input by the clinician or data collector of the patient, which can be used to treat the pathogenic factors of foodborne diseases. In the initial assessment and screening of suspected foodborne disease pathogens, it is more advanced, scientific, standardized, efficient, high-quality and fast. It is also more targeted for follow-up treatment.

2 Related Works

Many researchers at home and abroad have conducted research on foodborne diseases, including surveillance, identification and outbreak prediction.

2.1 Methods Based on Surveillance Platform Data

In the global food safety strategy formulated by the World Health Organization in 2002, the establishment of a food safety infrastructure should give priority to the monitoring of foodborne diseases. At present, many countries have carried out foodborne disease surveillance. The purpose of the foodborne disease surveillance system is to identify and control foodborne disease outbreaks; identify susceptible populations, high-risk foods, and unhealthy food operating procedures; clarify the foodborne transmission routes of specific pathogens; assess the impact of foodborne diseases; reduce the hazards of foodborne diseases; formulate food safety assessment plans; study the traceability and early warning strategies of foodborne disease outbreaks. A number of foodborne surveillance data in the United States are used to estimate the number of foodborne diseases in the United States every year [6], and the research results are widely cited by all countries studying foodborne diseases in the world [7]. Other developed countries, such as the United Kingdom, Denmark, Canada, Australia and so on, have a strong food borne disease monitoring system [8]. In some countries, the monitoring scope covers the whole world pathogen monitoring from farmland to dining table, especially the application of microbial source tracking technology. The construction of foodborne disease monitoring network in China is relatively late, and the traditional passive monitoring mode is still dominant for a long time. In recent years, China has established the national foodborne disease monitoring and reporting system and other platforms to classify, store, monitor and statistically analyze the foodborne disease monitoring data collected nationwide, and gradually improve the foodborne disease surveillance network system [9, 10]. Although there are some researches on foodborne diseases based on surveillance platform data, most of them focus on statistical analysis, only a few people use the data for disease cluster analysis and outbreak prediction [11], and have not yet put forward suggestions on using monitoring platform data to identify pathogens. Therefore, there is still a need for an accurate identification method of foodborne pathogens based on surveillance platform data.

2.2   Methods Based on Rapid Detection Technology

In recent years, many methods for detecting food-borne pathogens have been developed to solve the problems of food safety and public health, especially with the increasing consumption of fresh food and short shelf life food, which makes the rapid detection technology more marketable. Many scholars have been innovating in different rapid detection technologies, and have greatly improved the detection time, sensitivity and accuracy. There are chromogenic medium technology, adenosine triphosphate luminescence method, immunology technology (enzyme-linked immunosorbent assay, Immunofluorescence, gold immunochromatography assay), molecular biology technology (polymerase chain reaction, amplification, gene chip, microfluidic chip, high-throughput sequencing), biosensor, flow cytometry, spectroscopy and mass spectrometry [12]. However, there are still some shortcomings: almost all rapid detection technologies have the shortcomings of insufficient sensitivity, so the process of enrichment and cultivation is needed to get the test results; in addition, the integration of artificial intelligence, gene editing, nanotechnology and other frontier disciplines into the rapid detection of food borne pathogens will also become the future development trend. Furthermore, it is urgent to develop a series of standards for rapid detection technology to make up for the lack of rapid detection standards.

2.3   Methods Based on Machine Learning

Traditional foodborne disease research mainly focuses on biology and medicine. In recent years, with the advent of the era of big data and artificial intelligence, there have been more and more researches using machine learning methods to solve medical problems. Compared with traditional statistical analysis methods, machine learning methods can handle larger and more complex data, and can obtain faster and more accurate results. Therefore, machine learning methods have become a new method to solve foodborne diseases. However, most of these studies focus on the analysis of a certain factor of foodborne diseases, or the identification and outbreak prediction of foodborne diseases [11,13–16], and there is a lack of comprehensive analysis and prediction of foodborne disease pathogens from the perspective of data. The work of this article fills up this gap to a certain extent.

3   Methods

3.1   Data Description

The data used in the experiment is provided by the Heilongjiang Provincial Center for Disease Control and Prevention. Excluding "unexplained" cases, a total of 32,511 foodborne disease samples are included, and each sample comes from the real case after desensitization.

Foodborne pathogenic factors can be roughly divided into the following 7 categories: bacterial (such as *Salmonella, Campylobacter, Enterohaemorrhagic Escherichia coli, Listeria, Vibrio cholerae*, etc.), viral (*Norovirus*, hepatitis A virus, etc.), parasitic (fish-borne trematodes, *Ascaris, Cryptosporidium, Entamoeba histolytica, Giardia,* etc.), chemical (heavy metals, pesticides), mycotoxins (mildew grains, etc.), animal toxins (tetrodotoxin, etc.), and plant toxins (poisonous mushrooms, sprouting potatoes, etc.).

As shown in Table 1, the pathogens in the data provided by CDC are mainly bacteria and viruses. If subdivided in these two categories, as shown in

Table 1. Categories of pathogens involved in the cases

Categories	Count, n	Percentage
Pathogenic bacteria	20968	64.50%
Virus	7902	24.31%
Chemical pollutants	1791	5.51%
Mycotoxins	8	0.02%
Parasites	0	0.00%
Poisonous animals	400	1.23%
Poisonous plants	1442	4.44%

Table 2, the pathogens are mainly *Salmonella, Norovirus, Vibrio parahaemolyticus, Staphylococcus aureus, Escherichia coli* and *Shigella*. Therefore, in the following work, we mainly focus on these 6 pathogens.

Table 2. Distribution of pathogens involved in the cases

Categories	Count, n	Percentage
Salmonella	8308	25.55%
Norovirus	7902	24.31%
Vibrio parahaemolyticus	3579	11.01%
Staphylococcus aureus	2178	6.70%
Escherichia coli	1307	4.02%
Shigella	1074	3.30%

A case data entry contains information about the patient's age, gender, occupation, time of illness, incubation period, and cause food information (including food type name, food processing and packaging method, place of food purchase and intake). The symptoms and signs of patients are as follows:

1. Systemic symptoms and signs: fever (°C), flushing, pale, cyanosis, dehydration, thirst, edema, weight loss, shivering, fatigue, anemia, swelling, insomnia, photophobia, mushy mouth, metallic taste, soapy/salty taste, excessive saliva, sagging feet/wrists, pigmentation, peeling, leucorrhea on nails;
2. Digestive system: nausea, vomiting (times per day), abdominal pain, constipation, tenesmus, diarrhea (times per day), stool traits;
3. Respiratory system: shortness of breath, hemoptysis, dyspnea;
4. Cardio-cerebrovascular system: chest tightness, chest pain, palpitations, shortness of breath;
5. Urinary system: decreased urine volume, back/kidney area pain, kidney stones, blood in urine;
6. Nervous system: headache, coma, convulsion, delirium, paralysis, speech difficulty, dysphagia, paresthesia, mental disorder, diplopia, blurred vision, vertigo, blepharoptosis, numbness of limbs, sensory disturbance without tip, abnormal pupil (enlargement, fixation, contraction), acupuncture sensation, convulsion;
7. Skin and subcutaneous tissue: itching, burning sensation, rash, bleeding point, jaundice.

3.2   Data Preprocessing

Data preprocessing is to ensure the quality of input data. It is the only link in the entire data mining process that requires manual processing, but it is an extremely important and indispensable link. In the above data, there are a large number of missing values, irregular items, and even error items. Therefore, the data needs to be cleaned and converted before mining to ensure that the model we build can correctly process these samples. Combining the characteristics of the sampling data, the training data set is processed as follows.

Excluding numerical variables (such as age, incubation period, fever temperature), categorical variables are elaborated as follows:

Convert gender variables into binary variables. Divide occupations into the following categories: 1. Children, 2. Students, 3. Workers, 4. Farmers, 5. Clerks, and 6. the Elderly. Food categories: 1. Meat and meat products 2. Vegetables and their products 3. Fruits and their products 4. Aquatic animals and their products 5. Infant foods 6. Milk and dairy products 7. Eggs and egg products 8. Beverages and frozen drinks 9. Packaged water (including bottled water) 10. Grain and its products 11. Beans and Soy products 12. Nut seeds and their products 13. Fungi and their products 14. Alcohol and their products 15. Candy, chocolate, honey and their products 16. Algae and their products 17. Oils and fats 18. Condiments. Processing or packaging methods: 1. Catering service industry 2. Home-made 3. Stereotyped packaging 4. Bulk (including simple packaging). Type of eating or buying place: 1. Family 2. Restaurant (hotel) 3. Food store 4. Street food 5. Unit canteens 6. School canteens 7. Construction site canteens 8. Farmers markets 9. Supermarkets 10. Retail stores 11. Rural banquets. For the main symptoms and signs, except fever (°C, divided into 3 levels, no fever, low fever, high fever), vomiting (times/day), diarrhea (times/day), stool characteristics (1. loose stool, 2. normal stool, 3. melena), the rest are dichotomous variables (yes/no).

At the same time, according to common sense, for a certain foodborne disease, only a few symptoms will appear, so only a few of the attributes are effective, and most of the remaining attributes are **no**. Therefore, we need to select a few representative attribute values, and avoid including all dozens of attributes into the model, otherwise it will bring about the redundancy of extremely irrelevant interference items, thereby reducing the accuracy of the model. By setting the threshold, the symptoms that occur rarely are filtered out. After such cleansing and transformation, these symptoms remain: fever, dehydration, thirst, chills, fatigue, sagging feet/wrists, pigmentation, nausea, vomiting, abdominal pain, diarrhea, stool traits, headache, dizziness. This result is consistent with our impression of foodborne diseases.

3.3 Model Building

As mentioned above, the research in this article is a multi-classification problem. The input parameters are various clinical indicators or characteristics, and the output result is one of these six pathogenic factors. Define the input as $X = x_i$, which contains 32511 variables $x_i (i = 0, 1, \ldots, 32510)$, and each x_i has 23 features, including 5 basic information of patients (age x_i^0, gender x_i^1, occupation x_i^2, time of illness x_i^3, incubation period x_i^4), 4 cause food information (food type name x_i^5, food processing and packaging method x_i^6, place of food purchase x_i^7 and place of food intake x_i^8), and 14 symptoms ($x_i^9, \ldots, x_i^{22}$ represent the 14 remaining symptoms that has been deleted at the end of the previous section). Output is defined as **Y**, which has six categories.y_i is one of $[1, 0, 0, 0, 0, 0]^T, [0, 1, 0, 0, 0, 0]^T, \ldots, [0, 0, 0, 0, 0, 1]^T$.

In the field of machine learning, commonly used multi-classification models are as follows: logistic regression, decision tree, Random Forest,Gradient Boosting Decision Tree, Multi-Layer Perceptual classifier, naive Bayes, support vector machine, etc. [19]. In this article, we choose three commonly used models for classification: RF, GBDT and MLP.

RF belongs to Bagging method in ensemble learning, which can solve the problem of weak generalization ability of decision tree [17]. RF is composed of many decision trees, and there is no correlation between different decision trees. It uses voting methods to get the final result. Each tree uses alternative sampling methods to obtain training data and sample features in a certain proportion. It can process high-dimensional data without feature selection.

Boosting method is a common statistical method. It can improve the performance of the model by changing the weight of training samples, learning multiple classifiers, and combining these classifiers linearly. Boosting method mainly uses linear combination of basis function and forward distribution algorithm [18].

Neural network is a computer system formed by several very simple processing units connected with each other in a certain way. The system processes information by the dynamic response of its state to external input information. Artificial neural network is a kind of information processing system which aims to imitate the structure and function of human brain. Back propagation is the most widely used algorithm for supervised learning using multilayer feedforward networks. MLP is an extension of neural network model. Its basic idea is to construct a multilayer neural network model by increasing the number of hidden layers. In general, the deep neural network model can be divided into three parts: input layer, hidden layer and output layer. In order to reduce the complexity of the model, MLP with only one hidden layer is used in this paper.

4   Experiments

4.1   Evaluation Criteria

Different from the two-category problem, the multi-category problem treats each category individually as "positive", and all other types are regarded as "negative". In the confusion matrix, the correctly classified samples are distributed on the diagonal from the upper left to the lower right. Among them, Accuracy is defined as the ratio of the number of samples correctly classified (on the diagonal) to the total number of samples. Accuracy measures the global sample prediction situation. For Precision and Recall, each class needs to calculate its Precision and Recall separately.

$$\text{Precision} = \frac{TP}{TP + FP}, \text{Recall} = \frac{TP}{TP + FN} \tag{1}$$

Generally, F-score has a parameter β, which is used to adjust the ratio of the two parts. The calculation formula with this parameter is:

$$F_\beta = \frac{(\beta^2 + 1) \times \text{Precision} \times \text{Recall}}{\beta^2 \text{Precision} + \text{Recall}} \tag{2}$$

When $\beta = 1$, it degenerates into a simple harmonic average, called F_1-score:

$$F_1 = \frac{2 \times \text{Precision} \times \text{Recall}}{\text{Precision} + \text{Recall}} \tag{3}$$

This indicator is widely used.

But this is a separate judgment for each category. If we want to evaluate the overall function of the recognition system, we must consider the comprehensive predictive performance of each category. In this paper, we adopt the simplest and most widely used *Macro-average method*. It directly adds up the evaluation indicators of different categories and averages them, giving all categories the same weight. Since this method treats each category equally, its value will be affected by the rare category. But for the research of this paper, this problem does not exist. For example, the definition of Macro-F_1 is as follows:

$$\text{Macro} - F_1 = \frac{1}{n} \sum_1^n F_i \tag{4}$$

4.2   Result

We divided 32,511 samples into training set and test set at a ratio of 7:3. The size of the training set is 22758 samples, and the size of the test set is 9753 samples. In order to adjust the parameters, we used the grid search method. Specifically, we estimated the range of several important parameters in the model and set the step size to obtain all possible values of these parameters. Select the parameter combination that obtains the best model result. In addition, we also used 10-fold cross-validation to improve the robustness of the model.

The classification results of the three models are shown in Table 3. It can be seen that the three models have good performance in the problem of pathogenic factor prediction. In comparison, the MLP model has better prediction accuracy, stronger generalization ability and simpler parameters. Relatively speaking, the GBDT model has many parameters and needs to be adjusted many times. However, decision tree algorithms like GDBT and RF can identify several factors that have a greater impact. In other words, they are highly interpretable, which is very much needed in the medical field. Through evaluating feature importance, it can be found that the patient's age, fever, frequency of vomiting, frequency of diarrhea, and time of illness have a greater contribution to the classification and prediction of foodborne disease pathogens. In summary, several models have good performance in classification algorithms, and it is also meaningful to study how to use multiple models to solve classification problems together.

Table 3. The classification results of 3 classification models

Model	Macro-P	Macro-R	Macro-F_1	Accuracy
Random forest	0.53	0.52	0.53	0.53
GBDT	0.56	0.57	0.55	0.57
MLP	0.61	0.60	0.61	0.59

4.3   Limitations

This study has certain limitations. The biggest limitation lies in the data set. Since the disease case data comes from the surveillance platform, the results are affected by the quality of the surveillance platform's data. The monitoring method mainly depends on factors such as whether the patient sees a doctor, whether the receiving doctor or the laboratory of the medical institution reports timely diagnosis and treatment. In fact, the number of disease cases collected on the monitoring platform is far lower than the actual value. Therefore, the monitoring platform's data is like "the tip of the iceberg" [20]. At the same time, it is easy to see that in this study, only the 6 pathogenic factors with the highest proportion in the data set were considered, while for the study of other pathogenic factors, due to the lack of data and representativeness, it needs to be further solved.

5   Conclusion

With the rapid development of information technology today, data analysis technology has been widely used in the field of public health, and it greatly facilitates the preliminary judgment of medical staff, and is of great significance to all aspects of society. We used data analysis to model the foodborne disease case data provided by the Heilongjiang Provincial Center for Disease Control and Prevention, and proposed a data-driven classification method for foodborne disease pathogens. Our best model achieves good classification accuracy for *Salmonella, Norovirus, Vibrio parahaemolyticus, Staphylococcus aureus, Shigella* and *Escherichia coli*. The model can use the patient data input by clinicians to calculate the most likely pathogenic factors of foodborne diseases, and can conduct rapid risk assessment. The data-driven approach reduces manual intervention and the difficulty of testing. Through communication with CDC personnel, the results are consistent with their previous work experience. Although the method we proposed cannot completely replace traditional laboratory tests, this method can help rapid identification, is also targeted for subsequent treatment, and has strong practical significance. The establishment of this model will greatly improve the timeliness and scientific nature of the investigation and treatment of food-borne diseases, thereby improving the low level of investigators when investigating food-borne diseases on-site using traditional manual methods. According to changes in living standards, the use of hospital population data and related clinical indicators to establish individual disease risk prediction models suitable for my country's national conditions is the focus and difficulty of future research. In addition, combining this model with data from the food-borne disease outbreak surveillance system can also be used to predict outbreaks.

References

1. WHO: Food safety in the 21st century.Foreign Medical Sciences (Section Hygine) **029**(001), 6–8, 14 (2002)
2. Kirk, M.D., Pires, S.M., Black, R.E., Caipo, M., Crump, J.A., Devleesschauwer, B., et al.: World Health Organization estimates of the global and regional disease burden of 22 foodborne bacterial, protozoal, and viral diseases, 2010: a data synthesis. PLoS Med. **12**(12), e1001921 (2015)
3. WHO: Food safety and foodborne illness. https://www.who.int/news-room/fact-sheets/detail/food-safety. Accessed 30 Apr 2020
4. Mei, L., Yuqi, Z.: Current situation and prospect of active surveillance of foodborne diseases in primary sentinel hospitals. Chin. Med. Guide **13**(10), 285–286 (2015)
5. Hui, L.: Discussion on the management of food safety accidents. Chin. J. Food Hygiene **23**(05), 446–449 (2011)
6. Centers for Disease Control and Prevention: Foodborne diseases active surveillance network. Morb. Mortal. Wkly Rep. **46**(12), 258–261 (1996)
7. Oliver, S.P.: Foodborne pathogens and disease special issue on the national and international Pulsenet network. Foodborne Pathogens Disease **16**(7), 439–440 (2019)
8. Hammerum, A.M., Heuer, O.E., Emborg, H.D., et al.: Danish integrated antimicrobial resistance monitoring and research program. Emerg. Infect. Dis. **13**(11), 1632–1639 (2007)
9. Shen, H.: Surveillance of foodborne diseases in China. Chin. Food Safe. Mag. **2015**(13), 46–48 (2015)
10. Wang, Y.: Current status of foodborne diseases and control measures. J. Commun. Med. **2007**(01), 59–61 (2007)
11. Xiao, X., et al.: Automated detection for probable homologous foodborne disease outbreaks. In: Cao, T., Lim, E.-P., Zhou, Z.-H., Ho, T.-B., Cheung, D., Motoda, H. (eds.) PAKDD 2015. LNCS (LNAI), vol. 9077, pp. 563–575. Springer, Cham (2015). https://doi.org/10.1007/978-3-319-18038-0_44
12. Wang, D., Liu, M., Yang, Y., et al.: Research progress on rapid detection of foodborne pathogens. Food Sci. 1–16 (2021). http://kns.cnki.net/kcms/detail/11.2206.TS.20210205.1508.012.html
13. Oldroyd, R.A., Morris, M.A., Birkin, M.: Identifying methods for monitoring foodborne illness: review of existing public health surveillance techniques. JMIR Public Health Surveill. **4**(2), e57 (2018) https://doi.org/10.2196/publichealth.8218
14. Wenxiu, P., Jiewen, Z., Quansheng, C.: Classification of foodborne pathogens using near infrared (NIR) laser scatter imaging system with multivariate calibration. Sci. Rep. **45**(1), 1–8 (2015)
15. Hanxue, W., Wenjuan, C., Yunchang, G., et al.: Machine learning prediction of foodborne disease pathogens: algorithm development and validation study. JMIR Med. Inform. **9**(1), e24924 (2021)

16. Teyhouee, A., McPhee-Knowles, S., Waldner, C., Osgood, N.: Prospective detection of foodborne illness outbreaks using machine learning approaches. In: Lee, D., Lin, Y.-R., Osgood, N., Thomson, R. (eds.) SBP-BRiMS 2017. LNCS, vol. 10354, pp. 302–308. Springer, Cham (2017). https://doi.org/10.1007/978-3-319-60240-0_36
17. Breiman, L.: Random forests. Mach. Learn. **45**(1), 5–32 (2001)
18. Friedman, J.H.: Greedy function approximation: a gradient boosting machine. Ann. Stat. **29**, 1189–1232 (2001)
19. Guangzhou, A., Kazuko, O., et al.: Comparison of machine-learning classification models for glaucoma management. J. Healthc. Eng. **2018**, 6874765 (2018)
20. Peigang, J., Gangqiang, D., Zhenhua, G.: Prevention and Emergency Treatment of Foodborne Diseases. Fudan University Press, Shanghai (2006)

Infrastructure for Data Science

Intelligent Storage System of Machine Learning Model Based on Task Similarity

Shuangshuang Cui[1], Hongzhi Wang[1,2(✉)], Yuntian Xie[1], and Haiyao Gu[1]

[1] Harbin Institute of Technology, Harbin, China
wangzh@hit.edu.cn
[2] Peng Cheng Laboratory, Shenzhen, China

Abstract. With the closer integration of database and machine learning, machine learning task in database can reduce the data transmission, thus dramatically boosting the runtime performance of the whole task. Moreover, if there is a chance of storing machine learning models involved in similar tasks in the system intelligently, the computation resource and time cost of repeated training will be greatly reduced. However, the intelligent storage system of machine learning model has not been developed yet. In order to achieve this goal, a method is proposed to measure the similarity of machine learning tasks. Second, the intelligent storage system of machine learning model was designed to manage models. Finally, it introduced the overall architecture and key technologies of intelligent storage system of machine learning model based on task similarity (ISSMLM), and describe three demonstration scenarios of the system. The results show the validity of the proposed method.

Keywords: DB4AI · Model management · Task similarity

1 Introduction

In recent years, with the rapid growth of data scale [1], the multi-task effect of artificial intelligence in the fields of image recognition [2], speech recognition [3], and intelligent question-answering [4] has exceeded the level of real human beings. Machine learning is the most effective way to realize artificial intelligence. The machine learning model storage system is used to manage the models that have been trained for machine learning tasks. The traditional machine learning process generally starts from loading data, extracting features, training the model, and then model prediction, that is, using the trained model to predict a specific task, and finally model evaluation. However, the machine learning model is only used once and then discarded. When the next machine learning task comes, the new model needs to be retrained. Such repeated training model will not only cause a large amount of overhead in memory resources, but also a large amount of overhead in time, making machine learning task processing efficiency very low [5]. Therefore, how to intelligently manage machine learning models and improve the reuse rate of models [6] has become a new and important research direction. In order to solve this problem, we propose a machine learning model intelligent storage

© Springer Nature Singapore Pte Ltd. 2021
J. Zeng et al. (Eds.): ICPCSEE 2021, CCIS 1451, pp. 119–124, 2021.
https://doi.org/10.1007/978-981-16-5940-9_9

system. Firstly, by verifying the similarity between two or more machine learning tasks, the reuse rate of machine learning model is improved, the training cycle of machine learning model is shortened, and the processing speed of the whole machine learning task is accelerated. The main characteristic of our system summarized as follows:

Intelligent Storage   Our system implements automatic management of machine learning models, which intelligently generates storage decisions by verifying the similarity between tasks greatly increasing the reuse rate of the models. It is suitable for intelligent recommendation, intelligent analysis and other occasions involved in machine learning, greatly improving the efficiency of data science teams.

User-friendly Interface   Our system provides a user-friendly interface. In our system, users only need to input some workloads and parameters, and then the system can automatically complete the management of the model.

The paper is organized as follows: We introduce the relevant work in Sect. 2.We introduce the architecture and implementation of the system in detail(Sect. 3); then we show two key technologies of the core function of the system, which are task similarity validation and model management(Sect. 4); Finally, we show the demonstration scheme of the system(Sect. 5).

2   Related Work

Many scholars have done a lot of work on how to store the model generated after machine learning training. Models are generally expressed as directed acyclic graphs between parameter matrices, which can be stored in a common format, ONNX [7]. Different types of machine learning frameworks may have different model storage formats. In addition to data, features and models, the training process also includes multiple types of information, such as task description, super parameters, loss function changes, etc., so it is necessary to form a unified storage scheme to reflect the correlation between data. ModelDB [8] takes the model as the center, associates and stores dependent training data, features, super parameters, evaluation nodes and other data, and supports viewing and tracking the data flow of the training process. At the same time, in order to facilitate the training process problem discovery and analysis. The machine learning training process is iterative, and multiple versions of the same data may be produced. First, a round of training for the same task usually involves iterating over input data to optimize the objective function, each iteration producing a version of the model. For training data, there is usually a large proportion of data repetition between multiple versions formed by multiple iteration training of the same task, so the storage cost can be optimized by combining full storage and incremental storage. DataHub [9, 10] borrows the idea of GitHub to manage the continuously changing multi-version data, supporting the operation of adding new versions after modification, merging data from multiple existing versions to form new versions and so on. At present, the machine learning data and model management main task centered way, difficult to achieve the global optimal computation and storage, with the wide application of machine learning, there will be a lot of training objectives and processes similar tasks, set up across the mission data associated with the model, implement the task resource reuse is an important approach to improve the overall training efficiency of the task. There have been preliminary studies in this regard [11, 12].

3 System Architecture

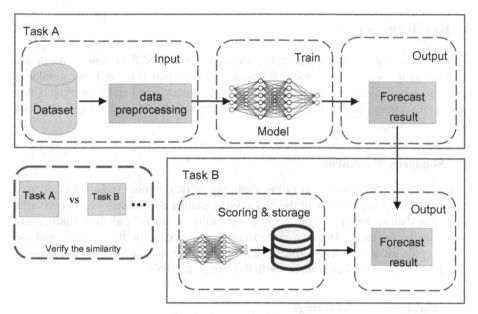

Fig. 1. System Architecture

In this section, we introduce the architecture of the entire system as shown in Fig. 1, which mainly consists of 5 modules.

- Input module: The input module of the initial task is responsible for reading and preprocessing the original data set.
- Training module: The training module is responsible for training the model with the preprocessed data.
- Similarity verification module: The similarity verification module is responsible for verifying whether two or more machine learning tasks are similar.
- Scoring and storage module: The scoring and storage module is responsible for scoring and evaluating the trained model and deciding whether to save it.
- Output module: The output module is responsible for outputting the results predicted by the corresponding model.

The core is the similarity verification module and the scoring and storage module, which are also involved in the key technologies of this system. In the verification similarity module, firstly, we extract the features of the two tasks, then perform one-hot encoding[13] on the features, use the similarity algorithm[14] to calculate the similarity of the tasks, compare with the set threshold, and draw a conclusion about whether they are similar. In the storage module, we store the model vector in the relational table, the parameters of the model are used as columns in the relational table, and the rows in the relational table represent the parameter information of the same model. Taking into account the need to complete the training process of the entire machine learning task in the system, our system is built based on the C/S architecture of Python and java.

4   Key Technologies

In this section, we will introduce the key technologies of ISSMLM system. We divide the intelligent storage system into two stages. The first stage is to verify the similarity between tasks, and the second stage is to score the models and store the models that exceed the set threshold. Next, we will separately introduce the key technologies involved in these two stages.

4.1   Similarity Verification

First of all, we study how to verify the similarity of two or more tasks. The difficulty lies in extracting the features of the tasks. First, extract the features of the two tasks (including data set features, Input & Output dimensions, data distribution features and task essential features), And then perform one-hot encoding on the features, and use similarity algorithms (such as cosine similarity, Euclidean distance, etc.) to calculate the similarity of the task. Then compare with the preset threshold.

4.2   Scoring and Model Storage

In the scoring module, we establish a multi-dimensional evaluation standard for the machine learning model, use the multi-dimensional evaluation standard to evaluate the model, and finally generate an evaluation result report to decide whether to save the model. Since we focus on the training and model management of machine learning tasks in the database, we store the model vector in the relational table of the database. The parameters of the model are used as columns in the relational table. The rows in the relational table represent the parameter information of the same model. In this way, besides realizing the reuse of machine learning models[15], it also realizes a deeper integration of database and machine learning.

5   Demonstration Scenarios

We plan to demonstrate our ISSMLM from the following three parts as shown in Fig. 2.

- Input task: The user uploads the dataset, and then performs model training after preprocessing.
- Task similarity verification: Click the similarity judgment button, and the system will automatically verify the task similarity and get the result.

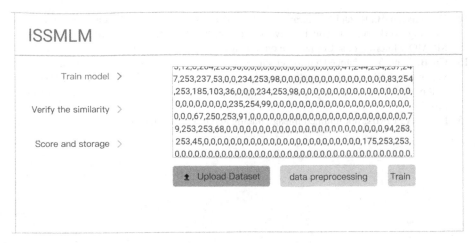

Fig. 2. System Demonstration

- Scoring and model storage: After similarity verification, the system scores the model, and the user can get the score and decide whether to save the model.

References

1. Du, X.Y., Lu, W., Zhang, F.: History, present, and future of big data management systems. Ruan Jian Xue Bao/J. Softw. **30**(1), 127–141 (2019). (in Chinese). http://www.jos.org.cn/1000-9825/5644.htm
2. Deng, J., Dong, W., Socher, R., Li, L.J., Li, K., Fei-Fei, L.: Imagenet: a large-scale hierarchical image database. In: 2009 IEEE Conference on Computer Vision and Pattern Recognition, pp. 248–255. IEEE, 20 June 2009
3. Lian, Z., Li, Y., Tao, J., Huang, J.: Improving speech emotion recognition via transformer-based predictive coding through transfer learning. arXiv. 2018 Nov:arXiv-1811
4. Devlin, J., Chang, M.W., Lee, K., Toutanova, K.: Bert: pre-training of deep bidirectional transformers for language understanding. arXiv preprint arXiv:1810.04805. 11 October 2018
5. Cui, J.W., Zhao, Z., Du, X.Y.: Data management technology for machine learning. Ruan Jian Xue Bao/J. Softw. (2021). (in Chinese)
6. Strubell, E., Ganesh, A., McCallum, A.: Energy and policy considerations for deep learning in NLP. arXiv preprint arXiv:1906.02243.2019, June 5
7. ONNX (2020). https://en.wikipedia.org/wiki/Open_Neural_Network_Exchange
8. Vartak, M., et al.: ModelDB: a system for machine learning model management. In: Proceedings of the Workshop on Human-In-the-Loop Data Analytics, pp. 1–3. 26 June 2016
9. Bhardwaj, A., et al.: Datahub: collaborative data science & dataset version management at scale. arXiv preprint arXiv:1409.0798. 2 September 2014
10. Miao, H., Li, A., Davis, L.S., Deshpande, A.: Modelhub: towards unified data and lifecycle management for deep learning. arXiv preprint arXiv:1611.06224. 18 November 2016
11. Smith, M.J., Sala, C., Kanter, J.M., Veeramachaneni, K.: The machine learning bazaar: harnessing the ML ecosystem for effective system development. In: Proceedings of the 2020 ACM SIGMOD International Conference on Management of Data, pp. pp. 785–800. 14 June 2020

12. Derakhshan, B., Rezaei Mahdiraji, A., Abedjan, Z., Rabl, T., Markl, V.: Optimizing machine learning workloads in collaborative environments. In: Proceedings of the 2020 ACM SIGMOD International Conference on Management of Data, pp. 1701–1716, 14 June 2020
13. Gu, B., Sung, Y.: enhanced reinforcement learning method combining one-hot encoding-based vectors for CNN-based alternative high-level decisions. Appl. Sci. **11**(3), 1291 (2021)
14. Eaton, C.: Using a similarity measure to investigate factors in failure classification schemes (2021)
15. Phani, A.: LIMA: fine-grained lineage tracing and reuse in machine learning systems (2021)

A Localization Algorithm in Wireless Sensor Network Based on Positioning Group Quality

Kaiguo Qian, Chunfen Pu, Yujian Wang, Shaojun Yu$^{(\boxtimes)}$, and Shikai Shen

School of Information Engineering, Kunming University, Kunming, China

Abstract. Localization is fundamental component for many critical applications in wireless sensor networks (WSNs). However, DV-Hop localization algorithm and its improved ones cannot meet the requirement of positioning accuracy for their high localization errors. This paper proposes a localization algorithm based on positioning group quality (LA-PGQ). The average estimate hop size was first corrected by link singularity and difference between the estimation hop length and true hop length among beacons, the best positioning group was constituted for unknown node by using node trust function and positioning group quality evaluation function to choose three beacons with best topological distribution. Third, LA-PGQ algorithm uses two-dimensional hyperbolic algorithm instead of the classical three-side method/least square method to determine the coordinates of nodes, which are more accurate. Simulation results show the positioning accuracy of LA-PGQ algorithm is obviously improved in WSNs, and the average localization error of LA-PGQ algorithm is remarkable lower than those of the DV-Hop algorithm and its improved algorithm and Amorphous, under both the isotropy and anisotropy distributions.

Keywords: Wireless sensor networks · Localization algorithm · Positioning group quality · Positioning accuracy

1 Introduction

As an advanced wireless network technology, wireless sensor networks (WSNs) are consisted of a large number of sensor nodes, which are densely deployed in close proximity to the monitoring environment and gather local data. Then, the sensor nodes send the data to a sink or a base station (BS) by using wireless multiple hop ad-hoc transmission techniques [1, 2]. WSNs have been used in wide application such as military applications, industrial control and wildlife monitoring with the advantages of rapid disposition, convenient networking and inexpensive data-collection [3]. Localization information of the sensor nodes must be clear in most of the applications and WSNs own core technology such as the networking topology control and the routing protocol [4]. However, it is impossible to install localization modules for all the nodes. Therefore, sensor network nodes must calculate their own localization according to node communication connection from the beacon nodes which have identified their location node. Many studies have been conducted to solve the localization problem in WSNs that can be divided

© Springer Nature Singapore Pte Ltd. 2021
J. Zeng et al. (Eds.): ICPCSEE 2021, CCIS 1451, pp. 125–139, 2021.
https://doi.org/10.1007/978-981-16-5940-9_10

into range-based and range-free [5] algorithms whether they directly measure the distance or angle among nodes by assembling measure hardware module or estimate the distance from beacon nodes through the topology relationship of sensors. The range-based localization algorithm firstly measures the distance or angle information which got through the following measurement technology: Received Signal Strength Indicator (RSSI) [6], Time of Arrival(TOA) [7], Time Difference of Arrival(TDOA) [8] and Angle of Arrival(AOA) [9]. However, they face two disadvantages. First, multipath attenuation reduces range accuracy。Second, due to the integrated battery module, sensors consume energy quickly and increase significantly overall cost of node deployment. By contrast, range-free localization algorithm [10] uses only the connectivity information among sensors. Thus, the range-free approach does not require any additional hardware for measurements so that it would be preferred In the case of large-scale WSN applications involving hundreds or thousands of sensors. Typical range - free location algorithms are distance vector-hop(DV-Hop) algorithm [11], Amorphous,Centroid Algorithm [12], Convex Optimization [13] and MDS-MAP [14]. Depending to the implementation details of this range-free localization algorithms, one of the most fatal drawbacks of the range-free algorithms is that the accuracy of the location estimation is somewhat poor, and it especially cannot locate accurately in anisotropy wireless sensor network because of the absence of any distance measurements between the nodes and collinear beacon nodes participating in the positioning.

In this paper, we propose a localization algorithm based on positioning group quality (LA-PGQ), which improves the DV-Hop algorithm. We firstly correct the average estimate hop size by link singularity and difference between the estimation hop length and true hop length. Best beacon groups are selected to localize the unknown nodes that avoid collinear beacons to participate in orientation. Finally, two-dimensional hyperbolic algorithm is used to calculate node position instead of the classic trilateration/least square method.This method has reduced the calculation complexity and improved the positioning accuracy in sensor network, including isotropy network and anisotropic network. To the best of our knowledge, the contributions of this paper are summarized as follows:

(1) Link singularity function is introduced to remove the beacon which deviates too much straight hop path in the process of the estimating hop size, which will increase the estimating accuracy of the average hop size.
(2) The difference size of the average estimated hop length and the actual hop length among beacon nodes are used to optimize the average estimated hop size of the whole network, which aims to make the average estimated hop size in anisotropic network to approach the hop size in homogeneous network.
(3) We firstly analyze positioning accuracy for different distributions of three reference beacon topology, and then build a quantitative model of positioning group quality to select the best positioning group which is consisted of three beacon nodes and one unknown node.
(4) In order to improve the positioning accuracy of nodes, the two-dimensional hyperbolic algorithm is used to calculate the position of nodes.
(5) Extensive simulation studies demonstrate the effectiveness of our proposed localization algorithm in WSNs.

The rest of the paper is organized as follows. Section 2 briefly introduces related research about range-free localization algorithms, DV-hop algorithm and its improved ones in WSNs. Section 3 present s the LA-PGQ algorithm. Section 4 shows simulation experiments with different sensor networks. Finally, Sect. 5 concludes the whole paper.

2 Related Works

In range-free localization, nodes used the accumulated hops distance in the shortest hop path in between two nodes to estimate the Euclidean distance between them. So, they can only provide coarse location information. However, range-free localization algorithm does not require hardware support which will reduce the cost. In the premise of keeping advantages of DV-Hop algorithm and ensuring the positioning accuracy, it is important to improve range measurement accuracy through taking full use of beacon nodes and connection relationship among nodes. Range-free localization algorithms are attracting more and more attentions in both academia and industry, since the distance between nodes is not essential to be measured. At present, a lot of research works focused on the study of improving the traditional DV-Hop algorithm which pay more attention to some influential factors, including the average every hop size of beacon nodes[15], collinearity [16, 17].

The Dv-hop algorithm includes three steps, firstly, the distance vectors including beacon ID, coordinates, and the variable Hops initialized to zero is flooded from respective beacon i. When sensor j receiving those vectors containing Hops, it checks the value of the Hops that it maintains about beacon i, if this value is less than the received one, then the latter is ignored. Otherwise, the sensor increments the Hops one, and then floods it in the network. In the second stage, every beacon i computes the average hop size from its perspective, using Eq. 1.

$$hopzise_i = \frac{\sum\limits_{j=1,j\neq i}^{m} \sqrt{(x_i - x_j)^2 + (y_i - y_j)^2}}{\sum\limits_{j=1,j\neq i}^{n} hop_{ij}} \tag{1}$$

Where m is the number of the beacon, (x_i, y_i) and (x_j, y_j) are coordinates of beacon i and beacon j, hop_{ij} is the hops from beacon i to beacon j. Then beacon i broadcasts its average hop size to all sensors over network using controlled flooding. After receiving an average hop size from beacon i, the unknown sensor u calculates the distance from the beacon i by the Eq. 2.

$$dis \tan ce_{ui} = hopsize_i \times hop_{ui} \tag{2}$$

Where hop_{ui} is the shortest hops between beacon i and unknown sensor u. In the third phase, sensor u uses the Least Square (LS) technique to determine its position. The DV-Hop method is easy to implement and its average localization error is in the order of 30%r, where r is the sensor communication radius [18], and it cannot locate it accurately in anisotropy network in which proposed NDSL approach [19], it divided into many sub-regions where the nodes density is relatively uniform and then corrects the single-hop distance for each beacon to locate unknown nodes. A lot of research works focused on the study of improving the traditional DV-Hop algorithm are conducted. In [20], QDV-Hop used the quadratic programming minimize the error to obtain better localization. However, quadratic programming requires special optimization tool box that increases computational complexity. References [21] establish the simultaneous equations with the ratio of distance and path length instead of calculating the average size of one hop, which removed the calculating processes of the average hop length, and reduced the complexity of the positioning algorithm. HDV-Hop [22] localization algorithm was proposed to meet localizing events in hostile environments which recommended placing the beacons in a circle or a semi-circle around the perimeter of the WSN, its purpose is to minimize the power of positioning process. In [23], authors firstly corrected the average hop size used distance estimation error, then used two-dimensional hyperbolic location algorithm in place of traditional triangulation algorithm for location estimation. A cluster-based architecture for range-free localization in references [24] was proposed to work on anisotropy network, which focused on reducing the communicating burden in the process of positioning. Reference [25] proposes algorithm based a reliable anchor node selection. These methods above improve the localization accuracy in WSNs effectively compared with DV-Hop.

3   Problem Identification and LA-PGQ Algorithm

3.1   Problem Identification

Analyzing the basic principle of traditional DV-hop algorithm and its' improved one in references in [23], its' main errors originated from three factors, the first is that used the straight distance instead of accumulated hop distance between two beacons and used the shortest hop path distance instead of straight distance from unknown sensor to beacon. The average hop size error will be increased when the shortest path seriously deviate the straight line between two beacons, so does hop distance from unknown sensor to beacon. It is shown as Fig. 1. The estimated shortest hop path distance is 5 hop size from beacon a to beacon f, but the actual distance is approaching 2 hop size.

The second is that in the process of polygonal positioning calculation, singular matrix in calculating will be made when beacon nodes keep collinear or approach collinear, the localization errors will be increased. As is shown in Fig. 1, when the unknown node u is positioned, it is not suitable for selection the beacon nodes d, e and f because those beacons approximate collinear distribution although they are very close to the node u, which the best choice of the referenced nodes should b,e,c,d and f. The third is that the remote beacon is selected to determine the location, large number of hop for remote beacon will increase distance estimation error, especially in the anisotropy network.

To ensure localization accuracy, this paper proposes LA-PGQ Algorithm. In this technique, the quantitative models of the singular hop path are firstly built to remove the beacons that make the shortest path seriously deviate the straight line during the process of average hop size calculated. Then average hop size is optimized to make it approach average hop length when nodes are uniform random distribution in anisotropic network. Secondly, Beacon node trust-rank functions and positioning group quality measures are built to choose three beacons that constitute the best positioning group for unknown

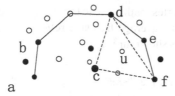

Fig. 1. Error originates from singular hop path and collinear localization

node position calculation. At last, we adopt two-dimensional hyperbolic function rather than the classic trilateration/least square method to determine the locations of unknown nodes, which makes it very closes to their actual locations.

3.2 The Average Hop Size (AHS)

3.2.1 Singular Path Beacon

In the second stage of the DV-Hop algorithm, the average hop size from every beacon i is calculated used all Euclidean distance that from beacon i to the other beacons include ones locate in the shortest path seriously deviate the straight line between two beacons, which causes the average hop size error increased. It is shown in Fig. 2, the shortest hop path 3 from beacon A to B, the hop path distance is 2 times of ctual distance. We define the shortest path that seriously deviate the straight line as singular path such as path 3 in Fig. 2.

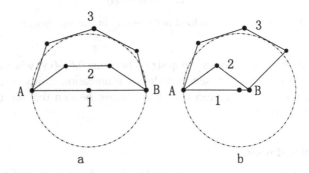

Fig. 2. Singular shortest hop path

We should remove these beacons located in singular path during the process of calculating the average hop size from every beacon i. So, the quantitative model of the singular path beacon is built to evaluate the deviation degree from straight line in Eq. 3.

$$q_s = \frac{d_{ij}}{hops_{ij}} \tag{3}$$

Where, d_{ij} is Euclidean distance from beacon i to beacon j, $hops_{ij}$ is the shortest hops from beacon i to j. The average hop size is related to the sensor density in homogeneous network because that the shortest path close to the straight line between nodes. Kleinrock and Silvester [26] deduce the average hop size relies on the numbers of the neighbor nodes, and the calculating is shown in Eq. 4.

$$HopSize_i = r(1 + e^{-n_{i\_local}} - \int_{-1}^{1} e^{-\frac{n_{i\_local}}{\pi}(ar\cos t - t\sqrt{1-t^2})}) \tag{4}$$

We make experiments to study the relationship between the average hop size and its local neighbor node density. The experimental hop size and the calculating results used in Eq. 4 are presented in Fig. 3.

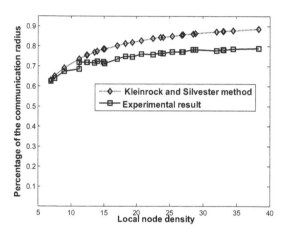

Fig. 3. Hop size related to local node density in homogeneous network

It is clearly shown that the average hop size is between 0.6r–0.9r when local neighbor nodes are more than 5, where r is sensor node communication radius. According to this conclusion, the threshold of q_s can be set to remove beacon that are unsuitable for estimating average hop size.

3.2.2 The Initial Hop-Size

Before beacon i calculates its' Hop-size$_i$ used in Eq. 1, beacon j calculates its' q_s from beacon i according to Eq. 3, and it does not participate in Hop-size calculation if the q_s is out of the threshold. By this way, all beacons have the initial Hop-size form itself respective.

3.2.3 Initial Hop-Size Optimization

After that all beacons get the average hop size, the hop distance form beacon j to beacon i can be calculated by Eq. 5.

$$d_{ej \to i} = Hop\text{-}size_i \times hops_{j \to i} \tag{5}$$

The estimated difference of Hop-size$_i$ is calculated by Eq. 6.

$$e_{hop\text{-}diffi} = \frac{\sum\limits_{j} \left(d_{ej \to i} - d_{true\, j \to i} \right)}{\sum\limits_{j} hops_{j \to i}} \tag{6}$$

The average hop size from the beacon i is optimized as following Eq. 7.

$$Hop\text{-}size_i = Hop\text{-}size_i + e_{hop\text{-}diffi} \tag{7}$$

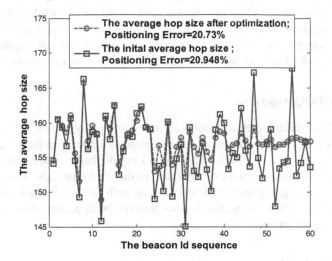

(a)Hop size in the isotropy network

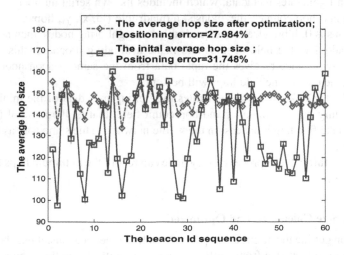

(b) Hop size in the anisotropy network

Fig. 4. The average hop size and the corresponding positioning error

In 1000 m*1000 m area, experiment is conducted to verify the effectiveness of this hop-size optimization. We deployed 300 nodes with isotropy network topology and anisotropy C-shape network topology. In those deployed nodes, 60 nodes are beacons, node communication radius r is set to 200 m. The average hop size after optimization and the corresponding positioning error using DV-Hop are shown in Fig. 4.

As shown in Fig. 4a, in the isotropy network, there is so little difference between initial Hop- size and the optimized Hop-size, their size locates between 145 m and 165 m, but the optimized Hop-size is smoother than the initial Hop-size. Those two Hop-sizes are used in the positioning of the other nodes, and there is almost the same positioning accuracy. However, from the Fig. 4b, in the C-shape anisotropy networks, all beacons initial hop-sizes are so big difference between 90 m and 165 m, while optimized Hop-sizes of all beacons are located in scope of the 140 m to 160 m, which are almost equals the Hop-size in isotropy networks. The positioning error of corresponding the optimized Hop-size is 27.984% that obvious lower than 31.748% corresponding the inital Hop-size.

3.3 The LA-PGQ Algorithm

The LA-PGQ Algorithm comprises of four non-overlapping stages. First, it employs quadruple of (id, x, y, hops) exchange so that all nodes get minimum hops to all beacons. In the second stage, hop-size from each beacon i is calculated, which is then deployed as a correction hop distance to the other nodes. In the third phase, the sensor u chooses the best positioning group and calculates the hop path distance to beacon in the best positioning group. When getting the distance to three beacons, the trilateration is used to determine sensor location.

3.3.1 The Quadruple Exchange

Each beacon i generates metadata, which includes its own serial number, the fields of localization and hop counter, they consist a quadruple (id, x_i, y_i, hops), among them, the initial hops is 0. It broadcasts the quadruple to neighboring nodes, then neighboring nodes will add 1 to the field of hops and make a flooding broadcast this quadruple. If every receiving node receives the tuple data from the same beacon node, then the minimum tuple data of the field hops will be kept.

Similarly, the unknown node generates the polling metadata of the minimum hop count, including its serial number and hop count field, that is, the tuple of (id; hops), and it will make flooding broadcast in the whole network in the same way as that of the beacon node.

Therefore, through this process, every node can record the minimum hops from itself to other nodes.

3.3.2 Hop-Size Calculated and Optimized

When beacon getting the minimum hops from itself to other beacons, it may be calculate and optimize the Hop-size from itself in accordance with the method shown in above Sect. 3.2.

3.3.3 Positioning Group Selection

1. *The Topological Distribution of Positioning Group*

   In positioning of the unknown nodes, the classical DV-Hop algorithm chooses all the beacon nodes as the localization referenced nodes include following two types of beacons, the first have large hops to the unknown node so that it introduces big errors in estimating hop path distance; the second are from collinear beacon nodes that causes low positioning inaccuracy. In practice, the unknown u can determine own location that only needs 3 beacons.

   Therefore, the unknown node u and the other three beacons are together are defined as positioning group, which is presented as $lg(u) = tri(u, \{s_i, i = 1, 2, 3\})$ and s_i beacon. The topology of lg(u) is shown as Fig. 5.

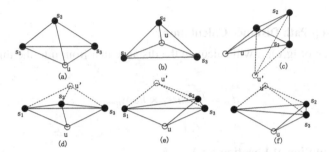

Fig. 5. The topology of positioning group lg(u)

   In Fig. 5a and b, the geometric distribution of three beacons for the lg(u) almost approach to regular triangle and the hops does not so big, which causes lower hop distance error. These geometric distribution of lg(u) is preferable to be selected to u. In Fig. 5c, although three beacon nodes close to triangular distribution, lg(u) is not suitable for positioning because that the u is far from three beacons, which will cause big hop distance error. In Fig. 5d and e, three beacons are collinear that does not meet the redundant rigid graph [27] so that it introduces poor positioning accuracy or even non-localization. One beacon is far from the others, which can be seen special case of colinearity in Fig. 5f. On the basis of the influence that topology of lg(u) to the positioning accuracy, we build beacon credibility model to remove remote hop path node and localization group quality evaluation function to choose three beacons to constitute a best lg(u).

2. *Beacon Credibility Model*

   For the unknown sensor u, beacon i credibility is calculated as Eq. 8.

$$w_u(i) = \sum_{1}^{k} \beta_i \times \frac{hop(i)}{hop_{\max}} + \left(1 - \sum_{1}^{k} \beta_i\right) \times \frac{d_e}{l} \tag{8}$$

Where $\beta_i = n_i/N$, n_i is the i-hop number of neighbors, de is estimated distance from unknown u to beacon i. It sets the threshold of $w_u(i)$ and chooses the beacons which $w_u(i)$ are smaller than the threshold to form the candidate beacon collection Ω.

3. *Localization Group Quality Metric Function*
   According to the impact that the topology of lg(u) puts on undecided sensor u in
   localization accuracy, the metric function of localization group quality could be built
   in Eq. 9

$$c_\Delta = \frac{2h_{min}}{\sqrt{3}l_{max}} \tag{9}$$

Where, l_{max} is the max edge length in triangle, h_{min} is corresponding edge high for
the triangle. If the numbers of the candidate beacon Ω is m, there will be C_m^3 positioning
group. Every c_Δ of lg(u) is calculated and the largest c_Δ corresponding positioning group
is selected to form the referenced beacons for the unknown sensor u.

3.3.4 The Hop Path Distance Calculation

After formation of lg(u), it will calculate the distance that from u to the three beacons
by Eq. 10.

$$d_{ui} = hopsize_i \times hops_{u \rightarrow i} \tag{10}$$

3.3.5 Computation of Location for U

For the selected lg(u), let the position u be (x_u; y_u), and the position of the beacon i is (x_i;
y_i), and dui denotes the hop distance from node u to beacon i, the position of unknown
node u is determined as Eq. 11.

$$\begin{cases} -2x_1x_u - 2y_1y_u + x_u^2 + y_u^2 = d_{u1}^2 - x_1^2 - y_1^2 \\ -2x_2x_u - 2y_2y_u + x_u^2 + y_u^2 = d_{u2}^2 - x_2^2 - y_2^2 \\ -2x_3x_u - 2y_3y_u + x_u^2 + y_u^2 = d_{u3}^2 - x_3^2 - y_3^2 \end{cases} \tag{11}$$

Let $A_i = x_i^2 + y_i^2$, $B_u = x_u^2 + y_u^2$

$$Z = [x_u, y_u, B]^T$$

$$G = \begin{pmatrix} -2x_1 & -2y_1 & 1 \\ -2x_2 & -2y_2 & 1 \\ -2x_3 & -2y_3 & 1 \end{pmatrix}$$

$$H = \begin{pmatrix} d_{u1}^2 - A_1 \\ d_{u2}^2 - A_2 \\ d_{u3}^2 - A_3 \end{pmatrix}$$

Then the matrix expression of Eq. 11 is determined as Eq. 12.

$$GZ = H \tag{12}$$

Then

$$Z = (G^T G)^{-1} GH \tag{13}$$

Therefore, we get the coordinate of the unknown node j as follows:

$$(x_u, y_u) = (Z(1), Z(2)) \tag{14}$$

4 Simulation Results and Analysis

In this section, based on the implementation of LA-PGQ algorithm, Simulations are conducted to evaluate the performance of the LA-PGQ algorithm and to compare with the DV-Hop and its' improved one in [23] and Amorphous algorithm [26]. In the 1000 m *1000 m network area which shown isotropic network in Fig. 6(a) and the anisotropic network (C-shape) shown in Fig. 6(b), we deploy 300 sensors that have an identical communication radius.

We use the average localization error (ALE) to evaluate the performance of the algorithm. It is calculated as follows:

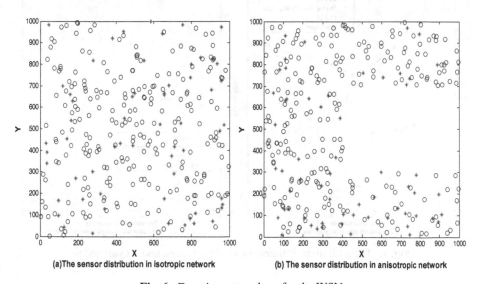

(a)The sensor distribution in isotropic network   (b) The sensor distribution in anisotropic network

Fig. 6. Experiment topology for the WSN

$$ALE = \frac{\sum\limits_{i} \frac{\sqrt{(\widehat{x_i}-x_i)^2+(\widehat{y_i}-y_i)^2}}{r_{comm}}}{n} \tag{15}$$

where $(\widehat{x_i}, \widehat{y_i})$ is the estimated position of sensor i by the respective localization algorithm, (x_i, y_i) is the actual position of sensor i, r is the communication radius,n is the number of sensor nodes.

4.1 Simulation Results with Varying the Ratio of the Beacon Nodes

When setting of communication radius 200 m, we have varied the values of the ratio of the beacon nodes from 10% to 60% of total nodes, and measured the average localization error at each 5% increment, which respective simulation results are shown in Fig. 7.

The figure shows that, as we increase the ratio of the beacon nodes, the ALE in LA-PGQ is lower than their counterparts in DV-Hop and Amorphous on average 6% and lower than the improve DV-Hop algorithm on average 4% in isotropic network. While in the anisotropic network (C-shape), the ALE in Amorphous, DV-Hop and its' improved one is more than 80, with improved greatly, the ALE in LA-PGQ is lower than 30%.This means that LA-PGQ has better localization accuracy than DV-Hop and its improvements under both the isotropic network and anisotropic network. In the meantime, we observe that the ALE in LA-PGQ has little change varied the values of the ratio of the beacon nodes, which illustrate that the number of beacons is put little impact on the positioning accuracy of LA-PGQ algorithm. So, we may deploy fewer beacons to reduce the costs network on premise of the positioning accuracy.

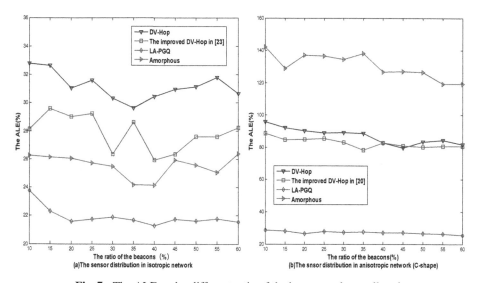

Fig. 7. The ALE under different ratio of the beacon nodes to all node

4.2 Simulation Results with Varying Communication Radius

We have fixed the ratio of the beacon nodes 20%, respectively, incremented the sensor communication radius from 150 m to 290 m, and measured the average localization error which respective simulation results are shown in Fig. 8 at every 20 m increment.

It can be shown from Fig. 8 that the LA-PGQ outperforms compared DV-Hop, Amorphous and its' improved algorithm because of the ALE in LA-PGQ is far lower than compared algorithm both in isotropic network and anisotropic network. Especially, in the heterogeneous network, the ALE in LA-PGQ is declined more than 30% compared with DV-Hop, Amorphous and its' improved one. Meanwhile, it's worth noting that the ALE in LA-PGQ is not significantly affected by the increase the sensor communication radius both in isotropic network and anisotropic network, therefore, we can dynamically adjust the sensor node transmission power so as to reduce energy consumption of sensor networks in the process of locating.

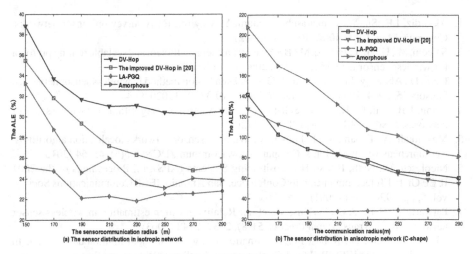

Fig. 8. The ALE under different communication radius

5 Conclusions

In this paper, we have surveyed the classical DV-Hop, Amorphous and its improved algorithms, which show their particular advantage for localizing the normal nodes that are not cost-effective and robust. However, its' positioning accuracy relies on the sensor distribution in area. On the one hand, it has high accuracy in isotropy wireless sensor network, but it can be further improved. On the other hand, it cannot locate normal nodes in anisotropy wireless sensor network with low accuracy. Focusing on DV-hop and improved one, we have proposed the LA-PGQ algorithm, which firstly corrects the Hop-size by removing the beacon node that does not locate on the straight path in the process of the estimating hop distance, than optimizes the Hop-size to make it in anisotropic network to approach the hop distance in homogeneous network. Then the beast positioning group is selected and the two-dimensional hyperbolic is used to locate normal node. So, in almost isotropy wireless sensor network and anisotropy wireless sensor network, the LA-PGQ algorithm performs better than the DV-Hop,Amorphous and improved algorithm in terms of lower localization error. Moreover, the deployed beacons and the communication radius of sensor put little impact on the localization errors. Therefore, in practical application, we can deploy fewer beacons to reduce the costs and dynamically adjust the sensor node transmission power so as to reduce energy consumption of sensor networks in the process of locating.

Acknowledgments. The authors are grateful to the anonymous reviewers for their comments. This work was supported by the Yunnan Local Colleges Applied BasicResearch Projects (2017FH001-059, 2018FH001-010, 2018FH001-061), National Natural Science Foundation of China (61962033).

References

1. Akyildiz, I.F., Su, W., Sankarasubramaniam, Y., Cayirci, E.: A survey on sensor networks. Commun. Mag. IEEE **40**(8), 102–114 (2002)
2. Saleem, M., Ullah, I., Farooq, M.: BeeSensor: an energy-efficient and scalable routing protocol for wireless sensor networks. Inf. Sci. **200**, 38–56 (2012)
3. Izadi, D., Abawajy, J.H., Herawan, S.G.T.: A Data fusion method in wireless sensor networks. Sensors **15**, 2964–2979 (2015). https://doi.org/10.3390/s150202964
4. Wang, F.B., Shi, L., Ren, F.Y.: Self-localization systems and algorithms for wireless sensor networks. J. Softw. Chin. **16**(5), 857–867 (2005)
5. Mesmoudi, A., Feham, M., Labraoui, N.: Wireless sensor networks localization algorithms: a comprehensive survey. Int. J. Comput. Netw. Commun. (IJCNC) **5**(6), 45–64 (2013)
6. Nieoleseu, D., Nath, B.: Ad-Hoc Positioning Systems (APS), In: Proceedings of the 2001 IEEE Global Telecommunications Conference. San Antonio: IEEE Communications Society, vol. 5, pp. 2926–2931 (2001)
7. Patwari, N., Hero, A.O., Perkins, M., et al.: Relative location estimation in wireless sensor networks. IEEE Trans. Signal Process. **51**(8), 2137–2148 (2003)
8. Girod, L., Estrin, D.: Robust range estimation using acoustic and multimodal sensing, In Proceedings of the IEEE/RSJ Int'l Conference on Intelligent Robots and Systems(IROS01), Maui:IEEE Robotics and Automation society, vol. 3, pp. 1312–1320, 2001.
9. Lazos, L., Poovendran, R.: POPE: robust position estimation in wireless sensor networks. In: Proceedings of the 4th IEEE International Conference on Information Processing in Sensor Networks, pp. 324–331 (2005)
10. Lee, J., Chung, W., Kim, E.: A new kernelized approach to wireless sensor network Localization. Inf. Sci. **243**, 20–38 (2013)
11. Niculescu, D., Nath, B.: DV-based positioning in AD Hoc networks. J. Telecommun. Syst. **22**(1), 267–280 (2003)
12. Bulusu, N., Heidemann, J., Estrin, D.: GPS-less low-cost outdoor localization for very small devices. IEEE Pers. Commun. **7**(5), 28–34 (2000)
13. Doherty, L., Pister, K.S., El Ghaoui, L.: Convex position estimation in wireless sensor networks. In: Proceeding of Joint Conference of the IEEE Computer and Communications Societies (INFOCOM 2001). Anchorage, pp. 1655–1663 (2001)
14. Shang, Y., Ruml, W., Zhang, Y.: Localization from mere connectivity. In: Proceeding of the 4th ACM International Symposium on Mobile Ad Hoc Networking and Computing (MobiHoc 2003). Annapolis, pp. 201–212 (2003)
15. Yi, X., Liu, Y., Deng, L., He, Y.: An improved DV-Hop positioning algorithm with modified distance error for wireless sensor network. In: Proceedings of the 2nd International Symposium on Knowledge Acquisition and Modeling (KAM 2009), pp. 216–218, December 2009
16. Poggi, C., Mazzini, G.: Collinearity for sensor network localization. In: Proceedings of the 58th IEEE Vehicular Technology Conference (VTC 2003), pp. 3040–3044, October 2003
17. Zhang, Y., Xiang, S., Fu, W., Wei, D.: Improved normalized collinearity DV-Hop algorithm for node localization in wireless sensor network. Int. J. Distribut. Sensor Netw. **10**, 14 (2014). Article ID 436891

18. Guang, W., Wang, S., Wang, B., Dong, Y., Yan, S.: A novel range-free localization based on regulated neighborhood distance for wireless ad hoc and sensor networks. Comput. Netw. **56**, 3581–3593 (2012)
19. Tang, Z., Zhang, J., Wang, L., Han, J., Fang, D., Wang, A.: NDSL: node density-based subregional localization in large scale anisotropy wireless sensor networks. Int. J. Distribut. Sensor Netw. **11**, 16 (2015). Article ID 821352
20. Kumar, S., Lobiyal, D.K.: Improvement over DV-Hop localization algorithm for wireless sensor networks. World Acad. Sci. Eng. Technol. **7**(4), 235–245 (2013)
21. Liu, Y., Luo, X.Y., Long, C.: Improved DV-hop localization algorithm based on the ratio of distance and path length. J. Inf. Comput. Sci. **9**(7), 1875–1882 (2012)
22. Safa, H.: A novel localization algorithm for large scale wireless sensor networks. Comput. Commun. **45**, 32–46 (2014)
23. Shi, W.R., Jia, C.J., Liang, H.H.: An improved DV-hop localization for wireless sensor networks. Chin. J. Sens. Actuators **24**(1), 83–87 (2011)
24. Manisekaran, S.V., Venkatesan, R.: Cluster-based architecture for range-free localization in wireless sensor networks. Int. J. Distribut. Sensor Netw. **2014**, 9 (2014). https://doi.org/10.1155/2014/963473.Article ID 963473
25. Woo, H., Lee, C., Oh, S.: Reliable anchor node based range-free localization algorithm in anisotropic wireless sensor networks. In: Proceedings of the 27th International Conference on Information Networking (ICOIN 2013), pp. 618–622, January 2013
26. Kleinrock, L., Silvester, J.: Optimum tranmission radii for packet radio networks or why six is a magic number. In: Proceedings of the IEEE National Telecommunications Conference, pp. 4.3.1–4.3.5 (1978)
27. Eren, T., Goldenberg, O.K., Whiteley, W.: Rigidity, computation and randomization in network localization. In: Proceedings of IEEE, vol. 4, pp. 2673–2684 (2004)

A Collaborative Cache Strategy Based on Utility Optimization

Pingshan Liu[1,2(✉)], Shaoxing Liu[2], and Guimin Huang[2]

[1] Business School, Guilin University of Electronic Technology, Guilin, China
[2] Guangxi Key Laboratory of Trusted Software, Guilin University of Electronic Technology, Guilin, China

Abstract. With the continuous development of network technology, the number of streaming media videos is growing rapidly. More and more users are watching videos through the Internet, which leads to the increasing huge server load and the increasing transmission cost across ISP domains. A feasible scheme to reduce transmission cost across ISP domains and alleviate the server load is to cache some popular videos in a large number of terminal users. Therefore, in this paper, in order to utilize the idle resources of the terminal peers, some peers with good performance were selected from the fixed peers as the super peers, which were aggregated into a super peer set (SPS). In addition, with the supply and demand relation of streaming videos among ISP domains, a mathematical model was formulated to optimize the service utility of ISP. Then, a collaborative cache strategy was proposed based on the utility optimization. The simulation results show that the strategy proposed can effectively improve the user playback fluency and hit rate while ensuring the optimal service utility.

Keywords: Peer-to-Peer · Super Peer Set · Collaborative cache management · ISP · Utility optimization

1 Instruction

With the increase of intelligent terminals and cloud services, a large amount of data will be generated. According to the "Global Mobile Data Traffic Forecast: 2016–2021" [1] released by Cisco on March 28, 2017 predicts that by 2021, the number of wireless devices connected to the network will reach 110 billion, and mobile data traffic will also rise from 440 million bytes per month in 2015 to 4.9 billion bytes per month in 2021.

In recent years, the issue of P2P caching has attracted many researchers [2–4]. The authors of [2] considered the caching capability of the user equipment. Users can obtain the requested streaming media content from BSs or themselves. And a low complexity sub-optimal method is proposed to reasonably place the content on the relay station and the user's equipment. The author of [3] proposed a P2P caching mechanism that supports broadcasting based on cloud architecture and the software defined network (SDN) technology. The authors of [4] pointed out that it is a good strategy to store content on storage devices (BS, user equipment) near the edge of the wireless network.

© Springer Nature Singapore Pte Ltd. 2021
J. Zeng et al. (Eds.): ICPCSEE 2021, CCIS 1451, pp. 140–153, 2021.
https://doi.org/10.1007/978-981-16-5940-9_11

It can effectively improve the QoE of the user to cache videos on the user side. However, the bandwidth and capacity of different devices are different. Watching devices can be divided into mobile devices (called mobile peer) and fixed devices (called fixed peer). Two types of buffer structures are shown in Fig. 1. Figure 1(a) is a buffer structure of the mobile peer, it includes only a playback buffer. Figure 1(b) is a buffer structure of the fixed peer, the structure includes a playback buffer, a supply buffer and a hard disk buffer. The supply buffer can store shared resources. The hard disk is a storage device that can store multiple complete streaming videos for a long time.

The authors of [5] proposed a new LFU-SIZE cache replacement algorithm. The algorithm preserved the most popular files, discarded rarely used files. The authors of [6] proposed cooperative willingness aware collaborative cache mechanism, the cache mechanism includes a D2D UE cache capability estimation method and a D2D UE service relationship perception method. The authors of [7] studied that the current video-on-demand (VoD) system distributes content through the content distribution network (CDN). And based on the game theory model, they found the best profit function by analyzing the cache investment under competition among multiple interconnected ISPs. By virtue of the high scalability of P2P overlay net, low deployment cost and the characteristics of CDN content storage and distribution, the network congestion problem has been effectively alleviated. However, with the number of user requests increases, the traffic of crossing ISP domains is also rising.

In this article, our main contributions are as follows:

1) We build a hybrid CDN-P2P architecture. In this architecture, we select high-performance fixed peers as super peers. Considering the weakness of a single super peer, the super peers are aggregated into a Super Peer Set (SPS).
2) According to the supply and demand relation among ISP domains, we design a mathematical model to achieve the optimal service utility. Furthermore, we propose a collaborative cache strategy based on utility optimization using the cache capabilities of CDN and SPS.

The remainder of the paper is arranged as follows: we summarize the related researches in literature in Sect. 2. Then in Sect. 3, a hybrid CDN-P2P network architecture is described, in addition, we introduce the selection of the super peer and the aggregation of the super peer set. In Sect. 4, a service utility optimization model is established, based on utility optimization, we propose a collaborative cache strategy. In Sect. 5, we conduct a lot of simulations to evaluate the relevant performance. The relevant conclusions and the future work are given in Sect. 6.

2   Related Work

Cache is a method of sacrificing storage resources in exchange for bandwidth. It can improve network performance to cache streaming videos in a location close to the user. Caching streaming videos at the edge can reduce bandwidth costs [8] and content transmission delays [9]. Therefore, under the constraints of architecture and system capacity, it is worth studying to find out the best routing and cache strategy. Considering the

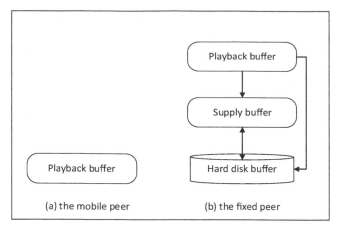

Fig. 1. (a) Cache structure of the mobile peer (b) Cache structure of the fixed peer

limited capabilities of a single client peer, the authors of [10] used long-term connected peer clusters to share cache. They treat the capacity of all peers in the cluster as a large capacity, which reduces duplication and improves cache space utilization. The authors of [11] proposed a fully cooperative local cache, categorize the MTs into different interest groups according to the preferences of MTs. In addition, they formulate cache decision problem as a weighted-sum utility maximization problem. In the case of partial cooperation and non-cooperation, with and without inter-group file sharing, the optimal cache distributions for two cases are derived. The authors of [12] proposed a distributed collaborative cache management scheme. Moreover, they extend the sharing mode category and proposed the sharing mode selection scheme by constructing a general social welfare maximization with efficient incentive mechanisms. The author of [13] studied how to replace the video copies based on available bandwidth limitations when the disk storage is full. And they proposed a BAB algorithm to achieve the best distribution of streaming video copies by a distributed method.

The authors of [14] studied the relationship between cloud CDN bandwidth costs and crossing ISP domains traffic in a hybrid environment of cloud CDN and P2P. Different from the game model among ISP domains, the authors of [15] proposed a peer-to-peer sharing model among ISPs. This model can reduce the service cost by coordinating the reciprocity and fairness of content among ISPs. The authors of [16] analyzed the basic issues of the interaction among ISP domains, and designed a novel inter-ISP domain traffic control scheme to make full use of network resource sharing. This approach explores the impact of ISP peering relationship and bandwidth allocation to maximize the total system revenue. Most of the existing researches only consider the issue of collaborative cache or crossing ISP domains, it does not comprehensively consider the issues of collaborative cache and crossing ISP domains. In our paper, considering that the fixed peers have a larger bandwidth and disk capacity, we choose the fixed peers with good performance as super peers. Then, the super peers are aggregated into Super Peer Set (SPS). According to the supply and demand relation among ISP domains, we design a mathematical model to achieve optimal utility. Base on the utility optimization, we propose a collaborative cache strategy based on the utility optimization.

3   System Architecture Construction

In this part, we introduce the hybrid CDN-P2P network architecture and the aggregation of the super peer set.

3.1   Hybrid CDN-P2P Architecture Model

The hybrid CDN-P2P network architecture is shown in Fig. 2. which is composed of a video source server, the multiple Internet Service Provider (ISP) domains, the tracker server peer, the cloud CDN, mobile peers, and super peer sets. The architecture has two layers of overlay networks where the first layer consists of cloud CDNs and the second layer includes super peer sets and normal peers. We consider a scenario of multiple ISPs domains. Cloud CDNs are distributed in different geographical locations. Cloud CDNs in different ISPs domains transmit streaming videos through backhaul links. The video source server stores all streaming video and distributes streaming videos to different Internet Service Provider (ISP) domains. The tracker peer stores information about cloud CDNs and super peer sets. The mobile peers can only request streaming videos from cloud CDN and SPS. When the supply of ISP domain is insufficient for the demand, the cloud CDN and SPS will collaboratively cache the idle resources from neighbor ISP domains.

3.2   The Aggregation of SPS

We know that the storage structure and bandwidth are different between the fixed peers and the mobile peers. In order to ensure the stable service status of the entire network, we select the fixed peers with good performance as super peers, which are aggregated into a Super Peer Sets (SPS). Each super peer in the SPS stores the information of neighbor super peers.

For evaluating the performance of the peers, we let $Ability_i$ to represent the strength of peer i. In general, we select super peers according to the online time, the bandwidth and capacity. But a fact is what there are some peers that will not contribute their remaining resources. If a peer does not intend to contribute their own bandwidth and capacity, even though the peer has a lot of available bandwidth and available capacity, it cannot contribute to the network. So, we add $Sharing$ attribute to the peer, which represents the sharing willingness of peers. The greater the $Sharing$, the peer is more willing to provide services to other peers. The calculation formula for $Ability_i$ of peer i is as follows:

$$Ability_i = \mu_1 OTime + \mu_2 Bw + \mu_3 Capacity + \mu_4 Sharing \tag{1}$$

In formula (1), we let $OTime$ represent the historical online time of the peer, Bw represent the bandwidth of the peer, $Capacity$ represent the disk storage capacity of the peer, and $Sharing$ represent the service willingness of the peer. $\mu_1, \mu_2, \mu_3, \mu_4$ represent weighting factor.

Algorithm 1
1   **For** each peer i **do**
2   Obtain the information $OTime, Bw, Capacity$ and $Share$ of peer i
3   Calculate $Ability_i$ of peer i according to （1）
4   **End for**
5   Sort $Ability_i$ in a non-increasing order
6   Select the first $\beta \cdot N$ peers as super peers
7   **For** each super peer sp **do**
8   Calculate $Dist$ between the super peer sp and other super peers k_n
9   Obtain the minDist$_{sp, k1}$ and the minDist$_{sp, k2}$
10   **If** $(N_{k_1} < \tau)$
11   Super peer sp joins SPS_{k_1}
12   **Else If** $(N_{k_2} < \tau)$
13   Super peer sp joins SPS_{k_2}
14   **Else** peer sp self-organize into a set
15   **End If**
16   **End For**

We let N_k donate the number of super peers in SPS_k and τ donate the maximum number of super peers in SPS. We use $\Phi = \{SPS_1, \ldots, SPS_L\}$ to denote the set of the SPSs. In Algorithm 1, the pseudo code for the aggregation of the Super Peer Set is given. From the line 1 to 4, the procedure first loops through each peer, obtains the peer's information of $OTime$, Bw, $Capacity$ and $Sharing$, then, calculates $Ability_i$ of peer i. In line 5, sorting the peer in descending order according to the value of $Ability_i$. The existing method for measuring the distance between peers is mainly based on the network ranging method of detection messages. From the line 7 to 15, super peer sp will sends detection packets to other super peers and gets the distance $Dist$ according to the RRT (Round Trip Time) of the detection packets. sp obtains peer k_1 with the smallest $Dist$, if N_{k_1} of SPS_{k_1} is not greater than τ, sp is added to SPS_{k_1}, otherwise, sp obtains peer k_2 with the $Dist$ next to last. If N_{k_2} of SPS_{k_2} is not greater than τ, sp is added to SPS_{k_2}. If the above two conditions are not met, sp will self-organize into a set.

4   Problem Formulation and Algorithm

In this part, we introduce a mathematical model for optimizing ISP services and a collaborative caching strategy based on utility optimization. The video source server has a batch of streaming videos. We use $F = \{f_1, f_2, \ldots, f_V\}$ to denote the set of the videos, and assume that the popularity of streaming videos follows a Zipf's distribution [18]. We use $\Psi = \{ISP_1, ISP_2, \ldots, ISP_G\}$ to represent a set of ISP domains. The symbols used in our article are listed in Table 1.

Fig. 2. The hybrid CDN-P2P network architecture

4.1 The Utility Optimization of ISP

In this part, we mainly introduce a mathematical model of utility optimization based on greedy algorithm. The model can effectively solve the issue of resources imbalance in ISP domains. According to the supply and demand relation of the resources among ISP domains, the service utility optimization problem is divided into several sub-optimization issues by the greedy algorithm. The global optimal solution is divided into optimal solutions of multiple sub-problems. We assume the resource required by an ISP domain as Q_v, greedy algorithm is used to optimize the allocation of resource Q_v. the formula for Q_v is as follows.

$$Q_v = \sum_{g=1}^{G} P_{ISPg} + P_{Source} \tag{2}$$

Table. 1. List of the symbol used in the paper

Symbol	Description
V	The number of streaming videos
M	The number of the segment of each video
N	The number of users
C	The disk capacity of the peer
Bw	The bandwidth of the peer
$utility$	The utility of the server
s_j	The size of segment j
$Replace_v$	The replacement index of video v
P_v	The popularity of video v
$freq_v$	The frequency of video v
$Secqec_{v,T}$	The number of valid requests for video v in a period T
$Scarcity_v$	The scarcity of video v
$Size_v$	The size of video v

In formula (2), P_{ISPg} represents the resources provided by neighbor *ISPg* domain, P_{Source} represents the resources supplied by the video source server. In each neighbor *ISPg* domain, there are the CDN storage and the hard disk in the SPS. Therefore, the P_{ISPg} subdivision formula is as follows:

$$P_{ISPg} = P_{ISPg}^{CDN} + P_{ISPg}^{SPS} \qquad (3)$$

In formula (3), P_{ISPg}^{CDN} represents that the requested resources are provided through cloud CDN of neighbor *ISPg* domain, P_{ISPg}^{SPS} represents that the requested resources are provided through SPSs of neighbor *ISPg* domain. The expression for Q_v is as follows.

$$Q_v = \sum\nolimits_{g=1}^{G} (P_{ISPg}^{CDN} + P_{ISPg}^{SPS}) + P_{Source} \qquad (4)$$

In formula (4), we make a detailed allocation of Q_v. Further, we formulate the service utility $Utility_{Q_v}$ of Q_v, the expression of $Utility_{Q_v}$ is shown as follow.

$$Utility_{Q_v} = \sum\nolimits_{g=1}^{G} (tSPSP_{ISPg}^{CDN} + tCDNP_{ISPg}^{SPS}) + tSourceP_{Source} \qquad (5)$$

In formula (5), *tSPS* indicate the cost of transmitting one unit of video over super peer set, *tCDN* and *tSource* indicate respectively the cost of transmitting one unit of video over CDN and the video source server.

By minimize $Utility_{Q_v}$, we finally get the utility optimization of ISP as $Utility$.

$$Utility = MinimizUtility_{Q_v} \tag{6}$$

Algorithm 2	
1	Obtain information of resources requested in ISP domain, called Q_v
2	Count neighbor ISP domains with idle resources, called G
3	**For (G)**
4	Get the resources of SPSs in P_{ISPg} domain, update $Q_v = Q_v - P_{ISPg}^{SPS}$
5	**If ($Q_v == null$)** skip to step 10
6	Get the resources of CDN in P_{ISPg} domain, update $Q_v = Q_v - P_{ISPg}^{CDN}$
7	**If ($Q_v == null$)** skip to step 10
8	**End For**
9	**If ($Q_v! = null$)** request the resources from the video source server
10	Compute $Utility$
11	**End Procedure**

The pseudo code of the utility optimal model is given in Algorithm 2. The tracker server gets supply and demand of ISP domain. Through the greedy algorithm, ISP first obtains the resources from SPSs in neighbor ISP domains, if the resources of SPSs is not enough, ISP obtains the satisfied resources from CDNs in neighbor ISP domains. If the resources of all neighbor ISP domains are insufficient to meet the demand, it is necessary to send a request to the video source server provider for the relevant resources. It needs less cost to get streaming videos from SPS than from CDN or source video provider. The greedy algorithm can request resources from SPS preferentially.

4.2 The Collaborative Cache Strategy Based on Utility Optimization

Through utility optimization model in Sect. 4.1, we obtain the optimal allocation of resources among ISP domains and design a collaborative cache strategy based on utility optimization. Under the limitations of available bandwidth and capacity, SPS and CDN cooperatively cache optimal streaming videos. A replacement index $Replace_v$ is defined to evaluate the importance of streaming videos in the cache. The evaluation parameters of $Replace_v$ are mainly the popularity, scarcity, number of valid requests and size of streaming video. The definition of $Replace_v$ is shown in the following formula.

$$Replace_v = \sigma \cdot P_v \cdot 1 - Scarcity_v + \beta \cdot Sucqec_{v,T} + \gamma \cdot Size_v \tag{7}$$

$$Constraints : \sum_{v=1}^{V} Size_v \leq C \tag{7a}$$

$$\sum_{qv=1}^{Q_v} \sum_{j=1}^{M} 8\frac{S_j}{t} \leq Bw \tag{7b}$$

In formula (7), we let P_v represent the popularity of the video v and $Scarcity_v$ represent the proportion of the streaming videos. $Sucqec_{v,T}$ represent the number of valid requests for video v. in a period T. $Size_v$ represent the size of video v. σ, β and γ represent normalization factors. In (7a), the size of the streaming videos in the peer cache is not greater than the total storage capacity C. And in (7b), the video transmission bandwidth of each peer is not greater than the initial bandwidth Bw.

$$Scarcity_v = \frac{Num_v}{\sum_{v=1}^{V} Num_v} \tag{8}$$

In formula (5), we let Num_v denote the number of copies of video v, and $\sum_{v=1}^{V} Num_v$ denote the number of copies of total streaming videos.

Algorithm 3	
1	**Initialize** pre-cached resources C_v
2	**For** each SPS, each CDN **do**
3	**If** $(C_v == null)$ skip to step 13
4	**If** $(Bw$ is enough) **then**
5	**If** $(Capacity$ is enough)
6	Cache video directly to disk
7	**Else**
8	Calculate $Replace_v$ according to Eq (7), replace the video with a minimum $Replace_v$
9	**End If**
10	**End If**
11	**End If**
12	**End for**
13	**End Procedure**

The pseudo code of collaborative cache strategy based on the utility optimization is given in **Algorithm 3**. The cache strategy is mainly divided into two phases. The first phase is that streaming videos are cached by using the disk capacity of SPS, the second phase is that cloud CDN store streaming videos. The cache method of two phases is similar, so we analyze only the first phase in detail. In line 3, the procedure first judges whether pre-cached resources is empty, if it is, the procedure skips to line 13, end the procedure, otherwise, judges whether the peer bandwidth in the SPS is sufficient in line 4, if it is not, start the next round of inquiry, otherwise, determine whether the disk capacity is sufficient, if it's enough, the peer cache directly the video in line 6, otherwise, the procedure skips to line 8, peers calculate $Replace_v$ according to formula (7), and replace the video with a minimum $Replace_v$.

5   Experiment Evaluation

5.1   Simulation Setting

In our simulation, we deploy a video source server, three network service providers {ISP1, ISP2 and ISP3}, three tracking servers {Tracker1, Tracker2, Tracker3} and 10,000 peers. Each ISP domain has a trace server. We initialize the bandwidth, disk capacity, and online time for each peer. The mobile peer has no hard disk buffer, so in order to distinguish the mobile peers and fixed peers, we set the *Capacity* of some peers to 0 as mobile peers. 10% of total fixed peers are selected as super peers. We set bandwidth of CDN as 20Mb, and the bandwidth range of the super peers is set to $\theta = \{1\,\text{Mb}, 2\,\text{Mb}, 4\,\text{Mb}, 6\,\text{Mb}\}$. In order to simulate real network conditions, some peers leave the network during the experiment. For a convenience, we assume that the size of each video is the same and the popularity of the video follows a Zipf's distribution law. Reference [17] show that *OTime*, *BW* and *Capacity* are equally important. However, while a sharing willingness the peer is small, its contribution to the network is small even though it has remaining bandwidth and capacity we set the weight of *OTime*, *BW* and *Capacity* to 1/5, and the weight of *Sharing* to 2/5. The cost of different transmission paths is also different, we set *tSPS* to 1 unit, *tCDN* to 6 units and *tSource* to 12 units. We set the running time of the simulation experiment to 24 h. Compared with the classical LRU and LFU algorithms, the simulation results show that the proposed scheme is better than the other two algorithms.

5.2   Simulation Results

In order to evaluate the performance of our proposed scheme, we used PeerSim to conduct related simulation experiments.

1) ISP Pressure Index

The ISP pressure index indicates the load of the server. The higher the ISP pressure index, the server load is bigger. It can be seen from Fig. 3 the cache size is between 4 and 7, the ISP pressure index drops rapidly in our scheme, but the LRU and LFU schemes decrease slowly. After the cache size reaches 10, the ISP pressure index gradually becomes stable. Compared with the other two solutions, the scheme our proposed has a better effect for reducing ISP service index. This is because our scheme can make full use of the limited cache resources of super peer. It can store more popular videos efficiently, thereby increasing the overall cache hit rate and reducing server pressure.

2) **Hit-Rate**

The Hit-Rate is the ratio of the number of videos successfully received by peers. In our collaborative cache strategy, the streaming videos are cached cooperatively through super peers in the SPS. As shown in Fig. 4, we can see that as the number of peers in the SPS increases, the Hit-Rate gradually increases. When the number of peers reaches 20, the hit rate tends to stabilize. And our proposed algorithm has a

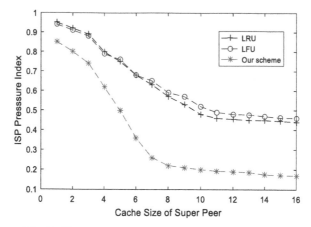

Fig. 3. The ISP pressure index with different cache size

higher hit rate at any point. Therefore, compared with the traditional LRU and LFU schemes, our proposed scheme can greatly improve Hit-Rate, thereby reducing the bandwidth costs of the backhaul link.

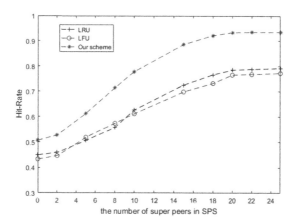

Fig. 4. The relationship between the number of super peers in SPS and the hit rate

3) **Continuity Index**

Continuity index indicates the smoothness of video playback. Continuity index is the proportion of the number of video segments received by the peer to the total number of video segment requested by the peer. If the continuity index of the peer is close to 1, it means that the peer viewing experience is excellent. From the CDF function graph in Fig. 5, we know that the traditional LRU and LFU schemes grow very fast when the continuity index is approximately equal to 0.8. However, our proposed scheme grows very quickly after the continuity index is equal to 0.9. Therefore, compared with the other two schemes, our proposed scheme can make more users have good QoE.

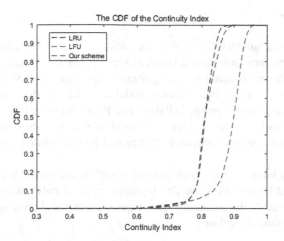

Fig. 5. The CDF of the Continuity Index

4) Proportion of Service

Proportion of service represents the proportion of super peers successfully served to the total super peers. The larger the service proportion of the super peers in the network is, the less costs the operators need to provide. We set different numbers of peers to detect the proportion of service. From Fig. 6, we can know that as the number of peers

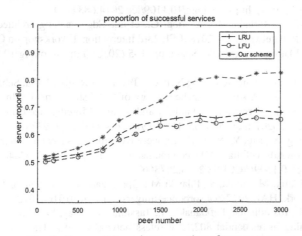

Fig. 6. Successful service proportion of super peers

increases, the proportions of the three schemes are also increasing. When the number of peers reaches 1500, the LRU and LFU algorithms tend to be stable. Our proposed scheme gradually stabilizes near 2000 peers and still has a small increase. Therefore, our proposed scheme is better than the traditional two algorithms.

6 Conclusion

In the paper, we first build a CDN-P2P hybrid architecture. In the architecture, the fixed peers with good performance were selected as super peers. Considering the insufficient service ability of the individual peer, we aggregate super peers into SPS. In addition, we design an ISP service utility optimization model according to the supply and demand relation of streaming videos among ISP domains. Based on the utility optimization, we propose a collaborative cache strategy. The simulation results show that our proposed strategy can improve the user playback fluency and hit rate while ensuring the optimal service utility.

In the current work, our research content is still based on the traditional network architecture. In order to follow the development trend of network technology in the future, we will research the streaming video transmission based on edge computing and construct the relevant algorithms.

Acknowledgement. This research was supported by the national key research and development program of China (No. 2020YFF0305301), the National Natural Science Foundation (61762029, U1811264).

References

1. Cisco: Cisco visual networking index: global mobile data traffic forecast update, 2016–2021 white paper [EB/OL]
2. Yang, C., Chen, Z., Yao, Y., Xia, B., Liu, H.: Energy efficiency in wireless cooperative caching networks. In: 2014 IEEE International Conference on Communications (ICC). Sydney, NSW, pp. 4975–4980 (2014). https://doi.org/10.1109/icc.2014.6884109
3. Nikoloudakis, Y., Markakis, E., Alexiou, G., et al.: Edge caching architecture for media delivery over P2P networks. In: 2018 IEEE 23rd International Workshop on Computer Aided Modeling and Design (CAMAD). Spain, pp. 1–5 (2018). https://doi.org/10.1109/CAMAD. 2018.8514935
4. Wessels, D.: Web Caching: Reducing Network Traffic. O'Reilly Media, Sebastopol (2001)
5. Zeng, Z., Zhang, H.: A study on cache strategy of CDN stream media. In: 2020 IEEE 9th Joint International Information Technology and Artificial Intelligence Conference (ITAIC), pp. 1424–1429 (2020). https://doi.org/10.1109/ITAIC49862.2020.9338805.
6. Zhang, P., Kang, X., Liu, Y., Yang, H.: Cooperative willingness aware collaborative caching mechanism towards cellular D2D communication. IEEE Access. 67046–67056 (2018). https://doi.org/10.1109/ACCESS.2018.2873662
7. Pham, T., Minoux, M., Fdida, S., Pilarski M.: Optimization of content caching in Content-Centric Network. HAL.2017. hal.sorbonne-universite.fr/hal-01016470v2
8. Poularakis, K., Iosifidis, G., Pefkianakis, I., Tassiulas, L., May, M.: Mobile data offloading through caching in residential 802.11 wireless networks. IEEE Trans. **13**, 71–84 (2016). https://doi.org/10.1109/TNSM.2016.2521352

9. Sengupta, A., Tandon, R., Simeone, O.: Fog-aided wireless networks for content delivery: fundamental latency tradeoffs. IEEE Trans. Inf. Theory **63**, 6650–6678 (2017). https://doi.org/10.1109/TIT.2017.2735962

10. Bok, K., Kim, J., Yoo, J.: Cooperative caching for efficient data search in mobile P2P networks. Wireless Pers. Commun. **97**(3), 4087–4109 (2017). https://doi.org/10.1007/s11277-017-4714-1

11. Guo, Y., Duan, L., Zhang, R.: Cooperative local caching under heterogeneous file preferences. IEEE Trans. Commun. **65** 444–457 (2017).https://doi.org/10.1109/TCOMM.2016.2620164

12. Wu, D., Zhou, L., Cai, Y., Qian, Y.: Collaborative caching and matching for D2D content sharing. IEEE Wirelss Commun. **25**, 43–49 (2018). https://doi.org/10.1109/MWC.2018.170 0325

13. Liu, P., Feng, S., Huang, G., Fan, J.: Bandwidth-availability-based replication strategy for P2P VoD systems. Comput. J. **57**, 1211–1229 (2014). https://doi.org/10.1093/comjnl/bxt071

14. Zhao, J., Wu, C., Lin X.: Locality-aware streaming in hybrid P2P-cloud CDN systems. Peer-to-Peer Netw. Appl. **8**, 320–335 (2015). https://doi.org/10.1007/s12083-013-0233-3

15. Mokryn, O., Akavia, A., Kanizo, Y.: Optimal cache placement with local sharing: an ISP guide to be benefit of the sharing economy. Int. J. Comput. Telecommun. Netw. Elsevier, North-Holland, United States (2020). https://doi.org/10.1016/j.comnet.2020.107153

16. Kim, S.: Cooperative inter-ISP traffic control scheme based on bargaining game approach. IEEE Access **9**, 31782–31791 (2021). https://doi.org/10.1109/ACCESS.2021.3058445

17. Liu, P., Fan, Y., Huang, K., Huang, G.: Super peer-based P2P VoD architecture for supporting multiple terminals. In: ICPCSEE 2020, CCIS 1257, pp. 389–404 (2020). https://doi.org/10.1007/978-981-15-798-3_28

18. Breslau, L., Cao, P., Fan, L., Phillips G., Shenker, S.: Web caching and Zipf-like distributions: evidence and implications: In: IEEE INFOCOM 99. Conference on Computer Communications, New York, USA, pp. 126–134 (1999). https://doi.org/10.1109/INFCOM.1997.749260

19. Stockhammer, T.: Dynamic adaptive streming over HTTP. In: ACM Press the Second Annual ACM Conference. San Jose, CA, USA, pp. 133–143 (2011). https://doi.org/10.1145/1943552.1943572

Research on Route Optimization of Battlefield Collection Equipment Based on Improved Ant Algorithm

Haigen Yang[1][✉], Wang Sun[1], Xiao Fu[2], Gang Li[1], and Luyang Li[1]

[1] Engineering Research Center of Wider and Wireless Communication Technology, Ministry of Education, Nanjing University of Posts and Telecommunications, Nanjing, China
[2] Information Construction and Management Office, Nanjing University of Posts and Telecommunications, Nanjing, China

Abstract. Accurate battlefield collection plays a crucial role in the end of the war. How to effectively improve the ability of accurate battlefield collection has become a hot issue of research. However, the existing support force is limited. To solve the problem, an improved ant algorithm is applied to the path optimization problem of battlefield collection equipment. A model for solving the collection path optimization problem of battlefield collection vehicles was designed, and an example was used to simulate calculations. The final results show that the algorithm is effective and practical, which improves the army's ability to accurately collect equipment in the modern battlefield.

Keywords: Ant algorithm · Battlefield collection · Path optimization · Simulation calculation

1 Introduction

As an important part of military logistics, battlefield collection refers to the collection of battlefield equipment at the end of training or combat [1]. Usually organized by the headquarters, a certain number of battlefield collection vehicles are assigned to recover the equipment within the battlefield, and deliver them to the logistics activities at the designated locations of the troops on time [2]. According to preliminary investigations and research, in the process of collecting equipment on the battlefield in my country, manual distribution of battlefield collection vehicles is the main focus. Based on experience, the path planning and scheduling of battlefield collection vehicles based on the location of the equipment is inefficient and costly [3]. With the development of the swarm intelligence algorithm, the ant algorithm can be used to solve the battlefield collection equipment path optimization problem [4], so this paper proposes an improved ant algorithm to solve the battlefield collection vehicle collection path optimization problem model, and use examples to simulate calculations [5].

J. Zeng et al. (Eds.): ICPCSEE 2021, CCIS 1451, pp. 154–162, 2021.
https://doi.org/10.1007/978-981-16-5940-9_12

2 The Optimization Model of the Battlefield Collecting Equipment Path

2.1 Problem Description

The problem of optimizing battlefield collection equipment path can generally be described as follows: multiple battlefield collection vehicles are assigned from a certain base center to reclaim equipment at multiple different locations. The location and weight of each equipment is fixed, and the carrying capacity of each battlefield collection vehicle is fixed. It is required to arrange the collection path of the battlefield collection vehicle reasonably so that the objective function is optimized. And meet the following conditions:

1. The sum of the equipment on each collection path does not exceed the load capacity of the battlefield collection vehicle;
2. The length of each collection path does not exceed the maximum travel path of the battlefield collection vehicle at one time;
3. Each equipment must be collected and can only be served by a battlefield collection vehicle.

2.2 Build Path Optimization Model

Let the center of the base be the letter w; $Eq = \{1,2,\ldots,n\}$ represents the equipment collection; i and j respectively represent the serial number of the base center or equipment; s represents the serial number of the battlefield collection vehicle; l represents the battlefield collection vehicle Total; q represents the load capacity of each battlefield collection vehicle; a_0 represents the fixed cost of the battlefield collection vehicle; a_{ij} represents the transportation cost of the battlefield collection vehicle from i to j; d_{ij} represents the distance from i to j; m_i represents the weight of the equipment; x_{ij} 0–1 variable that indicates whether the battlefield collection vehicle is from battlefield i to j; x_{ijs} indicates whether the battlefield collection vehicle s is 0–1 variable from equipment i to equipment j; y_{is} represents whether the equipment i is completed by battlefield collection vehicle s 0–1 Variable; and set the total distance as Z_1 and the total cost as Z_2.

Objective function:

$$\min Z_1 = \sum_{i=0}^{n}\sum_{j=0}^{n}\sum_{s=1}^{l} a_{ij}x_{ijs} + a_0 s \tag{1}$$

$$\min Z_2 = \sum_{(i,j)\in\{0\}\cup N} d_{ij}x_{ij} \tag{2}$$

Objective function:

$$\sum_{s=1}^{l} y_{is} = \begin{cases} 1, i=1,2,\ldots,n \\ l, i=0 \end{cases} \tag{3}$$

$$\sum_{i=0}^{n} m_i y_{is} \leq q \tag{4}$$

$$\sum_{i=0}^{n} x_{ijs} = y_{js} \tag{5}$$

$$\sum_{j=0}^{n} x_{ijs} = y_{is} \tag{6}$$

$$\sum_{i=0}^{n} \sum_{j=0}^{n} x_{ij} \leq n+1 \tag{7}$$

Among them, formula (1) reasonably arrange battlefield collection vehicles to minimize the total cost; formula (2) requires reasonable arrangement of the delivery path of battlefield collection vehicles to make the total path the shortest; formula (3) ensures that each large piece of equipment in the collection process Can only accept the collection service of one battlefield collection vehicle, and all the collection tasks are completed by one battlefield collection vehicle; formula (4) is the load constraint of the battlefield collection vehicle; formula (5) and formula (6) guarantee that there are only A battlefield collection vehicle arrives and departs from a certain large equipment location; formula (7) ensures that the battlefield collection vehicle starts and stops at the center of the base.

3   Improved Ant Algorithm

Italian scholar Marco Dorigo first proposed the ant algorithm in 1992 [6], which was inspired by the process of worker ants going out to find food sources. This paper proposes an improved ant colony algorithm through the pheromone incremental dynamic update strategy, which uses the number of iterations and the current optimal solution to dynamically and adaptively update the pheromone, which can speed up the convergence while ensuring that the ants can find the optimal solution. Shorten the convergence time [7].

3.1   Principles of Ant Algorithm

The ant colony algorithm optimizes and solves the problem by simulating the process of finding food sources in the real ant colony. The behavioral characteristics of ants mainly include the following three aspects: one is to mark the path, when the ant colony goes out for food, it will leave a special mark "pheromone" on each path it walks; the other is indirect communication, ants In the process of looking for food sources, you can perceive the strength of pheromone to guide your direction of movement, and always choose the path with more pheromone; the third is cluster optimization, and the concentration of pheromone will gradually decrease over time Through these features, the entire ant colony will move towards a shorter path and finally reach the food source [8]. In the ant colony algorithm, a certain number of ants are used to find the best. Each ant builds its own solutions and candidate solutions from the initial state according to the constraints of the problem, and solves the problem through the characteristics of the problem and its own behavior characteristics. Make adjustments to the feasible solutions of ants. The cooperative search of ants makes the entire ant colony update and evolve repeatedly, making it easier to approach or find the optimal solution in the optimization problem. The quality of the solution found by each ant in the algorithm is different. There are better solutions and poor solutions. By establishing different solutions, the ant colony algorithm has a global advantage [9].

3.2 The Basic Model of the Improved Ant Algorithm

Based on the principle of ant algorithm, suppose there are n battlefield points and the corresponding ant colony has m ants. When $b_i(t)$ is the time t, the expression of the number of ants at i is $m = \sum_{i=1}^{m} b_i(t)$. The probability of the ant transferring path from i to j at time t is

$$\rho_{ij}^k(t) = \begin{cases} \dfrac{\left[\tau_{ij}(t)\right]^\alpha \cdot \left[\eta_{ij}(t)\right]^\beta}{\sum\limits_{s \in J_k(i)} \left[\tau_{is}(t)\right]^\alpha \cdot \left[\eta_{is}(t)\right]^\beta} & j \in J_k(i) \\ \\ 0, j \notin J_k(i) \end{cases} \tag{8}$$

$\tau_{ij}(t)$---concentration heuristic factor, its value is the pheromone concentration between equipment i and j at time t;

$\eta_{ij}(t)$---path heuristic factor, also known as visibility, its value is generally the reciprocal of the distance dij between equipment i and j;

α---The relative importance of residual pheromone on path (i,j);

β---The relative importance of path heuristics;

$J_k(i)$---is all currently available target equipment of Ant K, that is, equipment that has not been visited yet.

After n times, after all the ants complete a tour, the path they traveled forms a solution. At this time, calculate the path L_k traveled by each ant, and save the shortest path $L_{min} = \min\{L_k | k = 1,2,...,m\}$. The pheromone left before will gradually decrease over time, so the longer the path the ant takes, the more time it takes, resulting in the faster the decrease of the pheromone left on the path. At this time, the amount of information on each path should be adjusted according to the following formula:

$$\tau_{ij}(t+n) = (1 - \rho)\tau_{ij}(t) + \Delta\tau_{ij}(t) \tag{9}$$

Among them: $0 < \rho < 1$ indicates the degree of attenuation of pheromone with the passage of time, and the concentration of pheromone on the path can be adjusted in time. $\Delta\tau_{ij}(t)$ represents the pheromone increment on the path (i, j) in this iteration process as follows:

$$\Delta\tau_{ij}(t) = \sum_{k=1}^{m} \Delta\tau_{ij}^k(t) \tag{10}$$

$\Delta \tau^k{}_{ij}(t)$ represents the amount of pheromone left on (i,j) by the kth ant in this iteration. In order to speed up the distinction between better paths and other paths, we use the pheromone incremental dynamic update strategy to dynamically and adaptively update the pheromone using the number of iterations and the current optimal solution. While ensuring that the ants can find the optimal solution, speed up the convergence speed and shorten the convergence time. The specific formula is as follows:

$$\Delta \tau_{ij}^{k}(t) = \begin{cases} \frac{Q}{NC}\,\mathrm{arccot}(L_k - L_{best}), \text{ the kth ant goes through the path } (i,j) \\ 0, \text{ otherwise} \end{cases} \tag{11}$$

Among them, Q/NC > = f; NC is the number of iterations; L_k is the length of the path taken by ant k in this cycle; L_{best} is the optimal solution algorithm found by the current entire ant colony in the iteration process.

The new pheromone incremental update rule is more in line with the law of ant pathfinding. At the beginning, the pheromone increment maintains a large value, which can quickly reduce the solution space containing the better solution to a certain range, and then increase with the number of iterations. The increment of growth pheromone gradually weakens, which is convenient for expanding the range of ants' path finding and jumping out of the local optimal solution. We can divide Eq. (11) into two parts, the first half of Q/NC makes the information increment change adaptively with the search process of artificial ants, and we set its minimum value f to avoid the excessive number of iterations. The pheromone increment is too small; the second half of the arccot() function adjusts the pheromone increment by comparing the distance between the current path found and the optimal solution in the entire iteration process [10].

3.3 Improved Ant Algorithm Process

The specific process of the ant algorithm is shown in Fig. 1. The process is as follows [11]:

(1) Initialize a parameter. Set the starting time $t = 0$, Nc = 0 (Nc represents the number of iterations), and set the maximum number of iterations to G, the pheromone concentration $\tau_{ij}(t) = c$ on each path, where c is a constant, set α, β, For the initial values of ρ and Q, m ants are placed at the central node of the base, and the initial time $\tau_{ij}(0) = 0$.

(2) The number of update cycles N_c increases by 1, the index number of the taboo table of ants is set to 1, and the number of ants $k = k + 1$.

(3) For each artificial ant, it is necessary to find out the nodes that it has not reached from the list, and select the next reasonable equipment according to the transition probability formula mentioned in the context.

(4) Change the index number in the taboo table, move the previously selected ant to the next new equipment, and include this equipment in the taboo table until the ant visits all the equipment and ends the ant's cycle.

(5) Calculate the optimal path length of this iteration. The optimal solution in the current iteration number is the optimal path for collecting equipment on the battlefield, and update $\tau^k{}_{ij}$ according to the pheromone concentration.

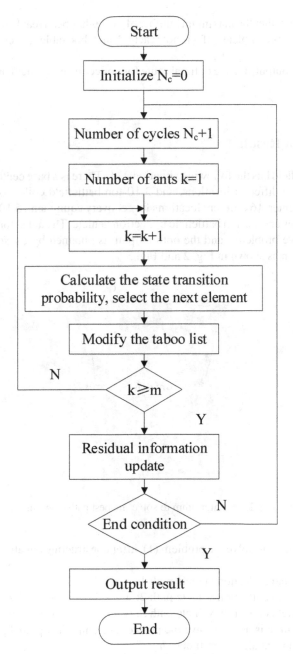

Fig. 1. Flow chart of ant algorithm

(6) Determine whether the maximum number of cycles has been reached, if it is reached, the collection is completed; if it is not reached, all taboo tables are cleared, and jump to (1).

(7) The result is output, and the optimal route for collecting equipment on the battlefield is drawn.

4 Simulation Result

Suppose a battlefield has the following situations: (1) There is a base center, 8 equipment to be recovered in different locations, and 3 10-ton battlefield collection vehicles. (2) There is a base center, 16 different locations for recovery equipment, 5 10-ton battlefield collection vehicles and 1 5-ton battlefield collection vehicle. The ant algorithm is applied to solve the above problems, and the optimal path is obtained by constructing the ant algorithm function as shown in Fig. 2 and Fig. 3.

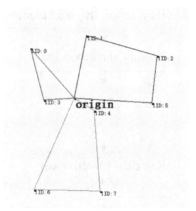

Fig. 2. Ant algorithm to solve the best path diagram

Use ant algorithm to solve the problem (1), after constructing ant algorithm function, the best path is:

10 ton battlefield collection vehicle path 1: w → 0 → 3 → w

10 ton battlefield collection vehicle path 2: w → 4 → 7 → 6 → w

10 ton battlefield collection vehicle path 3: w → 1 → 2 → 5 → w

The ant algorithm is used to solve the problem (2), and the optimal path is obtained by constructing the ant algorithm function:

10 ton battlefield collection vehicle path 1: w → 6 → 0 → 4 → w

10 ton battlefield collection vehicle path 2: w → 10 → 15 → 11 → w

10 ton battlefield collection vehicle path 3: w → 5 → 2 → 3 → w

10 ton battlefield collection vehicle path 4: w → 1 → 8 → 9 → w

10 ton battlefield collection vehicle path 1: w → 14 → 12 → 13 → w

5 ton battlefield collection vehicle path 1: w → 7 → w

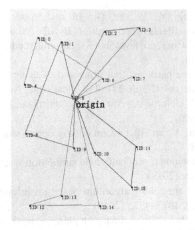

Fig. 3. Ant algorithm to solve the best path diagram

5 Conclusion

Accurate battlefield collection has become more and more important after the end of the war. How to effectively improve the ability of accurate battlefield collection has become a hot issue of research. However, the existing support force is limited. On the existing basis, optimizing the precise battlefield collection decision-making, especially the path selection of the battlefield collection of large pieces of equipment, is an important means to improve the battlefield collection ability. This article uses an improved ant algorithm to optimize the route and make it easily found and quickly searched, so as to reasonably arrange the driving route of the battlefield collection vehicle and improve the efficiency of the guarantee.

Acknowledgment. This work was supported by the Research Program of the Basic Scientific Research of National Defense of China under Grant JCKY2019210B005, JCKY2018204B025, and JCKY2017204B011, the Key Scientific Project Program of National Defense of China under Grant ZQ2019D20401, the Open Program of National Engineering Laboratory for Modeling and Emulation in E-Government, Item number MEL-20-02, and The Foundation strengthening project of China under Grant 2019JCJZZD13300.

References

1. Gao, K., Sui, K.: Explore and grasp the characteristics of war evolution in the intelligent age. People's Liberation Army Daily, 2021-05-27 (2007)
2. Mingchang, X.: Thoughts on improving military logistics guarantee capability. China Storage Transp. **11**, 198–199 (2020)
3. Bin, Z., Yao, L., Xiao, Q.: Research on online path planning algorithm of battlefield transportation. Firepower Command Control **45**(01), 79–84 (2020)
4. Deng Q., Xue Q., Chen, L., Chen, J.: Application of a hybrid path planning method in armored vehicle CGF. Ordnance Equip. Eng. J. **39**(07), 120–122+150 (2018)

5. Wang, Y., Guo, K., Fang, Y., Ye, Y.: Design and application of fuzzy neural net-work based on cluster intelligence optimization[J/OL]. Aviation weapons: 1–7[2020-12-13]. https://cc0eb1c56d2d940cf2d0186445b0c858.vpn.nuist.edu.cn/kcms/detail/41.1228.TJ.20200629.0903.001.html.I

6. Jacobs, S., Bean, C.P.: Fine particles, thin films and exchange anisotropy. In: Rado, G.T., Suhl, H. (eds.) Magnetism, vol. III, pp. 271–350. Academic, New York (1963)

7. Dorigo, M., Maniezzo, V., Colorni, A.: Ant system: optimization by a colony of cooperating agents. **26**(1), 29–41 (1996)

8. Raka, J., Milan, T., Stefan, V.: an efficient ant colony optimization algorithm for the blocks relocation problem (2018)

9. Wen, B.: Design and implementation of battlefield simulation system based on Unreal Engine. Beijing Forestry University (2019)

10. Xiao, J., Li, L.: Adaptive ant colony algorithm based on information entropy adjustment. Comput. Eng. Des. **31**(22), 4873–4876 (2010)

11. Jing, Y., Jin, Z., Liu, G.: A three-dimensional path planning method for farmland level navigation based on improved ant colony algorithm. Trans. Chin. Soc. Agri. Mach. **51**(S1), 333–339 (2020)

12. Luo, Z., Liu, X.: Research on optimization of logistics distribution route based on ant colony algorithm. J. Chongqing Technol. Bus. Univ. (Nat. Sci. Edition) **37**(04), 89–94 (2020)

Parallel Region Reconstruction Technique for Sunway High-Performance Multi-core Processors

Kai Nie[1] , Qinglei Zhou[2] , Hong Qian[3] , Jianmin Pang[1] , Jinlong Xu[1](✉) , and Yapeng Li[1]

[1] Information Engineering University, Zhengzhou 450001, Henan, China
[2] Zhengzhou University, Zhengzhou 450001, Henan, China
[3] Jiangnan Institute of Computing Technology, Wuxi 214083, Jiangsu, China

Abstract. The leading way to achieve thread-level parallelism on the Sunway high-performance multicore processors is to use OpenMP programming techniques. In order to address the problem of low parallel efficiency caused by high thread group control overhead in the compilation of Sunway OpenMP programs, this paper proposes the parallel region reconstruction technique. The parallel region reconstruction technique expands the parallel scope of parallel regions in OpenMP programs by parallel region merging and parallel region extending. Moreover, it reduces the number of parallel regions in OpenMP programs, decreases the overhead of frequent creation and convergence of thread groups, and converts standard fork-join model OpenMP programs to higher performance SPMD model OpenMP programs. On the Sunway 1621 server computer, NPB3.3-OMP and SPEC OMP2012 achieved 8.9% and 7.9% running efficiency improvement respectively through parallel region reconstruction technique. As a result, the parallel region reconstruction technique is feasible and effective. It provides technical support to fully exploit the multi-core parallelism advantage of Sunway's high-performance processors.

Keywords: Sunway high-performance multi-core processors · OpenMP programming technique · Parallel domain reconstruction technique

1 Introduction

Compiler design and development is a complex and difficult task, and one of the crucial aspects is compiler optimization. With the rapid development of multi-core processor design technology, the basic computing power of CPUs has been continuously improved, but the necessary way to fully exploit it to enhance the application performance of information systems is to generate efficient target code to match the multi-core processor chips through compiler optimization techniques. The main way to fully utilize the performance advantages of multi-core processors is to start multiple threads to execute programs in parallel on the multi-core processors. Currently, the main way to implement

© Springer Nature Singapore Pte Ltd. 2021
J. Zeng et al. (Eds.): ICPCSEE 2021, CCIS 1451, pp. 163–179, 2021.
https://doi.org/10.1007/978-981-16-5940-9_13

multi-threading on Sunway's high-performance multi-core processors with symmetric shared storage structure is the OpenMP programming technique. For this reason, the compilation and optimization of OpenMP programs is the core task of the Sunway Multi-Threaded Parallel Compilation System that accompanies the Sunway High Performance Multi-Core Processors. The major goal of OpenMP program compilation optimization is to express parallelism in the program as efficiently as possible, including translation-time optimization, run-time optimization, and thread-level optimization [1–3]. At present, the focus of OpenMP program compilation optimization is mainly on thread-level optimization.

There are two main models for implementing thread-level parallelism in OpenMP programming techniques: the standard fork-join model and the more complex SPMD (Single Program Multi-Data) model. Fork-join model is flexible, easy to handle parallel regions in a program, relatively simple to program and supports incremental development. The SPMD model uses a collaborative multi-threaded execution approach with less thread control overhead, but it requires the programmer to consider distributing workload and data among threads, which is complex to program. At present, the fork-join model is used in the Sunway multi-threaded parallel compilation system, and the threads are created and ended more frequently, which is in an inefficient state in terms of parallelism expression. To address this shortcoming, this paper proposes the parallel region reconstruction technique. The parallel region reconstruction technique combines scalar analysis, dependency analysis, and automatic parallelization techniques such as data and computation division information to merge multiple individual parallel domains in fork-join model into a large parallel region as a small SPMD zone to obtain a hybrid OpenMP program in fork-join and SPMD models. It reduces the control overhead of threads and improves the efficiency of parallelism expression of thread-level code without increasing the programmer's workload.

Implementing thread-level parallel execution of scientific computing programs on Sunway high-performance multi-core processors is the main means to improve the efficiency of program operation. The parallel domain reconstruction technique proposed in this paper can further compile OpenMP programs into more efficient thread-level parallel code, and provide technical support to fully utilize the advantages of multi-core parallelism of Sunway's high-performance processors.

2 Related Works

The programming standard of OpenMP programs strictly adheres to the fork-join model, where each parallel domain must go to shared memory to read data before starting execution. Neither the small and medium-sized SMP architectures nor the currently popular CMP architectures can avoid the problem of data division and distribution. For this reason, the fork-join model of OpenMP programs is not a good model for parallel programming. A large number of researchers have been working on optimizing OpenMP programs using an improved programming model on different platforms.

Naoyuki Onodera et al. pointed out through optimization studies of OpenMP programs such as MPCG (Multigrid Preconditioned Conjugate Gradient) on GPU platform that OpenMP programs can achieve coarser granularity by merging some parallel regions of parallelism, thus improving the overall performance of the program [4, 5]. Soumitra Pal et al. gave some specific OpenMP program optimization methods for jacobi programs on Intel Xeon processors, where they merged the parallel regions of OpenMP programs by operations such as variable privatization and removal of redundant synchronization to improve the running efficiency of the programs [6–8]. Shigehisa Satoh et al. illustrate the use of SPMD mode to eliminate redundant synchronization, expand the program parallel region, and improve the performance of executable programs from the perspective of OpenMP compiler optimization [9–11]. Zhu et al. showed earlier tests on a C64 multicore simulator that as the number of threads increases, the compilation directives "#pragma omp parallel for" and "#pragma omp parallel" are comparable in cost, but both are much more expensive than "#pragma omp for", with a difference of about 100 times, and their results also confirm that the parallel cost of OpenMP programs after parallel domain merging is much lower than that of OpenMP programs without merging and extension [12]. These results point out that merging parallel regions in OpenMP programs is a more efficient optimization method, but they do not give a general method for merging parallel domains in OpenMP programs. In addition, the research in the area of thread-level optimization on the recently released new generation of Sunway high-performance multi-core processors is still in a gap. Therefore, based on the summary of previous optimization methods, this paper seeks a programming model with high parallel efficiency and avoiding the complexity of hand-written parallel code for the thread-level parallel execution criteria adopted on Sunway high-performance processors.

3   Parallel Region Reconstruction Technique

The parallel region reconstruction technique converts fork-join model OpenMP programs into SPMD model OpenMP programs to improve the performance of parallel programs, including parallel region merging and parallel region extending. Next, we first analyze and explain the advantages and disadvantages of fork-join model and SPMD model in OpenMP programming technique, then we illustrate the specific practices of parallel region merging and parallel region extending in Sunway multi-threaded parallel subsystem through program examples one by one, and finally we give the implementation method of parallel region reconstruction technique.

3.1   OpenMP Programming Model and Problem Analysis

The fork-join model is the standard for OpenMP programming techniques and is currently the default parallel programming model for the Sunway multi-threaded parallel compilation system. When a program starts to execute, there is only one main thread executing serially. When a parallel domain is encountered, the master thread derives a set of slave threads, and the code in the parallel domain is executed in parallel in different slave threads, and after the derived slave threads are executed, they will exit or hang, and finally only the master thread continues to execute. In contrast, in the SPMD model, all threads execute the entire program, the threads are always active, and the serial part is either repeated or restricted to single-threaded execution. When a parallel loop is encountered, threads execute the corresponding part of the iteration based on the thread number, and synchronization needs to be inserted when there is a dependency between threads. Figure 1 and Fig. 2 show two parallel execution models of the OpenMP model.

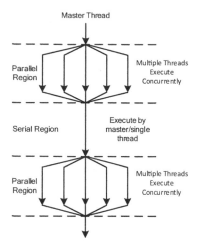

Fig. 1. Fork-join model.

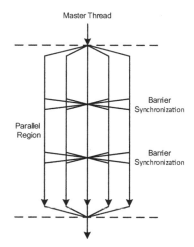

Fig. 2. SPMD model

OpenMP programs in fork-join programming model are parallelized for the corresponding loops, which are simple to program and easy to develop incrementally. The disadvantages are the high cost of starting and synchronizing multiple threads, the impact of serial part of the code on parallel performance, the poor scalability, the inability to perform cross-loop optimization, and therefore the inefficient use of Cache. In contrast, the SPMD model OpenMP program combines parallel loops into one large parallel domain, reducing the overhead of multiple thread group creation and merging, while the compiler can analyze and eliminate redundant synchronization for more adequate data partitioning and efficiently improve Cache utilization. However, OpenMP programs in SPMD mode have a larger parallel domain, the transformation of shared and private variables can appear more complex, and data correlation analysis as well as thread synchronization and loop optimization have greater difficulty, making them difficult to program.

3.2 Parallel Region Merging

Merging of parallel regions means that a gate barrier synchronization is inserted between two adjacent individual parallel regions (if there is a data dependency between the two parallel regions), and then the two parallel regions are directly merged into one large parallel region as a small SPMD region. Figure 3 shows a code example of merging adjacent parallel regions. In the Sunway compiler, the compiler directive "parallel", "for", "sections" and "single" all have an implicit barrier synchronization at the end.

```
#pragma omp parallel //parallel region 1
{
    #parallel omp for
        for (i=0;i<100;i++)
            a[i] = i;
}

#pragma omp parallel //parallel region 2
{
    #parallel omp for reduction(+: sum)
        for (i=0;i<100;i++)
            sum += a[i];
}
```

```
#pragma omp parallel //parallel region
{
    #parallel omp for
        for (i=0;i<100;i++)
            a[i] = i;

    #parallel omp for reduction(+: sum)
        for (i=0;i<100;i++)
            sum += a[i];
}
```

(a) Before Merger   (b) After Merger

Fig. 3. Example of parallel regions merging

In order to ensure that the semantics of the program are the same before and after merging, it is necessary to ensure that the order of execution of each thread and the order of updating the data seen by each thread are consistent. The former can be achieved by the barrier synchronization operation in OpenMP, and the latter can be achieved by the flush operation of OpenMP's directive to ensure the consistency of the visible data. In addition, the merging of parallel regions needs to solve two problems: (1) the conflict of variable data attributes encountered during the merging process. (2) The handling of the serial part of the program during the merging process.

Variable Data Attribute Conflict Processing. In multi-threaded programs, variable data attributes include both shared and private. A shared variable is a variable that is shared by all threads in the parallel domain. There is only one copy of the variable in the thread group, and modifications to the variable are visible to all threads. Private variables have a copy of the variable for each thread in the thread group, and modifications to the variable do not affect the value of the variable in other threads. In the process of extending parallelism through parallel domain merging, it may involve merging conflicts of different attributes of the same variable, and affect the correctness of the program.

Figure 4 shows an example of a conflict of data attributes encountered by a parallel region merge. The variable k is declared as a shared variable in parallel region 1 and as a private variable in parallel region 2. Due to the different data attributes of variable k, these two parallel regions cannot be simply merged. Before merging, the conflicting data attributes of variable k must be resolved first. To address this issue, this paper uses the standard OpenMP compiler-guided statements "private", "firstprivate" and "lastprivate", modify the data attributes of the shared variables, and then try to merge them again.

```
int k=1, sum=0, mul=1;

#pragma omp parallel shared(k) //parallel region 1
{
    #parallel omp for reduction(+: sum)
    for (i=0;i<10;i++) {
        sum += (k+i);
    }
}

#pragma omp parallel firstprivate(k) //parallel region 2
{
    #parallel omp for reduction(*: mul)
    for (j=0;j<10;j++) {
        mul *= (k+j);
    }
}
```

Fig. 4. Example of variable data attribute conflict

During the compilation of the example program in Fig. 4, the definition-use chain analysis of the variable k reveals that the shared variable k is used in parallel region 1, while the definition is in a previous program segment in parallel region 1. Therefore, to make the variable k private in parallel region 1, it is necessary to make the previous definition of k visible inside parallel region 1, and it is necessary to introduce this definition into parallel region 1 before it is used, which can be achieved by the OpenMP directive "firstprivate", as shown in Fig. 5. After modifying the data attributes of variable k, further merging of parallel domains 1 and 2 can be achieved. Figure 6 shows the merging of parallel regions 1 and 2 after modifying the data attributes of variable k.

```
int k=1, sum=0, mul=1;

#pragma omp parallel firstprivate(k) //parallel region 1
{
    #parallel omp for reduction(+: sum)
    for (i=0;i<10;i++) {
      sum += (k+i);
    }
}

#pragma omp parallel firstprivate(k) //parallel region 2
{
    #parallel omp for reduction(*: mul)
    for (j=0;j<10;j++) {
      mul *= (k+j);
    }
}
```

Fig. 5. Variable data attributes modification example

```
int k=1, sum=0, mul=1;

#pragma omp parallel firstprivate(k) //parallel region
{
    #parallel omp for reduction(+: sum)
    for (i=0;i<10;i++) {
      sum += (k+i);
    }

    #parallel omp for reduction(*: mul)
    for (j=0;j<10;j++) {
      mul *= (k+j);
    }
}
```

Fig. 6. Parallel region merging after data attribute modification

Privatizing variables blocks out their original definitions, both within the parallel domain and between the parallel domain and its surrounding programs, potentially violating the previous definition-use chains, a problem that has been studied in depth [13–15]. In order to satisfy all definition-use relationships prior to variable privatization, this paper requires that the shared variables undergoing privatization are not defined in the parallel region. In this paper, we determine whether the shared variables in a parallel domain can be privatized under this premise. If this condition cannot be satisfied, then the conflict of variable attributes for parallel domain merging cannot be eliminated, and further parallel domain merging cannot be performed. If the shared variables that need to be privatized are not redefined in the parallel domain, then privatization is attempted to further parallel domain merging.

Serial Program Processing. When there are several serial programs between two parallel domains, the merging of parallel regions cannot be implemented directly. In this case, you can constrain the execution so that the serial programs are included in the parallel region, and then implement the merging of parallel regions, which can be achieved by using the OpenMP directive clause "master" or "single".

```
int sum=0,mul=0;

#pragma omp parallel //parallel region 1
{
    #pragma omp for reduction(+: sum)
    for(i=0;i<10;i++)
        sum += i;
}

i++; mul++; //serial region

#pragma omp parallel //parallel region 2
{
    #pragma omp for reduction(*: mul)
    for(j=0;j<10;j++)
        mul *= i+j;
}
                            //serial region 2
printf("sum=%d, mul=%d.\n", sum, mul);
```

(a)Before merger

```
int sum=0,mul=0;

#pragma omp parallel //parallel region
{
    #pragma omp for reduction(+: sum)
    for(i=0;i<10;i++)
        sum += i;

    #pragma omp single //single-threaded execution
    {
        i++; mul++;
    }

    #pragma omp for reduction(*: mul)
    for(j=0;j<10;j++)
        mul *= i+j;

    #pragma omp master //master-threaded execution
        printf("sum=%d, mul=%d.\n", sum, mul);
}
```

(b)After merger

Fig. 7. Serial statements processing

Figure 7 shows a program example of serial statements processing during merging. The serial statement is executed with the OpenMP constraint "single" and "master" to merge with the parallel statement and further merge more parallel regions. However, this also increases the synchronization operations and introduces new additional overhead, with the possible benefit that the code of the serial domains can be used as new workloads in the parallel regions, or can be further transformed into "section" type parallel tasks depending on the characteristics of the serial regions code, thus providing more load balancing options. This provides more load balancing options.

3.3 Parallel Region Extending

There are three main cases of parallel domain extending: loop parallel region extending, control flow parallel region extending, and parallel region extending across function boundaries.

Loop Body Parallel Region Expanding. If all statements in a loop are contained in a parallel domain, the parallel domain can be extended beyond that loop. This is a more efficient optimization model, and if a parallel domain is extended beyond a loop of N iterations, the number of parallel domain creation and joins will be reduced by N−1. Figure 8 shows a simple example of a loop parallel region extending, where the parallel domain corresponding to loop j can be extended beyond loop i. Before the parallel domain extension, the parallel domain is created and converged 100 times, while after the extension the parallel domain is created and converged only 1 time, a reduction of 99 times.

```
for(i=0;i<100;i++)
#pragma omp parallel for reduction(+: sum)
  for(j=0;j<100;j++)
    sum += i+j;
```

(a)Before merger

```
#pragma omp parallel private(i)
for(i=0;i<100;i++)
 #pragma omp for reduction(+: sum)
  for(j=0;j<100;j++)
    sum += i+j;
```

(b)After merger

Fig. 8. Schematic diagram of the loop parallel region extending code

Control Flow Parallel Region Extending. When a parallel region appears in a branch of a selection structure, the entire selection structure can be contained in a large parallel region. For example, the if-then-else structure in Fig. 9 can be turned into a large parallel region by using parallel region extending. Note that if the selection structure contains serial branches, the serial branches must be protected or computationally privatized.

```
if(is_init_A){
  #pragma omp parallel for
    for(i=0;i<100;i++)
      A[i] = i;
} else {
  #pragma omp parallel for
    for(i=0;i<100;i++)
      B[i] = i;
}
```

(a)Before merger

```
#pragma omp parallel
{
  if(is_init_A){
    #pragma omp for
      for(i=0;i<100;i++)
        A[i] = i;
  } else {
    #pragma omp for
      for(i=0;i<100;i++)
        B[i] = i;
  }
}
```

(b)After merger

Fig. 9. Schematic diagram of the control flow parallel domain extending code

Parallel Region Extending Across Function Boundaries. Parallel region extending is usually not enough if it can be done only inside the function. It has been possible to extend the parallel region to the entire function body by means of parallel region merging. Parallel region extending across function boundaries means that the parallel region of a specific function is raised around a call to a specific function, thus gaining a greater chance of parallel region merging.

```
const int N=100;
void init_arry(int* a){
    #pragma omp parallel for
    for(int i=0;i<N;i++)
        a[i]=i;
}

int main(){
    int a[N];
    init_arry(a);
    printf("a[55] is %d\n", a[55]);
    return 0;
}
```

(a)Before merger

```
const int N=100;
void init_arry(int* a){
    #pragma omp for
    for(int i=0;i<N;i++)
        a[i]=i;
}

int main(){
    int a[N];
    #pragma omp parallel
    {
        init_arry(a);
    }
    printf("a[55] is %d\n", a[55]);
    return 0;
}
```

(b)After merger

Fig. 10. Schematic diagram for parallel region extending across function boundaries

Figure 10 illustrates an example of parallel region merging across function boundaries. Parallel region extension across function boundaries further extends the scope of the parallel region and thus has the potential to further improve the structure of the program and increase the granularity of parallel computation. However, parallel region extending across function boundaries completely changes the stack structure of threads, and this practice implicitly privatizes all local variables within the function body. Therefore, for such extending to not change the correctness of the program, the legitimacy of privatizing these variables needs to be confirmed by data flow analysis, and the derivation of data attributes needs to be optimized to eliminate unnecessary duplicate privatization.

Extending across function boundaries is based on the results of compiler interprocedural analysis and requires the ability to access both the called function as well as at least one call point of this function. Parallel domain extension of the called function implies modifying the function body in a way that is visible to all callers, so this extension can be accomplished by finding all call points of the function in the entire program and inserting the derivative clause. When the function is exposed as an external function or when it is not possible to determine all its call points through interprocedural analysis, or when there is no guarantee that the function will be extended across function boundaries to benefit all call points, this paper uses multiple versions of the function, i.e., the function is copied, and the copied version is extended across function boundaries while retaining the original version. This avoids modifying all call points and eliminates the need for global inter-procedural analysis.

3.4 Implementation of Parallel Region Reconstruction Technique

On the basis of parallel region merging and parallel region extending, the algorithm for constructing SPMD parallel zones in this paper adopts a bottom-up recursive invocation approach. Initially, each parallel loop is regarded as a SPMD region, and parallel region merging operation is performed at the same level starting from the innermost loop, and if all statements at the same level are contained in a parallel region, parallel region extending is performed to the outer level until a serial statement that cannot be merged is encountered.

```
#pragma omp parallel //parallel region
{
    ......
    #pragma omp for  //parallel region merging
        for ( ......)
        {
            ......
        }
    ......
    #pragma omp single/master  //executing serial statements by constraints
    {......}
    ......
    #pragma omp sections
    {
        #pragma omp section
        {......}
        ......
        #pragma omp section
        {......}
    }
    ......
    if( ......) { //parallel region extending
        #pragma omp for
        ......
    }else if (......){
        ......
    ......
    }else {
        ......
    }
    ......
}
```

Fig. 11. Parallel region reconstruction after the code framework

Parallel region reconstruction combines multiple parallel regions and the serial code between parallel regions into one SPMD region, while expanding the parallel regions in the loop body, control flow, and function body, it expands the whole program into one large parallel region, as shown in Fig. 11.

The parallel performance and scalability of the program can be significantly improved by expanding the parallel scope as much as possible through the parallel region reconstruction technique to realize the SPMDization of fork-join programs, reduce the number of parallel regions in the program, and lower the overhead of thread group start and termination. In addition, by expanding the scope of parallel regions, some redundant synchronization operations can also be eliminated. In some programs, redundant synchronization operations can also be a major factor in program performance degradation. Finally, expanding the parallel range can also improve the storage behavior of the program. There is a large class of programs that are very sensitive to the characteristics of the Cache, such as BLAS (Basic Linear Algebraic Subroutines) library operations for dense matrices, where whether the matrix is already in the Cache has a very significant impact on the performance of the program.

4   Experimental Evaluation

The work in this paper is derived from an optimization project under the Sunway multi-threaded parallel compilation system, where the parallel region reconstruction technique has been integrated in the prototype system.

4.1   Experimental Environment, Benchmarks and Test Content

The experimental environment is a high-performance server with Sunway 1621 shared memory architecture, equipped with Sunway multi-threaded parallel compilation system, CSC Kirin operating system, and Sunway computer storage management system, etc. Table 1 shows more detailed configuration information. The benchmarks are

Table 1. Experimental environment.

Configuration items	Value
CPU	Sunway 1621 CPU (Core3A @ 2 GHz × 16)
L1 cache	32 GB
L2 cache	32 KB
L3 cache	32 MB
Memory	32 GB
Hard disk	1 TB
OS	Zhong Biao Neoshine
Compiler	swgcc, swg++ & swgfortran

NPB3.3-OMP and SPEC OMP2012, which test the performance of the OpenMP parallel system of the high-performance server. The test scale is B and Ref respectively.

The performance improvement effect of parallel region reconstruction technique on multi-threaded parallel compilation subsystem is tested. On the Sunway 1621 server machine, the original Sunway multi-threaded parallel compilation system and the Sunway multi-threaded parallel compilation system integrated with parallel region reconstruction technique were used to compile and run the OpenMP programs in the NPB3.3-OMP and SPEC OMP2012, check the compilation and running results, compare the running time, and calculate the runtime efficiency gains. The runtime efficiency gains is calculated as:

$$\eta = \frac{t_1 - t_2}{t_1} \times 100\% \tag{1}$$

where t_1 is the target code runtime generated by the original Sunway multi-threaded parallel compilation system, t_2 is the target code runtime generated by the Sunway multi-threaded parallel compilation system with integrated parallel domain reconstruction technology, and η is the runtime efficiency gains.

4.2 Experimental Results and Analyses

The CPU of Sunway 1621 computer integrates 16 Core3A processors, so to fully test the performance improvement effect of parallel region reconstruction technique, the runtime of 16 threads is chosen to calculate the runtime efficiency gains in this paper. Table 2 and Table 3 show the compilation and running results, number of parallel regions of OpenMP programs in NPB3.3-OMP and SPEC OMP2012 before and after parallel region reconstruction, respectively. Table 4 and Table 5 show the runtime efficiency gains of OpenMP programs in NPB3.3-OMP and SPEC OMP2012, respectively.

Table 2. NPB3.3-OMP result statistics

Benchmark	Compile and run[1]	Number of parallel regions	
		before reconstruction	After reconstruction
BT	✓	12	10
CG	✓	7	6
DC	✓	1	1
EP	✓	3	2
FT	✓	8	5
IS	✓	7	4
LU	✓	12	9
MG	✓	11	7
SP	✓	13	10
UA	✓	60	40

[a] ✓: Compile and run correctly. ×: Compile or run error.

The compilation and running results in Table 2 and Table 3 show that the Sunway multi-threaded parallel compilation system with integrated parallel region reconstruction technique passes all the correctness tests (compilation passes and running results are correct) of NPB3.3-OMP and SPEC OMP2012. The reason is that the technique in this paper mainly focuses on intermediate process optimization and adopts a conservative approach, which has less impact on the target machine description file and the base library of the system at the back end, and therefore does not affect the correctness of the whole compilation system.

Table 3. SPEC OMP2012 Result Statistics

Benchmark	Compile and RUN[a]	Number of parallel regions	
		before reconstruction	After reconstruction
350.md	√	30	18
351.bwaves	√	29	21
352.nab	√	16	10
357.bt331	√	13	11
358.botsalgn	√	1	1
359.botsspar	√	1	1
360.ilbdc	√	5	4
362.fma3d	√	125	94
363.swim	√	12	11
367.imagick	√	190	120
370.mgrid331	√	15	13
371.applu331	√	10	8
372.smithwa	√	4	3
376.kdtree	√	4	4

[a] √: Compile and run correctly. ×: Compile or run error.

The results in Table 4 and Table 5 show that the more significant performance improvements are BT, FT, SP, UA, IS, MG, and LU in NPB3.3-OMP, and 367, 370, 376, 360, and 372 in SPEC OMP2012. The parallel regions in the BT, SP, UA, 370 and 360 programs are relatively concentrated, and there are fewer conflicts of variable data attributes between parallel regions. The compilation process merges more parallel regions to ensure the data continuity between adjacent parallel regions, and the running efficiency is improved significantly. There are not many parallel regions in FT, IS, 367, and 376, but the parallel regions in the hot loop segments of these programs can be extended, and the parallel region extensions greatly reduce the management control overhead of the thread group. Inter-thread communication in MG, LU and 372 core loop segments is implemented through shared variables, so there are more thread synchronization operations. After the parallel region is refactored or extended to remove

Table 4. Runtime efficiency promoted on NPB3.3-OMP

Benchmark	Runtime (s)		Runtime efficiency gains (%)
	before reconstruction	After reconstruction	
BT	58.56	54.05	7.7
CG	16.14	14.36	11.0
DC	235.06	230.36	2.0
EP	7.95	7.31	8.1
FT	9.51	8.53	10.3
IS	0.48	0.42	12.5
LU	47	42.3	10.0
MG	3.08	2.84	7.8
SP	67.98	61.59	9.4
UA	65.53	58.78	10.3
AVG	—	—	**8.9**

Table 5. Runtime efficiency promoted on SPEC OMP2012

Benchmark	Runtime (s)		Runtime efficiency gains (%)
	before reconstruction	After reconstruction	
350.md	18013	16791	6.8
351.bwaves	10225	9509	7.0
352.nab	6337	5940	6.3
357.bt331	3655	3363	8.0
358.botsalgn	4269	4098	4.0
359.botsspar	13056	12098	7.3
360.ilbdc	37045	33741	8.9
362.fma3d	4131	3801	8.0
363.swim	3030	2815	7.1
367.imagick	8996	7957	11.5
370.mgrid331	4624	4115	11.0
371.applu331	4196	3909	6.8
372.smithwa	3908	3578	8.4
376.kdtree	3223	2901	10.0
AVG	—	—	**7.9**

redundant synchronization, the data continuity between parallel regions is improved and the performance is more improved. The rest of the programs are less efficient to run, but the reconstruction of the parallel region expands the parallelism and improves the overall storage behavior of the program, providing opportunities for further other optimizations, so that the program performance is still ultimately improved.

The final results show that the overall efficiency improvement of NPB3.3-OMP is better than that of SPEC OMP2012, because the program of SPEC OMP2012 is larger and more complex, contains more function calls with more side effects that are difficult to analyze, and has more complex read and write operations on the properties of data variables, which makes it difficult to merge more effective parallel regions. For example, although 350 and 362 merged a large number of parallel regions, several hot functions in their core loop segments had more complex function call relationships or too many complex operations on private variables that prevented merging, and therefore could not achieve significant runtime efficiency gains.

5 Conclusion

In order to improve the running efficiency of thread-level parallel programs on Sunway high-performance multicore processors, this paper proposes the parallel region reconstruction technique. The parallel region reconstruction technique expands the parallel range of parallel regions in OpenMP programs by parallel region merging and parallel region extending, reduces the number of parallel regions in OpenMP programs, decreases the control overheads such as frequent creation and merging of thread groups, and converts fork-join model OpenMP programs into SPMD model OpenMP programs. The parallel region reconstruction technique is integrated in the Sunway multi-threaded parallel compilation system. To test the performance improvement effect of the parallel region reconstruction technique, this paper chooses NPB3.3-OMP and SPEC OMP2012 benchmarks to conduct experiments on a Sunway 1621 server computer. The experimental results show that the parallel region reconstruction technique does not affect the correctness of the Sunway multi-threaded parallel compilation system and the performance improvement is obvious. As a result, the method in this paper is feasible and effective, and can provide technical support for giving full advantage to the multi-core parallelism of Sunway high-performance processors.

References

1. Tiotto, E., Mahjour, B., Tsang, W.: OpenMP 4.5 compiler optimization for GPU offloading. IBM J. Res. Dev. **3**(5), 1–11 (2020)
2. Neth, B., Scogland, T.R.W., Strout, M.M., de Supinski, B.R.: Unified sequential optimization directives in OpenMP. In: Milfeld, K., de Supinski, B., Koesterke, L., Klinkenberg, J. (eds.) IWOMP 2020. LNCS, vol. 12295, pp. 85–97. Springer, Cham (2020). https://doi.org/10.1007/978-3-030-58144-2_6
3. Mosseri, I., Alon, L.O., Harel, R., Oren, G.: *ComPar*: optimized multi-compiler for automatic OpenMP S2S parallelization. In: Milfeld, K., de Supinski, B., Koesterke, L., Klinkenberg, J. (eds.) IWOMP 2020. LNCS, vol. 12295, pp. 247–262. Springer, Cham (2020). https://doi.org/10.1007/978-3-030-58144-2_16

4. Onodera, N., Idomura, Y., Hasegawa, Y.: GPU acceleration of multigrid preconditioned conjugate gradient solver on block-structured Cartesian grid. In: Proceedings of International Conference on High Performance Computing in Asia-Pacific Region, pp. 120–128 (2021)
5. Pereira, F.H., Verardi, S.L.L., Nabeta, S.I.: A fast algebraic multigrid preconditioned conjugate gradient solver. Appl. Math. Comput. **179**(1), 344–351 (2006)
6. Pal, S., Pathak, S., Rajasekaran, S.: On speeding-up parallel Jacobi iterations for SVDs. In: Proceedings - 18th IEEE International Conference on High Performance Computing and Communications, 14th IEEE International Conference on Smart City and 2nd IEEE International Conference on Data Science and Systems, pp. 9–16 (2016)
7. Yang, X., Mittal, R.: Efficient relaxed-Jacobi smoothers for multigrid on parallel computers. J. Comput. Phys. **332**, 135–142 (2017)
8. Kudo, S., Yamamoto, Y., Bečka, M., Vajteršic, M.: Performance of the parallel one-sided block Jacobi SVD algorithm on a modern distributed-memory parallel computer. In: Wyrzykowski, R., Deelman, E., Dongarra, J., Karczewski, K., Kitowski, J., Wiatr, K. (eds.) PPAM 2015. LNCS, vol. 9573, pp. 594–604. Springer, Cham (2016). https://doi.org/10.1007/978-3-319-32149-3_55
9. Cervini, S.: System and method for efficiently executing single program multiple data (SPMD) programs, US7904905 B2, US (2011)
10. Intel Corporation: Architecture and method for data parallel single program multiple data (SPMD) Execution: US,US20200104139[P], 4 February 2020
11. Sprenger, S., Zeuch, S., Leser, U.: Exploiting automatic vectorization to employ SPMD on SIMD registers. In: Proceedings - IEEE 34th International Conference on Data Engineering Workshops, pp. 90–95 (2018)
12. Zhu, W., del Cuvillo, J., Gao, G.R.: Performance characteristics of OpenMP language constructs on a many-core-on-a-chip architecture. In: Mueller, M.S., Chapman, B.M., de Supinski, B.R., Malony, A.D., Voss, M. (eds.) IWOMP -2005. LNCS, vol. 4315, pp. 230–241. Springer, Heidelberg (2008). https://doi.org/10.1007/978-3-540-68555-5_19
13. Stelle, G., Moses, W.S., Olivier, S.L.: Implementing OpenMP tasks with tapir. In: Proceedings of LLVM-HPC 2017: 4th Workshop on the LLVM Compiler Infrastructure in HPC - Held in conjunction with SC 2017: The International Conference for High Performance Computing, Networking, Storage and Analysis, pp. 1–12. OpenMPIR (2017)
14. Bouraoui, H., Castrillon, J., Jerad, C.: Comparing dataflow and OpenMP programming for speaker recognition applications. In: PARMA-DITAM 2019 - Proceedings: 10th Workshop on Parallel Programming and Run-Time Management Techniques for Many-Core Architectures - 8th Workshop on Design Tools and Architectures For Multicore Embedded Computing Platforms, pp. 1–6 (2019)
15. Scogland, T.R.W., Gyllenhaal, J., Keasler, J., Hornung, R., de Supinski, B.R.: Enabling region merging optimizations in OpenMP. In: Terboven, C., de Supinski, B., Reble, P., Chapman, B., Müller, M. (eds.) IWOMP 2015. LNCS, vol. 9342, pp. 177–188. Springer, Cham (2015). https://doi.org/10.1007/978-3-319-24595-9_13

Thread Private Variable Access Optimization Technique for Sunway High-Performance Multi-core Processors

Jinying Kong[1] ⓘ, Kai Nie[2] ⓘ, Qinglei Zhou[1] ⓘ, Jinlong Xu[2(✉)] ⓘ, and Lin Han[1]

[1] Zhengzhou University, Zhengzhou 450001, Henan, China
[2] Information Engineering University, Zhengzhou 450001, Henan, China

Abstract. The primary way to achieve thread-level parallelism on the Sunway high-performance multicore processor is to use the OpenMP programming technique. To address the problem of low parallelism efficiency caused by slow access to thread private variables in the compilation of Sunway OpenMP programs, this paper proposes a thread private variable access technique based on privileged instructions. The privileged instruction-based thread-private variable access technique centralizes the implementation of thread-private variables at the compiler level, eliminating the model switching overhead of invoking OS core processing and improving the speed of accessing thread-private variables. On the Sunway 1621 server platform, NPB3.3-OMP and SPEC OMP2012 achieved 6.2% and 6.8% running efficiency gains, respectively. The results show that the techniques proposed in this paper can provide technical support for giving full play to the advantages of Sunway's high-performance multi-core processors.

Keywords: Sunway high-performance multi-core processors · OpenMP programming technique · Privileged instruction-based thread-private variable access technique · Sunway 1621 processor

1 Introduction

Compiler design and development is a complex and difficult task, and one of the crucial aspects is compiler optimization. With the rapid development of multi-core processor design technology, the basic computing power of CPUs has been continuously improved, but the necessary way to fully exploit it to enhance the application performance of information systems is to generate efficient target code to match the multi-core processor chips through compiler optimization techniques. The main way to fully utilize the performance advantages of multi-core processors is to start multiple threads to execute programs in parallel on the multi-core processors. Currently, the main way to implement multi-threading on Sunway's high-performance multi-core processors with symmetric shared storage structure is the OpenMP programming technique. For this reason, the compilation and optimization of OpenMP programs is the core task of the Sunway Multi-Threaded Parallel Compilation System that accompanies the Sunway High Performance Multi-Core Processors. The major goal of OpenMP program compilation

J. Zeng et al. (Eds.): ICPCSEE 2021, CCIS 1451, pp. 180–189, 2021.
https://doi.org/10.1007/978-981-16-5940-9_14

optimization is to express parallelism in the program as efficiently as possible, including translation-time optimization, run-time optimization, and thread-level optimization [1–3]. At present, the focus of OpenMP program compilation optimization is mainly on thread-level optimization.

In order to ensure a thread-safe process, thread-private variables are widely used during thread execution, such as the "__thread" attribute variables in the base library, including "errno" in the error reporting mechanism, heap operations in virtual memory allocation control variables, character handling for different character sets, etc. Therefore, the way of accessing thread-private data has a significant impact on the performance of thread-level parallel programs, and is one of the core issues of compilation optimization that many researchers are currently concerned about. At present, the implementation of thread private variables in the Sunway multi-threaded parallel compilation system uses the thread-local storage technique, which requires the support of the operating system, the runtime library and the compiler. Therefore, there are serious portability problems with high requirements for the kernel of the operating system, the compiler and runtime library versions, and other support tools (e.g., binutilities). At the same time, access to thread private variables needs to be done using the interface provided by the OS kernel, which usually requires a runtime mode switch, is slow and has a large impact on the performance of the application. To address these flaws, this paper proposes a thread private variable access technique based on the privileged instructions of the Sunway processor in the Sunway multi-threaded parallel compilation system, which concentrates the implementation of thread private variables entirely at the compiler level, and the compiler uses the privileged instructions provided by the Sunway processor to read the start address information of thread private variables without entering the core of the operating system, making the number of thread private variable access cycles greatly reduced. The number of thread private variable access cycles is greatly reduced, and the portability problems caused by platform environment version upgrades are also avoided.

Implementing thread-level parallel execution of scientific computing programs on the Sunway high-performance multicore processor is the primary means to improve the efficiency of program operation. The thread-private variable access technique based on privileged instructions proposed in this paper can further compile OpenMP programs into more efficient thread-level parallelism code and provide technical support to fully exploit the advantages of Sunway high-performance multi-core processors.

2 Related Works

There are two major ways to handle thread-private variables in general: one is to assign thread-private variables on a dynamic stack in a similar way to variable declarations in functions; the other is to extend variables corresponding to multiple threads and access them using thread identifiers (IDs) [4–7]. The former requires providing the storage addresses of thread-private variables in the stack as parameters for data transfer between threads, and the programs become more complicated [8–11]. The latter requires the operating system to provide support for thread IDs, and even if the operating system provides the corresponding support, it is not applicable to the case where the number of threads changes dynamically, and the access efficiency is low [12–15]. The implementation of thread private variables in the Sunway multi-threaded parallel compilation system belongs to the latter, which not only has low access efficiency during operation, but also requires high versions of platform operating systems, runtime libraries and other tools, which often brings serious portability problems. Therefore, this paper investigates improving the efficiency of thread private variable access while avoiding the impact of changes in the platform environment.

3 The Privileged Instruction-Based Thread-Private Variable Access Technique

This section first introduces the current techniques for implementing thread private variables in the Sunway multi-threaded parallel compilation system, then points out the problems in its practical application, and finally gives the techniques for accessing thread private variables based on privileged instructions.

3.1 Technique Realization Mode of Thread Private Variables in Sunway and Its Problems

The implementation of thread private variables in the current OpenMP programming techniques for the Sunway platform is usually supported by a combination of thread local storage (TLS) techniques provided by the operating system, runtime libraries and compilers, as shown in Fig. 1.

Fig. 1. Implementation mechanism of threaded private variables in Sunway

In the thread-local storage mechanism, thread-private variables are implemented, managed and allocated by various parts of the entire software environment. Thread-private variables can be used in the same way as other types of variables in multiple compiler modules, and when special data sections (e.g. ".tls" sections) are generated during code generation, the compiler (including the linker Linker) organizes the thread-private variables into these special data sections and generates the corresponding special code to access these thread-private variables. The operating system provides special data structures that support threads to manage threads, including access to thread-local data, which is typically achieved by maintaining a thread pointer. The thread pointer directly accesses the thread-related data structures to get information about the thread's private variables. Runtime libraries, in particular the runtime loader, do the allocation of thread private variables when the code module is loaded (e.g., the main program is loaded and the runtime library is loaded). Based on the information provided inside the program, memory is allocated to these private variables, the internal data structures are modified, and the corresponding parts of the static program image are modified to access these memories. Also, when these modules are unloaded, the runtime library needs to free the allocated thread-local storage.

The implementation of TLS techniques for accessing thread-private variables requires the use of an interface provided by the operating system kernel, which usually implies a switch in runtime model, which results in a relatively high access overhead. At the same time, TLS techniques for storing and accessing thread private variables require the support of the entire software environment of the platform, which has high require-ments for the kernel, compiler (SWGCC), runtime libraries (glibc), and other support tools (binutilities) of the platform operating system, and not all versions can support the full functionality of TLS techniques well, which usually serious portability issues often arise.

3.2 Threaded Private Variable Access Techniques Based on Privileged Instructions

Due to the costly and cumbersome access to thread-private variables in traditional tech-niques, this paper concentrates the implementation of thread-private variables entirely at the compiler and base library levels. The starting address information of thread private variables is read directly in the compiler using the privileged instructions provided by the Sunway processor (e.g., rtid for reading thread IDs) to achieve direct access to thread private data without the need to enter the operating system processing, which reduces the number of access cycles of thread private variables significantly, as shown in Fig. 2.

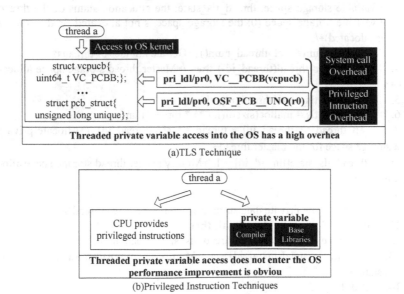

(a)TLS Technique

(b)Privileged Instruction Techniques

Fig. 2. Thread private variable access optimization

The implementation of thread-private variables based on privileged instructions at the base library level and in the compiler is almost identical, so this paper only describes the implementation in the compiler. To fully implement the storage and access of thread-private variables in the compiler, four main issues need to be solved: (1) storage allocation of thread-private variables; (2) access of thread-private variables; (3) data transfer between thread-private variables; and (4) release of thread-private variables.

(1) Storage allocation and access of private variables for privileged threads.
Because thread-private variables need to be shared between different parallel regions, thread-private variables cannot be allocated in stack space like local variables. In addition, sometimes the number of threads cannot be determined at compile time, so thread private variables cannot be allocated storage space like static variables. In this paper, we put thread private variables in a special field in the executable file, and when a new thread is started, we allocate a block of space from the heap of the program process large enough to copy the contents of the special field. In general, the master thread allocates an array to store the addresses of the private variables owned by the slave threads, and then each slave thread allocates the storage space needed for its own local private variables. This is done as in Algorithm 1.

Algorithm 1 Thread private variable storage space allocation

Input: thread id /*Thread ID*/
Output: p thread tls /*Pointer to store thread private data*/
1: **function** thread_locate_store_allocate(thread_id)
2:    **static int*** p_thread_tls, thread_tls_size, thread_tls_state /*p_thread_tls: pointer to the array of thread private variables, thread_tls_size: the size of the thread private variable storage space, thread_tls_state: the allocation status of the thread private variable storage space (0: the storage space is not allocated, 1: the storage space is allocated)*/
3:    **int** thread_num = get_thread_num() /*Get the number of threads*/
4:    **if** (is_master_thread(thread_id)) **then** /*Master thread storage space allocation*/
5:     **if** (!thread_tls_state[thread_id]) **then**
6:      p_thread_tls = malloc(sizeof(int *) * thread_num);
7:       p_thread_tls[thread_id] = thread_tls_size[thread_id]; /*Allocate private data storage space for the master thread*/
8:       thread_tls_state[thread_id] = 1 /*Modify master thread storage space allocation status*/
9:     **end if**
10: **else** /*Allocate private data storage space for the slave thread*/
11:   **if** (!thread_tls_state[thread_id]) **then**
12:    p_thread_tls[thread_id] = thread_tls_size[thread_id]
13:     thread_tls_state[thread_id] = 1 /*Modify slave thread storage space allocation status*/
14:   **end if**
15: **end if**
16: **return** p_thread_tls /*Returns a pointer to the thread's private data*/
17: **end function**

The basic idea of Algorithm 1 is that the compiler declares multiple adjoint variables for each thread-private variable to allocate the thread-private data storage space. Each thread private variable requires three global variables to maintain, an array pointer to the thread private data, a size corresponding to the thread private variable's allocated storage space, and a status flag to indicate whether the thread private variable storage space has been allocated. In the case of a master thread, the compiler first allocates an array to store the addresses of all thread-private variables, then allocates the required storage space for the thread-private variables and modifies the thread-private variable storage space allocation status flag. In case of sub-threads, the compiler first allocates the thread private variable storage space and then assigns the address to the corresponding element of the thread private variable storage array. Finally, a pointer to the array storing the thread private variable block is returned. Thus, the thread private variables can be accessed by adjoint variables.

(2) Data transfer and release between thread private variables of privileged level. Because the thread private variables are only visible within threads, special directives are needed to transfer data between thread private variables, and there are no corresponding shared variables for thread private variables, so the data transfer only takes place between thread private variables. The OpenMP standard provides "COPYPRI-VATE" and "COPYIN" clauses for copying data from one thread's private variables to other threads and from the master thread to the slave threads, respectively, both of which can be implemented through special functions. These can be implemented with special functions. For example, the "_COPYPRIVATE" clause can be translated by the function "__rtl_copy_private", and "_COPYIN" is also function, except that some of the parameter variables are replaced with ones that indicate whether the thread is the master thread or not.

4 Experimental Evaluation

The work in this paper is derived from an optimization project under the Sunway multi-threaded parallel compilation system, where the technique of threaded private variable access based on privileged instructions has been integrated in a prototype system.

4.1 Experimental Environment, Benchmarks and Test Content

The experimental environment is a high-performance server with Sunway 1621 shared memory architecture, equipped with Sunway multi-threaded parallel compilation system, CSC Kirin operating system, and Sunway computer storage management system, etc. Table 1 shows more detailed configuration information. The benchmarks are NPB3.3-OMP and SPEC OMP2012, which test the performance of the OpenMP parallel system of the high-performance server. The test scale is B and Ref respectively.

Table 1. Experimental environment.

Configuration items	Value
CPU	Sunway 1621 CPU (Core3A @ 2 GHz × 16)
L1 cache	32 GB
L2 cache	32 KB
L3 cache	32 MB
Memory	32 GB
Hard disk	1 TB
OS	Zhong Biao Neoshine
Compiler	swgcc, swg++ & swgfortran

The test is to evaluate the performance improvement of thread private variable access technique based on privileged instruction in multi-threaded parallel compilation subsystem. On the Sunway 1621 server computer, we compile and run the OpenMP programs in NPB3.3-OMP and SPEC OMP2012 using the original Sunway multi-threaded parallel compilation system and the Sunway multi-threaded parallel compilation system integrated with the technique of this paper, respectively, to check the compilation and running results, compare the running time, and calculate the runtime efficiency gains. The runtime efficiency gains is calculated as:

$$\eta = \frac{t_1 - t_2}{t_1} \times 100\% \tag{1}$$

where t_1 is the target code runtime generated by the original Sunway multi-threaded parallel compilation system, t_2 is the target code runtime generated by the Sunway multi-threaded parallel compilation system with integrated threaded private variable access based on privileged instructions technique, and η is the runtime efficiency gains.

4.2   Experimental Results and Analyses

The CPU of the Sunway 1621 computer integrates 16 Core3A processors, so in order to fully test the performance improvement effect of the threaded private variable access technique based on privileged instructions, the runtime of 16 threads is chosen in this paper to calculate the runtime efficiency gains. Table 2 and Table 3 show the test results of NPB3.3-OMP and SPEC OMP2012, respectively.

Table 2. NPB3.3-OMP result statistics

Benchmark	Compile and run[a]	Runtime (s)		Runtime efficiency gains (%)
		TLS	Privileged instructions	
BT	✓	58.56	53.82	8.1

(*continued*)

Table 2. (*continued*)

Benchmark	Compile and run[a]	Runtime (s)		Runtime efficiency gains (%)
		TLS	Privileged instructions	
CG	√	16.14	14.96	7.3
DC	√	235.06	218.14	7.2
EP	√	7.95	7.50	5.6
FT	√	9.51	8.43	11.4
IS	√	0.48	0.45	6.4
LU	√	47	42.91	8.7
MG	√	3.08	2.81	8.9
SP	√	67.98	63.77	6.2
UA	√	65.53	61.73	5.8
AVG	—	—	—	**6.2**

[a] √: Compile and run correctly. ×: Compile or run error.

The overall results in Tables 2 and 3 show that the performance improvement of the thread-private variable access technique based on privileged instructions is more obvious on SPEC OMP2012 than on NPB3.3-OMP. By comparing the source codes of the scientific computing programs in the two test sets, it is found that the programs in SPEC OMP2012 are more complex and use more thread-private variables in order to avoid some thread concurrency safety issues and improve thread flexibility, so the speed of thread-private variable access has a greater impact on SPEC OMP2012 performance. In addition, some of the programs are not large in size running time is relatively short (e.g., FT and MG, etc.), and the access time of thread private variables accounts for a higher proportion of the running time of the whole program, so a more obvious acceleration effect is achieved.

Table 3. SPEC OMP2012 result statistics

Benchmark	Compile and run	Runtime (s)		Runtime efficiency gains (%)
		TLS	Privileged instructions	
350.md	√	18013	16410	8.9
351.bwaves	√	10225	9898	3.2
352.nab	√	6337	5716	9.8
357.bt331	√	3655	3549	2.9
358.botsalgn	√	4269	4068	4.7

(*continued*)

Table 3. (*continued*)

Benchmark	Compile and run	Runtime (s)		Runtime efficiency gains (%)
		TLS	Privileged instructions	
359.botsspar	√	13056	12116	7.2
360.ilbdc	√	37045	35600	3.9
362.fma3d	√	4131	3829	7.3
363.swim	√	3030	2794	7.8
367.imagick	√	8996	8393	6.7
370.mgrid331	√	4624	4175	9.7
371.applu331	√	4196	3839	8.5
372.smithwa	√	3908	3603	7.8
376.kdtree		3223	2991	7.2
AVG				**6.8**
	—	—	—	

5 Conclusion

In order to improve the execution efficiency of multi-threaded programs on the Sunway processor, this paper proposes a thread-private variable access technique based on privileged instructions. This technique concentrates the access to thread private variables entirely at the compiler level, and the programmer directly reads the start address information of thread private variables using the privileged instructions provided by the Sunway processor without the need of OS kernel processing, which makes the number of thread private variable access cycles greatly reduced. The privileged instruction-based thread private variable access technique is integrated in the Sunway multi-threaded parallel compilation system. To test the performance improvement effect of this technique, this paper selects the NPB3.3-OMP and SPEC OMP2012 test sets to conduct experiments on the Sunway 1621 server computer. The experimental results show that the threaded private variable access technique based on privileged instructions does not affect the correctness of the Sunway multi-threaded parallel compilation system and has significant performance improvement. So the method in this paper is feasible and effective, and can provide technical support to fully exploit the advantages of Sunway's high-performance multi-core processors.

References

1. Tiotto, E., Mahjour, B., Tsang, W.: OpenMP 4.5 compiler optimization for GPU offloading. IBM J. Res. Dev. **3**(5), 1–11 (2020)
2. Neth, B., Scogland, T.R.W., Strout, M.M., de Supinski, B.R.: Unified Sequential optimization directives in OpenMP. In: Milfeld, K., de Supinski, B., Koesterke, L., Klinkenberg, J. (eds.) IWOMP 2020. LNCS, vol. 12295, pp. 85–97. Springer, Cham (2020). https://doi.org/10.1007/978-3-030-58144-2_6

3. Mosseri, I., Alon, L.O., Harel, R., Oren, G.: *ComPar*: optimized multi-compiler for automatic OpenMP S2S parallelization. In: Milfeld, K., de Supinski, B., Koesterke, L., Klinkenberg, J. (eds.) IWOMP 2020. LNCS, vol. 12295, pp. 247–262. Springer, Cham (2020). https://doi.org/10.1007/978-3-030-58144-2_16
4. Schreter, I.: Systems and methods for accessing thread private data (2008)
5. Wei, P.F., Brylinski, M.: Accelerated structural bioinformatics for drug discovery. In: High Performance Parallelism Pearls: Multicore and Many-Core Programming Approaches, pp. 55–72 (2015)
6. Lin, Y., Chakrabarti, G., Marathe, J., Kwon, O., Sabne, A.: System and method for translating program functions for correct handling of local-scope variables and computing system incorporating the same (2008)
7. Marathe, V.J., Byan, S., Seltzer, M.I., Mishra, A., Trivedi, A.: Efficient memory management for persistent memory (2019)
8. Bratanov, S.V.: Method of concurrent instruction execution and parallel work balancing in heterogeneous computer systems, US (2019)
9. Greenwood, S.R., Peterson, K.R., Schreiber, B.L.: Thread private memory storage for multithread digital data processors (1991)
10. Chen, F., Ganglin, Y., Shen, S., Ye, X., Yang, F., Wang, K.: Parallelization and optimization of RMC for criticality computing based on the heterogeneous architecture of the Sunway Taihu Light supercomputer. Ann. Nucl. Energy **11**(145), 1–12 (2020)
11. Shirakihara, T.: Method and apparatus for managing thread private data in a parallel processing computer, US(1996)
12. Gerofi, B., Takagi, M., Ishikawa, Y.: Toward operating system support for scalable multithreaded message passing. In: Proceedings of the 22nd European MPI Users' Group Meeting, pp. 21–23 (2015)
13. Hori, A., Takagi, M., Si, M., Dayal, J., Ishikawa, Y., Gerofi, B., Balaji, P.: Process-in-process: techniques for practical address-space sharing. In: HPDC 2018 - Proceedings of the 2018 International Symposium on High-Performance Parallel and Distributed Computing, pp. 131–143 (2018)
14. Coon, B.W., Lindholm, J.E.: System and method for grouping execution threads, US (2007)
15. Kadir, A., Cevdet, A.: Exploiting locality in sparse matrix-matrix multiplication on manycore architectures. IEEE Trans. Parallel Distrib. Syst. **28**(8), 2258–2271 (2017)

Basic Theory and Techniques for Data Science

Anti-obfuscation Binary Code Clone Detection Based on Software Gene

Ke Tang, Fudong Liu$^{(\boxtimes)}$, Zheng Shan, and Chunyan Zhang

State Key Laboratory of Mathematical Engineering and Advanced Computing,
Zhengzhou 450001, China

Abstract. Information technology facilitates people's lives greatly, while it also brings many security issues, such as code plagiarism, software in-fringement, and malicious code. In order to solve the problems, reverse engineering is applied to analyze abundant binary code manually, which costs a lot of time. However, due to the maturity of different obfuscation techniques, the disassembly code generated from the same function differs greatly in the opcode and control flow graph through different obfuscation options. This paper propose a method inspired by natural language processing, to realize the semantic similarity matching of binary code in basic block granularity and function granularity. In the similarity matching task of binary code obtained by different obfuscation options of LLVM, the indicator reaches 99%, which is better than the existing technologies.

Keywords: O-LLVM · Anti-obfuscation · Code cloning detection · Machine learning · Software gene

1 Introduction

In recent years, the booming software brings convenience but also many great challenges, such as code plagiarism [5], malicious code raging and other threats to software copyright and software security. However, due to the maturity of different obfuscation techniques, the disassembly code generated from the same function differs greatly in the opcode and control flow graph through different obfuscation options. Although obfuscation technology can protect software copyright to a large extent, it also makes it difficult to detect code plagiarism and malicious code variants [6]. Previous research in resisting obfuscation technology [5] focus on block semantic equivalence, and can not handle the modern obfuscation techniques well. Therefore, it is of great significance to establish an anti-aliasing method of binary code similarity detection [2]. Inspired by natural language processing, we propose a novel scheme for detecting binary code similarity with resisting modern obfuscation techniques. Data for this study were

Supported by the Foundation of National Natural Science Foundation of China (No. 61802435).

J. Zeng et al. (Eds.): ICPCSEE 2021, CCIS 1451, pp. 193–208, 2021.
https://doi.org/10.1007/978-981-16-5940-9_15

collected by compiling open source code with different obfuscation options of LLVM. After data preprocessing, it is used as the input of the model. After training, the corresponding high-dimensional vector is obtained and used to evaluate the similarity of the binary code.

The contributions of this paper are as follows:

1. Different from the reverse engineering of the binary file to the assembler, this article analyzes the binary code from a forward perspective, that is, compiling from the source code to the assembler. Both have the same effect, but the latter can significantly reduce the workload;
2. The random walk algorithm [7] is used to convert the program control flow graph (CFG) into a sequence, which can convert the control flow graph into a sequential assembly code sequence, cleverly bypassing the graph matching algorithm, and effectively reducing the computational complexity, Improve efficiency;
3. Use the natural language processing (NLP) method to process the assembly code to obtain the vector containing the semantic information of the assembler, so that it can effectively detect the similarity of the binary code while resisting the confusion options or compiler optimization options.

The remainder of the paper proceeds as follows: Sect. 2 reports related researches in the field of code cloning detection at home and abroad. Section 3 presents the main details of the similarity comparison technique proposed in this article. Section 4 shows the experimental results. Section 5 analyzes the deep-seated reasons for the experimental results. The last section summarizes the full paper.

2   Related Work

In recent years, code cloning detection technology has become a hot spot for researcher scholars at home and abroad. It has high demand in the field of computer security and software protection, such as code plagiarism detection [3,4], patch comparison analysis [19], code search [20], program understanding [21], malicious family classification [22], vulnerability detection technology [23–25], etc. With rapid development in recent years, code cloning detection algorithms have gradually increased, but the detection granularity for binary code and the detection effect are both also significantly different. In exploring the development process of code cloning detection technology, we found that the main detection algorithms can be divided into the following five categories:

1. Text-based code clone detection technology.
   The text-based code clone detection method is mainly to regard the binary code fragments as ordinary text, and compare the similar distance between the texts (such as Longest Common Subsequence, LCS) [1] as the basis for the similarity of the binaries, which ignores the grammatical structure characteristics of the programming language, and the effect is not good in the experiment.

2. Code clone detection based on program dependency graph isomorphism [9]. Many researchers construct a Program Dependence Graph (PDG) [8] for binary programs based on the structural properties of programs and the control dependencies between codes, then use this as a basis for similarity between programs. This text-based code cloning detection method has a high accuracy rate for the completely copied code, but due to the high flexibility of the programming language, for example, in the C language, various keywords such as for, while, do-while, etc. are all expressed cycle programs, so they can replace each other, which will cause the poor accuracy.

3. Token-based code clone detection technology [10–13]. The token-based code cloning detection technology first extracts the token sequence of the program, and uses the token sequence as the basis for comparing the similarity of functions. This method is more flexible than the text-based method and has a wider application range. However, this scheme also does not consider the structure of the program in the source code and ignores the execution sequence of the code in the program, so the efficiency is not satisfactory.

4. Code clone detection technology based on AST [15–17]. In this detection technology, an Abstract Syntax Tree (AST) corresponding to the source code is generated. The syntax tree contains the logical structure information and semantic information of the program, and the code similarity measurement is converted into the similarity calculation of the syntax tree, which is more accurate. The advantage of this method is that the AST contains the structural features of the program, but the computational complexity is relatively high due to the need to traverse the tree. In addition, programs with similar functions may have very different structures, so this solution is difficult to resist current code obfuscation techniques.

5. Code clone detection technology based on program measurement [18, 26].

The method of program measurement is to extract the typical characteristics of the source code, convert the program into a certain set of attributes, and then use the similarity between the attribute sets as the basis of program similarity to detect the cloned code. Commonly used calculation methods for comparing the similarity between attribute sets usually use Euclidean distance, cosine similarity, etc. This detection method relies on the selection of the attribute set, and it is difficult to ensure efficiency.

Through the above research analysis, the text-based detection method can effectively detect the situation where the code is highly coincident, and there is no restriction on the language of the program, but it is slightly insufficient for the language with flexible semantics. Although the token-based detection method is more in-depth than the text method, it still does not fully contain the structure information of the program. The detection method based on the program dependency graph, for a software with tens of thousands of lines of code, the scale of the graph data generated by the algorithm is very huge, and the number of clone pairs can reach tens of millions, which makes the calculation cost too high and requires a lot of computing resources. Although the method based on the AST combines program structure and semantic information, and can effectively detect cloning methods such as code insertion and code deletion, it also needs to traverse the tree structure, so the execution efficiency is not high. The metric-based method extracts different feature metric values to simplify it as much as possible and effectively avoids complex calculations, it still has the problem of insufficient detail in the detection granularity and low accuracy.

3 Method

3.1 Overview

The whole experiment first uses the common source code on GitHub (such as OpenSSL), and applies various compile obfuscation options to compile the source code by O-LLVM compiler, and obtain the assembler as our data set. Then extract the program control flow graph (CFG) from the assembler, and then apply the concept of software gene to divide it into gene blocks so that we can extract the code sequence later. After dividing each node in CFG into independent gene blocks for instruction normalization, random walk algorithm is used to traverse the nodes in the graph to obtain the gene sequence as our training set. After that, machine learning algorithms are used to train our training set, mainly using natural language processing (word2vec) to embed the assembly instructions, and then use doc2vec to semantically embed the assembly sequence of the basic block of the function to extract the semantic information of the function. Finally, the trained model can achieve good performance in anti-aliasing code clone detection. Finally, the accuracy of the trained model is evaluated. The general process is shown in Fig. 1.

3.2 Basic Block Semantic Embedding

The similarity comparison of basic blocks is the basis of binary similarity analysis. Basic blocks are the basic components of binary functions. By comparing basic blocks, the similarity between two functions can be directly compared. It is widely used in code clone detection field. This section mainly solves the problem of semantic similarity comparison of basic blocks. Natural language processing is used to extract the semantic information of basic blocks, automatically encode basic blocks into vectors containing semantic information, and express the basic blocks in a mathematical way. This section mainly includes the extraction of basic blocks, which are divided into gene blocks, and then semantic embedding. The experimental results prove that the method is scientific and reasonable, and can compare the similarity between basic blocks well.

Preprocessing. Obfuscator-LLVM [28] is a multi-platform LLVM compilation suite that can ensure the security of the software through code obfuscation. It mainly contains three obfuscation methods: instruction replacement, fake control flow, and control flow flattening.

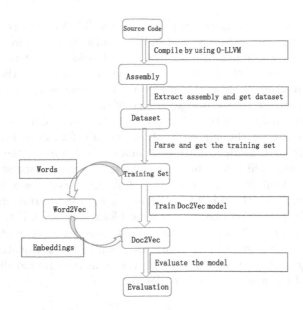

Fig. 1. The general process in this research.

We use the three main obfuscation techniques mentioned above to compile the source program. Some free and widely used open source projects have been selected as our data sources, such as OpenSSL, LibTomCrypt, and Libgmp. The detailed information is shown in Table 1. Then we can compile the source code to generate assembly files by O-LLVM.

Table 1. The description of dataset in the experiment.

Source	Description	Files	Functions
OpenSSL	Open secure socket layer protocol	1256	16070
LibTomCrypt	Common cryptographic algorithm library	537	1157
Libgmp	Open source mathematical operation library	970	4373

Then we can traverse all the assembler files and parse the contents of the assembler files. The parsing process is as follows: First, we create an ordered dictionary object to store the contents of all assembler files, the key is the file name of each assembler file, and the value is a new ordered dictionary object. In this new object, the key is the function name in current assembler file, and the value is still an ordered dictionary object. In this object, the key is the label in current function, that is, the identification of a basic block, the value is a list, which contains the assembly instructions in current basic block. After the analysis of all assembler files is completed, all assembler codes are saved in this structure. Figure 2 provides the division process of basic blocks.

Gene Block Segmentation. Software genes [31–33] are code fragments divided on the basis of the functions of the assembler. Learning from the concept of software genes, we divides the original basic blocks into "gene blocks". The control flow in each gene block is executed sequentially, and only the last instruction is a jump instruction or a ret instruction. The gene blocks are connected to each other according to the logical structure of the program.

Through the analysis of the assembly files, the main data is stored in the structure of an ordered dictionary. Next, we mainly focus on each basic block in the structure and divide it into gene blocks, and remove the empty basic blocks by the way, which will be converted to the gene data structure. The specific process of subdivision is roughly as follows: traverse each instruction in the basic blocks, and if the current instruction is a jump instruction, the current gene block ends, otherwise, the current gene block continues. If there are other instructions behind the basic block, then a new gene block is created, and then repeat the above process. The schematic diagram of the segmented gene block is shown in Fig. 3. After traversing all the values in the ordered dictionary, the segmentation of the gene block is completed.

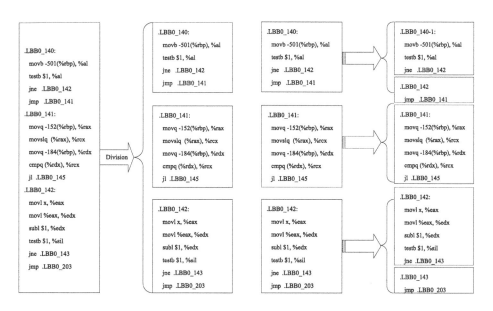

Fig. 2. Basic blocks division process.   **Fig. 3.** Gene segmentation process.

Instruction Normalization. After getting the gene data structure, we need to process the data more deeply. An instruction contains one opcode and zero, one or more operands. In the instruction, the number of the opcode is limited, but the operand is complex and diverse, such as the immediate number contains various integers, and the register contains various registers such as %eax, %ebx,

%edx, etc., the memory address also contains expressions of various addressing modes, etc. In order to eliminate the side-effect caused by the operands to the model during the training process, we need to normalize each instruction to ensure the quality of model for reducing interference with necessary information. The normalization rules are mainly as follows: Registers such as %eax, %ebx, %edx, etc. are all replaced by "REG", immediate data are all replaced by "IMM", and all accessed memory addresses are replaced by "ADDRESS", in the call instruction Use "FUNC" to replace the function name, "VAR" to replace the variable name, and "label" to replace the label in the assembly. The specific standardization process is shown in Fig. 4. After replacement, we save the extracted data to files as data set. Each function is saved as two files. One file (*.edge) saves all the nodes in the function and the connection relationship between the nodes, and the another file (*.node) saves the assembly instruction sequence of the gene block corresponding to each node.

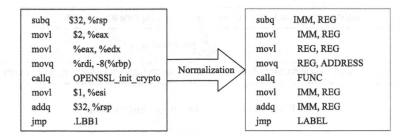

Fig. 4. The process of instruction normalization.

Basic Block Vectorization. In this paper, we learn from the natural language processing method and use the Word2Vec [29] model for processing. Word2Vec [20] is a set of machine learning models that can generate word vectors. The model is a shallow two-layer neural network used to train word to learn semantic information of words. The Word2Vec model can map any word to a specified fixed-length high-dimensional feature vector. It has two architectures: bag-of-words (CBOW) and skip-gram. In the CBOW architecture, the model predicts the current word from the surrounding window of upper and lower words, which does not consider the order between words. In the skip-gram architecture, the model uses a fixed-size window and predicts the current word based on the context in the window. Both architectures can represent the input words as fixed-length feature vectors, but the CBOW model has two obvious weaknesses: it will lose the order between words in the sentence and ignore the semantic information contained in sentences.

We regard assembly instructions as words, gene blocks, which contains a sequence of assembly instructions, as sentences, and gene sequences, which contains multiple gene blocks, as paragraphs. The order of each assembly instruction sequence is particularly important, because it contains the functions implemented by the code. If the CBOW model is used, the important information will be ignored, which is unbearable to us, so we use skip-gram architecture to train word vectors. Finally, we need to make the obtained word vector have such characteristics: the Euclidean distances of the word vectors obtained by similar words are also similar. In this way, we can preserve the semantic information of the word as much as possible in this process, so that the final vector can contain the function information of the function, which is used as the basis for comparing the similarity between functions. In the training stage, we have two ways, one is to divide the assembly instructions into single words further (as shown in Fig. 5), as the input of the model, another is to take a complete assembly instruction (as shown in Fig. 6) as a word for model training. And Next, we use both two methods to obtain the word vector, and compare its advantages and disadvantages according to the experimental results.

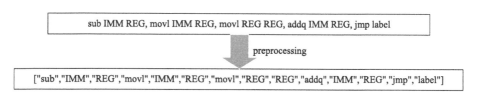

Fig. 5. Each word in the assembly instruction.

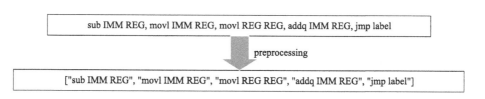

Fig. 6. A complete assembly instruction.

3.3   Semantic Embedding of Function Control Flow Graph

This section aims to solve the similarity comparison algorithm of assembly functions, which is composed of several gene blocks. Then we utilize random walk to obtain ordered instruction sequences as training set, and then use Doc2Vec for training to extract the semantics of assembly instruction sequences. After obtain the mathematical vector containing the semantic information of the function, we can calculate the cosine similarity to realize the similarity comparison between the assembly functions. At last, we use the p@n test to analyze the experimental results. Among them, in the task of comparing the similarity between the functions with the "sub" obfuscated option and the functions without the obfuscated options, the p@3 index reached 97%, which proves that our method can effectively extract the function semantic information, which can effectively perform code cloning detection on the basis of resisting confusion.

Control Flow Graph Serialization. In the CACompare [27] algorithm, authors used random input sequence to analyze the I/O behavior of assembler. Random input can perform random walk [13,14] algorithm on the effective execution flow to simulate program execution order. Inspired by this, we can use the random walk algorithm on the basis of the function control flow graph. We select any node in the graph, start from this node, and then follow the direction of the program control flow, randomly choose a node directly connected to the given node as the next node, this process is repeated until a specific condition is reached: such as a fixed sequence length or the end of the program.

In order to limit the length of the sequence obtained by random walk, we adopt truncated random walk algorithm. In this research, we set the longest length of the random sequence is 10. If the length of a certain execution path in the function exceeds 10, it woulf be truncated. In this way, the control flow graph can be converted into a series of assembly sequences and as the training data.

Semantic Extraction of Assembly Sequence. Doc2vec [30] is an unsupervised machine learning model that can map variable length text (such as a sentence, or a paragraph, or even an article) into a fixed-length feature vector. It can obtain the vector representing from the training by predicting words in the article, and a large number of studies have shown that the paragraph vector generated in this way can make up for the shortcomings of other representation techniques such as the bag of words model. In this model, although the paragraph vector is randomly initialized, it still can represent the semantic information contained in the text to some extent after the Doc2vec model is trained.

Inspired by this, we apply the Doc2Vec model to the semantic extraction of assembly sequences. In this experiment, we directly use the Doc2Vec model in gensim, because the Word2Vec model is implicitly called for word embedding inside the model, so we don't need to train the Word2Vec model to obtain word vectors.

4 Experiments

4.1 Explore Instruction Segmentation Methods

The two instruction segmentation methods mentioned above are used to train instructions to obtain their corresponding feature vectors. In the training process, we introduced t-SNE [34] to observe the training results more clearly and

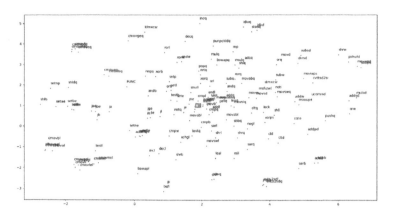

Fig. 7. Distribution of each word's vector.

compare the pros and cons between them. t-SNE is a relatively common high-dimensional data visualization tool. It can convert high-dimensional data into two-dimensional or three-dimensional data through model training, and then use the matplotlib package to visualize the converted data, so that we can intuitively see the similarity between the word embedding vectors obtained from the figure. As shown in Fig. 7 and Fig. 8, the effects of the vectors obtained by the two segmentation methods are respectively. As Fig. 7 puts that the distance between the points corresponding to each word is relatively uniform, no matter the meaning of the words is the same (such as "jle", "jl") or very different (such as "popq"). The difference in distance between words is not very obvious, so this representation method does not reflect the semantic information of words well. From Fig. 8, we can clearly find that instructions are no longer evenly distributed, but several instructions are clustered together and separated from other instructions. For partial enlargement of the figure, we can see that the distance between instructions with similar semantic informantic mapped to points in the two-dimensional space is also adjacent. For example, the instruction "cmovlel REG REG" and the instruction "cmovbel REG REG" have similar meanings in semantics. The distance between the word vectors obtained by the Word2Vcc model training is also relatively close, so we can claim that using a complete instruction as a word can express the semantic information corresponding to the instruction to a certain extent.

From the results of the two exploration experiments above, we can see that the results obtained by using the two instruction segmentation methods for word embedding are very different, and the word vector distribution obtained by the method of dividing each instruction for word embedding is relatively uniform. Which means that the word vector does not contain the semantic information of the assembly instruction well, and the feature vector obtained by using a complete instruction as a word for word embedding is more concentrated in the two-dimensional distribution of instructions with similar functions. This shows

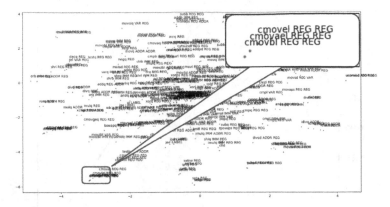

Fig. 8. Distribution of each instruction's vector.

that the word vector better contains the semantic information of the program instructions. It also demonstrates that there is a certain degree of scientificity and rationality by using NLP method to process the instruction sequences.

4.2 Model Training

In the model training process, in order to study the impact of the dimension of the word vector on the experimental results, we set the word vector dimension to 25, 50, 100, 150, 200 for experimental tests respectively, and evaluate results based on similarity indicators. At the same time, we randomly select a small part of the training data as experimental data when train the Doc2Vec model fot saving time and reducing experimental costs. After training, we use the trained model to test. The test results are shown in Fig. 9. The horizontal axis of the graph represents the vector dimension, and the vertical axis represents the similarity between the function vectors.

From the experimental results of the above line chart, it displays that the similarity between the function vectors changes as the vector dimension changes. For a function, there is an optimal dimension value, which makes the confused and unconfused The function have the highest similarity. In this experiment, it can be seen that when the dimension of the word vector is 150, the similarity between the vectors obtained by the similarity function is highest, so we choose 150 as the dimension of the word vector to train the model.

By using the previous case to select the best training parameters to train the Doc2Vec model, we get a vector representing the semantics for each node sequence. But in a function, there will be multiple node sequences. When comparing the similarity between two functions, in fact, it is comparing the similarity between multiple vectors. In this situation, we can have two options: one is to sum these vectors and take the average mathematically, and the another is to directly concatenate these vectors. These two algorithms have their own advantages and disadvantages, and we have to compare the scientific nature of these

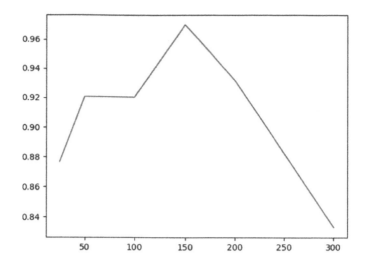

Fig. 9. The similarity index with different dimensions.

two processing methods through experimental research. After repeated experiments, it is found that concatenate vectors of the function directly can achieve higher accuracy. Therefore, in the following tests, we will directly use this method to calculate the similarity of functions.

Then we set out to train the model on our data set. We randomly divide the data set into two parts, with 80% as the training set and the rest as the test set.

4.3   p@n Test

The p@n test is defined as follows: Given a pair of similar functions A and B, then we randomly select 99 additional functions that are different from each other, together with B, a total of 100 functions, calculate the similarity with A, and sort them according to the similarity, and repeat the procedure for many times to get the probability that B ranks in the top n. In our experiment, the similar function we chose is the assembly function obtained by compiling the same function through different obfuscation options of the O-LLVM compiler. For example, a function is an assembly function obtained through any one of the three obfuscation options, and the remaining 100 functions to be compared are assembly functions obtained without any obfuscation options.

Test 1: In this test, the source code of LibTomCrypt is used. The functions in the test set are all compiled with the same obfuscation options (sub, fla, bcf), and the number of functions is 100. In the test process, we calculate the similarity between obfuscated functions and non-obfuscated functions, and sort all functions by similarity. Repeat the experiment many times, we will be able to gain p@1, p@3 and p@10 respectively. The results are shown in Table 2.

Table 2. Test result with LibTomCrypt.

Options	p@1	p@3	p@10
sub	0.74	0.97	0.99
fla	0.24	0.34	0.71
bcf	0.42	0.59	0.79
Average	0.47	0.63	0.83

Table 3. Test result with LibGmp.

Options	p@1	p@3	p@10
sub	0.79	0.98	1.00
fla	0.36	0.47	0.56
bcf	0.50	0.71	0.84
Average	0.55	0.72	0.80

Test 2: In this test, we use the source code of LibGmp instead. The test strategy is exactly as same as that of test 1. The results are shown in Table 3.

It can be seen from the experimental results that the Doc2Vec model works best for the "sub" obfuscation option of the O-LLVM compiler, the p@10 test is very close to 1. However, we can also find that the detection effect of the "fla" obfuscation option is poor relatively. So in future research, we can consider improving the internal algorithm structure of the Doc2Vec model, or proposing a new model architecture based on the model, making it resistant the "fla" confusion option and improving the similarity.

5 Discussion

According to the results mentioned above, it indicates that the similarities between different obfuscation options with the same function are quite different. The underlying reasons may be as follows:

For the "sub" obfuscation option, the goal of this obfuscation technique is only to replace simple instruction with a functionally equivalent but more complex instruction sequence. It only performs instruction replacement, remaining the general workflow of the entire function unchanged, so the effect on the similarity comparison of the function is limited, and the similarity will naturally be higher.

For the "bcf" obfuscation option, this obfuscation technique modifies the call relationship diagram of the program by adding a new basic block, and then jump to the original basic block by conditional jump instruction. Therefore, although this obfuscation technique changes the instruction flow of the function to a certain extent, the changed control flow graph is only in a partial range and does not change the overall control flow graph. Therefore, the detection effect of the function is only slightly affected by using this obfuscation technique.

For the "fla" obfuscation option, this obfuscation technique can completely flatten the control flow chart of a function and completely change the control flow chart of the program. Although some basic blocks are the same after the change, the control flow graph of the program is very different from before, which may be a major reason for low detection similarity.

6   Conclusions

The aim of the present research was to assess the similarity of binary codes against obfuscation options by using NLP techniques. The results of this investigation prove that our method can deal with modern obfuscation techniques well. From the result, we can see that the Doc2Vec model works best for the "sub" obfuscation option of the O-LLVM compiler, and the p@10 index is almost close to 1, which shows that our model has good performance against "sub" obfuscation options. However, the detection effect of the "fla" confusion option is poor.

In future research, we can consider improving the internal algorithm structure of the Doc2Vec model, or propose a new architecture based on the model to make it more resistant to "fla" confusion option to improve performance. Through this research, we find that the paragraph vector has a certain applicability in detecting the similarity of binary codes. We can guess that focusing on the generation process of the paragraph vector seems to have a bright future.

References

1. Allison, L., Dix, T.I.: A bit-string longest-common-subsequence algorithm. Inf. Process. Lett. **23**(5), 305–310 (1986). https://doi.org/10.1016/0020-0190(86)90091-8
2. Hu, Y., Zhang, Y., Li, J., Wang, H., Li, B., Gu, D.: BinMatch: a semantics-based hybrid approach on binary code clone analysis. arXiv:1808.06216 [cs], August 2018. http://arxiv.org/abs/1808.06216. Accessed 28 Mar 2021
3. Jhi, Y.-C., Wang, X., Jia, X., Zhu, S., Liu, P., Wu, D.: Value-based program characterization and its application to software plagiarism detection. In: Proceeding of the 33rd International Conference on Software Engineering - ICSE 2011, Waikiki, Honolulu, HI, USA, p. 756 (2011). https://doi.org/10.1145/1985793.1985899
4. Zhang, F., Jhi, Y.-C., Wu, D., Liu, P., Zhu, S.: A first step towards algorithm plagiarism detection. In: Proceedings of the 2012 International Symposium on Software Testing and Analysis - ISSTA 2012, Minneapolis, MN, USA, p. 111 (2012). https://doi.org/10.1145/2338965.2336767
5. Luo, L., Ming, J., Wu, D., Liu, P., Zhu, S.: Semantics-based obfuscation-resilient binary code similarity comparison with applications to software plagiarism detection. In: Proceedings of the 22nd ACM SIGSOFT International Symposium on Foundations of Software Engineering - FSE 2014, Hong Kong, China, pp. 389 400 (2014). https://doi.org/10.1145/2635868.2635900
6. Lindorfer, M., Di Federico, A., Maggi, F., Comparetti, P.M., Zanero, S.: Lines of malicious code: insights into the malicious software industry. In: Proceedings of the 28th Annual Computer Security Applications Conference on - ACSAC 2012, Orlando, Florida, p. 349 (2012). https://doi.org/10.1145/2420950.2421001
7. Metzler, R., Klafter, J.: The random walk's guide to anomalous diffusion: a fractional dynamics approach. Phys. Rep. **339**(1), 1–77 (2000). https://doi.org/10.1016/S0370-1573(00)00070-3
8. Ferrante, J.: The program dependence graph and its use in optimization. ACM Trans. Program. Lang. Syst. **9**(3), 31 (1987)
9. Simko, T.J.: Cloneless: Code Clone Detection via Program Dependence Graphs with Relaxed Constraints. California Polytechnic State University, San Luis Obispo, California (2019)

10. Kamiya, T., Kusumoto, S., Inoue, K.: CCFinder: a multilinguistic token-based code clone detection system for large scale source code. IIEEE Trans. Software Eng. **28**(7), 654–670 (2002). https://doi.org/10.1109/TSE.2002.1019480
11. Wu, Y., et al.: SCDetector: software functional clone detection based on semantic tokens analysis. In: 2020 35th IEEE/ACM International Conference on Automated Software Engineering (ASE), pp. 821–833, September 2020
12. Zou, Y., Ban, B., Xue, Y., Xu, Y.: CCGraph: a PDG-based code clone detector with approximate graph matching. In: Proceedings of the 35th IEEE/ACM International Conference on Automated Software Engineering, Virtual Event Australia, pp. 931–942, December 2020. https://doi.org/10.1145/3324884.3416541
13. Krinke, J.: Identifying similar code with program dependence graphs, pp. 301–309, February 2001. https://doi.org/10.1109/WCRE.2001.957835
14. Fang, C., Liu, Z., Shi, Y., Huang, J., Shi, Q.: Functional code clone detection with syntax and semantics fusion learning. In: Proceedings of the 29th ACM SIGSOFT International Symposium on Software Testing and Analysis, Virtual Event USA, pp. 516–527, July 2020. https://doi.org/10.1145/3395363.3397362
15. Baxter, I.D., Yahin, A., Moura, L., Sant'Anna, M., Bier, L.: Clone detection using abstract syntax trees. In: Proceedings. International Conference on Software Maintenance (Cat. No. 98CB36272), Bethesda, MD, USA, pp. 368–377 (1998). https://doi.org/10.1109/ICSM.1998.738528
16. Lazar, F., Banias, O.: Clone detection algorithm based on the abstract syntax tree approach. In: 2014 IEEE 9th IEEE International Symposium on Applied Computational Intelligence and Informatics (SACI), pp. 73–78, May 2014. https://doi.org/10.1109/SACI.2014.6840038
17. Buch, L., Andrzejak, A.: Learning-based recursive aggregation of abstract syntax trees for code clone detection. In: 2019 IEEE 26th International Conference on Software Analysis, Evolution and Reengineering (SANER), Hangzhou, China, pp. 95–104, February 2019. https://doi.org/10.1109/SANER.2019.8668039
18. Xue, H., Venkataramani, G., Lan, T.: Clone-slicer: detecting domain specific binary code clones through program slicing. In: Proceedings of the 2018 Workshop on Forming an Ecosystem Around Software Transformation - FEAST 2018, Toronto, Canada, pp. 27–33 (2018). https://doi.org/10.1145/3273045.3273047
19. Brumley, D., Poosankam, P., Song, D., Zheng, J.: Automatic patch-based exploit generation is possible: techniques and implications. In: 2008 IEEE Symposium on Security and Privacy (SP 2008), Oakland, CA, USA, May 2008, pp. 143–157 (2008). https://doi.org/10.1109/SP.2008.17
20. Mahinthan, C., Xue, Y., Xu, Z., Liu, Y., Cho, C., Tan, H.B.K.: BinGo: cross-architecture cross-OS binary search, pp. 678–689, November 2016. https://doi.org/10.1145/2950290.2950350
21. Hu, Y., Zhang, Y., Li, J., Gu, D.: Cross-architecture binary semantics understanding via similar code comparison. In: 2016 IEEE 23rd International Conference on Software Analysis, Evolution, and Reengineering (SANER), Suita, pp. 57–67, March 2016. https://doi.org/10.1109/SANER.2016.50
22. Ming, J., Xu, D., Jiang, Y., Wu, D.: BinSim: trace-based semantic binary diffing via system call sliced segment equivalence checking. In: Proceedings of the 26th USENIX Conference on Security Symposium, USA, August 2017, pp. 253–270. Accessed 28 Mar 2021
23. Pewny, J., Garmany, B., Gawlik, R., Rossow, C., Holz, T.: Cross-architecture bug search in binary executables. In: 2015 IEEE Symposium on Security and Privacy, pp. 709–724, May 2015. https://doi.org/10.1109/SP.2015.49

24. Eschweiler, S., Yakdan, K., Gerhards-Padilla, E.: discovRE: efficient cross-architecture identification of bugs in binary code, February 2016. https://doi.org/10.14722/ndss.2016.23185

25. Feng, Q., Zhou, R., Xu, C., Cheng, Y., Testa, B., Yin, H.: Scalable graph-based bug search for firmware images. In: Proceedings of the 2016 ACM SIGSAC Conference on Computer and Communications Security, Vienna Austria, pp. 480–491, October 2016. https://doi.org/10.1145/2976749.2978370

26. Toomey, D.: Code Similarity Comparison of Multiple Source Trees, May 2008

27. Hu, Y., Zhang, Y., Li, J., Gu, D.: Binary code clone detection across architectures and compiling configurations. In: 2017 IEEE/ACM 25th International Conference on Program Comprehension (ICPC), Buenos Aires, Argentina, May 2017, pp. 88–98 (2017). https://doi.org/10.1109/ICPC.2017.22

28. Junod, P., Rinaldini, J., Wehrli, J., Michielin, J.: Obfuscator-LLVM - software protection for the masses, May 2015, pp. 3–9 (2015). https://doi.org/10.1109/SPRO.2015.10

29. Rong, X.: word2vec parameter learning explained, November 2014

30. Lau, J.H., Baldwin, T.: An empirical evaluation of doc2vec with practical insights into document embedding generation. arXiv:1607.05368 [cs], July 2016. http://arxiv.org/abs/1607.05368. Accessed 29 Mar 2021

31. Kirat, D., Vigna, G.: MalGene: automatic extraction of malware analysis evasion signature. In: Proceedings of the 22nd ACM SIGSAC Conference on Computer and Communications Security, Denver Colorado USA, October 2015, pp. 769–780 (2015). https://doi.org/10.1145/2810103.2813642

32. Zhao, B., Shan, Z., Liu, F., Zhao, B., Chen, Y., Sun, W.: Malware homology identification based on a gene perspective. Front. Inform. Technol. Electron. Eng. **20**(6), 801–815 (2019). https://doi.org/10.1631/FITEE.1800523

33. Liu, F., Zhang, P., Hou, Y., Wang, L., Shan, Z., Wang, J.: Malware analysis platform based on software gene for cyberspace security practice teaching. In: 2020 IEEE 2nd International Conference on Computer Science and Educational Informatization (CSEI), Xinxiang, China, pp. 140–143, June 2020. https://doi.org/10.1109/CSEI50228.2020.9142516

34. Maaten, L.V.D., Hinton, G.: Visualizing data using t-SNE. J. Mach. Learn. Res. **9**, 2579–2605 (2008)

The Construction of Case Event Logic Graph for Judgment Documents

Congyao Zhang and Shiping Tang[✉]

Beijing Institute of Technology, Beijing, China
{3120181066,simontangbit}@bit.edu.cn

Abstract. The construction of a case event logic graph for the judgment document can more intuitively retrospect the development of the case. This paper proposes a joint model of event extraction and relationship recognition for judgment documents. By extracting the case information in the judgment document, a case event logic graph was constructed. The development process of the case was shown, and a reference was provided for the analysis of the context of the case. The experimental results show that the proposed method can extract events and identify the relationship between events, and the F1 value reaches 0.809. The case event logic graph reveals the development context of the case accurately and vividly.

Keywords: Event logic graph · Judgment document · Event extraction · Relationship recognition

1 Introduction

With the progress of judicial reform, the publicity of judgment documents has become an essential part of implementing the people's right to know, participate, express, and supervise. However, judgment documents are textual information with complex content and are not conducive to the people's understanding of the case. Converting it into structured data is conducive to the display of content and conducive to computer processing. Knowledge graph and event logic graph can achieve this. In the judgment document, a case contains multiple interrelated events. These events and their interrelationship constitute the development process of the entire case. Compared with the knowledge graph that describes the static properties of the entity, it describes the events and the chain between events. The event logic graph of type dependence obviously can better express the content of the case. The research content of this paper is to extract the relevant information of the case from the judgment document, extract the events contained in the case, and identify the relationship between the events, forming an event logic graph describing the development of the case, and simply and vividly describing the development process of the case.

© Springer Nature Singapore Pte Ltd. 2021
J. Zeng et al. (Eds.): ICPCSEE 2021, CCIS 1451, pp. 209–217, 2021.
https://doi.org/10.1007/978-981-16-5940-9_16

2 Related Work

Professor Ting Liu [1] first proposed the concept of event logic graph in 2017. He defined the event logic graph as "a logically directed graph of event evolution describing the succession and causal relationship between events". Because the event logic graph represents the evolution of events, people use the event logic graph in various fields [5–7]. In 2019, Han Zhu [8] conducted research on the causality of aviation safety accidents based on the event logic map. In 2020, Zhongbao Liu [9] and others developed an event logic graph visualization system platform for "Historical Records". Feng Jun [10] et al. extracted the events from the event clauses after extracting the causality to construct the urban waterlogging event logic graph and analyzed the causes of waterlogging. Qiankun Shi [11] constructed a consumer event logic graph to identify and predict consumer intentions. Chao Wu [12] sorted out event elements to construct an emergency event logic graph. In addition to these, the event logic graph has also been applied to the prediction of online public opinion events, the identification and prediction of consumer intentions, and teaching behavior research.

3 Construction Method

3.1 Research Framework

The construction of the event logic graph of judicial cases includes four stages: data collection and processing, event extraction, identification of the relationship between events, and construction of the event logic graph (see Fig. 1).

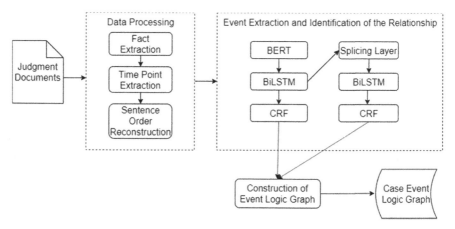

Fig. 1. Flow chart of construction of event logic graph.

This paper downloads the judgment documents needed for the research from the Peking University Law website, divides the judgment documents into sections, finds the description part of the facts of the case, reorders the paragraphs according to time, and marks them. Extract the events in the case and their attributes (time, place, event implementer, event bearer) from the judgment document, and identify the causal relationship between the events. Take the extracted events as nodes and the relations of events as edges to construct a case event logic graph.

3.2 Construction of Event Logic Graph

The difficulty in constructing the event logic graph lies in the extraction of events and the identification of relations between events. There are two main ways to construct a general event logic graph. One is to identify the sentence that contains the event relationship through the event relationship template (such as the causal relationship template), divide the sentence into two event clauses, and then directly use the event clause as the event or use the event extracted from the event clause; the other is to first extract the event in the text and then identify the event relationship of the sentence containing multiple events. The joint models are widely used in Named Entity Recognition [13–15]. In this paper, the method of the joint model is proposed to extract events and the relationship between events at the same time.

Data Processing. The data required for the construction of the judicial case event logic graph is the case judgment document. With the progress of judicial reform and the promotion of judicial openness, the judgment document is already available on the Internet. This paper is to download the required judgment document from the Peking University Law website for different reasons. Although the judgment document is in the form of a text, its writing has a specific structural standard, which can more accurately extract the relevant elements of the case. The purpose of this paper is to present the case in the form of an event logic graph. The rule-based method can be used to extract paragraphs describing the facts of the case, extract the time points in the sentences, and then rearrange these sentences in chronological order. A text describing the case in chronological order is constructed to prepare for subsequent processing.

Event Extraction. The extraction of events includes the extraction of events and event attributes. The attributes that this paper cares about mainly include time, place, implementer, and receiver. In this paper, the method based on sequence annotation is used to extract the event itself and event attributes simultaneously.

The labeling of events uses the BIEOS labeling mode. Taking into account that the detailed description of the judgment document makes the sentence contain a large number of words with low repetition (mostly company names, place names, etc.), resulting in poor word segmentation effects, this paper uses character rather than word input to label. The specific marking method is in Table 1.

Recognition of the Relationship Between Events. In a case, the relationship between events mainly includes sequential relationship and causal relationship. The sequential relationship is used to indicate the sequence of events in a case, while the causal relationship indicates the direct causes between events.

Table 1. Labels with their meanings

Label	Meaning	Label	Meaning
B-event	The beginning of the event	B-time	The beginning of time
I-event	The inner of the event	I-time	The inner of time
E-event	The end of the event	E-time	The end of time
S-event	Single event	B-address	The beginning of place
B-subject	The beginning of the implementer	I-address	The inner of place
I-subject	The inner of the implementer	E-address	The end of place
E-subject	The end of the implementer	B-object	The beginning of the receiver
O	Other character	I-object	The inner of the receiver
R	Result event	E-object	The end of the receiver

The judgment document uses strict time terms when describing the process of the case. Through the time point, the events can be sorted by time, and the timeline of the sequential relationship can be obtained. The specific identification is divided into two aspects. One is to obtain an overall event line of the case. The second is to make an event line with each person as the implementer.

The research focus of event relationship recognition focuses on the recognition of the causal relationship. At present, there are two common causality recognition methods, one is a pattern matching method based on causal words, and the other is a text classification method based on deep learning. Both methods focus on the situation that the events that constitute causality are in the same or adjacent sentences. In fact, the events that constitute causality are not always together in the text. There is a strong causal relationship between Guo's "serious illness" event and his "death" event in the "first-instance civil judgment on disputes between Liu, Guo, etc. and Sun, etc. on the right to life, health, and physical rights". Since the two events are far apart and there are many other dispute events between them, their causal relationship will be ignored in the construction of many event logic graphs. However, in this case, their causal relationship is more important than some other relationships. Therefore, this paper proposes to splice the event and the result of event extraction into the sentence sequence to identify whether events in the sentence result from the spliced event. This way can effectively solve the problem of identifying long-distance causality, and the same applies to one-cause-multi-effect and multi-cause-one-effect.

4   Model

This paper adopts the event and relationship joint extraction model, which comprises five modules: BERT, BiLSTM, CRF, splicing layer, and relationship recognition layer (see Fig. 2). First, use the BERT model to obtain the word vector and extract the essential features of the text. Then use the BiLSTM deep learning context feature information to perform named entity recognition. The CRF layer processes the BiLSTM output sequence according to the adjacent labels, combined with the state transition matrix in the CRF, to obtain an optimal global sequence. The splicing layer splices the extracted events and the results recognized by the CRF layer to the sequence obtained by the BiLSTM layer, forming a new input and sending it to the next layer. The relationship recognition layer obtains the causal relationship between events through BiLSTM and CRF in turn.

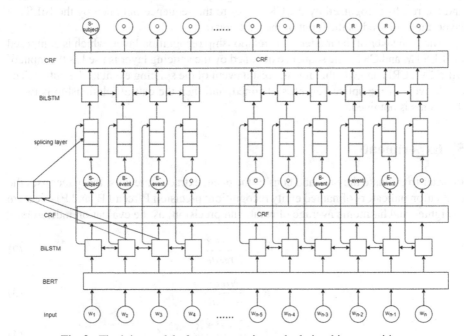

Fig. 2. The joint model of event extraction and relationship recognition.

The first layer of the model uses the pre-trained BERT model to initialize the input text to obtain a word vector containing text features.

The second layer of the model is the BiLSTM layer. The word vectors obtained in the first layer are used as the input of each time step of BiLSTM. The forward and backward hidden state sequences at the corresponding time are spliced to obtain the encoding result of the corresponding word $h_t = (\overrightarrow{h}_t, \overleftarrow{h}_t)$.

The third layer of the model is the CRF layer, which is modeled by the conditional probability $P(y|x)$ to get the label of the input text.

$$P(y|x) = \frac{1}{z(x)} \exp(\sum_{i,m} \lambda_m \mu_m(y_{i-1}, x, i) \cdot \sum_{i,n} \beta_n t_n(y_i, x, i)) \tag{1}$$

Among them, i represents the index of the current node in the sequence. m, n represents the total number of feature functions on the current node i. t_n represents the node feature function, which is only related to the current position. μ_m represents the local feature function, which is only related to the current position and the previous node position. β_n and λ_m respectively represent the weight coefficients corresponding to the feature functions t_n and μ_m, which are used to measure the trust of the feature function. $z(x)$ is the normalization factor.

$$z(y|x) = \sum_y \exp(\sum_{i,m} \lambda_m \mu_m (y_{i-1}, x, i) \cdot \sum_{i,n} \beta_n t_n (y_i, x, i)) \qquad (2)$$

The fourth layer of the model is the splicing layer. After splicing the extracted events and the results recognized by the CRF layer to the sequence obtained by the BiLSTM layer, they are used as the input of the next layer.

The fifth layer of the model is the relationship recognition layer, which is composed of BiLSTM and CRF. The sequence obtained by the splicing layer is used as the input of BiLSTM+CRF, and whether it is the result event of the splicing event as the output. The result event of the spliced event is obtained, and then the causal relationship between the events is obtained.

5 Experiment

In order to evaluate the pros and cons of the event extraction results, this paper uses the evaluation indicators of named entity recognition: precision P, recall R, and F1-measure (weighted and harmonic average of recall and precision) as the evaluation indicators.

$$P = \frac{Ncorrect}{Nresult} \qquad (2)$$

$$R = \frac{Nreal}{Nresult} \qquad (3)$$

$$F1 = \frac{2 * P * R}{P + R} \qquad (4)$$

Ncorrect is the number of words that are correctly labeled and indeed belong to the event. *Nresult* is the number of words that are labeled as an event. *Nreal* is the number of words that are actually an event in the test set.

In this paper, 1000 judge documents are used as training data, and 100 judge documents are used as test data. Table 2 shows the results of event extraction applying different models.

Table 2. Results of event extraction

Model	Precision	Recall	F1-measure
BERT+BiGRU+CRF	0.748	0.786	0.767
BERT+BiLSTM	0.687	0.855	0.762
BERT+CNN+LSTM	0.753	0.786	0.769
The joint model	0.791	0.828	0.809

The experimental results show that the method proposed in this paper has achieved good results in event extraction experiments. It is only inferior to the BERT+BiLSTM model in recall and is significantly better than other models in precision and F1-measure. In general, The method proposed in this paper helps to improve the effect of event extraction.

Taking "first-instance civil judgment on disputes between Liu, Guo, etc. and Sun, etc. on the right to life, health, and physical rights" as an example, the extracted events and the relationships between the events constitute the case event logic graph as Fig. 3.

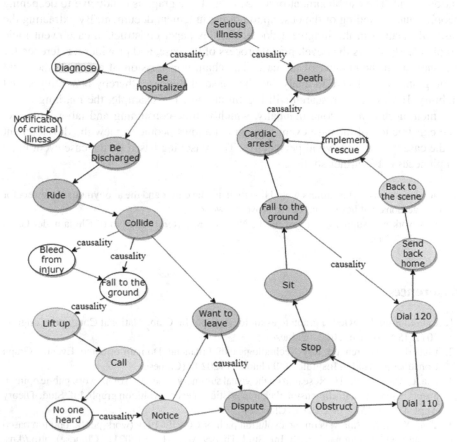

Fig. 3. Event logic graph constructed from "first-instance civil judgment on disputes between Liu, Guo, etc. and Sun, etc. on the right to life, health, and physical rights". (Color figure online)

The development process of this case can be seen from the figure. In the figure, the events initiated by different implementers are represented by different colors. The plaintiff uses the red node, the defendant uses the blue node, and the co-initiated event uses the purple node. Events implemented by others or no one implemented are white nodes. It can be seen that the direct cause of the defendant's death was cardiac arrest, and the indirect cause was the previous illness. A complete combing of the case context through the event logic graph revealed that the plaintiff is not at fault in the defendant's death. This result is consistent with the finding of the judgment document, and It is more convenient for people to feel intuitive.

6 Conclusion

Judgment documents are the final products of trial activities. Their publicity is of great significance to constructing a social credit system and the promotion of judicial education for all. The establishment of a case event logic graph is conducive to deepening people's understanding of the description of the judgment document. By extracting the case information in the judgment document, this paper constructs a case event logic graph, clearly shows the development process of the case, and provides a reference for the analysis of the context of the case. The chain transmission of events in the event logic graph is conducive to analyzing the cause of the case, thereby benefiting social stability. However, the research still has limitations. For example, the marking of the judgment document is done manually, which is time-consuming and labor-intensive. The construction of the case event logic graph allows people to view the development of the case from a more vivid perspective. The reasoning based on the case event logic graph needs to be further studied.

Acknowledgments. The authors would like to thank the editor and the anonymous reviewers for their suggestions that have helped us improve the work.

This work was supported in part by the National Key R&D Program of China under Grant 2018YFC0830104.

References

1. Liu, T.: From knowledge graph to event logic graph. In: China National Computer Congress 2017, 15 November 2017. (Chinese)
2. Liao, K.: Research on Key Technologies of Financial Domain-oriented Eventic Graph Construction. Harbin Institute of Technology (2020). (Chinese)
3. Xia, L., Chen, J., Yu, H.: Research on the visual summary generation of network public opinion events based on multi-dimensional characteristics of event evolution graph. Inf. Stud. Theory Appl. **43**(10), 157–164 (2020). (Chinese)
4. Tian, Y., Li, X.: Analysis on the evolution path of COVID-19 network public opinion based on the event evolutionary graph. Inf. Stud. Theory Appl., 1–13 (2021). (Chinese). http://kns.cnki.net/kcms/detail/11.1762.G3.20210208.1330.009.html
5. Bai, L.: Event Evolution Graph Construction in Political Field. University of International Relations (2020). (Chinese)
6. Chen, P.: Research on the Method of Constructing Event Evolutionary Graph of Housing Price Changes. Harbin Institute of Technology (2020). (Chinese)
7. Shan, X., Pang, S., Liu, X., Yang, J.: Research on internet public opinion event prediction method based on event evolution graph. Inf. Stud. Theory Appl. **43**(10), 165–170+156 (2020). (Chinese)
8. Zhu, H.: Research on Causality of Aviation Safety Accident Based on Event Evolutionary Graph. Civil Aviation University of China (2019). (Chinese)

9. Liu, Z., Dang, J., Zhang, Z.: Research on automatic extraction of historical events and construction of affair atlas in "historical records". Libr. Inf. Serv. **64**(11), 116–124 (2020). (Chinese)
10. Feng, J., Wang, Y., Wu, W., et al.: Construction method and application of event logic graph for urban waterlogging. J. Hohai Univ. (Nat. Sci.) **48**(6), 479–487 (2020)
11. Shi, Q.: Research on Key Technologies of Consumption Intention Identification and Prediction Based on Event Logic Graph. Harbin Institute of Technology (2020). (Chinese)
12. Wu, C.: Design and Implementation of Event Graph Platform for the Field of Emergency. University of Electronic Science and Technology of China (2020). (Chinese)
13. Miwa, M., Bansal, M.: End-to-end relation extraction using LSTMs on sequences and tree structures (2016)
14. Demeester, T., Develder, C., Deleu, J., et al.: Joint entity recognition and relation extraction as a multi-head selection problem. Expert Syst. Appl. **114**, 34–45 (2018)
15. Zheng, S., Wang, F., Bao, H., Hao, Y., Zhou, P., Xu, B.: Joint extraction of entities and relations based on a novel tagging scheme. In: ACL 2017 - 55th Annual Meeting of the Association for Computational Linguistics, Proceedings of the Conference (Long Papers), vol. 1, pp. 1227–1236 (2017)

Machine Learning for Data Science

Improved Non-negative Matrix Factorization Algorithm for Sparse Graph Regularization

Caifeng Yang[1], Tao Liu[1(✉)], Guifu Lu[1], Zhenxin Wang[1], and Zhi Deng[2]

[1] College of Computer and Information, Anhui Polytechnic University, Wuhu, China
liutao@ahpu.edu.cn
[2] School of Computer Science, Northwestern Polytechnic University, Xi'an, China

Abstract. Aiming at the low recognition accuracy of non-negative matrix factorization (NMF) in practical application, an improved spare graph NMF (New-SGNMF) is proposed in this paper. New-SGNMF makes full use of the inherent geometric structure of image data to optimize the basis matrix in two steps. A threshold value s was first set to judge the threshold value of the decomposed base matrix to filter the redundant information in the data. Using L_2 norm, sparse constraints were then implemented on the basis matrix, and integrated into the objective function to obtain the objective function of New-SGNMF. In addition, the derivation process of the algorithm and the convergence analysis of the algorithm were given. The experimental results on COIL20, PIE-pose09 and YaleB database show that compared with K-means, PCA, NMF and other algorithms, the proposed algorithm has higher accuracy and normalized mutual information.

Keywords: Image recognition · Non-negative matrix factorization · Graph regularization · Basis matrix · Sparseness constraints

1 Introduction

With the development of image recognition technology [1], it has been applied in navigation, military detective, face recognition, biomedical and other fields. Image recognition [2] is a series of operations such as image data acquisition, preprocessing, segmentation, feature extraction and classification, and finally identifying the target we need. In practical application, it is generally faced with the problem of large data dimension. NMF is a typical matrix factorization technique. The goal is to find two non-negative matrices, which are not only non-negative, but also much smaller than the original matrix in dimension. The product of these two matrices can effectively approximate the original matrix. NMF learning image local features, with natural non-negativity, not only can effectively reduce the dimension of data compression, but also meet the requirements of image representation, more and more people pay attention to it.

Lee and Seung et al. [3] first proposed NMF algorithm, pointed out that the algorithm can be used for image recognition, and gave the derivation process of iterative formula in the paper [4] published in the same year, based on two kinds of objective functions, and proved convergence. This is the first work of NMF, and also provides a new and

© Springer Nature Singapore Pte Ltd. 2021
J. Zeng et al. (Eds.): ICPCSEE 2021, CCIS 1451, pp. 221–232, 2021.
https://doi.org/10.1007/978-981-16-5940-9_17

effective method for image recognition. After that, many experts and scholars have made various improvements on NMF algorithm in practical application, Li et al. [5] defined an objective function to impose local constraints and proposed a local NMF algorithm. Liu et al. [6] combined with the knowledge of sparse coding, formed a sparse NMF algorithm, which makes the new data more sparse than the original NMF. Zhou Jing et al. [7] improved the iterative rule, which made the expression of the new objective function simple and the calculation more simple. Cai et al. [8] combined with graph structure, explicitly considered local invariance, obtained a matrix factorization algorithm preserving graph structure, and proposed the classical Graph Regularization Non-negative Matrix Factorization(GNMF). Li et al. [9] added the constraint of low rank on the basis of graph regular NMF to remove the interference noise, and then obtained a new improved algorithm of low dimensional NMF. Wang Xiaohua and others [10] improved the two-dimensional NMF algorithm in order to greatly reduce the dimension of data, which shortened the calculation time and improved the recognition rate. Xu Huimin et al. [11] proposed a GNMF algorithm with sparse discrimination for image feature extraction, in which the sparse constraint is to use $L_{2,1}$ norm and optimize the objective function to improve the classification accuracy. Long et al. [12] integrated geometric structure and discriminant information into NMF model, and used Laplacian information and supervised labeling information of graph to learn projection matrix in new model. Du et al. [13] proposed a GNMF algorithm with convex smooth $L_{3/2}$ norm constraint by using convex smooth $L_{3/2}$ norm as sparse constraint to constrain low dimensional features. Qiu Feiyue et al. [14] reconstructed the updating rule by manifold learning and $L_{2,1}$ norm, which made the algorithm stable under different noises. Yang et al. [15] not only considered the global and local features of manifold structure of image clustering, but also improved the convergence of GNMF algorithm, thus proposed a symmetric manifold GNMF method. Li et al. [16] combined the knowledge of multi view data clustering, constructed a multi-layer NMF model, and used graph regularization to extract more abstract representations, so as to improve the performance of the algorithm. Li Xiangli et al. [17] proposed a new GNMF method with kernel to solve the problem that the traditional NMF method can't use the label information of data, ignore the inherent geometric structure of data, and can't deal with the nonlinear situation of data. These methods can improve the application effect of the original NMF to a certain extent, but there are still some shortcomings, such as too much data redundancy, unable to highlight the features required for recognition, insufficient sparsity and high algorithm complexity.

Aiming at the application of NMF in image recognition, a new improved algorithm (New-SGNMF) is proposed: The image recognition effect can be improved by adding the spectrum theory to NMF and retaining the inherent geometric structure of image data. Adding threshold judgment to the NMF base matrix is a simple and effective optimization process, without increasing the complexity of calculation. The processed base matrix can filter out the redundant information in the process of image recognition, better extract image features, and reduce unnecessary interference. Using L_2 norm to continue the sparse constraint on the basis matrix after threshold optimization processing, further improve the sparsity of image data, and effectively control the sparsity for different application scenarios. In addition, the objective function of New-SGNMF algorithm is given, and the multiplicative rule of New-SGNMF algorithm is derived, and the convergence of New-SGNMF algorithm is analyzed theoretically.

2 NMF, GNMF Algorithm

NMF algorithm has been used in image recognition since it was proposed. Its basic principle is to find two non-negative matrices whose product is equal to or about equal to the original matrix. In the actual image recognition process, given an image data matrix $X = [x_1, x_2, \cdots, x_N] \in R^{M \times N}$ of $M \times N$, the purpose of NMF is to find the coefficient matrix $U = [u_{ik}] \in R^{M \times K}$ and the base matrix $V = [v_{jk}] \in R^{N \times K}$, whose product can satisfy the following formula:

$$X \approx U \times V^T \qquad s.t. \, U \geq 0, \, V \geq 0 \tag{1}$$

Lee and Seung et al. [3, 4] give the objective function of NMF as follows:

$$O_1 = ||X - UV^T||_F^2 = \sum_{i,j} \left(x_{ij} - \sum_{k=1}^{K} u_{ik} v_{jk} \right)^2 \tag{2}$$

Where $\|\bullet\|_F^2$ is the Frobenius norm of the matrix. In general, $K < < M$, $K < < N$, the original image data matrix X is compressed. The following rules are given in reference [4]:

$$u_{ik} \leftarrow u_{ik} \frac{(XV)_{ik}}{(UV^T V)_{ik}} \tag{3}$$

$$v_{jk} \leftarrow v_{jk} \frac{(X^T U)_{jk}}{(VU^T U)_{jk}} \tag{4}$$

In the actual image recognition application of NMF, the inherent geometric structure of image data is ignored when NMF processes data. However, these inherent geometric structures are very useful for practical application. The idea of graph regularization is embedded into the original NMF to form GNMF to maintain the local manifold structure in the image data and improve the application effect.

According to the graph theory [18], assuming that two image data points x_i and x_j are adjacent in the original space, the low dimensional representations x_i and x_j under the new basis are also adjacent. The graph composed of original image data points is defined as graph G. by using 0–1 weighting method, the weight matrix W can be obtained:

$$W_{ij} = \begin{cases} 1, & x_i \in N_p(x_j) \; or \; x_j \in N_p(x_i) \\ 0, & others \end{cases} \tag{5}$$

Where, x_i represented by $N_p(x_i)$ is the p adjacent points of. The definition of Laplacian matrix $L = D - W$, $D_{ii} = \sum_j W_{ij}$ is a diagonal matrix, its terms are W rows (or columns).

Cai et al. [8] gave the objective function of GNMF as follows:

$$O_2 = ||X - UV^T||_F^2 + \lambda Tr\left(V^T LV \right) \tag{6}$$

Where $Tr(\bullet)$ is the trace of the matrix and $\lambda \geq 0$ is the graph regularization parameter. The new iteration rules are as follows:

$$u_{ik} \leftarrow u_{ik} \frac{(XV)_{ik}}{(UV^TV)_{ik}} \tag{7}$$

$$v_{jk} \leftarrow v_{jk} \frac{(X^TU + \lambda WV)_{jk}}{(VU^TU + \lambda DV)_{jk}} \tag{8}$$

3   New-SGNMF Algorithm Design

In the actual image recognition, we hope to extract the main features of the original image, that is, the features extracted from the reduced image are needed in the process of image recognition. However, although the original NMF algorithm can effectively compress the image and reduce the dimension, it can't accurately extract the features needed in the actual image recognition. For example, there are redundant data, ignoring some important features, excessive dimension reduction and so on, an improved NMF algorithm, New-SGNMF, is proposed. Combined with the spectrum theory, the intrinsic geometric structure of the data is preserved. Set threshold, optimize base matrix and filter redundant information. In order to improve the effect of image recognition, L_2 norm is used to constrain the basis matrix sparsely.

3.1   Algorithm Model of New-SGNMF

Based on GNMF, the threshold is set for the matrix, and the redundant information between image data is filtered by threshold judgment. Select a threshold value S, for each column of the base matrix U, compare it with the threshold value S, and set the value less than the threshold value S to a number s greater than 0, so as to obtain a new base matrix U_{new}, and X obtains a new U_{new} in the new base projection, and the objective function is as follows:

$$O_3 = ||X - U_{new}V_{new}^T||_F^2 + \lambda Tr(V_{new}^T L V_{new}) \tag{9}$$

Sparse Coding [19] is an unsupervised learning algorithm, which has been widely concerned since it was proposed, and it has been proved to have good application effect in image recognition. In practical applications, for a given image data, sparse coding can capture the high-level semantics in the image data by finding a base set, which is called a dictionary. Sparse coordinates are learned through the dictionary to produce a sparse representation of the data. Using sparse coding technology, it is integrated into GNMF after threshold processing to improve the sparsity of image data representation, which can not only save storage space, but also improve the quality of decomposition. Sparse constraint can be realized by L_p norm constraint. The constraint object can be coefficient matrix V and base matrix U. Different norms and constraint objects can be selected for different application scenarios and data. In this paper, L_2 norm is used for sparse constraint of base matrix U to improve the sparsity of data. The specific operation is to integrate the sparse constraints into the objective function (9) to obtain a new New-SGNMF objective function:

$$O_4 = ||X - U_{new}V_{new}^T||_F^2 + \lambda Tr(V_{new}^T LV_{new}) + \beta(||U_{new}||_2) \tag{10}$$

Among them, $||\bullet||_2$ is the L$_2$ norm, β is the sparse constraint parameter, which has the function of adjusting the sparse constraint. By minimizing this objective function, we can find the local minimum.

3.2   Algorithm Solution of New-SGNMF

In this paper, the method of constructing Lagrange function is used to solve the algorithm, and the new iteration rules of New-SGNMF are obtained. The detailed process is as follows:

$$\begin{aligned} O_4 &= \left\| X - U_{new}V_{new}^T \right\|_F^2 + \lambda Tr(V_{new}^T LV_{new}) + \beta(||U_{new}||_2) \\ &= Tr((X - U_{new}V_{new}^T)(X - U_{new}V_{new}^T)^T) + \lambda Tr(V_{new}^T LV_{new}) + \beta Tr(U_{new}^T U_{new}) \\ &= Tr(XX^T) - 2Tr(XV_{new}U_{new}^T) + Tr(U_{new}V_{new}^T V_{new}U_{new}^T) \\ &\quad + \lambda Tr(V_{new}^T LV_{new}) + \beta Tr(U_{new}^T U_{new}) \end{aligned} \tag{11}$$

Let $\Psi = \{\varphi_{ik}\}$, $\Phi = \{\phi_{jk}\}$, and $\varphi_{ik} \geq 0$, $\phi_{jk} \geq 0$, construct Lagrange function:

$$\begin{aligned} F(U_{new}, V_{new}, \Psi, \Phi) &= Tr(XX^T) - 2Tr(XV_{new}U_{new}^T) \\ &+ Tr(U_{new}V_{new}^T V_{new}U_{new}^T) + \lambda Tr(V_{new}^T LV_{new}) + \beta Tr(U_{new}^T U_{new}) \\ &+ Tr(\Psi U_{new}^T) + Tr(\Phi V_{new}^T) \end{aligned} \tag{12}$$

The partial derivatives of function $F(U_{new}, V_{new}, \Psi, \Phi)$ for U_{new} and V_{new} are as follows:

$$\frac{\partial F}{\partial U_{new}} = -2XV_{new} + 2U_{new}V_{new}^T V_{new} + 2\beta U_{new} + \Psi \tag{13}$$

$$\frac{\partial F}{\partial V_{new}} = -2X^T U_{new} + 2V_{new}U_{new}^T U_{new} + 2\lambda LV_{new} + \Phi \tag{14}$$

Using Karush Kuhn Tucker (KKT) condition [20] $\varphi_{ik}u_{ik} = 0$, $\phi_{jk}v_{jk} = 0$, the following equations can be obtained:

$$-(XV_{new})_{ik}u_{ik} + (U_{new}V_{new}^T V_{new})_{ik}u_{ik} + \beta(U_{new})_{ik}u_{ik} \tag{15}$$

$$-(X^T U_{new})_{jk}v_{jk} + (V_{new}U_{new}^T U_{new})_{jk}v_{jk} + \lambda(LV_{new})_{jk}v_{jk} = 0 \tag{16}$$

The solution equation is obtained:

$$u_{ik} \leftarrow u_{ik} \frac{(XV_{new})_{ik}}{(U_{new}V_{new}^T V_{new} + \beta U_{new})_{ik}} \tag{17}$$

$$v_{jk} \leftarrow v_{jk} \frac{(X^T U_{new} + \lambda WV_{new})_{jk}}{(V_{new}U_{new}^T U_{new} + \lambda DV_{new})_{jk}} \tag{18}$$

3.3 Convergence Analysis of the Algorithm

In this section, we will explain the convergence of the algorithm. We can prove that the New-SGNMF algorithm is convergent by constructing auxiliary functions. The following definitions and lemmas will be used in the specific process.

Definition: Let an auxiliary function of function $O(u)$ be $G(u, u')$ and have, $G(u, u') \geq O(u)$, $G(u, u) = O(u)$.

Lemma: Under the condition that $G(u, u')$ is an auxiliary function of $O(u)$, $O(u)$ is nonadditive under the iteration of $u^{t+1} = argmin_u G(u, u^t)$.

From the above definitions and lemmas, it can be obtained that:

1: $G(u, u')$ gets the minimum $(u = u^{t+1})$.

2: $O(u^{t+1}) \overset{def}{\leq} G(u^{t+1}, u^t) \overset{min}{\leq} G(u^t, u^t) \overset{def}{=} O(u^t)$.

Further, it can be seen that there is only $O(u^{t+1}) = O(u^t)$ if u^t is $G(u, u')$ and has a local minimum. If the function $O(\bullet)$ has a continuous minimum field, and u^t satisfies this requirement, then the derivative $\nabla O(u^t) = 0$ is true. And then we have sequence $O(u_{mim}) \leq \cdots \leq O(u^{t+1}) \leq O(u^t) \leq \cdots \leq O(u^1) \leq O(u^0)$ converges to point $u_{min} = argmin_u O(u)$. Therefore, for the objective function (10) in this paper, an auxiliary function $G(u, u')$ can be constructed, and the iteration rules (17) and (18) can satisfy $u^{t+1} = argmin_u G(u, u^t)$.

3.4 New-SGNMF Algorithm Steps

According to all the above descriptions, the algorithm steps proposed in this paper are summarized as follows:

1: Input image data matrix X;
2: According to the rules of formula (7) and (8), X is decomposed to obtain the basis matrix U;
3: Setting parameters p, λ;
4: Select the values of parameters S and s, optimize the basis matrix U, and get the new basis matrix U_{new};
5: The projection of X onto the new basis matrix U_{new} gives us a new V_{new};
6: The L_2 norm constraint is applied to the basis matrix U_{new}, and parameter β is set;
7: It can be calculated as (17), (18);
8: Output the final result: AC and NMI.

4 Experimental Analysis

4.1 Experimental Database

In order to verify the effectiveness of the New-SGNMF proposed in this paper, COIL20 object image database and PIE-pose09 and YaleB face image databases are selected to carry out experiments on MATLAB2018. The three image databases are now open image databases, which are provided by different institutions and have important research value. The number of images, people and objects included are different. Table 1 shows the detailed information of the three image databases.

Table 1. Details of three image databases.

Dataset	Size(M)	# of classes(K)	Attitude/Angle degree	Light conditions	Expression change
COIL20	1440	20	5	1	
PIE-pose09	859	68	13	64	4
YaleB	5850	10	9	64	1

4.2 Comparison of Different Algorithms

The effectiveness of the improved algorithm in image recognition is verified by the visual contrast experiments of different algorithms. Here, seven algorithms, K-means, PCA (Principal Component Analysis), NMF, SC (Spatial Coding), GraphSC (Graph Regularized Spatial Coding), GNMF and New-GNMF, are selected to compare with the New-SGNMF algorithm. Among them, New-GNMF represents the method of threshold optimization for the base matrix U on the basis of GNMF. Different algorithms are described as follows:

1) K-means: It is a simple classical clustering algorithm. Based on the objective function of the data prototype, it uses the extreme value of the function to get the final iterative rules. The algorithm operates on the original matrix without extracting and utilizing the information contained in the original matrix. In the process of simulation experiment, the other seven algorithms are combined with K-means to achieve the purpose of image clustering recognition.

2) PCA: It is a widely used multivariate statistical method, which selects a small number of representative data from a large number of data through linear transformation. In the application of image recognition, PCA can extract the main components of the data set through linear transformation, and then analyze these important variables to achieve the purpose of dimensionality reduction.

3) NMF: Its purpose is to find two non-negative matrices, and the dimension of these two matrices is much smaller than that of the original matrix. In image recognition, it can not only extract the required application features and reduce the dimension, but also meet the requirements of image representation.

4) SC: The algorithm belongs to image statistics method, which has the advantages of adaptive orientation, local space and directivity. SC can produce sparse representation, represent data as a linear combination of a few basic vectors, and extract the main features of image data set.

5) GraphSC: Integrating graph regularization into SC can not only produce sparse representation, but also preserve the geometric structure of data. It is a combination of manifold learning and SC, which can find a more suitable image representation method than SC, and effectively improve the image recognition effect.

6) GNMF: The graph theory is combined with the original NMF to construct the neighborhood graph and preserve the intrinsic geometric structure of the data. In the experiment, the number of nearest neighbors p is set to 5, and the regularization parameter λ is set to 100.

7) New-GNMF: Considering not only the intrinsic geometric information of the image data, but also the redundant information of the data decomposed by GNMF, threshold processing is added to the base matrix, and the secondary information is set to s to filter the interference information in the process of image recognition. In the experiment, the number of nearest neighbors p is set to 5, the regularization parameter λ is set to 100, the threshold S of base matrix on COIL20 is set to 0.03, s is set to 0.00001, the threshold S of base matrix on PIE-pose09 and YaleB is set to 0.7, s is set to 0.00001.

8) New-SGNMF: On the basis of New-GNMF, we continue to use norm to impose sparse constraints on the basis matrix U, which can not only filter redundant information, but also produce more effective sparse representation and improve the quality of the algorithm. In the experiment, the number of nearest neighbors p is set to 5, the regularization parameter λ is set to 100, the threshold S of base matrix on the three databases is set to 0.02, and for sparse parameters β, COIL20, PIE-pose09 and YaleB are set to 2, 0.02 and 0.2 respectively.

4.3 Evaluation Criteria Used

To visually represent the effectiveness of the algorithm presented in this paper in practical applications, two commonly used measures are used, the first being AC (accuracy) and the second NMI (normalized mutual information). AC is defined as follows:

$$AC = \frac{\sum_{i=1}^{n} \delta(gnd_i, map(z_i))}{n} \tag{19}$$

Where, n represents the size of the data used in the experiment, gnd_i represents the actual label, if $a = b$, then $\delta(a, b) = 1$, in other cases $\delta(a, b) = 0$, z_i represents the sample label, which is different from gnd_i. It is the label obtained by the algorithm in this paper. From sample X, $map(z_i)$ represents a replacement function, which is an optimal result and a replacement function mapping from label z_i to gnd_i.

NMI is also a useful measurement method. Based on information theory, NMI is defined as follows:

$$MI = (C, C') = \sum_{c_i \in C, c_j' \in C'} p(c_i, c_j') log \frac{p(c_i, c_j')}{p(c_i)p(c_j')} \tag{20}$$

$$NMI\left(C, C'\right) = \frac{MI\left(C, C'\right)}{max(H(C), H(C'))} \tag{21}$$

Where, MI said mutual information, can reflect a data contained in another data information, C said known data, generally is a class, data from C' said algorithm in this paper, p(·) said probability, $p(c_i)$ said the probability is the probability of sample belongs to C, $p(c_j')$ said the probability is the probability of sample belongs to C', $p(c_i, c_j')$ said the probability is the probability of belonging to C and C' at the same time, $H(\cdot)$ according to entropy, $H(C)$ and $H(C')$, respectively, C and C' entropy.

4.4 Experimental Result

Table 2 and Table 3 show AC and NMI in COIL20 object image database, Table 4 and Table 5 show AC and NMI in PIE-pose09 face image database, and Table 6 and Table 7 show AC and NMI in YaleB face image database. In order to randomize the experiment, each given cluster number K is tested and run 10 times, and the performance results of each cluster number of different methods are reported in the table (take the average value, and keep two decimal places).

The experimental results may be different under different hardware configurations: i7-7700 3.6 GHz CPU; The RAM is 8 GB; 1T hard disk drive; 2G display card. The operating system is Windows 10.

Table 2. AC results on COIL20 (%).

K	1	3	5	10	20	23
K-means	47.71	52.15	60.83	60.49	60.49	60.49
PCA	46.18	51.94	61.81	61.11	61.32	61.32
NMF	51.67	51.67	71.60	65.63	66.74	66.74
SC	31.39	31.18	31.18	38.82	37.08	37.08
GraphSC	52.26	57.99	67.50	67.50	67.50	67.50
GNMF	66.81	64.03	72.78	71.88	72.22	72.22
New-GNMF	66.94	65.76	79.37	79.37	79.37	79.37
New-SGNMF	69.17	69.37	82.36	82.36	82.36	82.36

Table 3. NMI results on COIL20 (%).

K	1	3	5	10	20	23
K-means	65.85	71.72	69.96	73.86	73.86	73.86
PCA	65.18	70.06	70.72	74.45	73.75	73.75
NMF	66.75	66.75	76.42	75.16	74.36	74.36
SC	39.76	41.13	41.13	49.23	53.47	53.47
GraphSC	73.82	75.40	77.93	77.93	77.93	77.93
GNMF	83.60	81.79	87.31	87.64	87.60	87.60
New-GNMF	83.79	82.08	89.47	89.47	89.47	89.47
New-SGNMF	83.84	84.31	90.52	90.52	90.52	90.52

From the above simulation results, it can be seen that the New-SGNMF algorithm proposed in this paper is effective by setting threshold and applying sparse constraint on the basis matrix. Compared with other methods such as K-means, PCA and NMF, the AC and NMI of New-GNMF and New-SGNMF are improved in three image databases, namely COIL20, PIE-pose09 and YaleB, which shows the superiority of the two-step

Table 4. AC results on PIE-pose09 (%).

K	1	3	5	10	20	23
K-means	19.67	17.34	17.34	20.16	20.16	20.16
PCA	19.24	17.83	17.83	18.20	18.20	18.20
NMF	26.04	23.96	23.96	26.16	27.08	27.08
SC	29.47	27.39	27.39	28.31	28.31	28.31
GraphSC	28.13	28.13	28.13	27.02	27.02	27.02
GNMF	40.56	39.40	39.40	35.11	35.11	35.11
New-GNMF	41.42	38.17	38.17	37.99	37.56	37.56
New-SGNMF	41.30	40.01	39.22	38.24	39.40	39.40

Table 5. NMI results on PIE-pose09 (%).

K	1	3	5	10	20	23
K-means	48.00	46.32	46.32	49.01	49.01	49.01
PCA	48.18	47.88	47.88	47.35	47.35	47.35
NMF	55.80	55.36	55.36	55.81	57.01	57.01
SC	55.68	55.71	55.71	55.19	55.19	55.19
GraphSC	58.51	58.51	58.51	57.31	57.31	57.31
GNMF	65.66	65.40	65.40	62.79	62.79	62.79
New-GNMF	66.26	64.76	64.76	64.04	64.03	64.03
New-SGNMF	65.81	66.09	63.84	64.57	65.95	65.95

Table 6. AC results on YaleB (%).

K	1	3	5	10	20	23
K-means	08.04	07.33	07.33	07.33	08.20	08.20
PCA	08.41	07.46	08.91	08.62	08.62	08.62
NMF	13.59	13.59	14.87	14.87	13.63	13.63
SC	16.20	16.61	17.73	16.98	16.98	16.98
GraphSC	15.33	14.66	14.66	15.29	15.29	15.29
GNMF	20.96	21.17	21.17	21.21	21.21	21.21
New-GNMF	23.49	23.49	23.49	24.77	24.77	24.77
New-SGNMF	24.45	24.45	25.93	25.93	25.93	25.93

Table 7. NMI results on YaleB (%).

K	1	3	5	10	20	23
K-means	08.81	08.27	08.27	08.27	09.56	09.56
PCA	09.64	09.29	10.62	11.26	11.26	11.26
NMF	26.28	26.28	26.36	26.36	24.27	24.27
SC	25.78	28.32	26.98	28.17	28.17	28.17
GraphSC	25.72	25.24	25.24	25.37	25.37	25.37
GNMF	33.72	33.26	33.26	32.74	32.74	32.74
New-GNMF	36.05	36.05	36.05	36.72	36.72	36.72
New-SGNMF	36.17	36.17	38.07	38.07	38.07	38.07

optimization proposed in this paper. Due to the differences of images in different databases, including lighting conditions, posture and so on, the performance effect of New-GNMF and New-SGNMF on the three databases is different, and the effect on COIL20 and YaleB is more obvious than that on PIE-pose09. For different cluster numbers, this paper selects K (1, 3, 5, 10, 20, 23), AC and NMI results are different, but the algorithm results are overall stable.

5   Conclusions and Future Work

Based on GNMF, a new improved algorithm, New-SGNMF, is proposed to solve the problem of redundant information. This method not only uses the atlas theory technology to retain the inherent geometric structure of image data, but also implements two-step optimization operation on the basis matrix. Firstly, it sets the threshold value to filter the redundant information. Secondly, it uses the L_2 norm technique to embed it into the objective function to realize the efficient representation of image data. And the specific process of solving the corresponding objective function and iterative formula is given, which clearly shows the specific operation steps and convergence analysis of the algorithm. Finally, the simulation experiments are carried out in COIL20 object image database and PIE-pose09 and YaleB face image database. AC and NMI are used as evaluation system, and the effectiveness of New-SGNMF is illustrated by comparing with K-means, PCA and NMF. However, in the actual operation process, it is found that there are still some shortcomings in the algorithm. There is no unified standard for the selection of threshold S and sparse parameters β in different databases. This problem will be further studied in the future work.

Acknowledgment. This work was supported by the National Natural Science Foundation of China (Grant No. 61501005), the Anhui Natural Science Foundation (Grant No. 1608085 MF 147), the Natural Science Foundation of Anhui Universities (Grant No. KJ2016A057), the Industry Collaborative Innovation Fund of Anhui Polytechnic University and Jiujiang District (Grant No. 2021cyxtb4), the Science Research Project of Anhui Polytechnic University (Grant No. Xjky2020120).

References

1. Wang, J., Liu, X.P.: Medical image recognition and segmentation of pathological slices of gastric cancer based on Deeplab v3+ neural network. Comput. Methods Programs Biomed. **207**, 106210 (2021)
2. He, X., Wang, Y., Zhao, S., et al.: 451 Clinical image identification of basal cell carcinoma and pigmented nevus based on convolutional neural networks. J. Investig. Dermatol. **140**, 59 (2020)
3. Lee, D.D., Seung, H.S.: Learning the parts of objects by non-negative matrix factorization. Nature **401**, 788–791 (1999)
4. Lee, D.D., Seung, H.S.: Algorithms for non-negative matrix factorization. In: Neural Information Processing Systems, Vancouver CANADA, pp. 556–562 (2000)
5. Li, S.Z., Hou, X., Zhang, H., et al.: Learning spatially localized, parts-based representation. In: Proceedings of the 2001 IEEE Computer Society Conference on Computer Vision and Pattern Recognition. CVPR 2001, pp. 207–212 (2001)
6. Liu, W., Zheng, N., Lu, X.: Non-negative matrix factorization for visual coding. In: 2003 IEEE International Conference on Acoustics, Speech, and Signal Processing, 2003. Proceedings. (ICASSP 2003), pp. 293–296 (2003)
7. Zhou, J., Huang, X.H.: Face recognition method based on sparse convex nonnegative matrix factorization with improved iteration step. J. Huazhong Univ. Sci. Technol. (Nat. Sci. Ed. **46**, 48–54 (2018). (in Chinese)
8. Cai, D., He, X., Han, J.: Graph regularized non-negative matrix factorization for data representation. IEEE Trans. Pattern Anal. Mach. Intell. **33**, 1548–1560 (2011)
9. Li, X.L., Cui, G.S., Dong, Y.S.: Graph regularized non-negative low-rank matrix factorization for image clustering. IEEE Trans. Cybern. **47**, 3840–3856 (2017)
10. Wang, X.H., Yang, Q.M., Yang, T.: Face recognition based on improved Gabor transform and nonnegative matrix factorization. Comput. Eng. Appl. **53**, 132–137 (2017). (in Chinese)
11. Xu, H.M., Chen, X.H.: Graph-regularized, sparse discriminant, non-negative matrix factorization. CAAI Trans. Int. Syst. **14**, 1217–1224 (2019). (in Chinese)
12. Long, X., Lu, H., Peng, Y., Li, W.: Graph regularized discriminative non-negative matrix factorization for face recognition. Multimed. Tools Appl. **72**(3), 2679–2699 (2013). https://doi.org/10.1007/s11042-013-1572-z
13. Du, S.Q., Shi, Y.Q., Wang, W.L.: L3/2 sparsity constrained graph non-negative matrix factorization for image representation. In: The 26th China Conference on control and decision-making, Changsha, Hunan, China, pp. 2963–2966 (2014)
14. Qiu, F.Y., Chen, B.W., Chen, T.M., et al.: Sparsity induced convex nonnegative matrix factorization algorithm with manifold regularization. J. Commun. **41**, 84–95 (2020). (in Chinese)
15. Yang, S., Liu, Y., Li, Q., Yang, W., Zhang, Y., Wen, C.: Non-negative matrix factorization with symmetric manifold regularization. Neural Process. Lett. **51**(1), 723–748 (2019). https://doi.org/10.1007/s11063-019-10111-y
16. Li, J.Q., Zhou, G.X., Qiu, Y.N., et al.: Deep graph regularized non-negative matrix factorization for multi-view clustering. Neurocomputing **390**, 108–116 (2020)
17. Li, X.L., Zhang, Y.: Discriminative and graph regularized nonnegative matrix factorization with kernel method. J. Front. Comput. Sci. Technol., 1–11 (2020). (in Chinese)
18. Yu, J.B., Zhang, C.Y.: Manifold regularized stacked autoencoders-based feature learning for fault detection in industrial processes. J. Process Control **92**, 119–136 (2020)
19. Ecke, G.A., Papp, H.M., Mallot, H.A.: Exploitation of image statistics with sparse coding in the case of stereo vision. Neural Netw. **135**, 158–176 (2021)
20. Guo, H.S., Zhang, A.J., Wang, W.J.: An accelerator for online SVM based on the fixed-size KKT window. Eng. Appl. Artif. Intell. **92**, 103637 (2020)

MA Mask R-CNN: MPR and AFPN Based Mask R-CNN

Sumin Qi[1](✉) (iD), Shihao Jiang[2] (iD), and Zhenwei Zhang[1] (iD)

[1] School of Cyberspace Security, Qufu Normal University, Qufu 273165, China
qixm@qfnu.edu.cn
[2] Beijing Imperial Image Intelligent Technology Co., LTD., Beijing 100089, China

Abstract. Multi-resolution parallel ResNet (MPR) and Attention FPN (AFPN) are presented based Mask R-CNN (MA Mask R-CNN) to achieve image instance segmentation, which aims at improving the feature extraction ability of the model, and efficiently increasing the detection and segmentation accuracy. MPR adds parallel branches to extract the feature before each down-sampling of ResNet to make full use of the features. Attention FPN (AFPN) learns the attention mechanism by adding attention modules to each layer of the FPN, so that each layer can adaptively adjust the weight, emphasize the effective information, and suppress the unimportant information. Results of the comparison of the proposed MA Mask R-CNN to the previous state-of-the-art models on COCO dataset show a wider variety of scenarios and complete labels. Our method in instance segmentation task performs better, compared with the state-of-the-art methods. mAP, mAP_{50} and mAP_{75} are improved by 1.1%, 1.6% and 1.3 respectively, to Mask R-CNN, and 0.4%, 1.3%, 0.4% respectively, to PANet, and 7.4%, 7.9%, 9.5% respectively, to FCIS. The average segmentation accuracy of small, medium, and large targets (mAP_S, mAP_M, mAP_L) is increased by 1.4%, 1.0%, and 1.1% to Mask R-CNN, respectively.

Keywords: Instance segmentation · Object detection · MA mask R-CNN · MPR · FPN

1 Introduction

Instance segmentation is an emerging computer vision task, which requires the pixelwise segmenting [1] of each object in an image and the distinction of each instance. Instance segmentation is an extension of two classical computer vision tasks of object detection and semantic segmentation [2]. The aim of object detection is to identify each object and its position with a rectangle frame, while the aim of semantic segmentation is to classify each pixel without differentiating each instance. Our work aims at improving instance segmentation framework.

Object detection models are mainly divided into two types: two-stage models and one-stage models. Two-stage models consist of two stages: one is to find all the regions of interest (ROIs) of the target, and the other is to classify the ROIs and adjust their positions slightly. One-stage models predict the position of ROIs and classify them simultaneously

© Springer Nature Singapore Pte Ltd. 2021
J. Zeng et al. (Eds.): ICPCSEE 2021, CCIS 1451, pp. 233–245, 2021.
https://doi.org/10.1007/978-981-16-5940-9_18

with higher speed, but lower accuracy, such as YOLACT [3], PolarMask [4], SOLO [5]. R-CNN [6] adopts the two-stage procedure, which is a powerful baseline system for object detection. In this model, ROIs are extracted by using selective search algorithm instead of sliding window in traditional target detection, which reduces computation greatly. CNN is used to extract features in ROIs instead of traditional feature operator, and SVM is used to classify the ROIs and optimize their positions. Fast R-CNN [7] extracts the feature map of the input image and gets the corresponding feature of ROIs to avoid the repeated feature extraction and eliminate the redundant calculation, and adds ROI pooling layer to obtain the features with fixed size. In Fast R-CNN deep networks are used for classification and regression of ROIs concurrently to simplify the training process and the storage space needed for training greatly. Faster R-CNN [8] replaces the selective search in Fast R-CNN with Region Proposal Network. It integrates feature extraction, ROIs extraction, classification and regression into a network model, which greatly improves the training and testing efficiency of the model. R-FCN [9] is fully convolutional network with almost all computation shared on the entire image, which achieved the higher speed than Faster R-CNN. Mask-R-FCN [10] applied a learning-based and decision-level strategy to fuse both sematic segmentation and object detection.

Fully Convolutional Network (FCN) framework is used for semantic segmentation [11], which built a fully convolutional networks with input of arbitrary size and correspondingly-sized output to achieve pixelwise semantic segmentation. DeepLab [12] improves the FCN framework with deep convolutional nets, atrous convolution and fully connected CRFs to achieve more accurate semantic segmentation. RefineNet [13] exploits the multi-scale feature for semantic segmentation and introduces chained residual pooling to capture rich background context in an efficient manner.

Taking the effectiveness of R-CNN as the driving force, many models in instance segmentation are based on segment proposals. DeepMask [14] introduces an approach based on a discriminative convolutional network to generate object proposals and then is classified by Fast R-CNN. SharpMask [15] improves the DeepMask with faster speed. Instance-Sensitive FCN [16] develops FCNs to produce instance-level segment candidates, which achieves competitive results of instance segment proposals. FCIS [17] presents a fully convolutional end-to-end solution for instance-aware semantic segmentation task to carry out mask prediction and classification simultaneously. Mask R-CNN [18] adds a branch to predict an object mask to extend Faster R-CNN in parallel with the existing branch for bounding box recognition, which simplified the original R-CNN procedure largely and is a conceptually simple, powerful and the most representative instance segmentation mode.

In Mask R-CNN, ResNet [19] and Feature Pyramid Networks (FPN) [20] is used for feature extraction. However, the sampled higher level feature of FPN is merged with only one of the lower features while the low level feature only undergo a small amount of convolution. Then, there will be a problem of information mismatch while directly merged and the feature information of each layer can't be fully utilized. In addition, the focus of each FPN layer is different (for example, the high-level feature should pay more attention to the larger target), but the weight of each layer in Mask R-CNN is the same, which makes it difficult to make the best of the effective information in each layer.

In view of the above problems, this paper intends to improve the backbone network of Mask R-CNN, and proposes Multi-resolution parallel ResNet (MPR) and Attention FPN (AFPN) based Mask R-CNN (MA Mask R-CNN) (Fig. 1), which will enhance the feature extraction ability of the model.

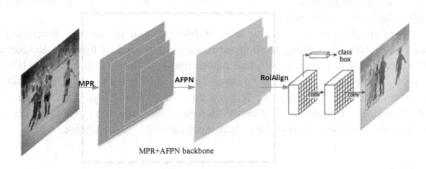

Fig. 1. MA mask R-CNN: MPR and AFPN based mask R-CNN.

Multi-resolution parallel ResNet (MPR) adds parallel branches to extract the feature before each down-sampling of ResNet, the goal of which is to make full use of the features of each level. Attention FPN (AFPN) learns the attention mechanism by adding attention modules to each layer of the FPN, so that each layer can adaptively adjust the weight, emphasize the effective information, and suppress the unimportant information.

We compare our method to the state-of-the-art methods in segmentation and detection task. Experiments results show that our model performs the best in instance segmentation and bounding box detection.

The contributions of our improved model are as follows.

(1) A multi-resolution parallel ResNet (MPR) is proposed to extract more features from each level of ResNet for segmentation effect enhancement of smaller targets.
(2) An attention FPN (AFPN) is proposed to emphasize the effective information and suppress the ineffective information for segmentation effect enhancement of the wanted targets.
(3) MPR and AFPN based Mask R-CNN is proposed for instance segments. Combining MPR with the AFPN, our model has improved segmentation accuracy of targets with different scales.

The rest of this paper is organized as follows. Section 2 introduces the related work. Section 3 introduces the proposed methods. Section 4 gives the experiment results and analysis. Section 5 concludes the paper.

2   Related Work

2.1   ResNet

Researchers have found that too deep network depth leads to over-fitting, which reduces the accuracy and increases the training error and test error of network. This phenomenon is called the Degradation problem.

ResNet attempts to solve degradation problem caused by network deepening by changing the network learning from features learning to residual learning. ResNet can further deepen the depth of convolution neural network. For example, Resnet-101 has 101 layers including one convolution layer on 7×7 ROI with stride 2, and then 33 residual modules. Each residual module has three convolution layers, and the last full connection layer. By introducing identity shortcut connection, the input of the residual module is directly transferred to the output layer for addition calculation.

2.2   FPN

FPN attempts to solve the multi-scale feature extraction by adding up-sampling operation and a small amount of convolution operation on the basis of the original network, in which the low-level features are fused with the up-sampled high-level features, so that each layer of features has appropriate resolution and corresponding semantic features. FPN improves the accuracy and ensures network speed without the extra overhead.

2.3   ResNet + FPN Backbone

Both Faster R-CNN and Mask R-CNN adopt ResNet + FPN backbone to extract multi-scale target features, shown as Fig. 2. The input of FPN is level 2–5 of ResNet output, written down as $\{I_2, I_3, I_4, I_5\}$, where the first level output is abandoned because of large resolution, large space occupation and low semantics. Through a top-down path, the higher level features are fused with the lower level features after up sampling. Each top-down fusion process consists of an up-sampling operation and a 1×1 convolution layer acting as a transverse connection, and then convolution is carried out in a 3×3 convolution kernel to eliminate the aliasing effect caused by up-sampling. The final features are marked as $\{P_2, P_3, P_4, P_5\}$.

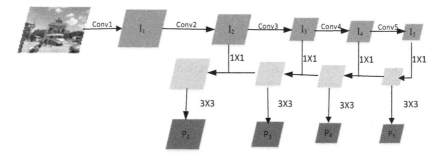

Fig. 2. ResNet+FPN backbone.

2.4 Attention Mechanism

Attention mechanism is an inherent mechanism of human beings to process information [21]. Human beings can quickly obtain the complex information and extract the parts that need to be focused on by allocating more brain resources to process the valuable information, and suppressing the unimportant information.

The goal of attention mechanism in deep learning is to reduce computation and improve the parallel efficiency without damaging the final experimental results, which is widely applied in the field of natural language processing (NLP) [22–24] and computer vision [25, 26]. It is mainly realized by generating masks. By multiplying the weighted mask with the original image, each position information of the image can be enhanced or suppressed to achieve the attention effect.

3   Our Method

Our method adopts the MPR+AFPN backbone, which mainly improves the feature extraction ability, as shown in Fig. 1. The image is first sent to MPR to produce a series of feature maps with different resolutions and the same depth. Then, the feature maps are transmitted to AFPN to obtain the final multi-level output features for subsequent detection and segmentation tasks.

3.1   Multi-resolution Parallel ResNet (MPR)

MPR proposed in this paper adds the multi-resolution parallel branch on the basis of the Resnet101, shown as Fig. 3. The main structure MPR is the same as resnet101, which consists of a head network composed of a 7×7 convolution kernel and a 3×3 pooling layer and 4 residual stages which are composed of 3, 4, 23, and 3 residual modules respectively. The head network changes the input image resolution to 1/4 of the original. In the last three residual stages, the features are down-sampled, expressed as $\{I_{33}, I_{44}, I_{55}\}$.

Our parallel branch attempts to retain the feature map with multi-resolution for accurate target feature, which carries a parallel 3×3 convolution operation before the down-sampling in each residual stage. Feature maps with different resolutions and the same depth can be extracted, expressed as $\{I_{32}, I_{43}, I_{54}\}$. Our parallel branch also adds an information interaction operation ⊛after each down sampling stage, in which the

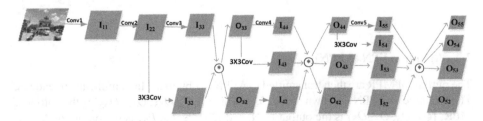

Fig. 3. MPR structure. ⊛is an interaction operation.

238   S. Qi et al.

features of each layer interact with the information of other layers. The last interaction operation outputs the fused feature maps, expressed as $\{O_{52}, O_{53}, O_{54}, O_{55}\}$.

To show the above interaction operation, we define serval operators as follows.

Definition 1: Let x, y, z be three feature maps, $\Downarrow$ is said to be a *down-sample operator* if x is down sampled to z with 3×3 kernel, stride 2, whose channels and resolution are the same as y, written as $z = x \Downarrow y$.

NOTE: *down-sample operator* $\Downarrow$ doesn't satisfies commutativity, but associative.

Definition 2: Let x, y, z be the three feature maps, $\Uparrow$ is said to be an *up-sample operator* if x is up sampled to z with linear interpolation operation, whose channels and resolution are the same as y, written as $z = x \Uparrow y$.

NOTE: up-sample *operator* $\Uparrow$ doesn't satisfies commutativity, but associative.

The feature maps with different resolutions must perform information interaction operation to fuse information from other layers as Fig. 3. Let $\{I_{m1}, I_{m2}, \ldots, I_{mn}\}$ be the input of information interaction, and $\{O_{m1}, O_{m2}, \ldots, O_{mn}\}$ be the output of the module, which is arranged in ascending order of resolution and n is the number of feature maps. Each output is an aggregation of input information, as shown in Eq. 1. Figure 4 can clearly show the operation procedure, where n is equal to 3, the inputs are at the left while the outputs are at the right. The number of channels of each branch is reduced to 1/4 so that the parameters and computation are basically the same as resnet101.

$$O_{mk} = \sum_{i=1}^{k-1} I_{mi} \Uparrow I_{mk} + I_k + \sum_{i=k+1}^{n} I_{mi} \Downarrow I_{mk} (0 \leq k \leq n). \tag{1}$$

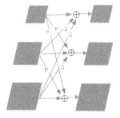

Fig. 4. Information interaction operation procedure. The inputs are at the left while the outputs are at the right.

3.2 Attention FPN (AFPN)

The outputs of MPR are then transmitted to AFPN to obtain the final multi-level features. The structure of AFPN is shown as Fig. 5, in which $\{O_{52}, O_{53}, O_{54}, O_{55}\}$ is the output of MPR, $\{O_2, O_3, O_4, O_5\}$ is the output FPN, and $\{P_2, P_3, P_4, P_5\}$ is the output of AFPN. An attention module, which can also be called attention operation Θ, is added to each layer of FPN to avoid losing some trivial position features, and make each layer focus on the features of the corresponding region, which improves the accuracy of target detection and segmentation tasks, shown as Eq. 2.

$$P_k = \Theta O_k (2 \leq k \leq 5) \tag{2}$$

What's more, the attention module is integrated with the entire network and is jointly trained with the entire network without additional supervision.

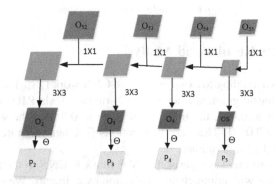

Fig. 5. AFPN structure. Θ is an attention operator.

For input features with dimensions $C \times H \times W$, the processing flow of a single attention operation (Fig. 6) is as follows:

(1) The input is pooled on the channel dimension C to aggregate the spatial features. Maximum pooling and average pooling are used respectively to obtain two feature maps with dimension $1 \times H \times W$, which represent the aggregation in the average way and the maximum way.

(2) The two obtained feature maps are connected into features map with dimension $2 \times H \times W$. And then they are sent to the subsequent attention network to get the attention feature with dimension $1 \times H \times W$ because the attention network consists of a series of 3×3 convolutional layers. Through the sigmoid function, the attention feature is mapped to the range of 0–1, which is an attention weight map. The weight map encodes the positions in the feature that need to be emphasized or suppressed. Finally, by multiplying the attention weight map and the original input element-by-element, a new feature containing attention mechanism is obtained.

(3) Because the attention weights between 0–1 weakens all features but important features, in order not to affect the representational ability of the feature map, a residual structure is added to the attention mechanism module, and the original input is added to the output feature.

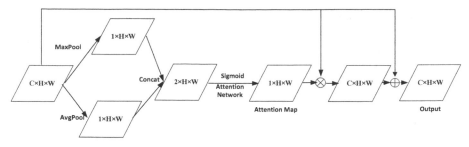

Fig. 6. A single attention operation.

4 Experiments Results and Analysis

We set parameters according to existing Mask R-CNN work [18]. The proposed mod-el is trained by stochastic gradient descent (SGD) on two Tesla P100 GPUs. The initial learning rate is set to 0.005, the momentum is set to 0.9, and the weight attenuation coefficient is set to 0.0001. There are 12 iterations. The learning rate is reduced to 0.1 times in the 8th and 11th iterations.

We perform a profound comparison of MA Mask R-CNN to the previous state-of-the-art models along with comprehensive ablation experiments. We choose the COCO dataset [27] for all experiments, which has 91 target categories and each category contains more than 5000 target instances except very few categories. The images in COCO dataset contain rich scenes, multiple objects and complex natural background, which makes the COCO dataset more difficult and challenging than other datasets. In addition, the targets in the data set are marked with borders, masks and other tags, which are suitable for instance segmentation task.

The standard COCO metrics are reported, including mAP (Mean Average Precision), mAP50 (mAP at an IoU of 0.5), mAP75 (mAP at an IoU of 0.75), and mAPS (mAP at small scale), mAPM (mAP at medium scale), mAPL (mAP at large scale), as shown in [28, 29]. As in previous work [18], 80k train images and a 35k val images are used for train, and the remaining 5k subset of val images are used for ablations. We perform a profound comparison of our model to the previous state-of-the-art models along with comprehensive ablation experiments.

4.1 Main Result

Our method is compared to Mask R-CNN in instance segmentation, which is visualized in Fig. 7. The model can correctly identify and segment targets of different sizes in complex scenes in the COCO dataset. What's more, our method can accurately identify and provide a more accurate segmentation mask, while Mask R-CNN Mistakes the background for targets.

The performance metrics of comparisons in instance segmentation task are shown in Table 1. All instantiations of our method outperform other methods, including FCIS [17], Mask R-CNN [18], PANet [30]. Our method in instance segmentation task performs the best. mAP, mAP50 and mAP75 are improved by 1.1%, 1.6% and 1.3% respectively, to

Fig. 7. Experiment results comparisons for instance segmentation. Top is results of mask R-CNN and bottom is results of our method.

Mask R-CNN, and 0.4%, 1.3%, 0.4% respectively, to PANet, and 7.4%, 7.9%, 9.5% respectively, to FCIS.

Table 1. Segment results on COCO test-dev vs. state-of-the-art on test-dev.

Model	mAP	mAP_{50}	mAP_{75}	mAP_S	mAP_M	mAP_L
FCIS	29.6	51.4	30.2	8.0	31.0	49.7
Mask R-CNN	35.9	57.7	38.4	19.2	39.7	49.7
PANet	36.6	58.0	39.3	16.3	38.1	53.1
Our method	**37.0**	**59.3**	**39.7**	**20.6**	**40.7**	**50.8**

Due to our MPR, the information contained in the low-level features is more abundant, so that the segmentation effects of smaller targets are more significantly improved. Attention feature maps are easier to train for larger targets, which can generate attention weights that are more in line with the distribution of real targets for large targets. Therefore, the AFPN can effectively improve the segmentation effects of large targets. Combining MPR with the AFPN, the model has improved segmentation accuracy of targets with different scales. The average segmentation accuracy of small, medium, and large targets (mAPS, mAPM, mAPL in Table 1) is increased by 1.4%, 1.0%, and 1.1% to Mask R-CNN, respectively.

4.2 Ablation Study

In order to further prove the effectiveness of each part of our mode proposed in this paper, ablation experiments are carried out on our MPR and AFPN. Each is applied to Mask R-CNN model separately and compared with the original Mask R-CNN, as shown in Table 2.

Of all the models in Table 2, the backbone with our MPR+AFPN performs the best. Compared to the original backbone with ResNet+FPN, the mAP of backbone with MPR, AFPN, MPR+AFPN have improved by 0.9%, 0.6%, and 1.1%, respectively. mAP at an IoU of 0.5 (AP_{50}) of backbone with MPR, AFPN, MPR+AFPN have improved by 0.8%, 0.7%, and 1.6%, respectively. mAP at an IoU of 0.75 (AP_{75}) of backbone with MPR, AFPN, MPR+AFPN have improved by 0.7, 0.8 and 1.3%, respectively. The effect of backbone with MPR only on small targets is more obvious, and the mAP_S is improved by 1.3%. The effect of backbone with AFPN only on larger targets is more significant, and the mAP_L has increased by 0.7%. It is worth noting that the mAP of backbone with MPR+AFPN on small targets, medium targets, and large targets have increased by 1.4%, 1.0% and 1.1%, respectively.

Table 2. Segment results of ablation experiments on COCO test-dev.

Backbone	mAP	mAP_{50}	mAP_{75}	mAP_S	mAP_M	mAP_L
ResNet+FPN	35.9	57.7	38.4	19.2	39.7	49.7
MPR	36.8	58.5	39.1	20.5	39.9	49.8
AFPN	36.5	58.4	39.2	19.5	40.1	50.4
MPR+AFPN	**37.0**	**59.3**	**39.7**	**20.6**	**40.7**	**50.8**

4.3 Bounding Box Detection Results

The proposed MA Mask R-CNN is also compared to the state-of-the-art object detection models including RentinaNet [9], Mask R-CNN [15], as is shown in Table 3. Our method in bounding box detection performs the best. The mAP^{bb}, mAP^{bb}_{50} and mAP^{bb}_{75} are improved by 1.8%, 2.5%, 1.5% respectively, compared to Mask R-CNN, and 2.1%, 4.3%, 2.5% respectively, compared to RentinaNet. The mAP of small, medium, and large targets (mAP^{bb}_S, mAP^{bb}_M, mAP^{bb}_L in Table 4) is increased by 1.7%, 1.7%, 2.1%, respectively, compared to Mask R-CNN, and 3.5%, 2.7%, 3.2% respectively, compared to RentinaNet.

As a further comparison, we train two versions of backbones with MPR only and with AFPN only. Both models perform better than the original backbone with ResNet and FPN. What's more, our method with both MPR and AFPN performs the best due to the efficiency of the two approaches. Finally, we notice that our method keeps a gap (points 4.2) between the mask mAP (37.0 in Table 1) and box mAP^{bb} (41.2 in Table 3). This shows that our method is more suitable for instance segmentation task.

Table 3. Object detection single-model results vs. state-of-the-art on test-dev.

Model	mAP^{bb}	mAP^{bb}_{50}	mAP^{bb}_{75}	mAP^{bb}_{S}	mAP^{bb}_{M}	mAP^{bb}_{L}
RentinaNet	39.1	59.1	42.3	21.8	42.7	50.2
Mask R-CNN	39.4	60.9	43.3	23.0	43.7	51.4
Our method	**41.2**	**63.4**	**44.8**	**25.3**	**45.4**	**53.4**

Table 4. Detection mask AP of ablation experiments on COCO test-dev.

Backbone	mAP^{bb}	mAP^{bb}_{50}	mAP^{bb}_{75}	mAP^{bb}_{S}	mAP^{bb}_{M}	mAP^{bb}_{L}
ResNet + FPN	39.4	60.9	43.3	23.0	43.7	51.4
MPR	40.7	61.9	44.5	25.0	44.2	51.7
AFPN	40.4	61.6	44.4	23.8	44.6	52.7
MPR + AFPN	**41.2**	**63.0**	**45.2**	**25.3**	**45.4**	**53.4**

5 Conclusions

In this paper we propose a multi-resolution parallel ResNet (MPR) and an attention FPN (AFPN), and use them as the backbones of Mask R-CNN to form MA Mask R-CNN and obtain the final multi-level output features for subsequent instance segment tasks. Our approach efficiently improves the feature extraction ability of the model. The results of a deep comparison of our method to the previous state-of-the-art methods on COCO dataset are shown. The proposed MA Mask R-CNN in instance segmentation and object detection task performs the best. Compared with Mask R-CNN, the standard COCO metrics including AP (averaged over IoU thresholds), AP_{50}, AP_{75}, and AP_S, AP_M, AP_L (AP at different scales) on instance segmentation task and bounding-box object detection task are increased respectively. Ablation experiments are carried out on our MPR and AFPN. The effect of our method with MPR only on small targets is more obvious, and the effect of our method with AFPN only on larger targets is more significant. Combining MPR with the AFPN, the model has improved detection and segmentation accuracy of targets on different scales.

Acknowledgments. This work has been supported by National Natural Science Foundation of China (No. 6160261) and acknowledge assistance and financial support from our school.

References

1. Arnab, A., Torr, P.H.S.: Pixelwise instance segmentation with a dynamically instantiated network. In: Proceedings of 30th IEEE Conference on Computer Vision and Pattern Recognition, pp. 879–888. IEEE (2017)

2. Watanabe, T., Wolf, D.F.: Instance segmentation as image segmentation annotation. In: Proceedings of IEEE Intelligent Vehicles Symposium, pp. 432–437. IEEE (2019)
3. Bolya, D., Zhou, C., Xiao, F., Lee, Y.J.: YOLACT: real-time instance segmentation. In: Proceedings of the IEEE International Conference on Computer Vision, pp. 156–9165. IEEE (2019)
4. Xie, E., et al.: PolarMask: single shot instance segmentation with polar representation. In: Proceedings of the IEEE Computer Society Conference on Computer Vision and Pattern Recognition, pp. 12190–12199. IEEE (2020)
5. Wang, X., Kong, T., Shen, C., Jiang, Y., Li, L.: SOLO: segmenting objects by locations. In: Vedaldi, A., Bischof, H., Brox, T., Frahm, J.-M. (eds.) ECCV 2020. LNCS, vol. 12363, pp. 649–665. Springer, Cham (2020). https://doi.org/10.1007/978-3-030-58523-5_38
6. Girshick, R., Donahue, J., Darrell, T., Malik J.: Rrich feature hierarchies for accurate object detection and semantic segmentation. In: Proceedings of the IEEE Computer Society Conference on Computer Vision and Pattern Recognition, pp. 580–587. IEEE (2014)
7. Girshick, R.: Fast R-CNN. In: Proceedings of IEEE International Conference on Computer Vision, pp. 1440–1448. IEEE (2015)
8. Ren, S., He, K., Girshick, R., Sun, J.: Faster R-CNN: towards real-time object detection with region proposal networks. IEEE Trans. Pattern Anal. Mach. Intell. **39**(6), 1137–1149 (2017)
9. Dai, J., Li ,Y., He, K., Sun, J.: R-FCN: object detection via region-based fully convolutional networks. In: Proceeding of Advances in Neural Information Processing Systems, pp. 379–387 (2016)
10. Zhang, Y., Chi, M.: Mask-R-FCN: a deep fusion network for semantic segmentation. IEEE Access **8**, 155753–155765 (2020)
11. Long, J., Shelhamer, E., Darrell, T.: Fully convolutional networks for semantic segmentation. In: Proceeding of IEEE Conference on Computer Vision and Pattern Recognition, pp. 1–3. IEEE (2015)
12. Chen, L., Papandreou, G., Kokkinos, I., Murphy, K., Yuille, A.L.: DeepLab: semantic image segmentation with deep convolutional nets atrous convolution, and fully connected CRFs. IEEE Trans. Pattern Anal. Mach. Intell. **40**(4), 834–848 (2018)
13. Lin, G., Milan, A., Shen, C., Reid, I.: RefineNet: multi-path refinement networks for high-resolution semantic segmentation. IEEE Trans. Pattern Anal. Mach. Intell. **42**(5), 1228–1242 (2017)
14. Chen, T., Lin, L., Wu, X., Xiao, N., Luo, X.: Learning to segment object candidates via recursive neural networks. IEEE Trans. Image Process. **27**(12), 5827–5839 (2018)
15. Pinheiro, P.O., Lin, T.-Y., Collobert, R., Dollár, P.: Learning to refine object segments. In: Leibe, B., Matas, J., Sebe, N., Welling, M. (eds.) ECCV 2016. LNCS, vol. 9905, pp. 75–91. Springer, Cham (2016). https://doi.org/10.1007/978-3-319-46448-0_5
16. Dai, J., He, K., Sun, J.: Instance-sensitive fully convolutional networks. In: Proceeding of IEEE Conference on Computer Vision and Pattern Recognition, pp. 534–549. IEEE (2016)
17. Li, Y., Qi, H., Dai, J., Ji, X., Wei, Y.: Fully convolutional instance-aware semantic segmentation. In: Proceeding of IEEE Conference on Computer Vision and Pattern Recognition, pp. 4438–4446. IEEE (2017)
18. He, K., Gkioxari, G., Dollar, P., Girshick, R.: Mask R-CNN. In: Proceeding of IEEE International Conference on Computer Vision, pp. 2980–2988. IEEE (2017)
19. He, K., Zhang, X., Ren, S., Sun, J.: Deep residual learning for image recognition. In: Proceeding of IEEE Conference on Computer Vision and Pattern Recognition, pp. 770–778. IEEE (2016)
20. Lin, T.Y., Dollar, P., Girshick, R., He, K., Hariharan, B., Belongie, S.: Feature pyramid networks for object detection. In: Proceeding of IEEE Conference on Computer Vision and Pattern Recognition, pp. 936–944. IEEE (2017)

21. Tsotsos, J.K.: A Computational Perspective on Visual Attention. MIT Press, Cambridge (2011)
22. Lin, L., Luo, H., Huang, R., Ye, M.: Recurrent models of visual co-attention for person re-identification. IEEE Access **7**, 8865–8875 (2019)
23. Toshniwal, S., Livescu, K.: Jointly learning to align and convert graphemes to phonemes with neural attention models. In: Proceeding of IEEE Spoken Language Technology Workshop, pp. 76–82 IEEE. (2016)
24. Yin, W., Schtze, H., Xiang, B., Zhou, B.: ABCNN: attention-based convolutional neural network for modeling sentence pairs. Trans. Assoc. Comput. Linguist. **4**, 259–272 (2016)
25. Hu, J., Shen, L., Sun, G.: Squeeze-and-excitation networks. IEEE Trans. Pattern Anal. Mach. Intell. **42**(8), 2011–2023 (2020)
26. Zhu, Y., Zhao, C., Guo, H., Wang, J., Zhao, X., Lu, H.: Attention CoupleNet: fully convolutional attention coupling network for object detection. IEEE Trans. Image Process. **28**(1), 113–126 (2019)
27. Lin, T.-Y., et al.: Microsoft COCO: common objects in context. In: Fleet, D., Pajdla, T., Schiele, B., Tuytelaars, T. (eds.) ECCV 2014. LNCS, vol. 8693, pp. 740–755. Springer, Cham (2014). https://doi.org/10.1007/978-3-319-10602-1_48
28. Bell, S., Zitnick, C. L., Bala, K., Girshick, R.: Inside-outside net: detecting objects in context with skip pooling and recurrent neural networks. In: Proceeding of IEEE Conference on Computer Vision and Pattern Recognition, pp. 2874–2883. IEEE (2016)
29. Lin, T.Y., Goyal, P., Girshick, R., He, K., Dollar, P.: Focal loss for dense object detection. IEEE Trans. Pattern Anal. Mach. Intell. **42**(2), 318–327 (2020)
30. Liu, S., Qi, L., Qin, H., Shi, J., Jia, J.: Path aggregation network for instance segmentation. In: Proceedings of the IEEE Computer Society Conference on Computer Vision and Pattern Recognition, pp. 8759–8768. IEEE (2018)

Generative Adversarial Network Based Status Generation Simulation Approach

Zhiru Chen[1,2(✉)], Zhi Zhang[1,2], Xi Zhao[1,2], and Liang Guo[3]

[1] State Grid ShanDong Marketing Service Center(Metrology Center),
Jinan 250000, Shandong, China
[2] Measuring Center, State Grid Shandong Electric Power Research Institute,
Jinan 250000, Shandong, China
[3] State Grid Shandong Electric Power Company, Jinan 250000, Shandong, China

Abstract. The generative adversarial network based methods have been applied in many fields for simulation data generation. For power equipment, due to the combined influences of multiple factors, how to generate reasonable simulation data that meets specific requirements has become a challenge. This paper proposes a power equipment status generation approach based on generative confrontation network. This approach considers the changing factors of power equipment and takes it as the conditional distribution of simulation data during training. The proposed approach is applied to the status generation of power equipment, and the rationality and effectiveness of the approach are verified through experiments.

Keywords: Generative adversarial network · Simulation data · Policy gradient

1 Introduction

For a long time, power grid companies have been committed to the development of automated collection of electric energy measurement. Up to now, the electricity consumption information collection system built by State Grid Co., Ltd. has achieved 500 million meters, and the number of collection and metering equipment connected to the largest provincial system is close to 50 million. With the advancement of the construction of the energy Internet, various emerging energy metering businesses develop continuously. It is necessary to establish a simulation environment for electric energy metering equipment with large-scale, system-level, complex environment, and multi-channel integration, which could support application and verification of new equipment and new technologies. However, due to the complexity of the on-site operating environment, how to simulate the impact of the operating environment on the operating status of the equipment has always been a major challenge.

Supported by Technology Project of State Grid (No. 5600-201955167A-0-0-00).

© Springer Nature Singapore Pte Ltd. 2021
J. Zeng et al. (Eds.): ICPCSEE 2021, CCIS 1451, pp. 246–255, 2021.
https://doi.org/10.1007/978-981-16-5940-9_19

The traditional simulation approaches for operation of electric energy metering equipment mainly include physical equipment simulation and digital simulation. Although the use of physical equipment can simulate the operating status of the equipment, it is difficult and costly to simulate various factors that affect the operating status of the equipment, such as temperature, humidity, and electromagnetic interference, especially when multiple factors are mixed and combined. In addition, physical device simulation approach is easily affected by the differences of individual devices. Moreover, simulation based on physical devices is difficult to scale. It is not appropriate to use physical equipment to simulate the operating status of equipment in a complex operating environment.

The digital simulation approach can display the operation index of the electric energy metering equipment under different environmental conditions by outputting the results, which is more suitable for simulating the operation status of the equipment under the complex operation environment. However, the traditional digital simulation is to generate data first, and then output the generated results through digital simulation equipment. Because the data is generated in advance in this way, it cannot be adjusted in real time in the running state, and cannot meet the needs of two-way interaction between the operator and the simulation equipment during the simulation of electricity consumption.

To solve these above issues, this paper proposes a generative adversarial networks based status generation simulation approach for electric energy metering equipment. According to the changing environment or human dynamic intervention, the proposed simulation approach could generate simulation data.

The main contributions of this paper include:

1) A generative adversarial networks based status generation simulation approach is proposed, which could take into account the changing environment or human intervention. The simulation status data could be generated according to the changing context.
2) Experiments demonstrate the effectiveness of the proposed simulation approach.

The structure of this paper is organized as follows: Sect. 2 gives the background and related work. The problem is given in Sect. 3. The proposed approach is presented in Sect. 4. Then Sect. 5 gives the experiments and analysis. Conclusion and future work is presented in Sect. 6.

2   Background and Related Work

This section gives the background of generative adversarial networks. The related work about data generation approaches are also presented here.

2.1   Background

Generative Adversarial Networks (GAN) [6] was proposed to generate fake data which is similar with ground truth realistic data. Two adversarial players are introduced and adversarial training is used to generate the similar data with the ground truth data. The two players are Generator G and Discriminator D. The Generator G generates data which is close to the realistic data distribution $p_d(x)$. The Discriminator D tries to distinguish the realistic data from fake data generated by G. The training of GAN is a minimax game between Generator G and Discriminator D. The objective of the training process is as shown in Eq. 1. After the training, the Generator G could generate fake data with close similarity to the ground truth data.

$$\min_G \max_D (\mathbb{E}_{x \sim p_d} log D(x) + \mathbb{E}_{z \sim p_z} log(1 - D(G(z)))) \qquad (1)$$

The z stands for the random noise, which is the input of Generator G, and p_z is the prior distribution of the noise data z.

For optimization of GAN, reinforcement learning [8] is often used as a general learning framework. The agent could learns how to take actions to maximize the cumulative rewards. Received rewards could make agent to learn a better policy to maximize the cumulative rewards from the environment. With the development of deep learning technology, the combination of deep learning and reinforcement learning is a hot research topic [5,15]. Policy gradients [9] are often used. The objective of policy gradient is to maximize the cumulative rewards over an episode of T, which is suitable for the optimization of status sequence generation for electric equipment.

Ensemble learning [3] could construct a stronger learner through the combination of a set of weak learners. The performance could be improved. Bagging, boosting and stacking are the popular ensemble learning methods.

2.2   Related Work

For the fake sequence data generation, Generative Adversarial Networks (GAN) [6] approaches are often used. In some situation, the data is conditional to some context. Then conditional generative adversarial [11] net was proposed to tackle this issue. In the training process, the extra information y could be combined into the both generator and discriminator as additional input layer. Reference [1] proposed a semi-supervised learning with context-conditional generative adversarial networks for images.

For improving the performance of GAN, Reference [14] utilized the step-wise evaluation to improve conditional sequence generative adversarial networks. EvolGAN [13] incorporated Evolution algorithm to improve the performance of the GAN and got the better outputs. A quality estimator and evolutionary methods were used to search the latent space of GAN. This could make the generator to generate high quality data, which had the diversity property. Penalty-based approach was also introduced into GAN for sequence generation [4]. Two-stage GAN [18] was proposed to generate radar image sequences.

For sequence generation task, there are many researches. SeqGAN [17] was proposed, which model the data generator as a stochastic policy in reinforcement learning. The generator differentiation problem could be bypassed by SeqGAN by directly performing gradient policy update. Extensive experiments demonstrated significant improvements over existing approaches. For tacking the specific optimization goal of task, OptiGAN [7] was proposed. Besides generating the similar fake data, the desired goals or properties specific to the task is considered. OptiGAN was applied in text and real-valued sequence generation and got better performance than baselines.

GAN has a good performance in fake data generation, and could be used in many different areas, such as transactions generation [2], wind power scenario generation [10] and so on. The GAN is the hot spot, which attracts a lot of researchers.

3   Problem Definition

3.1   Influencing Factors Analysis

The main factors affecting the operation of electric energy metering equipment can be considered from the following aspects, as shown in Table 1.

Table 1. Influencing factors for electric equipment status.

	Influencing factors
1	Installation environment
2	Temperature
3	Humidity
4	Vibration
5	Chemical Gas
6	Dust
7	Other natural environmental factors

1) Installation Environment
   Common installation environments for electric energy metering equipment include placement in the open-air, placement in a distribution cabinet, and placement in a safety protection system. Among them, the equipments placed in the open-air is more sensitive to the environment.
2) Temperature
   During the operation of smart energy metering equipment, temperature has a greater impact on measurement accuracy and equipment operating status.
3) Humidity
   The effect of humid air on maintaining the operating accuracy of electrical energy metering equipment is obvious. It could affect the accuracy of the equipment.

4) Vibration

Under the vibrating environment, the probability of failure of various components in the electric energy metering equipment increases.

5) Chemical Gas

For some relatively special working environments and operating areas of electric energy metering equipment, the device is inevitably exposed to the working environment of chemical gases, which ultimately leads to a significant drop in the operating stability of the equipment.

6) Dust

The influence of dust usually acts on the outside of the electric energy metering equipment, and a certain amount of heat will be generated during the operation of the equipment.

7) Other Natural Environmental Factors

Other natural environmental factors include lightning, wind, haze in the natural environment. Theoretically, electric energy metering equipment is relatively less affected in a non-open air environment.

To sum up, the status generation of electric equipment could be affected by many context information, as shown in Table 1. These changing context and environment information should be considered into the simulation data generation process to get a more similar and rational sequence data.

3.2   Status Generation for Electric Equipment

These changing context involves the installation environment, temperature, humidity, vibration, chemical gas, dust, and other natural environmental factors. The status sequence of electric equipment is conditioned to these context information. In different execution context, the status change of electric equipment is different. For this reason, the status generation process should consider the context information. That is, the status should be generated considering the conditional distribution. Then generative adversarial network based status generation approach incorporate the conditional probability. When feeding the random noise or real data into the GAN, the conditional distribution information should also be considered.

The electric equipment may face many different contexts. Not all contexts are considered and trained. The status executed in these neglected context should be considered carefully to improve the similarity of the generated fake data. Then the ensemble learning approach could be adopted.

4   Proposed Approach

A dynamic generation method of electric energy metering equipment status based on a generative countermeasure network is proposed. On the basis of traditional digital simulation, the operating environment factors can be changed in real time, and the digital simulation equipment can dynamically change its operating status according to the environmental factors and the actions of the operators, so as to achieve a high degree of restoration of the on-site operating status. Considering the changing context, the generative adversary network based status generation architecture is shown in Fig. 1.

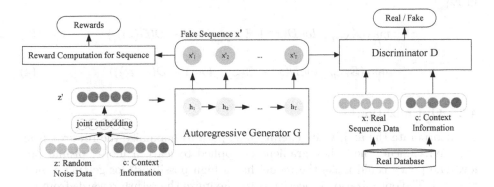

Fig. 1. Generative adversary network based status generation architecture.

4.1 Conditional Information Modeling

For electric equipment, the status is conditional to the execution environment, which is changing over time. For this reason, the generator and discriminator should be conditioned on these changing context information c in Fig. 1, which could be embedded from installation environment, temperature, humidity, vibration, chemical gas, dust, and other natural environmental factors.

These conditional information c could be feeding into both the discriminator and generator as additional input layer. For generator, the prior input noise data is z. However, the distribution of z is conditioned into the context information c. Then the noise data z and related information c is combined into a joint hidden representation. For discriminator D, the context information c should be considered.

Embedding technology could be used in the joint combination of noise information z and context information c. The combination is as shown in Eq. 2.

$$z' = E(z, c) \tag{2}$$

4.2 Adversarial Training

A neural autoregressive model Bi-RNN is employed as a generator, which could map combination of the input noise data z and the context information c into a fake sequence, which is mimic the realistic data from the real database. For given specific goal, policy gradient RL is incorporated for training. The overall training architecture is as shown in Fig. 1.

The model could be trained by alternatively update the discriminator D and the generator G. When training one of two players, the other one is fixed. When update discriminator D, the generator G is fixed. That is, the parameters of the generator is fixed. Then the optimization goal is as shown in Eq. 3. When update generator G, the discriminator D is fixed. And the optimization goal is as shown in Eq. 4.

$$\max_{D}(\mathbb{E}_{X \sim p_{d,c}} log D(x) + \mathbb{E}_{z \sim p_{z,c}} log(1 - D(G(z)))) \tag{3}$$

$$\max_{G}(\mathbb{E}_{x \sim p_{d,c}} log D(x) - \mathbb{E}_{z \sim p_{z,c}} log(1 - D(G(z)))) \tag{4}$$

4.3   Optimization

To incorporate the ability to generate fake data to maximize the rewards from the environment, the policy gradient is applied to learn a more optimal policy. The policy could make the model has a high possibility for getting a more rewards. The optimization objective is to maximize the return rewards from the environment, as shown in Eq. 5. These rewards are from an episode of T time steps.

$$J(\theta) = \mathbb{E}[\sum_{t} U_t log\pi(A_t|S_t, \theta)] \tag{5}$$

In the optimization objective, π is the policy. A_t and S_t are the action and state at time t, and $t = [0, 1, ..., T-1]$.. θ are the parameters of π and U_t is the return rewards at time t.

The REINFORCE algorithm [16] is used to search the parameter space and get the optimum one for policy G by gradient ascent of the gradient of J. The choice of U_t could be customized in different specific tasks.

The final total loss to train the generator G with adversarial learning and policy gradients could be as shown in Eq. 6.

$$\max_{G}(\mathbb{E}_{x \sim p_{d,c}} log D(x) + \lambda \mathbb{E}_{z \sim p_{z,c}} log(D(G(z))) + \alpha \mathbb{E}_{X_{P_G}}[\sum_{t} U_t log\pi(A_t|S_t, \theta)]) \tag{6}$$

Based on the input random noise data z, realistic data x and the related context information c, the proposed approach train the generator G in a principled adversary way. The proposed approach could generate similar fake data according to the context environment.

5   Experiments

5.1   Training Data and Baselines

The proposed approach is evaluated in the production data from a power company. The electric equipment is deployed in different positions with different context information. The performance and the status of electric equipment is affected by the environment. The realistic status data and related context information are extracted from these electric equipment.

LSTM are a frequently used generation model. Sequence data, such as text, could be generated through LSTM. LSTM is used as a baseline. OptiGAN [7] are used to generate sequence data in text generation and air-craft trajectory generation. These approaches are used as baselines in this paper.

5.2 Analysis

The negative log-likehood (NLL) [12] is used to evaluate the diversity of the generated data of the proposed approach.

$$NLL_{gen} = -\mathbb{E}_{x_{1:T} \sim P_d} log P_{G_\theta}(x_1, ..., x_T) \qquad (7)$$

P_d and P_{G_θ} are the real data and generated data distributions. The generated data would be more similar to the empirical data distribution, when the NLL has a lower value.

LSTM, OptiGAN and the proposed approach are performed in the electric equipment data from production. The experiments are done in Tensorflow platform. The experiments results are shown in Table 2.

Table 2. NLL comparison.

	NLL
LSTM	1.370 ± 0.125
OptiGAN	1.237 ± 0.132
Proposed approach	$\mathbf{1.138 \pm 0.118}$

6 Conclusion

For electric equipment status generation, a generative adversarial networks based status generation simulation approach is proposed in this paper. The context and conditional situation are incorporated into the training process of GAN based approach. For some rare situation, resemble learning is introduced to get a better simulation data.

However, there are many unforeseen conditions. For contexts that have not been seen and analyzed, ensemble learning is used to generate a more reasonable simulation data, which is the future work.

References

1. Denton, E.L., Gross, S., Fergus, R.: Semi-supervised learning with context-conditional generative adversarial networks. CoRR abs/1611.06430 (2016)
2. Doan, T., Veira, N., Keng, B.: Generating realistic sequences of customer-level transactions for retail datasets. In: Tong, H., Li, Z.J., Zhu, F., Yu, J. (eds.) 2018 IEEE International Conference on Data Mining Workshops, ICDM Workshops, Singapore, 17–20 November 2018, pp. 820–827. IEEE (2018)
3. Dong, X., Yu, Z., Cao, W., Shi, Y., Ma, Q.: A survey on ensemble learning. Front. Comput. Sci. 14(2), 241–258 (2019). https://doi.org/10.1007/s11704-019-8208-z
4. Duan, M., Li, Y.: Penalty-based sequence generative adversarial networks with enhanced transformer for text generation. In: 2020 International Joint Conference on Neural Networks, IJCNN 2020, Glasgow, United Kingdom, 19–24 July 2020, pp. 1–6. IEEE (2020)
5. François-Lavet, V., Henderson, P., Islam, R., Bellemare, M.G., Pineau, J.: An introduction to deep reinforcement learning. Found. Trends Mach. Learn. 11(3–4), 219–354 (2018)
6. Goodfellow, I.J., et al.: Generative adversarial nets. In: Ghahramani, Z., Welling, M., Cortes, C., Lawrence, N.D., Weinberger, K.Q. (eds.) Advances in Neural Information Processing Systems 27: Annual Conference on Neural Information Processing Systems 2014, 8–13 December 2014, Montreal, Quebec, Canada, pp. 2672–2680 (2014)
7. Hossam, M., Le, T., Huynh, V., Papasimeon, M., Phung, D.: Optigan: generative adversarial networks for goal optimized sequence generation. In: 2020 International Joint Conference on Neural Networks, IJCNN 2020, Glasgow, United Kingdom, 19–24 July 2020, pp. 1–8. IEEE (2020)
8. Johnson, J.D., Li, J., Chen, Z.: Reinforcement learning: An introduction: R.S. sutton, A.G. barto, MIT press, cambridge, MA 1998, 322, pp. 205–206, ISBN 0-262-19398-1. Neurocomputing 35(1-4) (2000)
9. Li, S., Wu, Y., Cui, X., Dong, H., Fang, F., Russell, S.J.: Robust multi-agent reinforcement learning via minimax deep deterministic policy gradient. In: The Thirty-Third AAAI Conference on Artificial Intelligence, AAAI 2019, The Thirty-First Innovative Applications of Artificial Intelligence Conference, IAAI 2019, The Ninth AAAI Symposium on Educational Advances in Artificial Intelligence, EAAI 2019, Honolulu, Hawaii, USA, 27 January–1 February 2019, pp. 4213–4220. AAAI Press (2019)
10. Liang, J., Tang, W.: Sequence generative adversarial networks for wind power scenario generation. IEEE J. Sel. Areas Commun. 38(1), 110–118 (2020)
11. Mirza, M., Osindero, S.: Conditional generative adversarial nets. CoRR abs/1411.1784 (2014)
12. Nie, W., Narodytska, N., Patel, A.: Relgan: Relational generative adversarial networks for text generation. In: 7th International Conference on Learning Representations, ICLR 2019, New Orleans, LA, USA, 6–9 May 2019 (2019) OpenReview.net
13. Roziere, B., et al.: EvolGAN: evolutionary generative adversarial networks. In: Ishikawa, H., Liu, C.-L., Pajdla, T., Shi, J. (eds.) ACCV 2020. LNCS, vol. 12625, pp. 679–694. Springer, Cham (2021). https://doi.org/10.1007/978-3-030-69538-5_41
14. Tuan, Y., Lee, H.: Improving conditional sequence generative adversarial networks by stepwise evaluation. IEEE ACM Trans. Audio Speech Lang. Process. 27(4), 788–798 (2019)

15. Wang, H., et al.: Deep reinforcement learning: a survey. Front. Inf. Technol. Electron. Eng. (2), 1–19 (2020). https://doi.org/10.1631/FITEE.1900533
16. Williams, R.J.: Simple statistical gradient-following algorithms for connectionist reinforcement learning. Mach. Learn. **8**, 229–256 (1992)
17. Yu, L., Zhang, W., Wang, J., Yu, Y.: Seqgan: sequence generative adversarial nets with policy gradient. In: Singh, S.P., Markovitch, S. (eds.) Proceedings of the Thirty-First AAAI Conference on Artificial Intelligence, February 4–9, 2017, San Francisco, California, USA, pp. 2852–2858. AAAI Press (2017)
18. Zhang, C., Yang, X., Tang, Y., Zhang, W.: Learning to generate radar image sequences using two-stage generative adversarial networks. IEEE Geosci. Remote. Sens. Lett. **17**(3), 401–405 (2020)

Research on Attitude Solving Algorithm of Towing Cable Based on Convolutional Neural Network Fusion Extended Kalman Filter

Bin Zheng, Xinran Yang[✉], and Haotian Hu

College of Instrumentation and Electrical Engineering,
Jilin University, Changchun 130061, China

Abstract. Marine seismic exploration is an important part of offshore oil and gas exploration, which requires accurate attitude information of submarine towing equipment. Conventional attitude solution algorithm or Kalman filter algorithm cannot satisfy the current requirements of high accuracy, high reliability, strong environmental adaptability and low cost. In view of the low accuracy and poor environmental adaptability of the traditional Kalman filter algorithm, this paper proposes a CNN-EKF fusion attitude calculation algorithm based on the study of the extended Kalman filter (EKF) model and the convolutional neural network (CNN) model. The system noise variance matrix (Q) and the observation noise variance matrix(R)of EKF were optimized by CNN, and the final solution results were obtained. Compared the traditional Kalman filtering model with the CNN-EKF fusion filtering model, experimental results shows that the algorithm improves the accuracy of attitude calculation and enhances the adaptive ability to the environment.

Keywords: Attitude solution · Extended Kalman filter · Convolutional neural network

1 Introduction

Marine resources are rich and have great development potential. At present, the common marine resource exploration methods include marine seismic exploration, marine geochemical exploration and marine electromagnetic exploration [1]. Among them, marine seismic exploration is the most widely used, especially in the distribution of deepwater oil [2]. In order to ensure the accuracy of seabed strata exploration, not only the stable operation of underwater equipment is needed, but also the attitude and position information of acquisition equipment is needed [3]. Therefore, towed cable attitude calculation is one of the research hotspots.

With the continuous development and maturity of MEMS technology and computer technology, accelerometers, magnetometers, gyroscopes and other attitude sensitive elements are widely used in various fields [4]. Accelerometer is susceptible to carrier vibration and motion acceleration; Magnetometer is susceptible to ferromagnetic material; Gyroscope has the characteristics of temperature drift, and the integration of errors will

© Springer Nature Singapore Pte Ltd. 2021
J. Zeng et al. (Eds.): ICPCSEE 2021, CCIS 1451, pp. 256–267, 2021.
https://doi.org/10.1007/978-981-16-5940-9_20

lead to the accumulation of errors [5]. Therefore, the separate and independent work of accelerometers, magnetometers and gyroscopes will produce large errors. It is necessary to integrate three kinds of sensors, fuse multi-sensor data, and filter out external noise so as to obtain attitude data with high accuracy, high precision and high reliability [6].

Attitudes of aircrafts, satellites, automatic vehicles and deep tow systems are essential control or monitoring targets. There are many multi-sensor data fusion methods for attitude calculation, among which Kalman filtering algorithm is the most widely used and mature attitude calculation algorithm [7]. Reference [8] introduces the application of wavelet denoising in attitude calculation. The key lies in the determination of threshold, decomposition level and the selection of wavelet basis function. If the selection of wavelet basis function is unreasonable, the accurate model cannot be obtained, which will lead to unsatisfactory filtering effect. Reference [9] used the extended Kalman filter algorithm for attitude calculation. The EKF prediction equation is nonlinear, and the model error will be introduced in linearization. There are many limitations in simply using EKF to improve the calculation accuracy. With the rapid development of neural network, it provides a new idea for attitude calculation. Neural network has strong fault tolerance and self-learning, self-organizing, self-adaptive ability, can analog complex nonlinear mapping, these characteristics and strong nonlinear processing ability, just satisfy the specific requirements of multi-sensor data fusion processing [10]. Reference [11, 12] introduced the BP neural network assisted Kalman filter algorithm for attitude calculation, which can reduce the model error of Kalman filter. Convolutional neural network model has excellent nonlinear fitting ability. In this paper, the filtering model of CNN-EKF fusion is used to calculate the attitude of towed cable. The CNN is used to train the system noise variance and the observation noise variance, and the output is used as the value of the prediction system noise variance matrix and the observation noise variance matrix of EKF, which can effectively improve the filtering effect of EKF and the adaptability to external input.

2 Attitude Angle Measurement Principle

2.1 Conversion of Coordinate System

There is a certain angle between the body coordinate system of the towed cable and the navigation coordinate system. It is assumed that the body coordinate system b is rotated by the navigation coordinate system n in the order of Z-X-Y axis, rotating θ, φ and ψ respectively, and the three rotations are all around a certain axis of the navigation coordinate system [13].

The towed cable is around the Z axis of the navigation coordinate system, and the body coordinate system θ is obtained by using the right-handed rotation angle θ as the pitch angle of the towed cable. The basic rotation matrix of rotating around Z axis is

$$C_\theta = \begin{bmatrix} 1 & 0 & 0 \\ 0 & \cos\theta & \sin\theta \\ 0 & -\sin\theta & \cos\theta \end{bmatrix} \tag{1}$$

The towed cable is around the X axis of the navigation coordinate system, and the body coordinate system is obtained by the rotation angle φ of the right-handed system. φ is the heading angle of the towed cable. The basic rotation matrix of rotating around X axis is

$$C_\phi = \begin{bmatrix} \cos\phi & -\sin\phi & 0 \\ \sin\phi & \cos\phi & 0 \\ 0 & 0 & 1 \end{bmatrix} \tag{2}$$

The towed cable revolves around the X axis of the navigation coordinate system, and the body coordinate system is obtained by using the right-handed rotation angle ψ, which is the rolling angle of the towed cable. The basic rotation matrix of fixed axis rotation around Y axis is

$$C_\varphi = \begin{bmatrix} \cos\varphi & 0 & -\sin\varphi \\ 0 & 1 & 0 \\ \sin\varphi & 0 & \cos\varphi \end{bmatrix} \tag{3}$$

The direction cosine matrix obtained by three rotations can be expressed as (4)

$$C_n^b = \begin{bmatrix} \cos\varphi\cos\theta & \sin\varphi\cos\theta & -\sin\theta \\ -\sin\varphi\cos\phi + \sin\phi\sin\theta\cos\varphi & \cos\varphi\cos\phi + \sin\varphi\sin\phi\sin\theta & \sin\phi\cos\theta \\ \sin\varphi\sin\phi + \cos\phi\sin\theta\cos\varphi & -\sin\varphi\cos\phi + \cos\phi\sin\theta\sin\varphi & \cos\phi\cos\theta \end{bmatrix} \tag{4}$$

2.2 Calculation of Attitude Angle by Quaternion Method

There are many parameters of the directional cosine matrix, nine differential equations need to be solved, and the amount of calculation is large. Quaternion attitude matrix differential equation only needs to solve four differential equations of first order, which is significantly less than the direction cosine matrix calculation [14]. Therefore, this paper uses quaternion method to solve attitude angle.

The sensor in the experimental part of this paper is the attitude sensor. In order to accurately capture the attitude information from three angles of pitch, roll and heading, three-axis gyroscope, three-axis magnetometer and three-axis accelerometer are used to solve the attitude angle relationship as follows:

Three attitude angles can be calculated by the combination of accelerometer and magnetometer. Accelerometer outputs three axis acceleration in real time and magnetometer outputs three axis geomagnetic intensity in real time. The output of the accelerometer is independent of the heading angle, so the pitch angle and roll angle can only be calculated by the accelerometer, and then the heading angle is calculated by the magnetometer.

The relationship between accelerometer, magnetometer and earth constant (g) is as follows:

Accelerometer output a_x, a_y, a_z:

$$\begin{bmatrix} a_x \\ a_y \\ a_z \end{bmatrix} = C_n^b \begin{bmatrix} 0 \\ 0 \\ -g \end{bmatrix} \tag{5}$$

Magnetometer output: m_x, m_y, m_z:

$$\begin{bmatrix} m_x \\ m_y \\ m_z \end{bmatrix} = C_n^b \begin{bmatrix} M \\ 0 \\ 0 \end{bmatrix} \tag{6}$$

So accelerometer, magnetometer attitude angle formula is as follows:

$$\theta = \arcsin\left(-\frac{a_x}{g}\right) \tag{7}$$

$$\varphi = \arctan\left(\frac{a_y}{a_z}\right) \tag{8}$$

$$\phi = \arctan\left(\frac{m_x \cos\theta - m_y \sin\theta \sin\varphi}{m_y \cos\varphi}\right) \tag{9}$$

Because the gyro has a singular effect on solving Euler angle [15], quaternion is used to represent the rotation of body axis system relative to reference axis system, quaternion q can be written as follows:

$$q = q_0 + q_1 i + q_2 j + q_3 k \tag{10}$$

In the formula q_0, q_1, q_2, q_3 are four real numbers, i, j, k are the bases of three imaginary units of quaternion. So the rotation matrix can be expressed as follows (11)

$$C_n^b = \begin{pmatrix} q_0^2 + q_1^2 - q_2^2 - q_3^2 & 2(q_1q_2 + q_0q_3) & 2(q_1q_3 - q_0q_2) \\ 2(q_1q_2 - q_0q_3) & q_0^2 - q_1^2 + q_2^2 - q_3^2 & 2(q_1q + q_2q_3) \\ 2(q_0q_2 + q_1q_3) & 2(q_2q_3 - q_0q_1) & q_0^2 - q_1^2 - q_2^2 + q_3^2 \end{pmatrix} = \begin{bmatrix} T_{11} & T_{12} & T_{13} \\ T_{21} & T_{22} & T_{23} \\ T_{31} & T_{32} & T_{33} \end{bmatrix} \tag{11}$$

Real-time update of quaternion can be obtained by quaternion differential equation, which is

$$\dot{Q} = \frac{1}{2}Q\omega \tag{12}$$

In the formula, is the angular velocity measured by three components of the gyroscope. The above expression is written in matrix form as follows:

$$\begin{bmatrix} \dot{q}_0 \\ \dot{q}_1 \\ \dot{q}_2 \\ \dot{q}_3 \end{bmatrix} = \frac{1}{2} \begin{bmatrix} 0 & -\omega_x & -\omega_y & -\omega_z \\ \omega_x & 0 & \omega_z & -\omega_y \\ \omega_y & -\omega_z & 0 & \omega_x \\ \omega_z & \omega_y & -\omega_x & 0 \end{bmatrix} \begin{bmatrix} q_0 \\ q_1 \\ q_2 \\ q_3 \end{bmatrix} \tag{13}$$

According to the Runge-Kutta method [16], the quaternion equation is updated to be:

$$q_k = \left\{ I + \frac{T}{2}\left[\Omega_{b_{(k-1)}}\right] \right\} q_{k-1} \tag{14}$$

$$\Omega_b = \begin{pmatrix} 0 & -\omega_x & -\omega_y & -\omega_z \\ \omega_x & 0 & \omega_z & -\omega_y \\ \omega_y & -\omega_z & 0 & \omega_x \\ \omega_z & \omega_y & -\omega_x & 0 \end{pmatrix} \tag{15}$$

According to the updated normalized quaternion, the rotation matrix (11) can be updated in real time, and three attitude angles can be determined:

$$\begin{pmatrix} \varphi \\ \theta \\ \phi \end{pmatrix} = \begin{pmatrix} -\arctan(T_{13}/T_{33}) \\ \arcsin(T_{23}) \\ -\arctan(T_{21}/T_{22}) \end{pmatrix} \tag{16}$$

3 CNN - EKF Fusion Filtering Model Algorithm

3.1 CNN Neural Network Model

Neural network has strong fault tolerance and self-learning, self-organizing, self-adaptive ability, can analog complex nonlinear mapping. CNN and other neural networks have advanced performance in various pattern recognition tasks, especially in the field of image recognition [17]. The convolution pooling layer can effectively collect the sensor information at multiple times and fuse the information into an effective and well-featured data set. Its sparse connection and weight sharing characteristics can reduce the network parameters and the complexity of the network model. The computational complexity is much lower than that of the hidden layer of the traditional neural network [18]. Therefore, the convolution pooling layer can greatly improve the computational efficiency.

3.2 CNN-EKF Fusion Filtering Model

The original data are measured by gyroscope, accelerometer and magnetometer, and the extended Kalman filter is improved by convolutional neural network. Firstly, the data are preprocessed. The original data are solved by quaternion method as roll angle, pitch angle and heading angle. Secondly, the CNN model is trained. The input layer is three angle data, Kalman gain and covariance matrix, and uses the seven hidden layer. The output layer is the system noise variance matrix (Q) and the observation noise variance matrix (R). Finally, the output results of the neural network are fed back to the filter to improve the filter and output the attitude angle of the optimal estimated value.

The CNN assisted extended Kalman filter process is as follows:

①The state equation of the system is established as:

$$X(k) = F(X(k-1)) + \Gamma W(k-1) \tag{17}$$

In the formula, X(k) is the state of the system at time k; F is the state transition matrix; W(k) is the white noise of the system; Γ is noise driven matrix.

The observation equation of the system is as follows:

$$Z(k) = H \bullet X(k) + V(k) \tag{18}$$

In the formula, Z(k) is the observation signal corresponding to X(k); V(k) is the measurement noise of observation; H is identity matrix.

②Kalman filter gain:

$$K(k) = P(k|k-1) \bullet H^{\wedge}T / (H \bullet P(k|k-1) \bullet H^{\wedge}T + R) \tag{19}$$

In the formula, K(k) is the Kalman gain; R is covariance matrix of observation noise.

③covariance matrix:

$$P(k|k-1) = F \bullet P(k-1|k-1) \bullet F^{\wedge}T + \Gamma Q \Gamma^{\wedge}T \tag{20}$$

In the formula, $P(k|k-1)$ is the corresponding covariance matrix, and it is also the estimation of $\hat{X}$ $(k|k-1)$covariance matrix. Q is the system noise.

④state update:

$$\hat{X}(k|k) = \hat{X}(k|k-1) + K(k)(Z(k) - H \bullet \hat{X}(k|k-1)) \tag{21}$$

In the formula, Z(k) is the observation matrix.The CNN training model is used to output Q and R, which are transmitted to the Kalman filter algorithm, and the state estimation value at time k is obtained through recursive calculation. The structure block diagram of CNN-EKF fusion filtering model is shown below (Fig. 1):

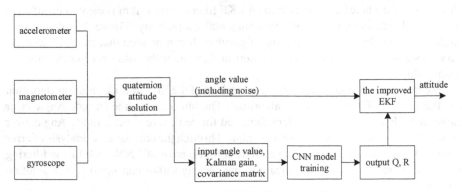

Fig. 1. Structure diagram of CNN-EKF fusion filter model.

4 System Verification

4.1 Training Sample Data Description

The experiment uses XSENS MTI-1 sensor, including three-axis accelerometer, three-axis magnetometer, three-axis gyroscope.

The experimental data come from a real project.The neural network model was trained with real sea trial data, and the sampling frequency was 100 Hz, and the experimental verification was carried out.

4.2   CNN Training Model

The activation function of CNN training model should be selected according to the input data. In attitude angle measurement, the output of the sensor is positive or negative, and there is a nonlinear relationship between the input and output of the sensor. Therefore, tanh is used as activation function in this algorithm, which has many advantages: input from −1 to 1; Value change sensitive; Output and input can maintain a nonlinear monotonic relationship of rise and fall; The function satisfies the gradient solution based on neural network; better fault tolerance; When the sensor is a trigonometric function, it satisfies the expression law [19]. Secondly, the relationship between multiple sensors and target values should be linear. Using nonlinear functions such as tanh reduces the efficiency of certain parameters. So we also use ReLU as activation functions for some parameters [20].

4.3   Test Results and Analysis

We compare the attitude calculation results of the traditional extended Kalman filter algorithm and the CNN-Kalman fusion filtering algorithm. Figure 2 is the roll, pitch and yaw calculated by the EKF algorithm. As it can see from the figure that the roll and yaw have jumped, and for the inertial device, this jump is not consistent with the inertial characteristics. Therefore, the traditional EKF filtering algorithm is easy to be affected by external factors and has poor environmental adaptability. Figure 3 is the attitude angle calculated by the fusion filtering algorithm. It can be seen that the attitude angle curve is stable without abnormal fluctuation, and the algorithm has strong environmental adaptability.

Figure 4 is the attitude Angle error comparison of the EKF attitude solution algorithm and the CNN-EKF fusion filtering algorithm. The blue curve is the attitude Angle error curve of EKF attitude solution algorithm, and the red curve is the attitude Angle error curve of CNN-EKF fusion filtering algorithm. Through the comparative analysis of error curves, it is intuitively seen that the attitude solution error of CNN-EKF fusion filtering algorithm is relatively small, indicating that the algorithm can improve the solution accuracy.

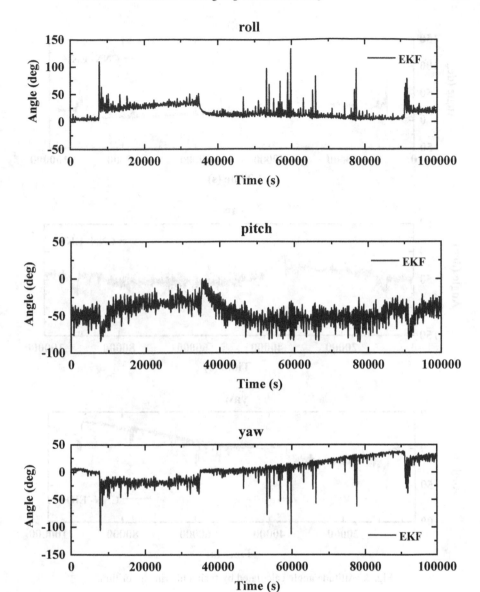

Fig. 2. Structure diagram of CNN-EKF fusion filter model.

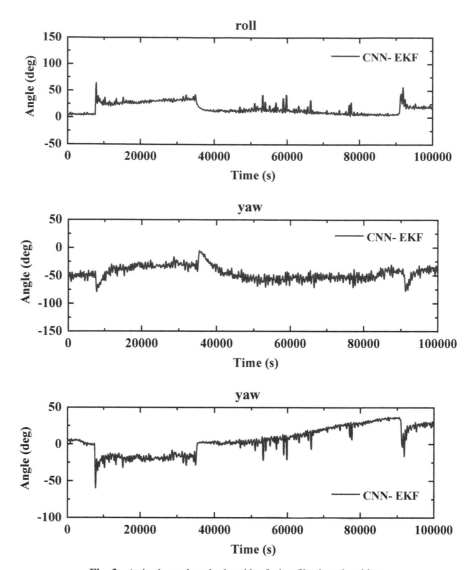

Fig. 3. Attitude angle calculated by fusion filtering algorithm.

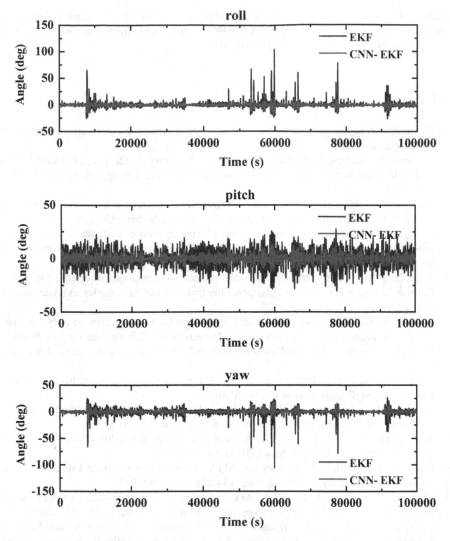

Fig. 4. Attitude angle errors calculated by EKF and CNN-EKF.

5   Conclusion

Aiming at the large noise interference caused by carrier jitter of deep sea towing equipment in sea test, the algorithm based on convolutional neural network and extended Kalman filter is used to solve the attitude of towing cable. The test results show that the attitude angle accuracy calculated by this algorithm is effectively improved and has stronger stability in a long time range. The effect of CNN-EKF fusion filter is more effective than that of traditional Kalman filtering, and this algorithm provides a valuable idea for the study of attitude solution and other issues.

Acknowledgements. The research was supported by the National Key Research and Development Program of China (Grant No. 2016YFC0303901), Southern Marine Science and Engineering Guangdong Laboratory (Zhanjiang) (ZJW-2019-04).

References

1. Fagen, P., Hui, F., Liang, P., Meixing, H., Gengen, Q.: Progress in electromagnetic exploration of gas hydrate. J. Prog. Geophys. **35**(2), 775–785 (2020)
2. Zhiqiang, W., Xunhua, Z., Weina, Z., Jianghao, Q., Xiaoqing, Z., Zhenxing, T.: Ocean bottom node (OBN) seismic exploration: progress and achievements. J. Prog. Geophys. **36**(1), 412–424 (2021)
3. Chunying, X.: Research on underwater topographic settlement mechanism and monitoring system based on MEMS 9-axis sensor array. PhD Thesis, Zhejiang University, China (2019)
4. Guanglin, Z.: Research and implementation of visualization of underwater robot attitude monitoring system based on LabVIEW. Master of Engineering Thesis, Dalian University of Technology, China (2019)
5. Jiajie, F.: Research on quadrotor attitude measurement system based on quaternion nonlinear filtering. Master of Engineering Thesis, Nanjing University of Aeronautics and Astronautics, China (2018)
6. Helong, W.: Research on key technologies of low-cost inertial/GNSS/vision integrated navigation for multi-rotor UAV. PhD Thesis, University of Chinese Academy of Sciences (Changchun Institute of Optics, Fine Mechanics and Physics, Chinese Academy of Sciences) (2020)
7. Xiling, S.: Research on attitude measurement and control technology of multi-rotor aircraft. PhD Thesis, North University of China (2020)
8. Guangwu, C., Xiaobo, L., Di, W.: Denoising algorithm of MEMS gyro signal based on improved wavelet transform. J. Electron. Inf. **5**, 1025–1031 (2019)
9. Yanguo, H., Yundong, F., Zuli, C.: Attitude calculation of quadrotor aircraft based on information fusion. J. Comput. Simul. **35**(9), 65–70 (2018)
10. Guruing, H., Banerjee, A.: Self-sensing SMA actuator using extended kalman filter and artificial neural network. J. Procedia Eng. **144**(25), 629–634 (2016)
11. Zhiqiang, D., Jieyu, L., Lixin, W., Xinsan, L., Qiang, S.: BPNN-assisted MEMS gyroscope data processing method for KF. J. Piezoelectrics Acoustooptics **42**(2), 284–288 (2020)
12. Chuxiong, Y., Jiahan, Y., Chengjun, G.: Research on attitude solution algorithm based on BP-Kalman fusion filter. In: Proceedings of the Ninth China Satellite Navigation Academic Conference–S10 Multi-source Fusion Navigation Technology, pp. 91–95 (2018)
13. Lei, D., Weijie, W., Ting, F.: An extended Kalman filtering method for fusion attitudes of multi-rotor aircraft. J. Anhui Univ. Technol. (Natural Science Edition). **35**(3), 240–248 (2018)
14. Yunjie, Q.: Research on inertial/satellite/vision integrated navigation technology of small UAVs. PhD Thesis, Harbin Institute of Technology, China (2020)
15. Wenwu, X.: Research on key technology of multi-functional stabilization platform based on MEMS inertial measurement unit. Master of Engineering Thesis, North University of China (2020)
16. Weixiong, Z., Shen, Y., Junqing, M., Litong, H., Shichen, F., Miao, W.: Research on calculation algorithm of position and pose of cantilever roadheader based on SINS. J. Coal Eng. **52**(9), 170–176 (2020)
17. Liang, C., Xiaoming, D., Mingquan, Z.: Convolutional neural networks in image understanding. J. Acta Automatica Sinica. **42**(9), 1300–1312 (2016)

18. Jiuxiang, G., Zhenhua, W., Kuen, J.: Recent advances in convolutional neural networks. J. Pattern Recognit. **77**(5), 354–377 (2018)
19. Chuanchao, P.: Research on path planning of indoor mobile robot based on deep reinforcement learning. Master of Engineering Thesis, China University of Mining and Technology (2020)
20. Yuanpan, Z., Guangyang, L., Ye, L.: A review on the application of deep learning in image recognition. J. Comput. Eng. Appl. **55**(12), 20–36 (2019)

Exploring Classification Capability of CNN Features

Shan Wang, Yue Wang$^{(\boxtimes)}$, Qilong Zhao, Zhijiang Yang, Weiyu Guo, and Xiuli Wang

School of Information, Central University of Finance and Economics, Beijing, China

Abstract. This paper explores the classification capability of features by three ways, respectively: decision tree/random forest, hierarchical clustering and Word-Net. To simulate the human judgment process, first, a decision tree is first constructed to reflect the importance of features. The model performs worse when top 5 features and top 10 features are used separately than when the all features are used, showing that the top k feature set omits some information that are important to classification. Second, hierarchical clustering is used to show the relationships between high-level features of different classes. The ward linkage method is selected to construct the hierarchical clustering tree. The parts of the adjacent classes with higher feature overlap are mapped back to the original image to visually show common features captured by the CNN network. Finally, to study the semantic value of neural network classification, the WordNet semantic structure is applied to fit the image classification process. However, results are relatively poor, demonstrating the inconsistency between the WordNet classification and machine learning classification.

Keywords: Hierarchical clustering · Decision tree · Random forest · Interpretability · CNN

1 Introduction

Image classification is to distinguish different categories of images according to their semantic information, which is an important basic problem in computer vision. Deep learning methods can learn hierarchical feature representation through supervised or unsupervised methods, among which CNN has made very good progress in the field of image classification. Compared with the traditional method, the CNN model image classification method can obtain a higher accuracy. The CNN model directly uses image pixel information for input, and uses convolution and pooling to extract features and obtain

Shan Wang and Yue Wang contributed equally to this work. This work is supported by: National Defense Science and Technology Innovation Special Zone Project (No. 18-163-11-ZT-002-045-04); Engineering Research Center of State Financial Security, Ministry of Education, Central University of Finance and Economics, Beijing, 102206, China; Program for Innovation Research in Central University of Finance and Economics; National College Students' Innovation and Entrepreneurship Training Program "Research on classification and interpretability of popular goods based on Neural Network".

© Springer Nature Singapore Pte Ltd. 2021
J. Zeng et al. (Eds.): ICPCSEE 2021, CCIS 1451, pp. 268–282, 2021.
https://doi.org/10.1007/978-981-16-5940-9_21

high-level abstraction. But it lacks interpretability because of its "black box" character-istics, so it faces many difficulties in the application of some advanced or sophisticated industries. Therefore, it is particularly significant to improve the interpretability of neu-ral network models and make them more easily understood and visualized. Our key research questions are:

- What is the performance using the decision tree and hierarchical clustering to classify images based on CNN features?
- Does CNN have the same semantic meanings with common sense?

2 Related Work

CNN has difficulty visualizing classification relationships and has been described as a "black box". As a result, their use is limited in areas such as medical image detection [1], public policy making [2], and biology [3]. Visual feature extraction is one of the foun-dations of content-based image classification. General visual features focus on texture, color, shape and so on. Some features of an image can be represented in many different ways [4]. Constructing high-level structural features and semantic features is an effec-tive method to detect image objects [5]. It is found that there are many misclassification and noise in the traditional prediction method. Zhao et al. used PCNN to explore the representation of high-level features of the hierarchy, and carried out a systematic study on five different layer configurations, which has the classification accuracy more than 90% [6]. Girish et al. tried to answer a number of questions related to interpretability by using unsupervised methods to identify features learned by CNN at different levels [7].

The unstructured format of images reduces the application of classification tech-niques. Clustering, as a method to better organize images, is of great significance for effective image retrieval [8]. It is possible for the clustering algorithm to use finite sym-bols to represent the visual features of an image. Hierarchical clustering is also used in the field of images, which divides the image into several non-overlapping areas by using the features of gray, color, texture and shape of the image, and makes these fea-tures appear similar in the same area to search for obvious difference between different areas [9]. According to the relationship between fuzzy boundary set and hierarchical image segmentation, Gomez et al. introduced the concept of fuzzy image segmentation by clustering hierarchically to form a more natural and consistent segmentation [10].

WordNet is also very useful for image classification. Jin et al. linked visual markers with knowledge-based WordNet keywords by using WordNet to prune unrelated key-words. The results showed that knowledge expansion based on the classical model can remove the irrelevant keywords and improve the accuracy of annotation [11]. Benitez et al. proposed an image classification method based on the knowledge found in anno-tated images. By using WordNet combined with a classifier based on the knowledge found and summarized from annotated images, it can achieve higher (up to 15%) or equivalent accuracy compared to a single classifier and a purely statistical learning clas-sifier structure [12]. ImageNet is an image database organized according to a WordNet hierarchy (currently only nouns) [13].

In order to explore the internal structure of neural network in the process of image classification, two methods can be adopted: one is structure visualization; the other is to extract decision rules by training decision trees. Lakkaraju and Himabindu (2017) introduced a black box explanation through the transparent approximation (Beta), which can learn (with optimality guarantee) a small set of compact decisions to explain the behavior of the black box model in a definite region of the feature space [14]. Zilke et al. proposed a new decision tree-based decomposition algorithm called "DeepRed", which is capable to extract rules from deep neural networks. The evaluation showed that it was able to exceed the preset benchmark in some tasks [15]. Andrew Silva et al. allowed for a gradient update over the entire tree that improves sample complexity affords interpretable policy extraction [16]. Through running the classification neural network as a decision tree, NBDT achieves good interpretability and classification effect, and improves the performance of all decision tree-based methods by about 14% [17].

In terms of image interpretability, general interpretability models have been developed, such as sensitivity analysis [18], Grad-CAM [19], LIME [20], etc. And deconvolution has excellent visualization effect. Matthew D Zeiler & Rob Fergus introduced the deconvolution visualization method, which provided insight into the function of the intermediate element layer of the Convolutional Neural Network, demonstrating impressive classification performance [21].

In this paper, decision tree is applied to simulate the image classification process of CNN, and key features are extracted from CNN to understand the application of features. A hierarchical clustering tree is constructed to show the distance and relationship between different classes, which helps to map similar features in different classes. In addition, a WordNet-based tree is used to compare the classification of images.

3   Decision Tree Interpretation Models

3.1   Input Features and Preprocessing

We select 10 classes numbered n02701002 (ambulance), n02704792 (amphibian), n02814533 (beach_wagon), n02950826 (cannon), n03272562 (electric_locomotive), n03345487 (fire_engine), n03478589 (half_track), n04008634 (projectile), n04389033 (tank), and n04467665 (trailer_truck) in the ILSVRC-2012 dataset [22], which are mainly vehicle targets. There are 1300 pictures in each category. We use all the pictures of them. In order to show the differences between classes, we also build an *others* class, which contains 1300 images that do not belong to the above 10 classes. Therefore, we obtain an image dataset of 11 classes. By using the VGG16 network to separate the high-level features of each image from three layers (24,26 and 28). We obtain the feature with the shape of (512,7,7) for each image. Then, we only keep the maximum value of each layer in the feature map. Now the output feature dimension of each image sample was (512 × 1), which is used as the original input of our method.

3.2   Decision Tree Performance

The decision tree is a commonly used model in machine learning, which has the advantage of good interpretability. In the classification process, a decision tree firstly chooses the best features to grow a split node based on either entropy or gini. The procedure is

Table 1. Two-classes decision tree's F1 classifying a target class and the others class by using all original features (eg. 24 + 26 refers to the features obtained from 24 and 26 layers)

Original features from some high layers							
Target Class Number	24	26	28	24+26	24+28	26+28	24+26+28
'n02701002',	0.901	0.913	0.92	0.893	0.906	0.903	0.897
'n02704792',	0.818	0.842	0.886	0.8	0.821	0.837	0.814
'n02814533',	0.928	0.942	0.95	0.914	0.931	0.932	0.913
'n02950826',	0.783	0.785	0.853	0.767	0.798	0.807	0.771
'n03272562',	0.922	0.925	0.953	0.894	0.91	0.929	0.903
'n03345487',	0.9	0.891	0.917	0.867	0.886	0.893	0.872
'n03478589',	0.875	0.899	0.941	0.858	0.887	0.895	0.876
'n04008634',	0.76	0.783	0.839	0.727	0.766	0.777	0.735
'n04389033',	0.825	0.866	0.907	0.819	0.841	0.865	0.831
'n04467665'	0.869	0.87	0.896	0.856	0.859	0.868	0.841
Mean F1 of 10 Classes	**0.8581**	**0.8716**	**0.9062**	**0.8395**	**0.8605**	**0.8706**	**0.8453**

Table 2. Two-classes decision tree's F1 classifying a target class and the others class by using top 10 features from some high layers (eg. 24 + 26 refers to the features obtained from 24 and 26 layers)

Top 10 features from some high layers							
Target class number	24	26	28	24 + 26	24 + 28	26 + 28	24 + 26 + 28
'n02701002',	0.894	0.892	0.908	0.903	0.902	0.9	0.904
'n02704792',	0.792	0.81	0.868	0.792	0.817	0.825	0.809
'n02814533',	0.931	0.909	0.948	0.923	0.938	0.923	0.921
'n02950826',	0.757	0.803	0.831	0.763	0.793	0.814	0.789
'n03272562',	0.878	0.894	0.938	0.895	0.901	0.919	0.906
'n03345487',	0.874	0.871	0.88	0.873	0.876	0.876	0.868
'n03478589',	0.87	0.867	0.913	0.869	0.895	0.897	0.889
'n04008634',	0.733	0.784	0.851	0.722	0.785	0.807	0.75
'n04389033',	0.794	0.837	0.88	0.802	0.837	0.855	0.834
'n04467665'	0.865	0.871	0.887	0.861	0.872	0.878	0.866
Mean F1 of 10 classes	**0.8388**	**0.8538**	**0.8904**	**0.8403**	**0.8616**	**0.8694**	**0.8536**

Table 3. Two-classes decision tree's F1 classifying a target class and the others class by using top 5 features from some high layers (eg. 24 + 26 refers to the features obtained from 24 and 26 layers)

Target Class Number	Top 5 features from some high layers						
	24	26	28	24+26	24+28	26+28	24+26+28
'n02701002',	0.893	0.893	0.919	0.903	0.899	0.89	0.894
'n02704792',	0.772	0.817	0.86	0.774	0.794	0.817	0.775
'n02814533',	0.908	0.902	0.921	0.906	0.918	0.908	0.9
'n02950826',	0.739	0.792	0.841	0.749	0.779	0.796	0.766
'n03272562',	0.863	0.894	0.931	0.87	0.889	0.896	0.869
'n03345487',	0.869	0.873	0.889	0.863	0.877	0.864	0.858
'n03478589',	0.859	0.869	0.912	0.862	0.884	0.884	0.879
'n04008634',	0.733	0.759	0.813	0.724	0.758	0.774	0.735
'n04389033',	0.77	0.82	0.861	0.775	0.815	0.835	0.791
'n04467665'	0.841	0.855	0.878	0.836	0.851	0.849	0.827
Mean F1 of 10 Classes	**0.8247**	**0.8474**	**0.8825**	**0.8262**	**0.8464**	**0.8513**	**0.8294**

pretty much like the natural mechanism for human beings to deal with decision-making problems. In this paper, we use scikit-learn to construct a decision tree for binary classification tasks. This decision tree is trained using the features, each feature is the max activation value in a feature map. The feature maps and the filters are corresponding to each other. By studying its decision path, we can easily know what features are considered most important in the situation. Furthermore, we compare top k features extracted by VGG-16 and the ones from the decision tree, and find that they overlap each other.

We mix samples of the target classes and "others" class from the same layer as the dataset. For instance, the "ambulance" class is chosen as a target class. Then, the whole dataset is split into two parts, 80% of which is used as the training set and the rest as the test set. The F1 values were printed to show the performance of the decision tree. Besides, features from different layers are also put together showing to what extent do different layers influence the result. On other counts, we extract and sort top k (k is 5 or 10 in the experiments) features according to the activation values in VGG-16. And these features are also used to feed the decision tree, just to test if the top k features are enough for training a well performed tree model. Results are shown in Table 1, Table 2 and Table 3.

It seems that top k feature has a slightly lower F1 score, and layer 28 provides the best performance in the binary classification tasks. Layer 24 performs the worst while layer 28 perform the best in the single layer test, especially in the top 5 feature set, showing lower-layer feature. And mixing with layer 24, layer 26 and 28 cannot perform well.

While binary classification tasks show good results and small differences between original features and top k features, we further conduct the multiple classification task shown in Table 4. The model performs the worst on the top 5 feature set. Although each of the results was not good enough, it shows different feature sets can have an influence on the output. Using the top k features may omit some features that are essential to classifying.

Similar conclusions can be found using random forest algorithm. For a clear exhibition, random forest algorithm, which is constructed by many decision trees, had been applied. The results are shown in Table 5. Apparently, the original feature set performs better than the top k feature sets. And fewer features we use, the worse performance we get. Top k features can be representative of the data in some degree, but the rest features can also contribute to the classification.

Table 4. Multi-classes decision tree's F1 of 11 classes (using original features of layer 28)

Features	F1
Original	0.60
Top 10	0.61
Top 5	0.55

Table 5. Multi-classes decision tree's F1 of 11 classes (using original features of layer 28)

Features	F1
Original	0.83
Top 10	0.74
Top 5	0.69

Furthermore, to figure out what features are the most important to binary classification, we use the ten target classes to test the binary classification performance of a decision tree. Data from top 5 feature set and layer 28 are been applied. Here, "feature" stands for features used for classifying, and "splitting condition" stands for the condition for a splitting node of the decision tree splits instances. A combination of the two presents a decision path, which is demonstrated in Table 6.

Table 6. Two-classes decision tree splitting features and conditions (using original features in layer 28)

Class	Features	Splitting condition
Ambulance	Filter 25, 143	Filter 25 > 10.848 and Filter 143 > 4.231; Filter 25 < = 10.848
Amphibious	Filter 318, 62	Filter 318 > 7.976 and Filter 62 > 1.522; Filter 318 < = 7.976
Beach_wagon	Filter 37, 343, 418	Filter 37 > 8.53 and Filter 418 > 2.766; Filter 37 < = 8.53 and Filter 343 < = 11.965
Cannon	Filter 6, 338, 289	Filter 6 > 9.126 and Filter 289 > 3.713; Filter 6 < = 9.126 and Filter 338 < = 10.126
Electric_locomotive	Filter 265, 57, 25	Filter 265 > 6.516 and Filter 25 > 6.441; Filter 265 < = 6.516 and Filter 57 < = 11.795
Fire_engine	Filter 25, 412	Filter 25 > 8.59 and Filter 412 > 3.682; Filter 25 < = 8.59
Halftrack	Filter 127, 183, 62	Filter 127 > 6.881 and Filter 62 > 4.98; Filter 127 < = 6.881 and Filter 183 < = 9.823
Projectile	Filter 6, 58	Filter 6 > 9.165 and Filter 58 > 1.378; Filter 6 < = 9.165
Tank	Filter 254, 472	Filter 254 > 7.709 and Filter 472 > 3.847; Filter 254 < = 7.709
Trailer_truck	Filter 292, 202, 40	Filter 292 > 7.607 and Filter 40 > 4.442; Filter 292 < = 7.607 and Filter 202 < = 5.31

In order to show how similar a decision tree and VGG-16 are in the way of classifying, top features extracted from the two models are put together (top features in a decision tree are based on feature importance, and top features in VGG-16 are based on concurrence frequencies of top activation features). From the comparison below, it's clear that some features overlap. For further proof of the high similarity in decisions the made by VGG-16 and the decision tree, *fidelity* is introduced. In here, fidelity measures the extent to which the decision tree approximates a VGG-16 model, which is the percentage that the predictions made by the decision tree trained using features extracted by a VGG-16 model is the same as that made by the VGG-16 model. We believe that to some degree, the way the decision tree classifies represents how VGG-16 is doing it. Thus, the decision path showed in Table 6 and the filters extracted in Table 7 can be used for enhancing the interpretability of the high-level features from the neural network.

Table 7. Comparing top features and fidelity of decision tree and VGG-16 (using layer 28 original features)

Class	Decision tree	VGG-16	Fidelity
Ambulance	Filter 25, 143	Filter 25, 59	0.92
Amphibious	Filter 318, 62	Filter 318, 145	0.88
Beach_wagon	Filter 37, 343, 418	Filter 418, 377, 37	0.92
Cannon	Filter 6, 338, 289	Filter 6, 464, 338	0.84
Electric_locomotive	Filter 265, 57, 25	Filter 25, 377, 254	0.93
Fire_engine	Filter 25, 412	Filter 25, 62	0.90
Half_track	Filter 127, 183, 62	Filter 62, 318, 183	0.92
Projectile	Filter 6, 58	Filter 6, 237	0.84
Tank	Filter 254, 472	Filter 472, 254	0.89
Trailer_truck	Filter 292, 202, 40	Filter 292, 254, 62	0.88

4 Hierarchical Clustering Models

4.1 Agglomerative Hierarchical Clustering Model

Hierarchical clustering takes each sample point as an original cluster, and then merges these atomic clusters until the expected number of clusters or other termination conditions are reached. There are four different methods of hierarchical clustering: ward linkage, average linkage, complete linkage and single linkage. The four methods have different measures of distance, and the calculation method is shown in Table 8.

Table 8. Four ways of calculating distance by hierarchical clustering

Method	Distance	Introduction
Ward	$ESS = \sum_{i=1}^{n} x_i^2 - \frac{1}{n}\left(\sum_{i=1}^{n} x_i\right)^2$	The increase in ESS was minimal after each merger
Average	$dist(C1, C2) = \frac{1}{|C1| \cdot |C2|} \sum_{P_i \in C1, P_j \in C2} dist(P_i, P_j)$	The distance between the classes is the average of the distances of all the points between the two clusters

(continued)

Table 8. (*continued*)

Method	Distance	Introduction
Complete	$dist(C1, C2) =$ $\max\limits_{P_i \in C1, P_j \in C2} dist(P_i, P_j)$	The distance between classes is the longest distance of sample points in the cluster
Single	$dist(C1, C2) =$ $\min\limits_{P_i \in C1, P_j \in C2} dist(P_i, P_j)$	The distance between classes is the shortest distance of sample points in the cluster

In this paper, the hierarchical clustering method is used to explore the correlation of high-level features of multiple classes of images. By comparing the distance between clusters, hierarchical cluster trees are generated. The generated hierarchical clustering tree visualizes the distance between the high-level features of the image set, measuring the relationship between classes from the point of view of a feature vector.

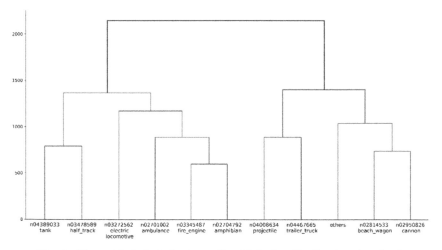

Fig. 1. Hierarchical clustering trees of multiple classification (ward method)

As is shown in Fig. 1, tanks and half-tracks are considered to be similar, and their graphic features are also quite similar in practice. Although some of these classifications are not accurate, hierarchical clustering trees can still provide an explanation for CNN's image classification. That is, similar images have similar high-level features on CNN. High-level features contain abstracted information about the image sample, which can also be semantically translated. Through the classification results of hierarchical clustering, the algorithm can view the overlapping feature regions between similar classes, and map the feature regions with high overlapping degree back to the original image by the deconvolution method to view hierarchical clustering to identify what CNN classification concerns.

4.2 Feature Visualization

In this paper we adopt the deconvnet method and a mapping method to visualize features. A deconvolution network is a convnet model that uses reverse layers (unpooling, rectification). We construct the deconvnet of VGG16 and assign the weights to the corresponding layers. We pass the feature maps as input to the deconvnet layer, then we get the mapping of the feature maps in the original image.

In the mapping method, we choose a threshold first. For each feature map of an input image, we set the pixels below the threshold to 0.4 and pixels above the threshold to 1, after that we multiply the feature map with the original image to obtain the output image. The output image shows the areas of the image whose activation is above a certain threshold.

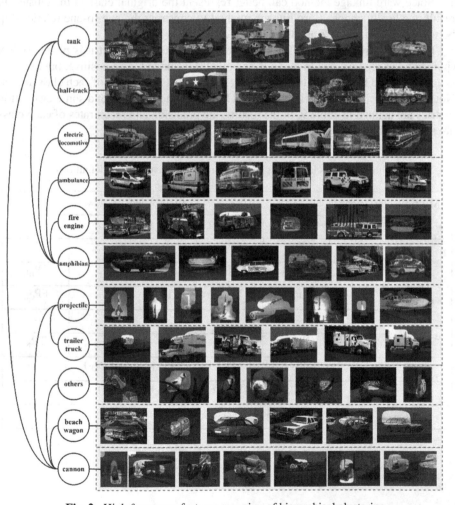

Fig. 2. High frequency features mapping of hierarchical clustering

We extract the high frequency characteristics of the hierarchical clustering and use the deconvolution method, and then map them back to the original image. The resulting image is shown in Fig. 2. Similar visualizations are found between adjacent classes. For instance, the caterpillar band region is identified for tank and half-track. The distinguishing characteristics of each classification are also well identified.

4.3 Clustering Result Evaluation

Clustering effect is evaluated to choose the best method by calculating precision and recall rate. As can be seen from the above figure, both *ward linkage* and *average linkage* methods failed completely. Ward linkage performs the best and complete linkage performs the second. None of the four approaches reaches the maximum at n_clusters = 11. Since ward linkage method can better represent the original class in the clustering result, we finally choose ward linkage method to construct the one-to-one relationship between the original class and the aggregate class.

In the clustering results, TP (judged to be positive, but in fact it is positive), TN (judged to be negative, but in fact it is negative), FP (judged to be positive, but in fact it is negative), and FN (judged to be negative, but in fact it is positive) of each class are set. Note that we evaluate each class (totally 11 classes) individually, and some evaluation metrics are defined in Table 9. In the results, the precision and recall rates of each class are shown in Table 10.

Table 9. Multi-classes classification precision, recall and F1 definitions

Multiple classification		Predicted Class				
		A	B	C	...	K
True Class	A	TP_A	FP_{AB}	FP_{AC}	...	FP_{AK}
	B	FP_{BA}	TP_B	FP_{BC}	...	FP_{BK}
	C	FP_{CA}	FP_{CB}	TP_C	...	FP_{CK}
	...	...	...	...	...	...
	K	FP_{KA}	FP_{KB}	FP_{EC}	...	TP_K
$Precison_A = TP_A/(TP_A+ FP_{AB}+ FP_{AC}+ ...+ FP_{AK})$						
$Recall_A = TP_A/(TP_A+FP_{BA}+FP_{CA}+...+FP_{KA})$						
$F1_A = 2Precison_A*Recall_A /(Precison_A+Recall_A)$						

Table 10. F1 of hierarchical clustering

Class	F1
n04389033	0.75
n02704792	0.21
n03272562	0.90
n02950826	0.68
n02701002	0.97
n03478589	0.64
n04008634	0.49
n02814533	0.62
n03345487	0.70
n04467665	0.84
Others	0.60

4.4 Compared with WordNet Classification

In order to further understand the semantic value and decision path of CNN image classification, we construct a WordNet inner product tree. WordNet is a semantic network covering a wide range of English words, which has a wide application in the fields such as image segmentation and recognition. WordNet inner product tree fills the gap of intermediate nodes on the basis of classification, making it possible to show the decision path from abstract class to concrete class.

According the classification using the predefined WordNet tree (Fig. 3), 75% of image samples from each of the 11 classes are used as the training set, and the remaining 25% of images are used as the validation set. In the classification using the WordNet tree [17], sample feature values are assigned to leaf nodes of the constructed tree, and the weighted average value of the child nodes is taken as the feature values of the parent node. For the samples in the validation set, the path from the parent node to the child node is selected by the method of inner product operation. After iteration, a specific category of leaf node can be finally reached.

WordNet trees are very different from hierarchical clustering trees. The former is entirely based on semantic characteristics, while the latter is mostly based on CNNs convolution features. In the WordNet tree, the categories of images are classified according to the function. Thus the characteristics of two subclasses of the same parent class may not be sufficiently similar. However, the structure of hierarchical cluster tree is based on the similarity of physical features, such as color, texture, size, etc. Quite different cognitive styles determine the difference of tree structure and classification results. The result of the WordNet inner product tree is shown in Table 11.

As can be seen from the results, WordNet's classification F1 value is quite unbalanced. Cannon, electric locomotive and projectile have achieved relatively high F1 value.

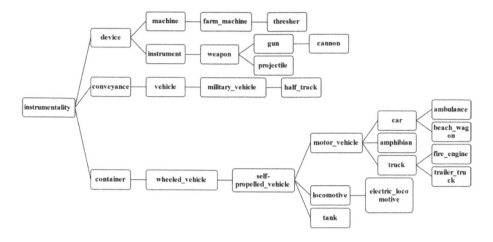

Fig. 3. WordNet semantic structure tree

Table 11. F1 of WordNet tree classification

Class	F1
Ambulance	0.59
Amphibian	0.08
Beach_wagon	0.60
Cannon	0.81
Electric_locomotive	0.98
Fire_engine	0.32
Half_track	0.97
Projectile	0.78
Tank	0.006
Trailer_truck	0.35

The classification of amphibian and tank is completely invalid. The process of classification from macro semantics to micro semantics is not reasonable. In the WordNet tree, the parent of two classes should be able to contain all the characteristics of the subclasses.

5   Conclusion

In this paper, we explore CNN classification rules by three ways: decision tree/random forest, hierarchical clustering and WordNet. (1) We construct decision trees to carry out binary classification and multiple classification. The model performs best using original feature set, worst on Top 5 feature set, indicating that top k features may omit some features that are essential to classifying. Similar conclusions can be found using random forest algorithm. Top features extracted from decision tree and VGG-16 have a high degree of overlap, showing that the decision tree provides good interpretability. (2) Hierarchical clustering is used to show relationships of different classes. The adjacent classes with higher feature overlap are mapped back to the original image, having similar recognition areas. (3) In terms of the semantic value of neural network classification, WordNet semantic structure is used to fit the image classification process. However, results are relatively poor, demonstrating the inconsistency between the WordNet classification and machine learning classification.

References

1. Litjens, G., Kooi, T., Bejnordi, B.E., et al.: A survey on deep learning in medical image analysis. Med. Image Anal. **42**, 60–88 (2017)
2. Brennan, T., Oliver, W.L.: Emergence of machine learning techniques in criminology: implications of complexity in our data and in research questions. Criminol Pub. Pol'y **12**, 551 (2013)
3. Angermueller, C., Pärnamaa, T., Parts, L., Stegle, O.: Deep learning for computational biology. Mol. Syst. Biol. **12**(7), 878 (2016)
4. Rui, Y., Huang, T.S., Chang, S.F.: Image retrieval: past, present, and future. J. Vis. Commun. Image Represent. **10**(1), 1–23 (1999)
5. Han, J., Zhang, D., Cheng, G., Guo, L., Ren, J.: Object detection in optical remote sensing images based on weakly supervised learning and high-level feature learning. IEEE Trans. Geosci. Remote Sens. **53**(6), 3325–3337 (2014)
6. Zhao, W., Du, S., Emery, W.J.: Object-based convolutional neural network for high-resolution imagery classification. IEEE J. Sel. Top. Appl. Earth Observations Remote Sens. **10**(7), 3386–3396 (2017)
7. Girish, D., Singh, V., Ralescu, A.: Unsupervised clustering based understanding of CNN. In CVPR Workshops, pp. 9–11 (2019)
8. Wang, L., Khan, L.: A New Hierarchical Approach for Image Clustering. Multimedia Data Mining and Knowledge Discovery, pp. 41–57. Springer, London (2007)
9. Singh, C., Murdoch, W.J., Yu, B.: Hierarchical interpretations for neural network predictions. arXiv preprint arXiv:1806.05337 (2018)
10. Gómez, D., Yáñez, J., Guada, C., et al.: Fuzzy image segmentation based upon hierarchical clustering. Knowl.-Based Syst. **87**, 26–37 (2015)
11. Jin, Y., Khan, L., Wang, L., Awad, M.: Image annotations by combining multiple evidence & wordnet. In: Proceedings of the 13th Annual ACM International Conference on Multimedia, pp. 706–715 (2005)
12. Benitez, A.B., Chang, S.F.: Image classification using multimedia knowledge networks. In: Proceedings 2003 International Conference on Image Processing, pp. III-613 (2003)
13. Deng, J., Dong, W., Socher, R., Li, et al.: Imagenet: a large-scale hierarchical image database. In: 2009 IEEE Conference on Computer Vision and Pattern Recognition, pp. 248–255 (2009)
14. Lakkaraju, H., Kamar, E., Caruana, R., Leskovec, J.: Interpretable & explorable approximations of black box models. arXiv preprint arXiv:1707.01154 (2017)
15. Zilke, J.R., Mencía, E.L., Janssen, F.: Deepred–rule extraction from deep neural networks. In: International Conference on Discovery Science, pp. 457–473 (2016)
16. Silva, A., Gombolay, M., Killian, T., et al.: Optimization methods for interpretable differentiable decision trees applied to reinforcement learning. In International Conference on Artificial Intelligence and Statistics, pp. 1855–1865 (2020)

17. Wan, A., Dunlap, L., Ho, D., et al.: NBDT: neural-backed decision trees. arXiv preprint arXiv: 2004.00221 (2020)
18. Li, J., Chen, X., Hovy, E., Jurafsky, D.: Visualizing and understanding neural models in nlp. arXiv preprint arXiv:1506.01066 (2015)
19. Selvaraju, R.R., Cogswell, M., Das, A., et al.: Grad-cam: visual explanations from deep networks via gradient-based localization. In: Proceedings of the IEEE International Conference on Computer Vision, pp. 618–626 (2017)
20. Ribeiro, M.T., Singh, S., Guestrin, C.: "Why should i trust you?" Explaining the predictions of any classifier. In: Proceedings of the 22nd ACM SIGKDD International Conference on Knowledge Discovery and Data Mining, pp. 1135–1144 (2016)
21. Zeiler, M.D., Fergus, R.: Visualizing and understanding convolutional networks. In: European Conference on Computer Vision, pp. 818–833 (2014)
22. Russakovsky, O., Deng, J., Su, H.: Imagenet large scale visual recognition challenge. Int. J. Comput. Vision **115**(3), 211–252 (2015)

Multi-objective Firefly Algorithm for Test Data Generation with Surrogate Model

Wenning Zhang[1,2(✉)] 📵, Qinglei Zhou[3] 📵, Chongyang Jiao[1] 📵, and Ting Xu[1,3] 📵

[1] State Key Laboratory of Mathematical Engineering and Advanced Computing, Zhengzhou 450000, China
[2] Zhongyuan University of Technology, Zhengzhou 450000, China
zhangwn@zut.edu.cn
[3] Zhengzhou University, Zhengzhou 450000, China

Abstract. To solve the emerging complex optimization problems, multi objective optimization algorithms are needed. By introducing the surrogate model for approximate fitness calculation, the multi objective firefly algorithm with surrogate model (MOFA-SM) is proposed in this paper. Firstly, the population was initialized according to the chaotic mapping. Secondly, the external archive was constructed based on the preference sorting, with the lightweight clustering pruning strategy. In the process of evolution, the elite solutions selected from archive were used to guide the movement to search optimal solutions. Simulation results show that the proposed algorithm can achieve better performance in terms of convergence iteration and stability.

Keywords: Firefly algorithm · Multi objective optimization · Surrogate model · Test data generation

1 Introduction

Software can be seen almost everywhere and and now is defining the world. As one of the fundamental and essential production activities, software testing is critical for software quality, counting about 40% budget of the total software development. Claimed by Harman and Jones in 2001 [1], the search based software engineering(SBSE) reformulates the classic software engineering problems as search optimization problems. Search based software testing (SBST) [2] is the most important branch of SBSE, which searches suitable test data required by testing targets defined by coverage criterions.

According to the targets covered in one run, the test data generation approaches can be divided into single target testing and multi targets testing. For the single target strategy, the efficiency is affected by some important factors, such as infeasible goals and the comprehensive relationship among them. To overcome these limitations, people proposed the whole suite approaches, which combine multi testing goals through the sum aggregation strategy [3, 4]. Although many works have proved its efficiency, it was not efficient for problems with non-convex region in the search space.

© Springer Nature Singapore Pte Ltd. 2021
J. Zeng et al. (Eds.): ICPCSEE 2021, CCIS 1451, pp. 283–299, 2021.
https://doi.org/10.1007/978-981-16-5940-9_22

It has been demonstrated that many multi objective optimization algorithms are more efficient than sum aggregation strategies [5, 6]. Motivated by the excellent performance of meta heuristic search algorithms, we convert the branch coverage criteria to a multi objective optimization problem.

In this paper, we analyze the properties of multi target test data generation, propose a novel fitness function presentation based on surrogate model and introduce multi firefly algorithm to solve the multi target test data generation, named MOFA-SM (Many Objective Firefly Algorithm with Surround Model).

2   Related Work

2.1   Multi Objective Optimization Algorithms

Optimization problem is one of the most common problems in engineering and scientific research. It has been proved that many objective optimization is more challenging because of the Pareto dominance relationships [7].

In the past decades, many excellent multi objective optimization algorithms have emerged and been successfully applied. Some traditional multi objective algorithms, such as PAES [8], NSGA [9], NSGA-II [10], SPEA2 [11], are efficient at optimization problems less than 3 targets, but poor in solving complex high dimensional multi objective optimization problems [12]. Based on decomposition strategy, Zhang et al. [13] proposed the multi objective evolutionary algorithm based on decomposition (MOEA/D). MOEA/D decomposes the total optimization goals into a number of sub goals and optimizes them simultaneously. By simulating typical behavior in nature, many meta heuristic search algorithms are proposed, genetic algorithm (GA), particle swarm optimization (PSO), firefly algorithm (FA) for example. These meta heuristic algorithms have attracted more attention and widely used to solve multi objective optimization problems, such as multi objective particle swarm optimization [14] and multi objective explosion optimization [15].

2.2   Test Data Generation Based on Multi Objective Optimization Algorithms

Researchers have shown the suitability of using multi objective meta heuristic algorithms in automatic software test data generation. Ahmed designed the multi objective genetic algorithm for covering multiple target paths in one run. Pooja Gopi [16] considered the branch coverage maximum and the test suite size reduction as two separate optimization goal. Zeng [17] put foreword a novel test preference based multi objective genetic algorithms (PMOGA) for test data generation based on the Pareto front sort strategy and local search strategy. Also, the Evosuite was analyzed to improve the efficiency. Concerning with the execution difficulty, Hong [18] defined the contact vector with the layer proximity theory and proposed multi objective optimization approach based on GA, where the sharing strategy between lay proximity was improved to cover difficult target paths. Annibale Panichella proposed the many objective sorting algorithm (MOSA) [5] and its improved dynamic many objective sorting algorithm (DynaMOSA) [6] for the path coverage criteria. The optimization targets were dynamically selected based on the analysis of the control flow and dependency relationship for test data generation. Targeting all definition-use pairs derived from control flow and data flow, Ji et al. [19] improved the multi objective genetic algorithm to search all optimal solutions simultaneously. The appropriate fitness function with dominance relations were defined for selecting definition-use nodes. Considering successful applications of intelligent algorithms and complexity of class integration test order, Zhang et al. [20] analyzed the convergence of some algorithms by the Markov process theory. They discussed two different multi objective optimization strategy and proved the efficiency of strategy with Pareto model. Liao et al. [21] proposed the multiple paths coverage based on ant colony algorithm.

Although there are some related literature of applying multi objective heuristic optimization in test data generation, some of them considered the path coverage as a single optimization problem and other domain specific objectives as other optimization questions. In this paper, we consider the path coverage as a multi objective directly. Then the novel improved multi objective firefly algorithm is proposed to generate test data for all path targets in one run.

3 Problem Definition

3.1 Problem Formulation

Generally speaking, any optimization problem can be converted to the maximum optimization problem or the minimum optimization problem. In this paper, we reformulate the path coverage as a multi objective minimum optimization problem, where the fitness function of each target path is a optimization objective function to be optimized. The smaller the fitness value is, the closer the input test data is to the target solution.

Supporting there are m target paths to be covered in the program under test(PUT) with n input parameters. Let $B = \{b_1, b_2, ...b_m\}$ be the set of target branches and $t = \{x_1, x_2, ...x_n\}$ be a test data. So the input space X can be named as n -dimension decision space X and their output space decided by their function value can be named as target space. The question then can be formulated to find a set of non dominated test case $T = \{t_1, t_2..., t_m\}$ to minimize the following(as in formulation (1) objective functions.

$$
\begin{cases}
min\, f_1(t) = al(b_1, t) + d(b_1, t) \\
\qquad\qquad \cdot \\
\qquad\qquad \cdot \\
\qquad\qquad \cdot \\
min\, f_m(t) = al(b_m, t) + d(b_m, t)
\end{cases}
\tag{1}
$$

Where $al(b_i, t)$ denotes the approach level of test data t from the testing path target b_i, and $d(b_i, t)$ denotes the branch distance of test data t from the testing path target b_i. Each $min\, f_i(t)$ represents a single optimization problem for a single path target. In the case of test data generation problem, we notice that one solution may cover several objectives because of the complexity dependency relationship among target paths. So the number of optimal solutions may less than the number of optimized objectives m.

For the multi objective problems, there is no single best solution exists but a set of optimal solutions where these objectives may be in direct conflict with each other. Some common concepts are described as following [22].

Definition 1 (Pareto dominance): supporting test data x and y are candidate solutions in the decision space, x Pareto dominates y (named $x \prec y$) if and only if x is better than y at least one objective and no worse in all other objectives, as in Eq. (2).

$$\forall i \in (1, ..., m), f_i(x) \leq f_i(y) \bigwedge \exists j \in (1, ..., m) \, f_j(x) < f_j(y) \qquad (2)$$

Definition 2 (Pareto optimal): if and only if there is no any other test data Pareto dominates test data t^*, *the t^* is* named as the Pareto optimal solution, or non dominated solution or non inferior solution.

Definition 3 (Pareto Optimal Set): the set of all Pareto optimal solutions in the population can be named as Pareto Optimal Set.

Definition 4 (Pareto Front): the set of the objective function values of the Pareto Optimal Set can be named as Pareto Front.

3.2 Firefly Algorithm

FA is a meta heuristic algorithm motivated by the idealized biological behavior and information interaction strategy of fireflies [23]. The less bright firefly will be attracted and move towards the brighter one. In search space of maximization problems, each firefly represents a candidate solution and its brightness can be simply set to its fitness function value.

Assuming there are N fireflies in D-dimensional space, the any two ith/jth firefly can be represented as $x_i = (x_{i1}, x_{i2}, ..., x_{iD})$, $i = 1,2,..., N$ and $x_j = (x_{j1}, x_{j2},..., x_{jD})$, $j = 1,2,...,N$ respectively. The mathematical description of FA can be described as follows.

Each firefly should be initialized as in (1)

$$x_{ij} = rand() * (U - L) \qquad (3)$$

where rand() is randomization function generating numbers between 0 and 1, U and L are upper bound and lower bound of the input space.

The distance between firefly i at x_i and firefly j at x_j can be defined as in (4).

$$r_{ij} = \|x_i - x_j\| = \sqrt{\sum\nolimits_{d=1}^{D} \left(x_{id} - x_{jd}\right)^2} \qquad (4)$$

where x_{id} is the dth spatial coordinate of the ith firefly and x_{jd} is the dth spatial coordinate of the jth firefly.

Considering the absorption and inverse square law, the light intensity can be defined as in (5) and the relative attractiveness can be defined as in (6).

$$I_{ij}(r_{ij}) = I_i e^{-\gamma r_{ij}^2} \qquad (5)$$

$$\beta_{ij}(r_{ij}) = \beta_0 e^{-\gamma r_{ij}^2} \qquad (6)$$

Where I_i is attractiveness of firefly i according to the encoded fitness function. γ is a given light absorption coefficient and β_0 is the initial attractiveness when $r = 0$.

The movement of firefly i attracted by firefly j is defined as in (7).

$$x_{id}(t+1) = x_{id}(t) + \beta(x_{jd}(t) - x_{id}(t)) + \alpha_i(t)\varepsilon \qquad (7)$$

where the second part is the attractiveness between two fireflies and the third part is the random walk.

3.3 Fitness Function with Surrogate Model

For automatic test data generation optimization problem, there are some typical fitness function widely used in previous research. The fitness function of one separate testing target path is defined as Eq. (8).

$$fitness(x) = approach\_level(x) + branch\_distance(x) \qquad (8)$$

Where x is input test data in decision space, $approach\_level(x)$ denotes the approach level of test data x from the testing target path [24], $branch\_distance(x)$ denotes the branch distance of test data x from the testing target path.

By self learning mechanism of each individual, the population becomes better and better during iterations and will cluster in the optimal solutions. However, too many calculations of fitness function consumes considerable compute resources. In the case of automatic testing data generation, each individual in the population will be decoded into actual input value to drive the program execution for fitness function value. Therefor, the calculation of fitness function becomes a important factor restricting the performance of optimization. If the fitness value can be approximately calculated or predicted, the execution efficiency can be greatly improved [25].

The intelligent optimization algorithms based on surrogate model use surrogate models instead of the original fitness functions, with the hope of reducing the computation consumption by repeatable precise calculation. Inspired by Yao's [26] research, the BP based surrogate model is constructed to simulate the precise fitness function, as shown in Fig. 1. As seen, the model has input layer, hidden layer and output layer with respective corresponding weight vector W and V. The detailed steps are described in Algorithm 1.

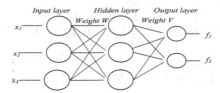

Fig. 1. The surrogate model based on BP

Algorithm1 : Surrogate model construction

Input: randomly initialized population
Output: the surrogate module
begin:
    Generate the initial population and calculate their exact fitness function value;
    Randomly initialized the weight vectors;
    Calculate output of each layer with the activation Sigmoid($f(x)=1/(1+e^{-x})$) function;
    Calculate the square error between the actual output and the expected output.if the error is less than the given threshold value or the training number reaches the maximum number, the training will stop. Otherwise , correct weight vector using the classical error back propagation algorithm.
    end

4 Multi Objective Firefly Algorithm with Surrogate Model

In the case of multi objective optimization algorithms, it is difficult to select offspring individuals because of too many non dominated solutions exists. Additionally, some actions should be taken to keep optimal solutions. Therefore, the offspring individual selection strategy and elite individual saving mechanism are critical for the optimization efficiency [27].

The population initialization, elite individuals archive and the offspring individuals selection strategies will be discussed in this section.

4.1 Chaotic Population Initialization

In the past decades, people discovered many chaotic maps applied to solve real world problems. Some popular chaotic maps include logistic map, tent map and Gauss map etc. Taking into the more evenly the chaotic sequence produced by tent map than logistic map, the chaotic sequence produced by tent map is used to initialize the whole population, as shown in Eq. (9) and mapped into the search space according to the Eq. (10).

$$\begin{cases} y_{n+1} = 4y_n^3 - 3y_n \\ -1 \leq y_n \leq 1, n = 0, 1, 2, ...N \end{cases} \tag{9}$$

$$\begin{cases} x_{id} = min_d + (1 + y_{id}) * \frac{max_d - min_d}{2} \\ \quad i = 1, 2, ..., N; d = 1, 2, ..., D \end{cases} \tag{10}$$

Where, some limitation on y_n ($y_n \in [-1,1]$ and $y_n \neq 0$) is required for the completeness of chaotic sequences in Eq. (9). In Eq. (10), the x_{id} denotes the dth dimensional position of x_i, y_{id} denotes the dth dimensional chaotic sequence, where min_d and max_d are lower bound and upper bound of parameters of programs under test.

The initialization process based on tent map can be summarized in Algorithm 2.

Algorithm 2 : Space mapping algorithm

Input: search space dimension D, population size N;
Output: initialized population
begin:
    Generate the first individual $Y= (y_1, y_2, ..., y_D)$ randomly;
    while($i < D$){
        Generate chaotic sequence of length N-1 iterated N-1 times according equation(9);
    }
    Map the chaotic sequence into search space according to equation(10)
end

4.2 Preference Based Non-dominated Sorting

Since meta heuristic search algorithms evolve and search optimal solutions with the population update at each iteration, the offspring individual selection mechanism is critical to the convergence of algorithms. The sorting mechanism based on Pareto non dominance relationship is popular in traditional multi objective optimization approaches and applicable in many real world problems. In the case of test data generation, the selection pressure of individuals increase because of the incomparable character of test data. If the preference of decision can be introduced into the selection process to assist individual selection, it will help to improve the efficiency of population selection [28, 29].

Once the preference mechanism is employed in automatic test data generation, we want more detailed information about uncovered target paths during the optimization process. According to the designed fitness function, we know that the lower the fitness value is, the closer to the target path. In other words, the candidate individual with lowest fitness value has the maximum probability of been the optimal solution for the target. So these special individuals should be selected as the offspring individuals for next generation.

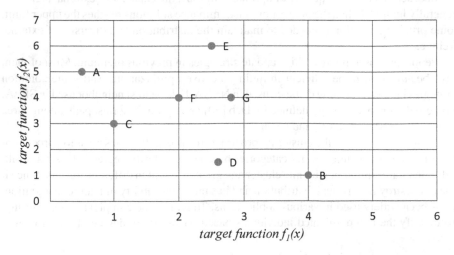

Fig. 2. Preference based non dominated sorting

Consuming the population distribution at a certain moment is illustrated in Fig. 2. According to the Pareto dominance definition, the four solutions A, B, C and D will be the Pareto optimal solutions, hence all of them will be sorted and selected. For test data generation, we prefer the solutions A and B, thus they will be ranked as the first level to the next generation. The preference based non dominated sorting strategy can be found in algorithm 3.

Algorithm3 : Preference based non dominated sorting

Input : the population T at the iteration t, the set of uncovered target B, population size N

Output : the non dominated solutions $F=\{F_0, F_1\}$

Begin :

$F = \{\}$

For($i=0; i<|B|,i++$){

　　Find the minimum fitness function value for each optimized function

　　Add the corresponding test data into the first level F_0;

}

Sort the rest test data based on domination relation and put all the non dominated test data in F_1

End

4.3 Archive Update Based on Lightweight Clustering

The archive strategy plays an important role in multi optimization process, which keeps track of the best solutions during interactions. With excellent genes, these individuals in archive contain more useful information to guide the optimization, called as elite individuals. As the output set of algorithms, its insertion and update mechanism is import to the optimization quality.

In order to ensure the quality of optimal solutions, the capacity of external archive is generally limited. When the number of non dominance solutions reaches the upper limit, some pruning operation is needed to maintain the distribution and diversity of external archive.

People proposed many archive update strategies in previous literature. Most of them have been proved to be efficient in multi objective optimization. And some common strategies include crowding distance used in NSGA-II, k-nearest neighbor used in SPEA, ε-pareto dominance strategy defined by Deb [30] and optimal shortest path among three points proposed by Xie [7], and so on.

Generally speaking, the clustering process is the population classification process of dividing individuals into several categories based on similarity degree [31]. As a result, individuals are similar in the same cluster and great different in different clusters. Since it doesn't destroy the original distribution and has the strong ability of feature recognition, it has been widely used in various applications. To reduce the complexity of clustering, we classify the non dominated individuals based on similarity degree of their function value.

4.4 Elite Individuals Learning

In the multi objective optimization algorithms, both the optimization convergence and diversity of Pareto optimal are the main considerations during the evolution process [32]. As for the firefly algorithm, each firefly will be attracted to any brighter one and update its position. The updated position is mainly affected by the current position, relative attractiveness and random disturbance. How the firefly algorithm evolves and search the optimal solutions? Maybe the fundamental cause lies on the learning ability of each firefly. It is simple in single optimization problem because of many brighter fireflies at each iteration. But the evolution strategy used successfully in single optimization algorithms are not efficient in the case of multi objective optimization.

Compared with single algorithms, there are more and more non dominated fireflies with the the population evolution in multi objective search space. Thus, the optimization process stops because there's less brighter fireflies and all individuals are similar, known as population premature. So it is necessary to employ some mechanism to activate the population. The elite individuals in external archive can be employed in view of their excellent genes. Based on the analysis, the position movement mechanism of fireflies can be updated as follows.

If firefly i and firefly j dose not dominate each other, the elite firefly g^* will be randomly selected from the external archive to promote the evolution process. The position update for firefly i and firefly j is defined in formula (11).

$$\begin{cases} x_i(t+1) = w_0 x_i(t) + (1-w_0)\beta_{g*i}(r_{g*i})(x_{g*}(t) - x_i(t)) + 2r_{g*i}\varepsilon \\ x_j(t+1) = w_0 x_j(t) + (1-w_0)\beta_{g*j}(r_{g*j})(x_{g*}(t) - x_j(t)) + 2r_{g*j}\varepsilon \end{cases} \tag{11}$$

When the dominance relation exists among fireflies, supposing firefly j dominates firefly i, the firefly i will be attracted and move toward firefly j and elite firefly g*. The position update for firefly i is defined in formula (12).

$$\begin{aligned} x_i(t+1) = x_i(t) + w_0\beta_{ij}(r_{ij})(x_j(t) - x_i(t)) \\ + (1-w_0)\beta_{g*i}(r_{g*i})(x_{g*}(t) - x_i(t)) + (r_{ij} + r_{g*i})\varepsilon \end{aligned} \tag{12}$$

Where, w_0 and ε both are random number between 0 and 1, r_{ij} denotes the Euclidean distance between firefly i and firefly j, β_{ij} denotes the relative attraction between firefly i and firefly j. By introducing the elite fireflies in optimization process, we hope the population becomes better under the excellent gene and the convergence rate will be improved.

5 Experiment Evaluation

In the field of test data generation, it has been proved that the multi-objective optimization algorithm is better than weighted aggregation optimization and iterative single path method. Therefore, this paper only carries out experiments in the range of multi-objective optimization algorithm. In order to verify the performance of the algorithm, the experiment is carried out on some common benchmark programs, and the following questions will be discussed.

RQ1: what is the accuracy of fitness function with surrogate model?
RQ2: How does the MOFA-SM perform compared to alternative approaches?
RQ3: How does the proposed algorithm depend on population size?

5.1 Experiment Setup

Algorithms for Experimental Analysis
Some typical alternative algorithms are selected from literature to verify their performance of multi objective test data generation. Their corresponding parameters are consistent with the reference as same as possible, as shown in Table 1.

Table 1. Algorithms for experimental analysis

Algorithm	Parameters	Reference
GA-M	Population size is set to 200, crossover probability is set to 0.95 and mutation probability is set to 0.8	Yao [33]
PSO-M	Population size is set to 200, acceleration factor is set to 2, maximum speed is 8 and the minimum is -8	Han [34]
HMOFA	$\alpha = 0.25, \beta_0 = 1, \gamma = 1.0$	Xie [7]
MOFA-SM	$\alpha = 0.25, \beta_0 = 1, \gamma = 1.0$	NA

Test Objects
ZDT functions and DTLZ functions are the existing frequently used data set, whose Pareto optimal front is known and predefined. These popular testing data set provides a quantifiable standard for the improvement of multi-objective optimization algorithm. In the area of automatic test data generation, there's no benchmark test data with any useful Pareto front. Most of researchers conduct experiment on some classical programs to efficiency verification. Although the scale of programs is limited, their input space dimensions and the nesting depth of target paths are various, ensuring the complexity of experiment, as shown in Table 2.

Experiment Design
With the assistant of fitness function, the practical problem can be transformed to optimization problem. Appropriate fitness functions can promote the optimization process and reflect the difference among individuals. The fitness function based surrogate model is used in this paper to reduce computation cost. In order to verify the similarity between the approximation value and exact value, several target paths in the program under test are selected and their separate fitness value in different ways are compared.

In order to verify the efficiency of algorithms for automatic test data generation, the experiment is conducted to calculate the average success rate of targets covering and the average convergence iterations, under the conditions of the same input space, population size and the same maximum iterations. Algorithms stop when the optimal solutions are found or reach the maximum iterations.

Table 2. Benchmark programs under test

Programs	Description
Triangle type	Given the input parameters x, y, z, the program checks whether these number form a triangle and further classify the triangle type equilateral, isosceles, or scalene
MaxAndMin	Given several numbers, the program find the maximum one and the minimum one
NumSort	Given three number, the program sort numbers in ascending order
Bubble sort	Given several numbers, the program sot them in ascending order
Insertion sort	Given several numbers, the program sot them in ascending order
Binary search	Given several numbers and the target key, the program check whether the key is in these numbers
Tax	Given a number representing the income, the program calculate the corresponding tax
GCD	Given two numbers, the program calculate the greatest common division between them, using the successive division method
Quadratic	Given three numbers, the program judge the roots type of the quadratic equation with one variable ($ax^2 + bx + c = 0$)

The efficiency of algorithms is closely related to the setting of parameters. It is meaningful to analyze the performance stability, which can be evaluated by changing some parameters. Additionally, each experiment was repeated 30 times independently and the average value of experimental results was used to reduce deviation caused by randomness.

5.2 Fitness Function Comparison

Some target paths of programs under test were selected to evaluate the approximation degree of the surrogate model. For each target, 100 test data are generated randomly and their fitness value with surrogate model and actual execution are collected. The approximate fitness value is calculated by surrogate model and the accurate fitness value is collected through the execution of the program driven by the input test data. The comparison result is showed in Fig. 3(a) and Fig. 3(b).

As seen from Fig. 3(a) and Fig. 3(b), the approximate value is similar to the accurate value from execution for the selected target paths. Although there is some sudden big difference happened in the comparison, most of the fitness function values are close. In the firefly optimization algorithm, all fireflies can be distinguished based on the approximate comparison, even if we do not know the exact fitness value collected from execution. On this basis, any firefly will be attracted and move to the better one in the optimization process. Hence, it is feasible to simulate the accurate value based on surrogate model.

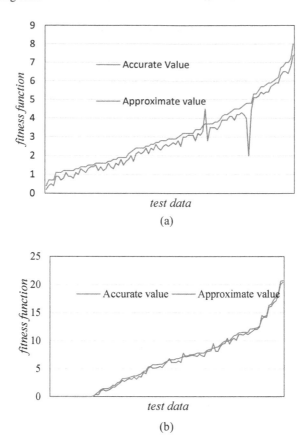

Fig. 3. (a) Fitness comparison of one function (b) Fitness comparison of another function

5.3 Efficiency

The target of the multi objective test data generation is to cover all reachable test targets in the program under test. So there are some typical measures should be collected to evaluate the efficiency of optimization algorithms, such as the success coverage rate and convergence iterations.

Success Coverage Rate

The success coverage rate is used to measure the effectiveness of algorithms in this paper. Each experiment was repeated 30 times independently to reduce deviation caused by randomness. During the execution, the coverage rate and the convergence iterations are collected for comparison. In each experiment, the coverage rate is calculated as the covered targets divided by the total targets within the maximum iterations. The coverage results are summarized in Table 3 and Fig. 4.

As seen from the experimental data, the success coverage rate of all algorithms is 100% for the program Tax. The rate is relatively low for program binary search. The average success coverage rate of the multi genetic algorithm is lower than other

Table 3. Success coverage rate

Algorithms	Triangle	MaxMin	NumSort	Bubble sort	Insertion sort	Binary search	Tax	GCD
GA-M	75	91	83	93	84	78	100	93
PSO-M	88	100	92	100	93	83	100	77
HMOFA	85	95	90	95	98	81	100	100
MOFA-SM	95	100	100	100	96	92	100	100

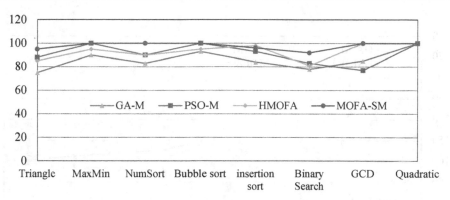

Fig. 4. Success coverage rate

algorithms. Generally, the proposed algorithm can search test data to cover targets with higher probability.

Convergence Iterations

Convergence iterations is an popular and important measurement to evaluate the performance of algorithms. Although there are different numbers of test targets in programs under test, the covering difficulty of each target are different. Some paths are easily covered in the early several runs of the optimization. In the automatic test data generation optimization, the number of test targets decrease along with the search. Sometimes, the convergence iterations are affected by the covering difficulty of some hard reachable paths. In this experiment, the iterations are recorded when all optimized targets are covered by the optimization process, and compared with each other. The results are shown in Table 4 and Fig. 5.

From the result, we can see the average convergence iterations of multi objective genetic algorithm is more than others. The number of iteration is 322 for triangle, 313 for quadratic and 357 for binary search, higher than other optimization algorithms. The performance of multi objective particle swarm optimization and hybrid multi-objective firefly algorithm are pretty even. Comparatively speaking, the average convergence iterations is lower than other algorithms.

Table 4. Convergence iterations

Algorithm	Triangle	NumSort	Bubble sort	Insertion sort	Binary search	GCD	Quadratic
GA-M	322	389	129	113	357	46	313
PSO-M	125	113	74	65	106	18	112
HMOFA	120	102	76	68	113	20	128
MOFA-SM	83	96	55	34	87	13	78

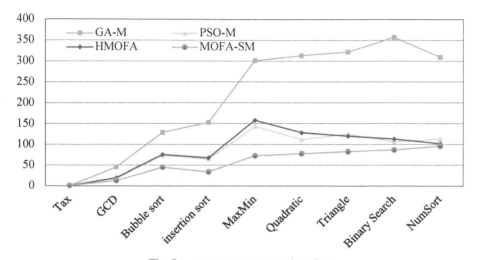

Fig. 5. Average convergence iterations

5.4 Parameters Analysis

Most of meta heuristic algorithms were proposed through simulating the amazing nature phenomena. They search and find the optimal solutions with a certain probability. Taking the triangle program as an example, we analyze the influence of population size to the performance of multi objective firefly algorithm.

To analyze the influence of population size to the performance, we focus on these experiments which can cover all test targets successfully in one run. Under the condition that the population size is set to 30, 50, 100, 150, 200, the algorithms search optimal solutions for all test targets and the execution data was recorded. Experiment data from 20 independent runs are shown in Table 5 and Fig. 6, including the minimum convergence iterations, the maximum convergence iterations and the average convergence iterations.

It can be seen from the experiment data that the average convergence generation decreases with the increment of population size and tends to stable when reaching a certain population size. In most cases, the proposed multi firefly algorithm can find all testing targets with less convergence iterations.

Table 5. Convergence iterations for triangle

Population size	30	50	100	150	200
Minimum	38	20	10	3	2
Maximum	100	95	87	75	78
Average	83	60	48	35	30

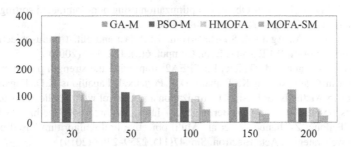

Fig. 6. Average convergence iterations at various size

6 Conclusion

The goal of automatic test data generation is to search a set of test data to maximize the testing target. According to the traditional search approach, the algorithm find solutions for one target at a time. For all target of program under test, the algorithm will be performed many times, consuming much more budget. Stemming from the excellent performance of multi objective optimization approaches, we tried to solve the automatic test data generation using multi meta heuristic search algorithm, where each target is considered as a independent optimization objective. Based on the analysis of test data generation optimization problem, the MOFA-SM is proposed to search optimal solutions for all test targets, which combing the preference strategy, the lightweight clustering archive update strategy and the elite individuals learning strategy. Experiment shows its fast convergence and strong stability. In future, we will focus on fitness function optimization, population variety improvement to improve its capability to solve various practical problems.

References

1. Harman, M., Jones, B.F.: Search-based software engineering. Inf. Softw. Technol. **43**(14), 833–839 (2001)
2. Harman, M., Yue, J, Zhang, Y.: Achievements, open problems and challenges for search based software testing. In: 8th IEEE International Conference on Software Testing, Verification and Validation (ICST 2015), Graz, Austria (2015)
3. Fraser, G., Arcuri, A.: Whole test suite generation. IEEE Trans. Software Eng. **39**(2), 276–291 (2013)
4. Fraser, G., Arcuri, A., Mcminn, P.: A memetic algorithm for whole test suite generation. J. Syst. Softw. **103**(2), 311–327 (2015)

5. Panichella, A., Kifetew, F.M., Tonella, P.: Reformulating branch coverage as a many-objective optimization problem. In: IEEE International Conference on Software Testing. IEEE (2015)
6. Panichella, A., Kifetew, F., Tonella, P.: Automated test case generation as a many-objective optimisation problem with dynamic selection of the targets. IEEE Trans. Softw. Eng. **44**, 1 (2018)
7. Xie, C.W., Xiao, C., Ding, L.X., Xia, X.W., Zhu, J.Y., Zhang, F.L.: HMOFA: a hybrid multi-objective firefly algorithm. J. Softw. **29**(4), 1143–1162 (2018)
8. Knowles, J.D., Corne, D.W.: Approximating the nondominated front using the Pareto archived evolution strategy. Evol. Comput. **8**(2), 149–172 (2000)
9. Srinivas, N., Deb, K.: Multi-Objective optimization using non-dominated sorting in genetic algorithms. Evol. Comput. **2**(3), 221–248 (1994)
10. Deb, K., Pratap, A., Agarwal, S., Meyarivan, T.: A fast and elitist multi-objective genetic algorithm: NSGA-II. IEEE Trans. Evol. Comput. **6**(2), 182–197 (2002)
11. Zitzler, E., Laumanns, M., Thiele, L.: SPEA2: improving the strength pareto evolutionary algorithm. In: Giannakoglou, K., Tsahalis, D.T., Periaux, J., Papailiou, K.D., Fogarty, T. (eds.) Evolutionary Methods for Design, Optimization and Control with Applications to Industrial Problems, pp. 95–100. Springer, Berlin (2002). https://doi.org/10.3929/ethz-a-004284029
12. Xie, C., Feilong, Z., Jianbo, L., et al.: Multi-objective firefly algorithm based on multiply cooperative strategies. Acta Electron. Sin. **47**(11), 2359–2367 (2019)
13. Zhang, Q., Hui, L.: MOEA/D: a multiobjective evolutionary algorithm based on decomposition. IEEE Trans. Evol. Comput. **11**(6), 712–731 (2008)
14. Coello, C.C.A., Pulido, G.T., Lechuga, M.S.: Handling multiple objectives with particles swarm optimization. IEEE Trans. Evol. Comput. **8**(3), 256–279 (2004)
15. Chengwang, X., Lei, X., Huairui, Z., et al.: Multi-objective fireworks optimization algorithm using elite opposite based learning. Acta Electronica Sinica **44**(5), 1180–1188 (2016)
16. Gopi, P., Ramalingam, M., Arumugam, C.: Search based test data generation: a multi objective approach using MOPSO evolutionary algorithm (2016)
17. Zhiqun, Z.: Test case generation based on multi-objective evolutionary algorithm. Hunan University (2016)
18. Dafei, H.: An improved evolutionary strategy for multi-target path coverage testing. Jiangxi University of Finance and Economics (2019)
19. Shunhui, J., Pengcheng, Z.: Test case generation approach for data flow based on dominance relations. Comput. Sci. **47**(9), 40–46 (2020)
20. Miao, Z., Shujuan, J., Yanmei, Z.: Research on multi-objective optimization in class integration test order. J. Chin. Comput. Syst. **38**(8), 1772–1777 (2017)
21. Weizhi, L., Xiaoyun, X., Xiaojun, J.: Test data generation for multiple paths coverage based on ant colony algorithm. Acta Electronica Sinica **48**(7), 1330–1342 (2020)
22. Chengwang, X., Weiwei, Y., Yingzhou, B., et al.: Many-objective evolutionary algorithm based on decomposition and coevolution. J. Softw. **31**(2), 356–373 (2020)
23. Yang, X.S.: Nature-inspired Metaheuristic Algorithms. Luniver Press, London (2008)
24. Dunwei, G., Yan, Z.: Novel evolutionary generation approach to test data for multiple paths coverage. Acta Electron. Sin. **038**(006), 1299–1304 (2010)
25. Gaoyang, L.: The research of swarm intelligence optimization algorithms based on surrogate model. Jilin University (2016)
26. Xiangjuan, Y., Dunwei, G., Bin, L.: Evolutional test data generation for path coverage by integrating neural network. J. Softw. **27**(4), 828–838 (2016)
27. Xiaoji, C., Chuan, S.: Multiobjective evolutionary algorithm based on hybrid individual selection mechanism. J. Softw. **30**(12), 3651–3664 (2019)
28. Xunxue, C., Chuang, L.: A preference based multi-objective concordance genetic algorithm. J. Softw. **16**(005), 761–770 (2005)

29. Dongdong, Y., Licheng, J., Maoguo, G., et al.: Clone selection algorithm to solve preference multi-objective optimization. J. Softw. **021**(001), 14–33 (2010)
30. Laumanns, M., Thiele, L., Deb, K., et al.: Combining convergence and diversity in evolutionary multi-objective optimization. Evol. Comput.**10**(3), 263–282 (2002)
31. Enze, Z.: Research on multi-objective particle swarm optimization algorithm and applications. Nanjing University of Science & Technology (2016)
32. Maoguo, G., Gang, C., Licheng, J.: Nondominated individual selection strategy based on adaptive partition for evolutionary multi-objective optimization. J. Comput. Res. Dev. **048**(004), 545–557 (2011)
33. Han, X., Lei, H., Wang, Y.S.: Multiple paths test data generation based on particle swarm optimization. IET Softw. **11**(2), 41–47 (2017)
34. Yao, X., Gong, D.: Genetic algorithm based test data generation for multiple paths via individual sharing. Comput. Intell. Neurosci. **2014**(3), 59–70 (2014)

Multimedia Data Management
and Analysis

Semantic Segmentation of High Resolution Remote Sensing Images Based on Improved ResU-Net

Songyu Chen, Qiang Zuo, and Zhifang Wang[✉]

Department of Electronic Engineering, Heilongjiang University, Harbin 150080, China
swangzhifang@hlju.edu.cn

Abstract. Image segmentation is an important basic link of remote sensing interpretation. High-resolution remote sensing images contain complex object information. The application of traditional segmentation methods is greatly restricted. In this paper, a remote sensing semantic segmentation algorithm is proposed based on ResU-Net combined with Atrous convolution. The traditional U-Net semantic segmentation network was improved as the backbone network, and the residual convolution unit was used to replace the original U-Net convolution unit to increase the depth of the network and avoid the disappearance of gradients. To detect more feature information, a multi-branch hole convolution module was added between the encoding and decoding modules to extract semantic features, and the expansion rate of the hole convolution was modified to make the network have a better effect on the small target category segmentation. Finally, the remote sensing image was classified by pixel to output the remote sensing image semantic segmentation result. The experimental results show that the accuracy and interaction ratio of the proposed algorithm in the ISPRS Vaihingen dataset are improved, which verifies its effectiveness.

Keywords: Semantic segmentation · ResU-Net · Atrous convolution

1 Introduction

With the continuous development of remote sensing technology, the semantic information contained in remote sensing images is becoming more and more abundant. Therefore, how to quickly and accurately extract the important semantic information from remote sensing images and carry out the later application and development is a very important research topic [1, 2].Semantic segmentation is an important step to achieve scene understanding, which generates predictions by assigning each pixel to a target category. It plays a vital role in many important remote sensing applications, such as natural disaster detection, urban planning, land cover mapping, etc. [3]. Therefore, people pay more and more attention to the semantic segmentation of ultra-high resolution remote sensing images.

Traditional machine learning methods employ manual features, such as random forests, support vector machines, markovrandom fields, and conditional random fields

© Springer Nature Singapore Pte Ltd. 2021
J. Zeng et al. (Eds.): ICPCSEE 2021, CCIS 1451, pp. 303–313, 2021.
https://doi.org/10.1007/978-981-16-5940-9_23

[4]. However, although the traditional algorithm is very efficient, it cannot meet the requirements of high precision. In recent years, as deep convolutional neural networks (DCNNs) have achieved advanced performance in the field of computer vision, a large number of DCNNs methods have been proposed to be applied to remote sensing semantic segmentation tasks [5–7]. Many studies have shown that deeper networks will have better performance. The proposal of Full Convolutional Network (FCN) [8] in the field of semantic segmentation has creative significance.FCN use output feature mapping and match the resolution of the input image through upsampling to achieve end-to-end training. However, the network encoding part adopts the VGG16 pre-training model to down-sample the input data 4 times, reducing it to 1/32 of the original, and there is a problem of rough predicting the fine edge feature map. In order to solve this problem, SegNet [9] and U-Net [10] networks have been proposed. The "encoding-decoding" type network represented by U-Net, such as R2U-Net [11] and CE-Net [12], benefit from the advantages of fewer parameters and easy training. In the process of feature decoding and restoration, it can be integrated with the encoding layer jump connection. The characteristics of semantic features and local detail features make the segmentation boundary gradient in the remote sensing fine segmentation image not obvious, the training set samples are less, and the shape size changes are gradually solved. However, when the receptive field becomes larger, some useful details of ground objects and image scenes will be deleted. Using hole convolution is the main method to solve this problem. When the size of the convolution kernel becomes larger, the receptive field is expanded without increasing the number of parameters and the size of the feature map. Subsequently, atrous convolution has been widely used in many algorithms, such as Deeplab [13] and PSPNet [14].

Many of the semantic segmentation methods currently used are pixel-level classification methods, which extract features from image objects and assign a class probability vector [15] to each pixel, such as cars, trees, or buildings. By contrast, using traditional methods such as random forests, the probability of each category is based only on the inherent characteristics of the spectrum. Information based on spectral features is less than based on object features. For example, judging the category of cars requires not only understanding its spectral features, but also understanding the changes in the features and the proportion in the image. It is known that cars are more likely to be surrounded by pixels belonging to the road and less likely to be surrounded by pixels belonging to buildings [16, 17].

In this paper, a method based on atrous convolution and ResU-Net (ARU-Net) is proposed for semantic segmentation. U-Net is taken as the backbone network, and the convolution unit of residual is used to replace the convolution unit of original U-Net, so as to train the deeper network and deepen the depth of network and extract deeper semantic information. A multi-branch atrous convolution module is added between the feature encoding module and the feature decoding module to perform multi-scale encoding of semantic features to realize the construction of an image segmentation model.

2   ARU-Net

In this paper, the ARU-Net method is an end-to-end deep convolutional neural network. This architecture not only combines the advantages of deep residual learning and U-Net, but also imports atrous convolution. The network of ARU-Net contains two sub-networks: one is the encoder, which realizes the recognition of image features by the network through 4 times of max-pooling; the other is the decoder, which restores the location information and detailed information of the image through up-sampling the extracted features. The encoder-decoder structure is mainly composed of a convolutional layer, an activation layer, a pooling layer, and an up-sampling layer. The convolution layer extracts the features of the input image, the activation layer adds non-linear factors, the pooling layer extracts the main features while reducing the resolution of the feature image, and the upsampling layer conducts deconvolution operation on the feature image to increase the resolution of the feature image. The encoder and the decoder realize the fusion of feature information between different levels through feature splicing, overcome the accuracy loss caused by upsampling, and optimize the classification result. The structure of ARU-Net is shown in Fig. 1.

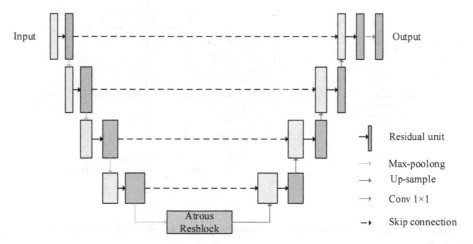

Fig. 1. ARU-Net structure.

In order to build a deeper network structure, ARU-Net uses a residual convolution unit, and solves the problem of degradation in the network, while reducing the computational cost. Atrousresblock acts as a bridge to expand the field of view of the filter and include a wider background. Artous convolution can arbitrarily expand the receptive field, obtain multi-scale information, and does not need to introduce additional parameters. At the same time, too much downsampling is not required, and the spatial resolution is not lost, which is conducive to retaining the characteristic information.

2.1 Residual Unit

Deepening the depth of the network can improve the performance of the network, but it is possible that the gradient disappears [18], that is, the accuracy will first rise and then reach saturation. Continuously increasing the depth will cause the model accuracy to decrease. Therefore, He et al. [19] proposed a deep residual learning framework to facilitate the training process and solve the degradation problem. The salient feature of the residual network architecture is the identical jump connection in the residual block, which allows us to easily train a deeper network architecture. The residual convolution unit can be expressed as:

$$y_l = h(x_l) + F(x_l, w_l)$$
$$x_{l+1} = f(y_l)$$
(1)

Where x_l and x_{l+1} are the input and output of the residual convolution block, $F(\cdot)$ is the residual function, $f(y_l)$ is the activation function, and $h(x_l)$ is the identity mapping function. The difference between the ordinary convolution unit and the residual convolution unit is shown in Fig. 2.

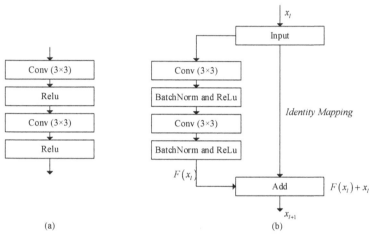

Fig. 2. Neural network convolution unit: (a) ordinary convolution unit; (b) Convolution unit of ResU-Net.

ARU-Net uses the complete activation remaining unit shown in Fig. 2 (b). In the deep residual, the jump connection unit helps to spread the information within the network without degradation and making the deep network easy to train. The design of neural network is improved by reducing parameters to improve the performance of semantic segmentation task.

2.2 AtrousResblock

At present, the key problem of semantic segmentation in existing models is that some information is lost during downsampling. The purpose of downsampling is to expand the receptive field so that each convolution output contains a larger range of information. However, in this case, the resolution of the image will continue to decrease, and the information contained will become more and more abstract. The partial information and detailed information of the system will gradually be lost [20]. Therefore, atrous convolution plays an important role in semantic segmentation.

Increasing the receptive field can fully understand the information between the context of the image and make the segmentation result more accurate. The size of the convolution kernel affects the size of the receptive field. The larger the size of the convolution kernel, the more detailed information contained in the extracted feature map. However, increasing the size of the convolution kernel will increase the model parameters and slow training. Artous convolution can increase the feature spatial resolution while increasing the receptive field without increasing the parameters [21]. The calculation formula of atrous convolution can be expressed as:

$$g_{i,j}(x_\ell) = \sum_{c=0}^{c_\ell} \theta_{k,r}^{i,j} * x_\ell^c \tag{2}$$

Where $g_{i,j}$ is the convolution operation of the output feature graph, x_ℓ is the feature graph belonging to the channel in i row and j column, and $\theta_{k,r}$ is atrous convolution, the size of the convolution kernel is k. The rate of expansion is r. After adding the atrous convolution, the size of the convolution kernel is $k + (k-1)(r-1)$, and when $r = 1$, it is equivalent to the standard convolution. When $r > 1$, it is atrous convolution, as shown in Fig. 3.

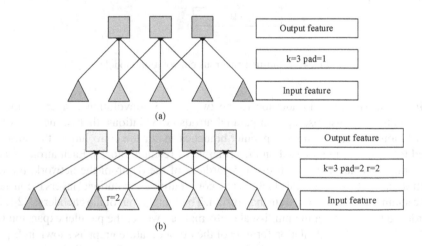

(a)

(b)

Fig. 3. Atrous convolution: (a) r = 1; (b) r = 2

Fig. 3 (a) is a standard convolution operation with the expansion rate, the corresponding receptive field is 3 × 3, and Fig. 3 (b) is atrous convolution. It can be seen that when $r = 1$, the input signal is sampled alternately. First, pad = 2 means that left and right are filled with 2 zeros, which keep the input and output feature maps the same size. Then, when $r = 2$, the input signal are sampled for convolution every 2 inputs. Therefore, there are 5 outputs at the output end, which makes the output feature maps increase. Atrous convolution expands the field of view of the convolution kernel to contain larger receptive field information. At the same time, it provides an effective mechanism to control the size of the receptive field and find precise positioning target information.

In order to achieve the fusion of multi-scale feature maps to adapt to the variability of remote sensing image scales [22], an atrous res-block is added between the encoding and decoding modules to encode high-level semantic feature maps. This module uses multiple parallel atrous convolutions with different expansion rates for the same feature map. The structure of the module is shown in Fig. 4, where $d_1 \ldots d_n$ represents different expansion rates.

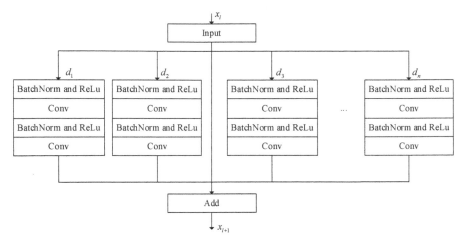

Fig. 4. Multi-branch atrous convolution module.

In ArtousResblock, in addition to the two 3 × 3 convolutional layers in the U-Net structure, we also use up to 3 parallel atrous convolutions, that is, the set of two convolutional layers has up to 4 parallel branches. After the convolution, the output is added to the initial input based on the principle of residual blocks. Each atrous branch is summed instead of fusion, because regardless of the depth of the network, the loss function shown by the residual blocks of two consecutive convolutional layers is constant, so the summation is easier to train than fusion. The final output of the encoder is 32 × 32. In order to better extract the multi-scale information, we set the parallel expansion rate as [1, 3, 15, 31]. The calculation formula of the output feature graph is shown in Eq. (3) to ensure that the output feature graph remains unchanged. The detailed information is shown in Table 1.

$$O = \left(\frac{i + 2p - k - (k - 1) * (r - 1)}{s} \right) + 1 \tag{3}$$

Where O is the size of the output feature map, k is the size of the convolution kernel, s is the step size, p is the number of pixels to be filled, and r is the expansion rate.

Table 1. The architecture of the proposed multi-branch Atrous convolution.

Name	Kernel Size	Stride	Expansion rate	Padding
Branch_1	3 × 3/3 × 3	1	1	1
Branch_2	3 × 3/3 × 3	1	3	3
Branch_3	3 × 3/3 × 3	1	15	15
Branch_4	3 × 3/3 × 3	1	31	31

3 Experiments and Analysis

3.1 Dataset

In order to test ARU-Net, the Vaihingen aerial image data of the public remote sensing image dataset ISPRS was selected [23]. Vaihingen is a small village with many independent buildings and small multi-storey buildings. The data set contains 16 labeled false-color aerial images (containing 3 channels of near-infrared, red, and green). The image size ranges from 1388 × 1281 pixels to 2995 × 3007 pixels, and the ground resolution is 0.09 m. The data set is classified into the five most common land cover categories, namely ground, buildings, low plants, trees, and cars. The data set does not contain cluttered backgrounds.

We divide the labeled dataset into two parts, of which 12 images are used for network training, and the remaining 4 images are used to test the performance of the network. Due to the small number of images in the original dataset, different sizes, and the size of a single image is too large to be directly sent to the network, due to the limitation of GPU memory, the original large training image is cropped into 8000 images with a size of 256 × 256 pixels. Datasets are allowed to overlap. The test image is cropped into small non-overlapping blocks of the same size. Horizontal flipping, vertical flipping, and 90-degree rotation are also used to expand the training set.

3.2 Evaluation Standard

In order to evaluate the proposed model, F1-score (AF), overall average accuracy (OA) and mIoU are used to evaluate the performance of semantic segmentation. The calculation formulas are as follows:

$$P = TP/(TP + FP) \tag{4}$$

$$R = TP/(TP + FN) \tag{5}$$

$$AF = 2PR/(P + R) \tag{6}$$

$$OA = \frac{TP + TN}{TP + FN + FP + TN} \tag{7}$$

$$mIoU = \frac{1}{k+1} \sum_{i=0}^{k} \frac{TP}{FN + FP + TP} \tag{8}$$

Where TP is True Positive, indicating that the predicted value and true value are 1. TN is True Negative, which means that the predicted value and true value are 0. FP is False Positive, indicating that the predicted value is 1 and the true value is 0. FN is False Negative, which means that the predicted value is 0 and the true value is 1.

3.3 Experiments and Analysis

The optimizer used in this experiment is SGD, momentum, initial learning rate, and minimum learning rate are set to 0.9, 0.01, and 0.001, respectively, and the batch normalization size is 6.

ARU-Net combines ResU-Net and Artous resblock. Since different expansion rates have a certain impact on the segmentation results, the expansion rates are set respectively [1, 3, 5], [1, 3, 15, 31], The experimental results are shown in Table 2 and Table 3.

Table 2. Semantic segmentation results of ISPRS vaihingen dataset.

Model	OA	AF	mIoU
FCN-8s	0.7860	0.6706	0.5134
U-Net	0.8363	0.7196	0.5712
ResU-Net	0.8475	0.7366	0.5748
R2U-Net	0.8432	0.7329	0.5703
ARU-Net + [1, 3]	0.8534	0.7474	0.5767
ARU-Net + [1, 3, 15]	0.8539	0.7472	0.5725
ARU-Net + [1, 3, 15, 31]	0.8524	0.7457	0.5878

It can be seen from Table 2 that the change of the expansion rate has a certain influence on the results of the experiment. The mIoU with expansion rate [1, 3, 15, 31] is increased by 1.53% compared with the other two settings. Although the results of OA and AF decreased slightly, the segmentation results of mIoU of trees and cars were significantly improved by 1% and 6.29%. According to the experimental results, the expansion rate [1, 3, 15, 31] is choosed as the final parameter. Compared with the original other method, the OA, AF and mIoU of Artousresblock with an expansion rate of [1, 3, 15, 31] have increased by at least 0.49%, 0.91% and 1.3%. The each category of mIoU has also been significantly improved. In particular, compared with mIoU and FCN-8S, U-Net, ResU-Net and R2U-Net, the mIoU increased by at least 4.57%,indi-cating that atrous convolution can effectively retain the characteristics of the information by expanding the receptive field information. For such small target objects with less detailed information, Artous resblock can better extract the semantic features of remote sensing images.

Table 3. The accuracy of each category is presented in the mIoU form.

Model	Surfaces	Building	Low veg	Tree	Car
FCN-8s	0.6919	0.7572	0.5848	0.5882	0.2881
U-Net	0.7751	0.7983	0.6375	0.6457	0.5056
ResU-Net	0.7955	0.8176	0.6667	0.6386	0.5306
R2U-Net	0.8022	0.8116	0.6494	0.6320	0.5268
ARU-Net + [1, 3]	0.8181	0.8333	0.6632	0.6261	0.5134
ARU-Net + [1, 3, 15]	0.8159	0.8312	0.6632	0.6365	0.4837
ARU-Net + [1,3,15,31]	0.8193	0.8266	0.6680	0.6465	0.5763

In order to make the experiment more scientific, this paper compares the segmentation results of the proposed methods and the experimental results is shown in Fig. 5. The first column in the figure is the local details of the input remote sensing image, and the second column is the local true label map. The method in this paper refines the boundaries of the classes in the segmented image, making the target edge closer to the real edge of the scene. Compared with FCN-8s, U-Net, ResU-Net and R2U-Net, the segmentation of buildings, trees, and small target objects is more accurate and reduces the phenomenon of false segmentation.

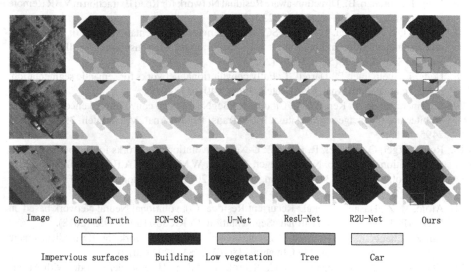

Image   Ground Truth   FCN-8S   U-Net   ResU-Net   R2U-Net   Ours

Impervious surfaces   Building   Low vegetation   Tree   Car

Fig. 5. Segmentation detail comparison.

4   Conclusions

This paper proposes an improved ResU-Net variant deep learning network archi-tecture combined with atrous convolution to achieve semantic segmentation of high-resolution remote sensing images. Artous Resblock is added between the encoder and the decoder, and multi-scale information is obtained by adjusting the expansion rate of the module, thereby constructing a network model suitable for remote sensing image segmentation and increasing the segmentation ability of small objects. Experiments were carried out on the Vaihingen dataset of ISPRS. The experi-mental results show that the overall mIoU is increased by 7.44%, 1.66%, 1.3% and 1.75%, compared to FCN-8s, U-Net, ResU-Net and R2U-Net, and the mIoU for the small target category of cars is increased by at least 4.57%.

References

1. Audebert, N., Le Saux, B., Lefevre, S.: Beyond RGB: very high resolution urban remote sensing with multimodal DeepNetworks. ISPRS J. Photogrammetry Remote Sens. **140**, 20–32 (2017)
2. Ma, J., et al.: Building extraction of aerial images by a global and multi-scale encoder-decoder network. Remote Sens. **12**(15), 2350 (2020)
3. Ma, L., Liu, Y., Zhang, X., et al.: Deep learning in remote sensing applications: A meta-analysis and review. ISPRS J. Photogrammetry Remote Sens.**152**, 166–177(2019)
4. Jiang, N., Li, J.: An improved semantic segmentation method for remote sensing images based on neural network. Traitement du Signal **37**(2), 271–278 (2020)
5. Wang, H., Wang, Y., Zhang,Q., et al.: Gated convolutional neural network for semantic segmentation in high-resolution images. Remote Sens. **9**(5), 446 (2017)
6. Ding, L., Lorenzo, B.: Direction-aware Residual Network for Road Extraction in VHR Remote Sensing Images. CoRR abs/2005.07232 (2020)
7. Li, H., Qiu, K., Chen, L., et al.: SCAttNet: semantic segmentation network with spatial and channel attention mechanism for high-resolution remote sensing images. IEEE Geosci. Remote Sens. Lett. **18**(5), 905–909 (2021)
8. Long, J., Shelhamer, E., Darrell, T.: Fully convolutional networks for semantic segmentation. IEEE Trans. Pattern Anal. Mach. Intell. **39**(4), 640–651 (2017)
9. Badrinarayanan, V., Kendall, A., Cipolla, R.: SegNet: a deep convolutional encoder-decoder architecture for image segmentation. IEEE Trans. Pattern Anal. Mach. Intell. **39**(12), 2481–2495 (2017)
10. Ronneberger, O., Fischer, P., Brox, T.: U-Net: convolutional networks for biomedical image segmentation. In: Navab, N., Hornegger, J., Wells, W.M., Frangi, A.F. (eds.) MICCAI 2015. LNCS, vol. 9351, pp. 234–241. Springer, Cham (2015). https://doi.org/10.1007/978-3-319-24574-4_28
11. Alom, M, Z., Hasan, M., et al.: Recurrent Residual Convolutional Neural Network based on U-Net (R2U-Net) for MedicalImageSegmentation. CoRR abs/1802.06955 (2018).
12. Gu, Z., Cheng, J., Fu, H., et al.: CE-net: context encoder network for 2D medical image segmentation. IEEE Trans. Med. Imaging **38**(10), 2281–2292 (2019)
13. Chen, L.-C., Zhu, Y., Papandreou, G., Schroff, F., Adam, H.: Encoder-decoder with Atrous separable convolution for semantic image segmentation. In: Ferrari, V., Hebert, M., Sminchisescu, C., Weiss, Y. (eds.) ECCV 2018. LNCS, vol. 11211, pp. 833–851. Springer, Cham (2018). https://doi.org/10.1007/978-3-030-01234-2_49
14. Cheng, G., Wang, Y., Xu, S., et al.: Automatic road detection and centerline extraction via cascaded end-to-end convolutional neural network. IEEE Trans. Geosci. Remote Sens. **55**(6), 3322–3337 (2017)

15. Chen, G., Zhang, X., Wang, Q., et al.: Symmetrical dense-shortcut deep fully convolutional networks for semantic segmentation of very-high-resolution remote sensing images. IEEE J. Sel. Topics Appl. Earth Obser. Remote Sens. **11**(5), 1633–1644 (2018)

16. Huang, H., et al.: UNet 3+: A Full-Scale Connected UNet for Medical Image Segmentation. CoRR abs/2004.08790 (2020)

17. He, C., Li, S., Xiong, D., et al.: Remote sensing image semantic segmentation based on edge information guidance. Remote Sens. **12**(9), 1501 (2020)

18. Shang, R., Zhang, J., Jiao, L., et al.: Multi-scale adaptive feature fusion network for semantic segmentation in remote sensing images. Remote Sens. **12**(5), 872 (2020)

19. Zhang, Z., Liu, Q., Wang, Y.: Road extraction by deep residual U-net. IEEE Geosci. Remote Sens. Lett. **15**(5), 749–753 (2018)

20. Wang, Y., Liang, B., Ding, M., et al.: Dense semantic labeling with atrous spatial pyramid pooling and decoder for high-resolution remote sensing imagery. Remote Sens. **11**(1), 20 (2019)

21. Chen,L ,C., Papandreou, G., Kokkinos, I., et al.:DeepLab: semantic image segmentation with deep convolutional nets, atrous convolution, and fully connected CRFs. IEEE Trans. Pattern Anal. Mach. Intell. **40**(4), 834–848 (2018)

22. Yang, M., et al.: DenseASPP for Semantic Segmentation in Street Scenes. CVPR, pp. 3684–3692 (2018)

23. Gerke, M. Use of the Stair Vision Library within the ISPRS 2D Semantic Labeling Benchmark (Vaihingen); Technical Report; University of Twente: Enschede, the Netherlands (2015)

Real-Time Image and Video Artistic Style Rendering System Based on GPU

Yang Zhao[1], Guowu Yuan[2(✉)], Hao Wu[2], Yuanyuan Pu[2], and Dan Xu[2]

[1] Department of Animation, Yunnan Normal University, Kunming 650500, China
[2] School of Information Science and Engineering, Yunnan University, Kunming 650091, China
yuanguowu@sina.com

Abstract. Aiming at the practical engineering application of video stylization, in this paper, a GPU-based video art stylization algorithm is proposed, and a real-time video art stylization rendering system is implemented. The four most common artistic styles including cartoon, oil painting, pencil painting and water-color painting are realized in this system rapidly. Moreover, the system makes good use of the GPU's parallel computing characteristics, transforms the video stylized rendering algorithm into the texture image rendering process, accelerates the time-consuming pixel traversal processing in parallel and avoids the loop processing of the traditional CPU. Experiments show that the four art styles achieved good results, and the system has a good interactive experience.

Keywords: Non-photorealistic rendering · GPU · Structure tensor · Line integral convolution · Kuwahara filter

1 Introduction

Artistic style rendering is an important research area of non-photorealistic rendering. It is one of the ways to simulate traditional art media on the computer. At the same time, it also makes image rendering more personalized.

With the wide use of smart phones, several image stylized rendering software, such as Prism and Dream-Art, have been designed and widely used.

These algorithms can be divided into two categories, which are physics-based simulation method and image-based processing method, and the latter is more widely used. The method splits an input video into frames, renders each frame, and then generates the video with specific artistic style through video compositing technology.

At present, researchers have developed all kinds of image-based stylized rendering technologies. But their studies mainly focus on cartoon rendering and oil painting rendering of input video, and no mature algorithms have been designed for other artistic styles such as watercolor painting and pencil drawing.

In addition, traditional image processing software needs artists to work through complex human-computer interaction to achieve the corresponding special effects. Although these software greatly simplify the rendering work of images, it is difficult for common users to directly generate stylized images with specific artistic characteristics.

J. Zeng et al. (Eds.): ICPCSEE 2021, CCIS 1451, pp. 314–338, 2021.
https://doi.org/10.1007/978-981-16-5940-9_24

In fact, artistic stylized rendering of videos has a wide range of application, but the shortcomings of the current methods are complexed implementing, long rendering time, high requirements for computer hardware, low degree of automation. It is necessary to design an algorithm to shorten the rendering period and even achieve the real-time performance is the hot spot in the current research field.

2 Related Work

Designing the video stylization algorithm is very important in NPR's research field. Video and Image stylization is the conversion and processing of the real images, so that the input image and video can be abstracted and stylized, which is convenient for users to intuitively understand. Video stylization algorithms have been applied to many fields, such as film and television production, advertising art and so on.

In 1997, Litwinowicz firstly proposed a rendering algorithm that processed the collected videos and images into impressionistic works [1]. By analyzing the video frames, the algorithm extracts the color features, and then selects different brushes to draw along the gradient field of the input video frames. The shortcoming of this algorithm is that the rendering speed is too slow.

In 2000, Hertzmann designed a new method to rendering the stylized video by considering the temporal coherence between frames [2]. Although the algorithm can obtain animation works with different styles such as oil painting and watercolor painting, the problem is that it consumes a lot of computing time to draw each frame in layers. In addition, according to the experimental results, the stylized video still has problems of jitter and flicker.

In 2002, Allison Klein et al. proposed to use dynamic programming method to form stylized videos with Mosaic effect from the collection of specified videos [3]. They propose to treat the video data as an entity of spatiotemporal data. By evaluating some functions, artists can interactively update the parameters in real time to complete the stylized effects of input video.

In 2002, Asean Agarwala designed a system called Snaketoonz. The system uses the active contour technology to track a set of manually specified contour lines, and then segment the video to obtain the cartoon animation effect [4]. Later, Aseem Agarwala et al. proposed a new technique called Rotoscoping [5]. It is essentially a video tracking algorithm, but by combining with manually defined strokes, the input video can be converted into animation with a variety of styles. The main problem of the above two technologies is that more human interactions reduce the automation of the system.

In 2004, Jue Wang et al. designed an animation rendering system to automatically generate cartoon style videos [6]. The basic idea of the system is that video is regarded as continuous data in time and space, and it can render cartoon style animation by means of the anisotropic mean-shift segmentation algorithm designed.

In 2006, Winnemoller et al. proposed an image-based cartoon stylized rendering system, which realized real-time cartoon rendering for the first time [7]. In 2007, Zhao Yang, Xu Dan et al. designed a real-time rendering system on this basis, which could render the input video into a stylized video with cartoon, watercolor and pencil drawing effects [8].

In 2011, Peter O'Donovan et al. designed an interactive system which can render video sequence to painting animation [9]. In this system, they proposed to use the greedy algorithm instead of calculating the optical flow vector field to track and guide the brush orientation in the video sequence.

None of the above video stylized rendering systems can achieve the real-time conversion and processing of the input video. In order to solve this problem, researches proposed to use GPU to accelerate rendering speed of images and videos in artistic style. However, the artistic style generated by these software systems is relatively not satisfactory, and which does not form a more systematic application framework.

In 2017, Qiaoyu Wang et al. designed an adaptive cartoon-like stylization algorithm for color video in real time [10]. They implement the algorithm parallel on a GPU having the Computer Unified Device Architecture (CUDA). The CUDA implement ion is able to process video in real time, running 45 to 180 times faster than the CPU. But the system they designed can't transfer the input image and video into different artistic style renderings.

In order to solve the above problems, this paper uses GPU based video processing technology to design and implement a video artistic stylized rendering system, which can process and transform the input video into specific artistic style in real time within a unified framework. According to the needs of users, the system can automatically render the input video into stylized video with special effects such as cartoon, Van Gogh's style oil painting, pencil drawing and watercolor painting.

3   System Framework

When GPU is used for video stylization, firstly, the input video needs to be output to the rendering texture frame by frame. Secondly, pixel shader is used for multi pass rendering. During the process, we save the intermediate results to the rendering texture, which will be used as the input texture for subsequent processing. Finally, the result of multi pass processing is output.

The system designed in this paper is mainly divided into three modules: (1) Video loading module, which is responsible for loading the original input video into the rendering texture; (2) Video stylization processing module, which contains four sub-modules, which can render the input video into four special effects of cartoon, oil painting, pencil drawing and watercolor painting. It can also be expanded to generate more artistic styles; (3) Video output module, which can output the result after rendering to the display device. The system framework is shown in Fig. 1.

4   Real-Time Video Stylization Algorithm

4.1   Structure Tensor

Firstly, we need to calculate the eigenvalues and eigenvectors of the current frame based on the structure tensor [11], which is used as the vector field to guide the orientation for the subsequent processing.

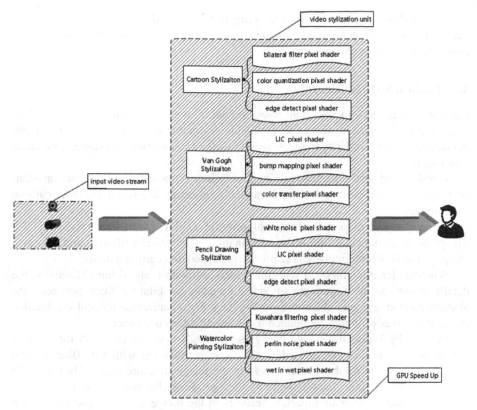

Fig. 1. System framework

Let $\mathbb{F}$ be the input image, $G_{\sigma,\,x}$ and $G_{\sigma,\,y}$ be the derivative of the normal distribution on the x axis and y axis respectively, it can be formalized into:

$$\mathbb{G}_\sigma(x) = \frac{1}{2\pi\sigma^2} exp\left(-\frac{x^2}{2\sigma^2}\right) \tag{1}$$

Convolve $\mathbb{F}$ using $\mathbb{G}_\sigma(x)$, we can get partial derivatives along the x and y axes, then the structure tensor can be expressed as:

$$J(\nabla F) = \begin{bmatrix} f_x \bullet f_x & f_x \bullet f_y \\ f_x \bullet f_y & f_y \bullet f_y \end{bmatrix} := \begin{bmatrix} E & F \\ F & G \end{bmatrix} \tag{2}$$

The non-negative eigenvalue of the structural tensor is:

$$\lambda_{1,2} = \frac{E + G \pm \sqrt{(E - G)^2 + 4F^2}}{2} \tag{3}$$

The direction of the feature vector is expressed as:

$$t = \begin{bmatrix} \lambda_1 - \mathbb{E} \\ -\mathbb{F} \end{bmatrix} \tag{4}$$

The local direction is defined as: $\theta = arg(t)$. Using the direction field of the feature vector as the input vector field, we can get some important applications of it in our designed real-time rendering system.

4.2 Cartoon Stylized Algorithm

Cartoon stylization is an important research field of image abstraction. The main characteristics of the cartoon works are: (1) highly abstract, with visual attraction; (2) universal color, color similarity in local areas; (3) edge enhancement, enhanced significant characteristic.

In order to get the abstraction style image, DeCarlo proposed a method based on attention perception, which mainly used Mean-shift clustering algorithm to render cartoon style images.

Wang Qiaoyu et al. proposed a new algorithm for color video cartoon stylizing [12]. The principle of the algorithm is the combination of Mean-shift filtering and Gaussian image pyramid, which improves the processing speed to a certain extent.

Winnemoller implements a real-time video abstraction algorithm [7]. Firstly, the details of low contrast region are smoothed by using the bilateral filter. Secondly, the abstract cartoon image is obtained by using the color quantization technology. Finally, the continuous edge image is obtained by Gaussian difference process.

Inspired by Winnemoller, we firstly use the DoG operator to detect the edge of the input video frame, then the edge image can be rendered with LIC (line integral convolution) guided by the direction field of the image structure tensor. By using this method, we can obtain the softened edge image. Secondly, the anisotropic bilateral filtering is used to filter unimportant features in the image. At the same time, color quantization technology is adopted to, we finally get the spatiotemporal continuity of visual abstraction image.

The main contribution of this paper is to implement the above algorithm using GLSL language, which can be implemented on GPU, and achieve the real-time performance, with good efficiency and stylized effect. Using GPU for real-time cartoon stylized rendering mainly includes the following steps:

(1) The input video frame is converted to the LAB color space for bilateral filtering;
(2) Color quantization is carried out on the frame of anisotropic bilateral filtering;
(3) DoG-LIC operator is used to detect the edge of the input video frame to obtain the softened edge image;
(4) The edge image is fused with the color quantization frame and converted to RGB color space to obtain the final artistic effect of cartoon.

Edge Preserve Smoothing Algorithm Based on Anisotropic Bilateral Filter

$$\mathbb{J}(x) = \frac{1}{W(y, x)} \sum_{y \in \Omega(x)} \mathbb{G}(x, y) \mathbb{H}(I(x), I(y)) I(y) \tag{5}$$

The bilateral filter consists of a low-pass filter $\mathbb{G}$ which represents the spatial relationship and a filter $\mathbb{H}$ that characterizes the relationship between pixel brightness.

Low pass filter $\mathbb{G}$ Defined as:

$$\mathbb{G}(x, y) = e^{-\frac{y-x^2}{2\sigma_d^2}}$$ (6)

Where: σ_d is the distance factor of the low pass filter, the larger of it the image is more blurred. $\|y\text{-}x\|$ is the Euclidean distance between point y and point x.

Range filter $\mathbb{H}$ Defined as:

$$\mathbb{H}(I(x), l(y)) = e^{-\frac{\|I(y)-I(x)\|^2}{2\sigma_r^2}}$$ (7)

Where: σ_r is the adjustment factor of the range filter; $\|I(y) - I(x)\|$ is the Euclidean distance between the color vectors of $I(y)$ and $I(x)$. Weight coefficient w(y, x)W(y, x) Defined as:

$$W(y, x) = \sum_{y \in \Omega(x)} \mathbb{G}(x, y)\mathbb{H}(I(x), I(y))$$ (8)

Bilateral filter is a local nonlinear image filter, which can smooth the image and keep the edge well. However, the directional signal of the image edge is not considered in the conventional bilateral filtering.

To solve this problem, Geusebroek et al. firstly proposed an improved anisotropic Gaussian filter, which could make the filter direction parallel to the edge direction during smoothing [13]. Inspired by this idea, by improving the Gaussian kernel of the bilateral filter and rotating it to the direction of the local structure tensor field of the input video frame, we can obtain the abstract effect while keep the edge features.

In the anisotropic bilateral filter designed in this paper, the value of the output pixel is still obtained by the weighted average value of the pixel in its neighborhood. Let $J(x,y)$ represent the pixel value of an image J at (x,y), then the expression of anisotropic bilateral filtering designed in this paper is (Fig. 2):

$$J(x, y) = \frac{\sum_{(i,j) \in W_{(x,y)}} \omega(i, j)M(\theta)}{\sum_{(i,j) \in W_{(x,y)}} \omega(i, j)}$$ (9)

The rotation matrix of the kernel filter defined as:

$$M\theta = \begin{bmatrix} cos\theta & -sin\theta \\ sin\theta & cos\theta \end{bmatrix}$$ (10)

Image Color Quantization Method

In order to obtain the abstract style of the input video frame, we use formula (11) for color quantization. The proposed color quantization algorithm is same to Winnemoller's method. The color quantization process can be defined as:

$$\mathbb{Q}(x) = q_n + bin_w \times tanh(\varphi_q \times (f(x) - q_n))$$ (11)

where: $\mathbb{Q}(x)$ is the color quantization image; bin_w is the quantization interval; q_n is the closest color pixel to the $f(x)$; parameter φ_q used to control the smoothness of the

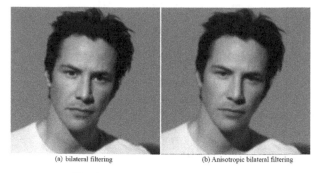

(a) bilateral filtering   (b) Anisotropic bilateral filtering

Fig. 2. Anisotropic bilateral filtering

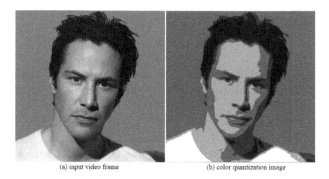

(a) input video frame   (b) color quantization image

Fig. 3. Color quantization

transition from one quantized color to another. With increase of the bin_w the color of the local area gradually converges (Fig. 3).

Image Edge Detection Operator

In cartoon stylization, line drawing is an important technology, which can reflect the outline features of one object. In order to facilitate the implementation on GPU and ensure the calculation speed, we use DoG operator to detect the edge of the input video frame like Winnemoller [7].

The second derivative $\nabla^2 G$ of DoG filter is obtained by Gaussian convolution. The idea is to use the difference of two Gaussian convolution filters with different standard deviations as the filter mask. The DoG filter can be expressed as:

$$G_\sigma(x) = \frac{1}{2\pi\sigma^2} exp(-\frac{\|x\|^2}{2\sigma^2}) \tag{12}$$

$$E(x, y) = G_{\varepsilon_1}(x) - t \times G_{\varepsilon_2}(x) \tag{13}$$

After the filtering image is obtained, the final edge effect can be obtained through binarization processing. The processing process can be expressed as:

$$Edge(x, y) = \begin{cases} 1 & E(x, y) > 0 \\ 1 + \tanh(\varphi_e \times E(x, y)) \ else \end{cases} \tag{14}$$

Line Integral Convolution.

To get the more softened edges, we use LIC method to took the edges detected by the DoG filter as the input texture, calculated the streamlines of the pixel points in the vector field along the direction of the structural tensor, and then convolved the streamlines corresponding to the edge values of the pixel points. According to the above principles, the discrete formula is obtained [14]:

$$h(t) = \int_{s_t}^{s+\Delta s_t} k(\tau)d\sigma, \, s_t = s_{t-1} + \Delta s_{t-1} \tag{15}$$

where, Δs_t is the step length at step t, s_t is the length of the streamline after step t. h_t represents the weight value, from which the pixel gray value of the output image can be obtained [14]:

$$F_{out}(x, y) = \frac{\sum_{t=0}^{l} F_{in}(p_t)h_t + \sum_{t=0}^{l'} F_{in}(p_t')h_t'}{\sum_{t=0}^{l} h_t + \sum_{t=0}^{l'} h_{t'}} \tag{16}$$

Where $\mathbb{F}_{out}(x, y)$ is the grayscale value at the pixel (x, y) of the output video frame, $\mathbb{F}_{in}(p_t)$ is the grayscale value at the pixel p_t of the edge image, p_t and p_t' is the pixel coordinate value along the positive and negative direction of the streamline respectively. l and l' is the number of integration steps in the positive and negative directions of the streamline. h_t is the weight, h_t' which represents the weight of the reverse streamline (Fig. 4).

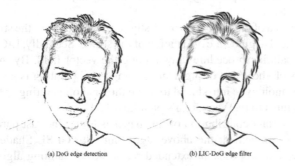

(a) DoG edge detection   (b) LIC-DoG edge filter

Fig. 4. LIC-DoG edge filter

Finally, we fuse the edge image with the abstracted image to get a cartoon style image (Fig. 5).

4.3 Van Gogh Stylization Algorithm

In a variety of stylization rendering algorithms, the simulation of oil painting style has been widely concerned by researchers. Image-based oil painting style rendering methods can be divided into two categories: physical simulation based rendering methods and

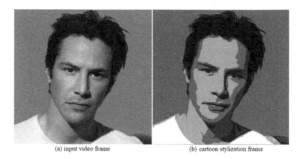

(a) input video frame   (b) cartoon stylization frame

Fig. 5. Cartoon stylization image

brush based rendering methods. How to generate art works with oil painting style in real time has gradually become a new research hotspot.

In 2003, Chung-Ming Wang et al. proposed to use LIC method to generate artistic images with oil painting style [15]. In 2006, Zhao Yang, Xu Dan et al. proposed a novel oil painting technology based on fluid dynamics simulation [16]. However, due to the large amount of computation, the above algorithms can't achieve the real-time artistic stylized conversion of the input video.

In 2016, Amir Semmo et al. presents an interactive system for transforming images into an oil paint look [17]. The main idea of this system is that it employs non-linear filtering based on the smoothed structure adapted to the main feature contours of the quantized image to synthesize a paint texture in real-time. Their system also introduces a generalized brush-based painting interface to locally adjust the level of abstraction of the filtering effects.

Based on the idea of the above paper, Firstly, we propose to use the structure tensor as the input vector field to guide the orientation of the stroke. Secondly, LIC method is used to simulate the painting procedure drawn along the vector field. By introducing local illumination model, the effect of pigment layering in oil painting is generated. Finally, color transfer technology is introduced to make the final oil painting special effect have similar color features to the Van Gogh's paintings.

The most important contribution of this paper is to use the fast parallel computing ability of GPU to implement the above algorithm with GLSL Shader Language. In addition, this section proposes an extended real-time oil painting algorithm based on multi-style brushes to achieve the stylized rendering of the input video.

Noise Map Generation Method

At present, Line Integral Convolution (LIC) is mainly used in texture-based vector field visualization. The basic principle is to select a noise image as the input texture after generating the original vector field, then filter the noise image along the tangent direction of the original vector field. The output texture can be used to describe the direction information of the vector field.

In order to generate noise textures while preserving the color and structure information of the input image, a pseudo-random perturbation method is designed in this section to increase the color variability.

Let $p_{in}(x, y)$ as the input image, $p_{out}(x, y)$ is the output image, (x, y) is the current pixel coordinate value, and par is the offset parameter. $rand(par)$ is defined as a pseudo-random function that generates random numbers in a specified range. The process of random disturbance is formalized as follows:

$$\mathbb{P}_{out} = \mathbb{P}_{in}(x + rand(par), y + rand(par)) \tag{17}$$

Van Gogh Stylization by Using LIC Method
We use LIC method to simulate the painting procedure drawn along the vector field. Taking the direction field of structure tensor mentioned above as the input vector field, the output video frame with Van Gogh oil painting style can be rendered by LIC operation on the noise image with random disturbance (Fig. 6).

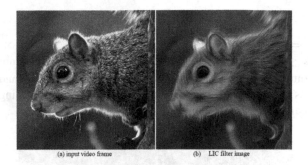

(a) input video frame   (b)   LIC filter image

Fig. 6. LIC filter image

Generate Oil Painting Layering Style Method
The algorithm designed by Litwinowicz and Hertzmann did not consider that Van Gogh's style oil paintings would present different visual perceptions under different lighting conditions. For this reason, Hertzmann proposed to extract the height field from the input image, apply the height field to the rendering brush, then use bump mapping technology to generate the oil painting layers [18]. Qian Wenhua et al. put forward image-based local illumination model in their research [19]. Inspired by the above ideas, this paper improves the simple lighting model and generates oil painting images through GPU in real time. The image-based lighting model designed in this section is formalized as follows:

$$\mathbb{P}_{i,j} = I_e k_e + I_p \left[k_d cos\theta + k_s cos^m \alpha \right] \tag{18}$$

where, $\mathbb{P}$ represents the image after illumination rendering; θ is the angle between the incident light L and the surface normal N; α is the angle between the line-of-sight direction V and the reflection direction R; I_e is the ambient light parameter; k_e is the reflection coefficient of ambient light; I_p represents the color channels; k_d and k_s represent specular reflection coefficient and diffuse reflection coefficient respectively.

In this paper, a method like Bump Mapping technology is designed to quickly generate output video frames with cascading visualization of oil painting. Define a three-dimensional parametric surface:

$$\mathbb{F}_{u,v} = [u, v, \mathbb{L}_{u,v}] \tag{19}$$

where u and v represent the coordinate position of the image; $\mathbb{L}_{u,v}$ represents the pixel value of the image at the point (u, v); Then the normal vector at point (u, v) can be expressed as:

$$\mathbb{N}_{(u,v)} = F_u \times F_v \tag{20}$$

The partial derivative of the image in the u and v directions is F_u, F_v; its value can be obtained by Sobel operator, then:

$$\mathbb{N}_{(u,v)} = [-F_u, -F_v, 1] \tag{21}$$

Let $N_{(u,v)_x} = -F_u N_{(u,v)_x} = -F_u$, $\mathbb{N}_{(u,v)_x} = -F_u$, $N_{(u,v)_z} = k \times 255 N_{(u,v)_z} = k \times 255 (0 \le k \le 1)$. Then, for a given light vector L, it can be obtained $\mathbb{N}_{(u,v)}L = cos\theta'$. By substituting Eq. (21) into the local light illumination model and acting on the rendering frames generated in real time, a stylized image with cascading visualization of oil painting can be obtained (as shown in Fig. 7).

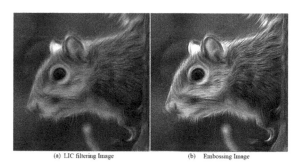

(a) LIC filtering Image   (b) Embossing Image

Fig. 7. Embossing image

Color Transfer Method

In order to make the output video frame after light rendering similar to other famous oil paintings such as Van Gogh's art works in visual perception. We use a linear mapping technology to make the color channels of the input video frame are consistent with the corresponding channels of the target image [20]. In order to reduce the correlation of each channel, we need to operate in the color space $l\alpha\beta$:

$$I_{l\alpha\beta} = \frac{\partial_{l\alpha\beta}B}{\partial_{l\alpha\beta}A}\left(I_{l\alpha\beta} - \mu_{l\alpha\beta}A\right) + \mu_{l\alpha\beta}B \tag{22}$$

(a) Target Style Image   (b) Color Transfer Image

Fig. 8. Color transfer between image

where, $\mu_{l\alpha\beta}A$, $\mu_{l\alpha\beta}B$ represents the mean value of image A and B on each channel in the color space $l\alpha\beta$; $\partial_{l\alpha\beta}A$, $\partial_{l\alpha\beta}B$ represents the variance value of image A and B on each channel in the color space $l\alpha\beta$ (Fig. 8).

In the design and implementation, we find that color transfer between images should be carried out according to formula (22). However, when using GPU to calculate the mean and standard deviation of a group of data, the traditional calculation method cannot be realized. Therefore, this section adopts the reduction method to solve this problem.

GPU Accelerated Painting Based on Multi-style Brush Model

In the actual oil painting process, the painter will choose different brushes for painting according to the actual visual attention. Although the LIC method proposed in this paper can effectively simulate the streamline characteristics of oil painting, it cannot describe the characteristics of oil painting with multi-style brushes. Therefore, in order to realize automatic and real-time generation of painting style by computer, it is necessary to further define and simulate the brush model.

Brush models can be extracted from various Van Gogh's style oil paintings. A brush object contains the following attributes: length, width, opacity, color, center point, and brush direction.

In our design, the center point of the brush is the position of the pen drop point. The color of the brush is the color value of the drop point; The direction of the brush is calculated by the structural tensor (Fig. 9).

Fig. 9. Brush model definition

In the process of stylizing, the calculation of the structure tensor can be completed by GPU, while the drawing of multi-style brush is still completed by the traditional CPU pipeline. GPU acceleration analysis will be introduced in Sect. 5 of this paper (Fig. 10).

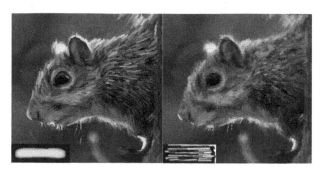

Fig. 10. Multi-style brush painting

Drawing Oil Painting Based on Visual Salience

When painters draw Van Gogh's style oil paintings, they often use brushes of different sizes to paint different areas hierarchically. In the process of painting, the region of interest is usually painted with a smaller brush to keep the detail information without loss, while other regions are painted with a larger brush. In order to simulate this process, it is necessary to define and calculate the brush size. Therefore, this section proposes a calculation model of brush size based on the guidance of image visual saliency.

Visual saliency is based on people's visual attention to detect and distinguish the object from other regions. In this paper, the spectral residual method proposed by Hou et al. [21] is used to calculate the visual saliency map. The advantage of this method is that it has high computational efficiency, can basically meet the requirements of real-time application, and can obtain the visual importance features in the image without using expensive eye tracker equipment. According to this method, a saliency map can be reconstructed to represent the saliency of each pixel in the original image, which is formally defined as:

$$S(x) = g(x) * F^{-1}\big[exp(R(f) + P(f))\big]^2 \tag{23}$$

Further define the brush size weight calculation model as follows:

$$weight(x) = \frac{k}{S(x)} \tag{24}$$

When the significance of the pixel is high, the weight of the region of visual interest is small. The rendering system we designed uses a small brush to draw this region, to highlight the detail features of the image. On the contrary, a larger brush is used to paint indicating that this area is not of interest to the people (Fig. 11).

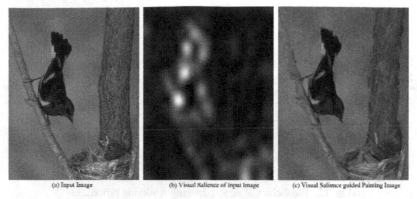

(a) Input Image   (b) Visual Salience of Input Image   (c) Visual Salience guided Painting Image

Fig. 11. Brush style rendering based on visual saliency guidance

4.4 Pencil Drawing Stylization Rendering Algorithm

Pencil drawing can express abstract information of the nature picture, which has the concise and accurate visual attraction. Using computer to simulate pencil drawing is an important research area of non-photorealistic rendering.

In 2012, Xiaoyang Mao et al. proposed an improved LIC method to transform an original image into a pencil drawing look [22], but the algorithm does not realize real-time GPU video style conversion.

In 2008, Xie Dangen and Zhao Yang et al. proposed a pencil filter generation algorithm based on convolution operator, and realized real-time pencil style rendering of video with GPU [23]. However, the limitation of this method is that rendering with convolution operator will produce filtering effect exceed the image boundary. In fact, the pencil texture will not go beyond the boundary, but naturally unfold along the tangent direction of the image boundary. In this section, LIC filtering along the direction of the input image structure tensor is proposed to improve the system, and the algorithm is extended to the implementation of GPU. The main principles of the algorithm are described in detail below.

The main steps of GPU based LIC pencil drawing stylized rendering algorithm are as follows:

Step 1: convert the original color space to LAB color space; Step 2: the brightness channel of the image is processed to generate the black noise; Step 3: the structure tensor vector field is calculated from the original image as the guide direction of pencil drawing;

Step 4: convolve the black noise image by LIC algorithm along the vector field to get pencil drawing texture image; Step 5: fuse LIC texture image and the edge to get the final paintings. The flow chart of pencil drawing rendering algorithm as show in Fig. 12.

Generation of Black Noise
This paper adopts the black noise generation method proposed by Xie Dangen [23], the function defined as follows:

$$I_{noise} = \begin{cases} 255, & if\, P \le TP \in [0.0, 1.0] \\ 0, & otherwise \end{cases} \tag{25}$$

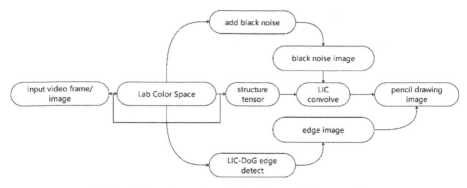

Fig. 12. Flow chart of pencil drawing rendering procedure

$$T = k\left[\frac{I_{input}}{255}\right] \tag{26}$$

The coefficient k is used to control how many black noises are added; P is a random number with the value of [0.0,1.0] (Fig. 13).

(a)input image   (b) black noise image

Fig. 13. Black noise process

After the black noise image is generated, the output video frame with pencil drawing style can be obtained by smoothing filtering along the direction of the structure tensor with the LIC method proposed in Sect. 4.2.4 (Fig. 14).

In addition, the current research on stylized simulation of pencil drawing is mainly focused on the algorithm design of gray pencil drawing. They propose a color pencil drawing algorithm, but it has a large amount of computation that cannot be applied to the field of real-time rendering.

In order to solve this problem, this paper proposes to convert RGB to Lab color space and add color information to the gray-scale pencil image generated by LIC method, so as to obtain the rendering effect of colored pencil. The algorithm processing effect is shown in Fig. 15. The efficiency of the algorithm will be discussed in detail in Sect. 5.

(a)input image   (b)pencil drawing image

Fig. 14. Pencil drawing stylization

(a) input image   (b) color pencil drawing image

Fig. 15. Color pencil drawing stylization

4.5 Watercolor Stylization Rendering Algorithm

Watercolor is an art form with a long history and widely used in various fields. Especially, in the process of digital animation, watercolor elements can enrich the form of expression and make the animation extremely artistic. However, due to the complexity of the drawing process and the variety of painting techniques, the use of computer to simulate the style of watercolor painting is not only complicated but also costly. At present, watercolor stylized rendering technology is mainly divided into two categories: one is based on physical simulation, the other is based on image filtering and processing.

This paper attempts to use GPU acceleration technology to simulate the style characteristics of traditional watercolor painting in real time, which is also one of the research hotspots in the field of non-photorealistic rendering.

Curtis et al. firstly proposed the watercolor style rendering method [24], which divides the canvas into three layers: shallow water layer, pigment deposition layer and water absorption layer. Based on the physical transmission of pigment and water on the paper, watercolor simulation is carried out. The effect of this method is realistic, but the calculation cost is high. Wang Miaoyi et al. proposed a GPU based real-time rendering and animation rendering algorithm for watercolor style, but it is mainly applied to the stylized rendering of 3D scene [25].

How to quickly render the input video with watercolor style and balance the rendering speed and rendering quality is a hot issue in the field of non-photorealistic research. This section we propose a more intuitive GPU based real-time video watercolor style rendering algorithm.

Firstly, the video frame with uniform color and similar to watercolor fading effect is obtained by Kuwahara filtering; Secondly, DoG filter is used to extract the edges, and LIC filter based on structure tensor is used to soften the extracted edges, so that the extracted edges are more coherent and smooth,; Thirdly, we use wet-in-wet technique to produce feather-like patterns along a region boundary; Finally, the video frame with watercolor rendering effect is obtained by mixing with paper texture. Through the above steps, it is possible to make real-time watercolor style paintings.

Kuwahara Filtering and Edge Detection

The traditional Kuwahara filtering algorithm is a nonlinear filtering method which preserves the edge feature [11]. The idea is to divide the filter kernel into N same neighborhood partitions, calculate the mean value and variance of each pixel in the partition, and select the mean value corresponding to the minimum variance as the filtering result. In this section, the filtering algorithm is used to process the input video to obtain the output frame with the effect of watercolor diffusion effect [10].

The Kuwahara filter divides the 3×3 size filter window into four parts, where $k \in 1, 2, 3$. Assume that the size of the square filter window is $(2n+1) \times (2n+1)$, the central pixel is (x_0, y_0), and the child window is θ_k. We calculate the local mean m_k and standard deviation δ_k of each sub-window respectively. For the central element (x_0, y_0), the Kuwahara operator defined as:

$$\xi(x_0, y_0) = \sum\nolimits_{k=0}^{3} m_k \times w_k \tag{27}$$

$$w_k = \begin{cases} 1, & \delta_k = min_k\{\delta_k\} \\ 0, & otherwise \end{cases} \tag{28}$$

The Kuwahara filter can keep the edges and corners of the image matrix well while smoothing the image, so as to simulate the effect of watercolor fading. At the same time,

(a) Input image   (b) Kuwahara filtering image   (c) LIC-DoG filter image

Fig. 16. Kuwahara filtering and edge detection

we use LIC-DoG filter designed in Sect. 4.2 to get the edge map of the input video frame. The algorithm processing effect is shown in Fig. 16.

Wet-In-Wet Technology
Wet-in-Wet is an important and frequently used technique in watercolor paintings. Miaoyi Wang et al. propose a technique based on image-filtering for simulating the wet-in-wet effect [26]. The main idea of this method is that, firstly randomly scatter some seeds around the boundaries, and then filter these areas with an ellipse-shaped kernel oriented along the normal vectors of a region boundary. Please refer to the reference [26] for detail implementation. The main contribution of our woks is to design and implement a GPU-based image watercolorization system with interactive performance (Fig. 17).

(a)Input image   (b)Wet-in-Wet image

Fig. 17. Wet-in-wet Process

Generation and Synthesis of Watercolor Texture
In actual painting, watercolor paper texture has a certain wrinkle. A simple random function can only generate a relatively rough noise image with a certain contrast, and such noise image usually does not exist in nature. In order to simulate these features of the watercolor paper, Therefore, we use Perlin noise function, a fractal algorithm, to generate a number of seemingly natural random noises. 2D Perlin noise is used to simulate the background watercolor paper texture. The main steps can be referred to in reference [26]. The result is shown in Fig. 18 after the edges are fused and combined with the watercolor paper texture.

The experimental pictures are taken from Wang Miaoyi's resources on the internet.

(a) input image   (b) watercolor stylization image

Fig. 18. Watercolor image fuse with edge and paper texture

4.6 Discussion Chinese Ink Stylization Method

Since the main characteristic of Chinese ink painting is to emphasize "writing spirit with shape", In order to achieve this effect, the algorithm designed in this paper retains the main shape features of the object in the input image, but leave out other details.

In the ink diffusion process, we took the lines extracted from the LIC-DoG method proposed in Sect. 4.2 as input, and then used a model similar to WLS (Weighted Least Square) for filtering and rendering. The ink style rendering algorithm discussed in this paper is essentially an image-based rendering method.

According to the characteristics of the original image and rice paper texture, the stylized rendering model of ink painting will propagate the value of the main shape position pixels of the object in the picture unevenly to its adjacent area, resulting in the effect of ink diffusion. By setting different model parameters, we can get different abstract image of Chinese ink painting style.

Inspired by Lingyu Liang's idea [27], let the input of the model be the line feature E extracted from the original image, and the output be the P after the diffusion of ink painting. The guiding feature vectors is G, then we can obtain the effect of ink diffusion by minimizing the following quadratic functional:

$$P = \underset{P}{\operatorname{argmin}} \left\{ \sum_z w(z)(P(z) - E(z))^2 + \sum_z \lambda \left(\frac{\|\frac{\partial P}{\partial x}\|^2}{\|\frac{\partial G}{\partial x}\|_p^\alpha + \varepsilon} + \frac{\|\frac{\partial P}{\partial y}\|^2}{\|\frac{\partial G}{\partial xy}\|_p^\alpha + \varepsilon} \right) \right\} \quad (29)$$

In the formula (29), the first term is the data term, which is used to ensure that the diffused image P is within the limit of the input line feature E. It makes the pixel value in the control area is as close as possible, and the non-uniform diffusion is carried out in the non-restricted area.

The second term is smooth term, which mainly contains three model parameters. Where is ε a small and the non-zero number is used to avoid zero denominator; the exponent α controls the change of P by adjusting the gradient change of G; λ is used to control the overall smoothness of P. Through iterative processing, we can get the rendering effect of Chinese ink style (Fig. 19).

(a) input image   (b) DoG filter image

(c) LIC-DoG filter image   (d) Chinese ink Painting Style image

Fig. 19. Chinese ink painting style achieved by using CUDA

Unfortunately, the solution based on this model involves the operation of large sparse matrix. Although we realize the solving operation of the above proposed model in real-time by using CUDA language, which provide the solution interface of large sparse matrix. However, how to use GLSL language to realize the calculation of the model and integrate it into the rendering system involved in this paper is still the problem to continue to study.

5   Analysis of Experimental Results

Aiming at the practical engineering application of video stylization, this paper proposes a GPU based video stylization algorithm, and implements a real-time video stylization rendering system. The system can make good use of the parallel computing ability of GPU to realize the fast conversion of the four artistic styles (cartoon, oil painting, pencil drawing and watercolor) of the input video, and provide users with a better interactive experience.

The hardware platform of the system operation environment designed in this paper is a laptop computer. The configuration of it is: Intel (R). Core (TM)i7-6700HQ processor, 2.6 GHz main frequency; 8.0 GB DDR3 memory; NVIDIA GeForce GTX960M video card; and 4G video memory. Software platform is Windows 10(X64) operating system, VC++ 2015, OpenGL, using GLSL language programming to implement the core algorithm. The experimental results are compared and the GPU parallel acceleration correlation algorithm is analyzed.

5.1 Comic-Style GPU Parallel Acceleration Analysis

In order to realize cartoon style, it is necessary to render the image abstractly, and the core process is edge preserving filtering. Figure 20(b, c) shows the effect of using the algorithm proposed in the literature [28, 29] to process the input image (a), and Fig. 20(d) shows the effect of using the anisotropic bilateral filter designed in this paper for edge preserving filtering. It can be seen from the result figure that although the edge information of the image is preserved in the filtering process in Fig. 20(b), some areas are also excessively blurred. The image is not very clear and cannot fully describe the artistic style of the cartoon image. Meanwhile, the algorithm in the literature is an iterative algorithm, and the final calculation time and processing effect are related to the number of iterations. In our experiments, the number of iterations is set to 10. Figure 20(c) is processed by image segmentation method. Although better abstraction effect is achieved, the computational complexity is high.

As can be seen from Fig. 20(d), this paper uses GPU for parallel acceleration to achieve real-time cartoon style rendering without losing abstract effect, Fig. 21 shows the comparison of calculation time obtained by different methods in this paper.

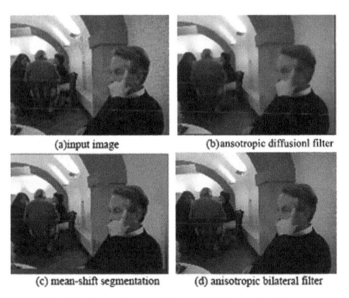

(a)input image   (b)ansotropic diffusionl filter

(c) mean-shift segmentation   (d) anisotropic bilateral filter

Fig. 20. Processing renderings of different algorithms

5.2 GPU Accelerated LIC Based on GLSL Architecture

With the development of GPU, it can deal with those highly parallel tasks, to complete the general computing tasks that should be completed by CPU. In the oil painting and pencil drawing algorithms proposed in this section, LIC algorithm is mainly used to convolve the input texture.

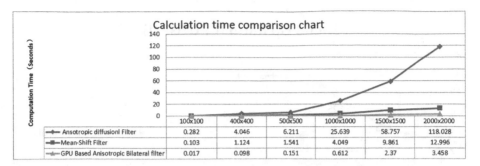

Fig. 21. Calculation time obtained by different methods

The traditional LIC algorithm needs to convolve each pixel of the input image in turn, which makes the complexity of the algorithm proportional to the size of the input image, and cannot achieve the purpose of real-time rendering. When using GPU parallelism, each input pixel can be allocated an execution unit to perform convolution operation, to improve the computational efficiency and achieve the purpose of real-time stylized rendering. This paper implements the GPU and CPU version of the algorithm. Figure 22 shows the efficiency comparison of the algorithm before and after GPU acceleration.

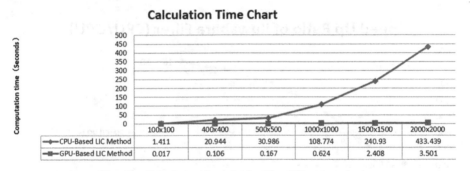

Fig. 22. Calculation time obtained by different methods

At the same time, this paper proposes a GPU accelerated rendering algorithm based on multi-style brush model. In the process of multi-style brush painting, the core program of parallel computing is the calculation process of structure tensor. This paper also implements the GPU and CPU version of the structure tensor calculation method to compare the efficiency of the algorithm. Figure 23 shows the change trend of algorithm speedup ratio after GPU acceleration.

5.3   GPU Accelerated Kuwahara Filtering Based on GLSL Architecture

When processing the input video with watercolor style, we mainly use the Kuwahara filter to abstract the frames. The traditional Kuwahara filter needs to calculate the mean and variance of the pixels in each region when smoothing the high contrast region of the

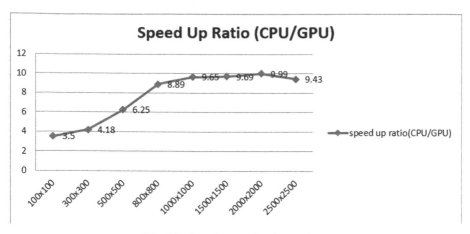

Fig. 23. Speed Up Ratio of CPU/GPU

image. In fact, it is a parallel processing procedure. In this paper, the Kuwahara filter is implemented by GPU, and the efficiency of the algorithm is compared with that of CPU. Figure 24 shows the change trend of algorithm speedup ratio after GPU acceleration.

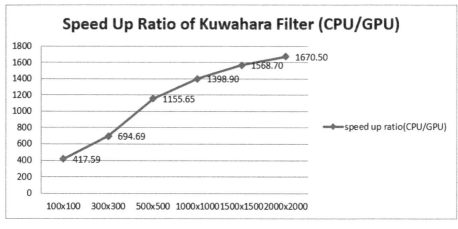

Fig. 24. Speed up ratio Kuwahara Filter (CPU/GPU)

6 Future Research Directions

Although this paper proposes system architecture, there are still a lot of works to be improved and optimized. The main improvement direction is the real-time stylization of Chinese ink painting. How to deal with the effect more realistically according to the physical model of ink diffusion needs further research.

In the future work, we intend to extend this technology to more complex 3D scene style rendering. Since different artists have different painting styles, how to use the parallel computing ability of GPU to make the computer quickly learn to simulate different styles of art paintings and expand the system function is also the key direction of future research work.

Acknowledgments. This work is supported by the Natural Science Foundation of China (Grant No.61761046, 62061049), the Application and Foundation Project of Yunnan Province (Grant No.202001BB050032, 202001BB050043, 2018FB100) and the Youth Top Talents Project of Yunnan Provincial "Ten Thousands Plan" (Grant No.YNWR-QNBJ-2018-329).

References

1. Litwinowicz, P.: Processing images and video for an impressionist effect. In: Proceedings of the 24th Annual Conference on Computer Graphics and Interactive Technique, pp. 407–414 (1997)
2. Hertzmann, A., Perlin, K.: Painterly rendering for video and interaction. In: Proceedings of the 1st International Symposium on Non-photorealistic Animation and Rendering (NPAR), pp. 7–12. ACM Press, New York (2000)
3. Klein, A.W., Grant, T., Cohen, F.: Video mosaics. In: Proceedings of the 2nd International Symposium on Non-photorealistic Animation and Rendering (NPAR), pp. 21–29. ACM Press, New York (2002)
4. Agarwala, A., Toonz, S.: A semi-automatic approach to creating cel animation from video. In: Proceedings of the 2nd International Symposium on Non-photorealistic Animation and Rendering (NPAR), p. 139. ACM Press, New York (2002)
5. Agarwala, A., Hertzmann, A., et al.: Keyframe-based tracking for rotoscoping and animation. ACM Trans. Graph. (TOG) 23(3), 584–591. Proceedings of ACM SIGGRAPH 2004 (S0730-0301) (2004)
6. Wang, J., Xu, Y., et al.: Video tooning. ACM Trans. Graph. (TOG) 23(3), 574–583. Proceedings of ACM SIGGRAPH 2004 (S0730-0301) (2004)
7. Winnemoeller, H., Olsen, S.C., Gooch, B.: Real-time video abstraction. ACM Transactions on Graphics (TOG) 25(3), 1221–1226. Proceedings of ACM SIGGRAPH 2006 (S1-59593-364-6)
8. Zhao, Y., Xu, D.: Automatic and real-time video stylization. In: Proceedings of 10th IEEE International Conference on Computer-Aided Design and Computer Graphics, pp. 505–508 (2007). (S978-1-4244-1578-6)
9. O'Donovan, P., Hertzmann, A.: AniPaint: interactive painterly animation from video. IEEE Trans. Vis. Comput. Graph. 18(3), 475–487 (2012). (S1077-2626)
10. Wang, Q., Chen, D., Li, S., Wu, Q., Zhang, Q.: An adaptive cartoon-like stylization for color video in real time. Multimedia Tools Appl. 76(15), 16767–16782 (2017)
11. Kyprianidis, J.E., Kang, H., et al.: Image and video abstraction by anisotropic Kuwahara filtering. Comput. Graph. Forum 28(7), 1955–1963 (2009)
12. Qiaoyu, W.: Research and implementation of cartoon-like stylization for color video image. J. Huaqiao Univ. Nat. Sci. 35(6), 659–664 (2014)
13. Geusebroek, J.-M., Smeulders, A.W.M., van de Weijer, J.: Fast anisotropic Gauss filtering. IEEE Trans. Image Process. 12(8), 938–943 (2003)
14. Cabral, B., Leedom, L.C.: Imaging vector fields using line integral convolution. In Proceedings of the 20th Annual Conference on Computer Graphics and Interactive Techniques, pp. 263–270 (1993)

15. Wang, C.-M., Lee, J.-S.: Using ILIC algorithm for an impressionist effect and stylized virtual environments. J. Vis. Lang. Comput. **14**, 255–274 (2003)
16. Yang, Z., Dan, X.: Oil style image generation via fluid simulation. J. Softw. **17**(7), 1571–1579 (2006)
17. Semmo, A., Limberger, D., Kyprianidis, J.E., Döllner, J.: Image stylization by interactive oil paint filtering. Comput. Graph. **55**, 1–16 (2016)
18. Hertzmann, A.: Fast paint texture. In: Proceedings of the 2nd International Symposium on Non-photorealistic Animation and Rendering (NPAR), New York, NY, USA, pp. 91 (2002). (1-58113-494-0)
19. Wenhua, Q.: Sketch artistic rendering based on significance map. J. Comput.-Aided Des. Comput. Graph. **27**(5), 915–923 (2015)
20. Reinhard, E., Ashikhmin, M., Gooch, B., Shirley, P.: Color transfer between images. IEEE Comput. Graphics Appl. **21**(5), 34–41 (2001)
21. Hou, X., Zhang, L., Saliency detection: a spectral residual approach. In: IEEE Conference on Computer Vision & Pattern Recognition, pp. 1–8 (2007)
22. Hata, M., Toyoura, M., Mao, X.: Automatic generation of accentuated pencil drawing with saliency map and LIC. Vis. Comput. **28**(6–8), 657–668 (2012)
23. Dang 'en, X., Yang, Z., Dan, X.: A method for generation of pencil filter and its implementation on GPU. J. Comput.-Aided Des. Comput. Graph. **20**(1), 26–31 (2008)
24. Curtis, C.J., Anderson, S.E., et al.: Computer-generated watercolor. In: ACM Transactions on Graphics (TOG) - Proceedings of ACM SIGGRAPH 1997(0-89791-896-7), pp. 421–430 (1997)
25. Miaoyi, W., Bin, W., Junhai, Y.: Real-time watercolor illustrations and animation on GPU. J. Graph. **33**(3), 73–79 (2012)
26. Wang, M., Wang, B., Fei, Y., et al.: Towards photo watercolorization with artistic verisimilitude. IEEE Trans. Vis. Comput. Graph. **20**(10), 1451–1460 (2014)
27. Liang, L., Jin, L.: Image-based rendering for ink painting. In: IEEE International Conference on Systems, Man, and Cybernetics, pp. 3950–3954 (2013)
28. Perona, P., Malik, J.: Scale-space and edge detection using ansotropic diffusion. IEEE Trans. Pattern Anal. Mach. Intell. **12**(7), 629–639 (1990)
29. Comaniciu, D., Meer, P.: Mean shift: a robust approach toward feature space analysis. IEEE Trans. Pattern Anal. Mach. Intell. **24**(5), 603–619 (2002)

Adaptive Densely Residual Network for Image Super-Resolution

Wen Zhao[✉]

Department of Computer Science and Technology, Heilongjiang University,
Harbin, China
2181419@s.hlju.edu.cn

Abstract. Many networks are designed to stack a large number of residual blocks, deepen the network and improve network performance through short residual connec-tion, long residual connection, and dense connection. However, without consider-ing different contributions of different depth features to the network, these de-signs have the problem of evaluating the importance of different depth features. To solve this problem, this paper proposes an adaptive densely residual net-work (ADR-Net) for the single image super resolution. ADRN realizes the evalua-tion of distributions of different depth features and learns more representa-tive features. An adaptive densely residual block (ADRB) was designed, combining 3 residual blocks (RB) and dense connection was added. It learned the attention score of each dense connection through adaptive dense connections, and the at-tention score reflected the importance of the features of each RB. To further en-hance the performance of ADRB, a multi-direction attention block (MDAB) was introduced to obtain multi-directional context information. Through comparative experiments, it is proved that theproposed ADRNet is superior to the existing methods. Through ablation experiments, it is proved that evaluating features of different depths helps to improve network performance.

Keywords: Deep learning · Single image super resolution · Multi-direction attention · Adaptive densely residual block

1 Introduction

In recent years, single image super resolution tasks have also achieved perfor-mance breakthroughs. The task of SISR [9] is to reconstruct a low resolution image (the deg-radation of a high resolution image) into a high resolution (HR) image. SISR is of great importance in many practical application scenarios, such as image recovery, satellite and aerial image, and video recovery.

The traditional interpolation [8] method is mainly characterized by fast speed, but it will cause the texture information and structure information to not be effectively restored, and it will feel very smooth to the naked eye. To further improve the struc-tural information, Allebach et al. [1] proposed edge-preserving

© Springer Nature Singapore Pte Ltd. 2021
J. Zeng et al. (Eds.): ICPCSEE 2021, CCIS 1451, pp. 339–349, 2021.
https://doi.org/10.1007/978-981-16-5940-9_25

interpolation and Baker's method based on edge priors [2]. To further improve the quality of restora-tion, some other methods have been proposed, such as neighborhood embedding [3], sparse coding [11], and local mapping function regression [10].

Dong et al. [6] applied the Convolutional Neural Network (CNN) to the SISR task for the first time, and the experimental results of stacking three layers of convolution greatly surpassed the traditional method. Kim et al. [16] further improved the net-work performance by increasing the network depth, and the depth of their proposed VDSR reached 20 layers. After these pioneering work, with the improvement of GPU performance, more and more methods [4,5,14,17, 18,21,27–30] are used to deepen the network to achieve the purpose of improving model performance.

Although the deep learning-based SISR methods have made good progress, there are still some unconcerned issues., although the SISR method of applying deep learning technology has made good progress, there are still some unconcerned issues. These methods do not take into account that the contributions of different depth feature map to the network are different, and the network has the problem of how to weigh the importance of different depth feature map.

Although the deep learning-based SISR methods have made good progress, there are still some unconcerned issues. These methods do not take into account that the contributions of different depth features to the network are different, and the network has the problem of how to weigh the importance of different depth features. Guo et al. [12] proposed a new adaptive dense connection module in the WCE image classification task, and the experiment proved that the adaptive dense connection module could effectively improve the accuracy of the model. Safarov et al. [23] proposed A-DenseUNet, which added adaptive dense connection to U-shape network, and the experiment proved that this technology could improve the segmentation accuracy. Xie et al. [26] proposed ADCSR and introduced adaptive dense connection.

Inspired by these work [12,23,26], this paper proposes a new adaptive densely re-sidual network (ADRNet) to realize the weight distribution of different depth features and learn more representative features. ADRN realizes the weight distribution of different depth features and learns more representative features. We designed an adap-tive densely residual block (ADRB), which combines 3 residual blocks (RB) and adds dense connection, and learns the attention score of each dense connection through adaptive dense connections, the attention score reflects the importance of the features of each residual block. To further enhance the performance of ADRB, we intro-duce a Multi-direction attention block to obtain multi-direction context information. Through comparative experiments, it is proved that our proposed ADRNet is superior to the existing methods. Through ablation experiments, it is proved that weighting features of different depths help to improve network performance.

To sum up, the main contributions of this paper are as follows:

– We propose an adaptive densely residual block (ADRB), which assigns weights to each residual block feature through adaptive dense connection.

The ablation ex-periment proves that weighting features of different depths can help improve net-work performance.
- We further introduce a Multi-direction attention block (MDAB) that can obtain contextual information in multiple directions and help enhance the performance of ADRB.
- We propose an adaptive densely residual network (ADRNet) for SISR. Through comparative experiments, it is proved that our proposed ADRNet is superior to the existing methods.

2   Related Work

Dong et al. [6] introduced the CNN to the SISR for the first time, and the experimental results of stacking three layers of convolution greatly surpassed the traditional method. Kim et al. [16] further improved the Peak Signal to Noise Ratio(PSNR) by increasing the network depth, their proposed VDSR had 20 layers. Dong et al. [7] introduced deconvolution in FSRCNN. Compared with SRCNN's early upsampling and then stacked convolutional layers, FSRCNN uses deconvolution at the end of the network to reduce computational costs and greatly improve the speed. The ESPCN proposed by Shi et at. [20] proved that the performance of sub-pixel convolutional in SISR task surpasses deconvolution layer. The EDSR proposed by Lim et al. [19] proved that simultaneously deepening and widening the network could enhance the nonlinear mapping ability of the network in the SISR task.

Huang et al. [25] proposed SRDenseNet, which introduced the dense connection of DenseNet [15] in the image classification task into the SISR task and aggregated the feature map of each Dense Block to the end of the network. After using deconvolution to upsampling the feature map, the network performance has been significantly improved. According to the characteristics that residual connections and dense connections can improve network performance, Zhang et al. [30] integrated the these connection to propose an RDN network, and the experimental results have been effectively improved. Haris et al. [13] proposed a D-DBPN with a feedback mechanism. By stacking multiple groups of upsampling units and downsampling units, and using dense connections to repeat the features of each up-sampling unit and down-sampling unit. The experimental results have better restored the texture structure information and won the championship of NTIRE2018 in the classic bicubic ×8 track. SRFBN [18] absorbed the advantages of D-DBPN and also used the method of dense connection.

3   Method

The adaptive densely residual network structure is shown in Fig. 1, the network structure consists of four parts: shallow feature extraction layer (SFEL), adaptive densely residual block (ADRB), upsampling block, and reconstruction layer. In the first step, the LR image is input to the SFEL, and the SFEL only uses a convolutional layer to extract the texture structure information of the LR image.

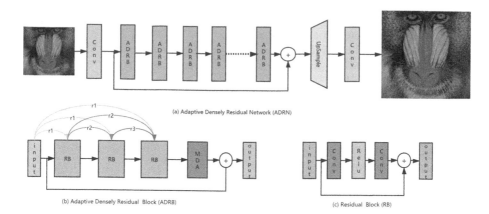

Fig. 1. (a) The structure of adaptive densely residual network (ADRNet), (b) the structure of adaptive densely residual Block (ADRB), and (c) the structure of residual block (RB).

The second step is input the feature maps obtained in the first step into N ADRB to further mine the deeper information of the LR image. The third step input the feature maps obtained in the first and second step are added through the long residual connection and then input to the upsampling block, in this paper, upsampling block used the sub-pixel convolution layer. The last step is restore the feature maps of upsampling block to an HR image of the RGB channel through one convolution layer.

3.1 Adaptive Densely Residual Block

As shown in Fig. 1(b), to train the network more effectively and weigh the importance of different depth features, we designed an adaptive densely residual block (ADRB), which combines 3 residual blocks (RB) and adds dense connection, and learns the attention score of each dense connection through adaptive dense connections, the attention score reflects the importance of the features of each residual block. Denote x_n is the nth RB feature map, and r_n is the weighted scalar of the corresponding dense connection. Then the adaptive dense connection Eq. (1) can be expressed as:

$$x_n = H_{RB}\left(\left[r_1 x_1, r_2 x_2, \cdots, r_{n-1} x_{n-1}\right]\right), \tag{1}$$

where $[[r_1 x_1, r_2 x_2, \cdots, r_{n-1} x_{n-1}]$ is the adaptive concatenation of the feature map generated by $1, 2, \cdots, n - 1th$ RB, $H_{RB}(\bullet)$ represents the RB, then input the resulting feature map into the next mentioned MDAB.

The ADRB can adaptively combine the feature information from the different residual blocks so that the feature map can be reused, ensure the maximum information flow between the residual block, and effectively aggregate useful information. The weights of all dense connections are initialized to 1, and

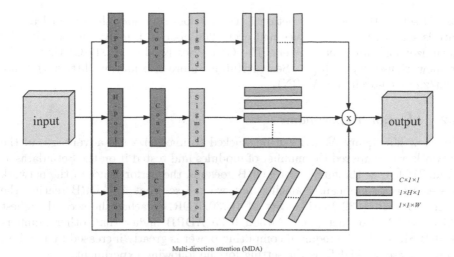

Multi-direction attention (MDA)

Fig. 2. The structure of multi-direction attention block.

iterative optimization is performed to learns the attention score of each dense connection through adaptive dense connections.

3.2 Multi-direction Attention Block

In CNN used attention mechanism can improve the performance of the network. In super resolution, the same concept can also be used to weight feature maps according to their relative importance. As shown in Fig. 2, to further improve the representation of the ADRB feature map, we introduce multi-direction attention block (MDAB). The MDAB can learn the context information of the C, H, and W directions of the feature map. The branches in each direction are composed of one global average pooling, one convolutional layer, sigmoid activation function. The activated feature map and the input feature map are subjected to a dot product operation to obtain a 3D attention feature map.

4 Experiment

4.1 Setting

We use the DIV2K dataset [24] to train the network. We use bicubic interpolation (BI) downsampling to generate low resolution images. We enhance the trainset by ran-domly rotating the images 90°, 180°, 270°, and horizontal flipping. We set patch-size of the low resolution image as 48 × 48, we use 5 open-source datasets for model testing, namely Set5, Set14, BSD100, Urban100, and Manga109, each dataset has different characteristics. PSNR and SSIM have measured the performance of the network. In the training process, the optimizer is Adam, the batch size is set to 16, the loss function is the L1 loss function, the initial learning rate is set to 10^{-4}, and reduce half every 200 epochs. Finally, we use the SGD optimizer for tuning parameters. The GPU uses Nvidia 2080Ti GPU, and the framework uses Pytorch [22]. See GitHub for more information (https://github.com/zhaowenlovehome/ADRN).

4.2 Ablation Experiments

To know how many ADRB can be stacked to maximize the advantages of the network, we compared the number of modules and tested it on the Set5 dataset. From Table 1, as the number of ADRB goes up, the performance of the network has also been significantly ameliorated, we can see that 35 ADRB reaches the highest PSNR = 32.38dB on Set (4×), 30 ADRB reaches the second highest PSNR = 32.37 on Set (4×). Although 35 ADRB higher than other numbers ADRB SR result, the required computing power is greatly increased by number, so we choose 30 ADRB as the setting for the following experiment.

To exhibit that the contributions of various depth feature maps to the network are different and the effectiveness of the MDAB. As shown in Table 2, we call the struc-ture contains 30 residual groups "base", contains 3 residual blocks as, a total of 90 residual blocks, "Dense Connection", "Adaptive Dense Connection", "Channel Atten-tion" and "Our Attention" are all components added to the "residual group".

Table 1. The performance of different numbers of ADRB on the network. On Set5 dataset (scale = 4) in 5.6×10^5 iterations.

Number	20	25	30	35
PSNR (dB)	32.01	32.25	32.37	**32.38**

(1)The PSNR of "base + Dense Connection" is higher than "base", which proves that the combination of residual connection and dense connection can contribute to the improvement of model performance. (2) Comparing the PSNR indicators of "base+ Dense Connection" and "base+Adaptive Dense Connection", it can be seen that the "Adaptive Dense Connection" we proposed is significantly higher than the unweighted "Dense Connection", in other words, the characteristics of different depths of the network The importance of the network model is different, which confirms our point of view. (3) By comparing "base+Adaptive Dense Connec-tion+Channel Attention" and "base+Adaptive Dense Connection", it can be conclud-ed that adding a channel attention block to the SISR task could improve the PSNR and SSIM. (4) By comparing the PSNR of "base+Adaptive Dense Connec-tion+Channel Attention" and "base+ Adaptive Dense Connection+ Multi-direction Attention", it testifies that we introduced multi-direction attention block is better than "Channel Attention", which also means more acquisition. The contextual information of the direction contributes to the improvement of the SR method.

Table 2. The performance of different components on residual group. On Set5 dataset (scale = 4) in 5.6×10^5 iterations.

	Base				
Dense connection		√			
Adaptive dense connection			√	√	√
Channel attention				√	
Our attention					√
PSNR (dB)	32.25	32.26	32.28	32.35	**32.37**

Table 3. Quantitative analysis of the network

Method	Scale	Set5	Set14	BSD100	Urban100	Manga109
		psnr/ssim	psnr/ssim	psnr/ssim	psnr/ssim	psnr/ssim
Bicubic	2	33.66/0.9299	30.24/0.8688	29.56/0.8431	26.88/0.8403	30.80/0.9339
EDSR [19]	2	38.11/0.9602	33.92/0.9195	32.32/0.9013	<u>32.93</u>/0.9351	39.10/0.9773
RDN [30]	2	<u>38.24</u>/0.9614	34.01/0.9212	<u>32.34</u>/0.9212	32.89/<u>0.9353</u>	39.18/0.9780
D-DBPN [13]	2	38.09/0.9600	33.85/0.9190	32.27/0.9000	32.55/0.9324	38.89/0.9775
SRFBN [18]	2	38.11/0.9609	33.82/0.9196	32.29/0.9010	32.62/0.9328	39.08/0.9779
OURS	2	38.22/0.9613	<u>34.12</u>/0.9215	<u>32.34</u>/0.9017	32.89/0.9351	<u>39.32</u>/0.9788
OURS+	2	**38.28/0.9616**	**34.26/0.9224**	**32.40/0.9023**	**33.21/0.9384**	**39.48/09791**
Bicubic	4	28.42/0.8104	26.00/0.7027	25.96/0.6675	23.14/0.6577	24.89/0.7866
EDSR [19]	4	32.46/0.8968	28.80/<u>0.7876</u>	27.71/0.7420	26.64/0.8033	31.02/0.9148
RDN [30]	4	32.47/0.8990	28.81/0.7871	27.72/0.7419	26.61/0.8028	31.00/0.9151
D-DBPN [13]	4	<u>32.47</u>/0.8980	28.82/0.7860	27.72/0.7400	26.38/0.7946	30.91/0.9137
SRFBN [18]	4	<u>32.47</u>/0.8983	28.81/0.7868	27.72/0.7409	26.60/0.8015	31.15/0.9160
OURS	4	32.46/<u>0.8986</u>	<u>28.83</u>/0.7873	27.75/0.7423	26.81/0.8066	31.10/0.9151
OURS+	4	**32.59/0.8998**	**28.92/0.7893**	**27.81/0.7437**	**27.02/0.8104**	**31.38/0.9177**

4.3   Quantitative Analysis

To reflect the superiority of our network, as shown in Table 3, we compare ADR-Net with the current best methods, including EDSR [19], RDN [30], D-DBPN [13], and SRFBN [18] methods. We bolded the first evaluation index and underlined the second. Like the EDSR [19] method, we use the self-ensemble method to enhance the performance of our model and mark it as "+". In general, our ADRNet is comparable to or better than other methods compared with all other methods. When the magnification is 2 (scale = 2), we achieved the best results in PSNR and SSIM on Set14, BSD100, and Manga109 datasets. and we achieved the second-best results on Set5 and Urban datasets. When the magnification is 4 (scale = 4), it achieved the first results in PSNR and SSIM on the BSD100 and Urban100 datasets, the first and second results in PSNR and SSIM on the Set14 dataset, and the second results on the Set5 and Manga109 datasets. Compared with other methods, we find that ADRNet performs particularly well on the datasets Set14 and BSD100, and can better restore texture and structure information on these datasets. Besides, our ADRNet+ surpasses other methods in PSNR and SSIM on all scaled sizes and all datasets.

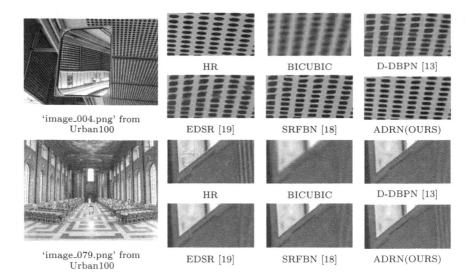

Fig. 3. Results visualization.

4.4 Visualization of Experimental Results

As shown in Fig. 3, we have enlarged the experimental results of various methods. From the figure, we can see that the experimental results of the interpolation method are very fuzzy. Other methods can not accurately reconstruct the image, and there are serious blur artifacts, the enlarged image is even a little distorted, and the high frequency detail information is not well restored. In contrast, our proposed ARDNet restores HR more clearly, the overall structure and texture information is better restored, the edges of objects in the image are restored more sharply, and the contrast is higher. As shown in the "image_004.png" Urban100 in the dataset, other methods cause excessive blurring and distortion in image details, while our method restores some of the detailed textures in comparison. The same "image_079.png" in the Urban100 dataset restored the texture on the wall, while the other methods restored it in a very fuzzy way and could not tell what it was.

As shown in Fig. 3, we have enlarged the experimental results of various methods. From the figure, the experimental results of the interpolation method are very fuzzy. Other methods can not accurately reconstruct the image, and there are serious blur artifacts, the enlarged image is even a little distorted, and the high-frequency detail information is not well restored. In contrast, our proposed ARDNet restores HR more clearly, the overall structure and texture information is better restored, the edges of objects are restored more sharply, and the contrast is higher. As shown in the "image_004.png" Urban100 in the dataset, other methods cause excessive blurring and distortion in image details, while our method restores some of the detailed textures in comparison. The same "image_079.png" in the Urban100 dataset restored the texture on the wall, while the other methods restored it in a very fuzzy way and could not tell what it was.

5 Conclusion

In this paper, we propose a new adaptive densely residual network (ADRNet) for image super-resolution. To solve the problem of insufficient feature utilization, we designed an adaptive densely residual block (ADRB), which combines 3 residual blocks (RB), adds dense connection, and learns the attention score of each dense connection through adaptive dense connections. The attention score reflects the im-portance of the features of each residual block. It makes full use of each residual block feature. To further enhance the performance of ADRB, we introduce a multi-directional attention block to obtain multi-directional context information. Through ablation experiments, it is proved that weighting features of different depths help to improve network performance, and MDA is more efficient than channel attention block. Through comparative experiments, it is demonstrated that ADRNet outbal-ances the existing methods.

References

1. Allebach, J., Wong, P.W.: Edge-directed interpolation. In: Proceedings of 3rd IEEE International Conference on Image Processing, vol. 3, pp. 707–710 (1996)
2. Baker, S., Kanade, T.: Limits on super-resolution and how to break them. IEEE Trans. Pattern Anal. Mach. Intell. **24**(9), 1167–1183 (2002)
3. Chang, H., Yeung, D.Y., Xiong, Y.: Super-resolution through neighbor embedding. In: Proceedings of the 2004 IEEE Computer Society Conference on Computer Vision and Pattern Recognition, 2004. CVPR 2004, vol. 1, pp. 275–282 (2004)
4. Chen, L., Guo, L., Cheng, D., Kou, Q., Gao, R.: A lightweight network with bidirectional constraints for single image super-resolution. Optik **239**, 166818 (2021)
5. Dai, T., Cai, J., Zhang, Y., Xia, S.T., Zhang, L.: Second-order attention network for single image super-resolution. In: 2019 IEEE/CVF Conference on Computer Vision and Pattern Recognition (CVPR), pp. 11065–11074 (2019)
6. Dong, C., Loy, C.C., He, K., Tang, X.: Image super-resolution using deep convolutional networks. IEEE Trans. Pattern Anal. Mach. Intell. **38**(2), 295–307 (2016)
7. Dong, C., Loy, C.C., Tang, X.: Accelerating the super-resolution convolutional neural network. In: European Conference on Computer Vision, pp. 391–407 (2016)
8. Duchon, C.E.: Lanczos filtering in one and two dimensions. J. Appl. Meteorol. **18**(8), 1016–1022 (1979)
9. Freeman, W.T., Pasztor, E.C., Carmichael, O.T.: Learning low-level vision. Int. J. Comput. Vis. **40**(1), 25–47 (2000)
10. Gu, S., Sang, N., Ma, F.: Fast image super resolution via local regression. In: Proceedings of the 21st International Conference on Pattern Recognition (ICPR2012), pp. 3128–3131 (2012)
11. Gu, S., Zuo, W., Xie, Q., Meng, D., Feng, X., Zhang, L.: Convolutional sparse coding for image super-resolution. In: 2015 IEEE International Conference on Computer Vision (ICCV), pp. 1823–1831 (2015)

12. Guo, X., Yuan, Y.: Triple ANet: adaptive abnormal-aware attention network for WCE image classification. In: Shen, D., et al. (eds.) MICCAI 2019. LNCS, vol. 11764, pp. 293–301. Springer, Cham (2019). https://doi.org/10.1007/978-3-030-32239-7_33

13. Haris, M., Shakhnarovich, G., Ukita, N.: Deep back-projection networks for super-resolution. In: 2018 IEEE/CVF Conference on Computer Vision and Pattern Recognition, pp. 1664–1673 (2018)

14. Hu, X., Mu, H., Zhang, X., Wang, Z., Tan, T., Sun, J.: Meta-sr: A magnification-arbitrary network for super-resolution. In: 2019 IEEE/CVF Conference on Computer Vision and Pattern Recognition (CVPR), pp. 1575–1584 (2019)

15. Huang, G., Liu, Z., van der Maaten, L., Weinberger, K.Q.: Densely connected convolutional networks. In: 2017 IEEE Conference on Computer Vision and Pattern Recognition (CVPR), pp. 2261–2269 (2017)

16. Kim, J., Lee, J.K., Lee, K.M.: Accurate image super-resolution using very deep convolutional networks. In: 2016 IEEE Conference on Computer Vision and Pattern Recognition (CVPR), pp. 1646–1654 (2016)

17. Lai, W.S., Huang, J.B., Ahuja, N., Yang, M.H.: Deep laplacian pyramid networks for fast and accurate super-resolution. In: 2017 IEEE Conference on Computer Vision and Pattern Recognition (CVPR), pp. 5835–5843 (2017)

18. Li, Z., Yang, J., Liu, Z., Yang, X., Jeon, G., Wu, W.: Feedback network for image super-resolution. In: 2019 IEEE/CVF Conference on Computer Vision and Pattern Recognition (CVPR), pp. 3867–3876 (2019)

19. Lim, B., Son, S., Kim, H., Nah, S., Lee, K.M.: Enhanced deep residual networks for single image super-resolution. In: 2017 IEEE Conference on Computer Vision and Pattern Recognition Workshops (CVPRW), pp. 1132–1140 (2017)

20. Liu, J., Zhang, W., Tang, Y., Tang, J., Wu, G.: Residual feature aggregation network for image super-resolution. In: 2020 IEEE/CVF Conference on Computer Vision and Pattern Recognition (CVPR), pp. 2359–2368 (2020)

21. Niu, Z.H., Lin, X.P., Yu, A.N., Zhou, Y.H., Yang, Y.B.: Lightweight and accurate single image super-resolution with channel segregation network. In: ICASSP 2021–2021 IEEE International Conference on Acoustics, Speech and Signal Processing (ICASSP) (2021)

22. Paszke, A., et al.: Automatic differentiation in Pytorch (2017)

23. Safarov, S., Whangbo, T.K.: A-denseunet: Adaptive densely connected unet for polyp segmentation in colonoscopy images with atrous convolution. Sensors **21**(4), 1441 (2021)

24. Timofte, R., et al.: Ntire 2017 challenge on single image super-resolution: methods and results. In: 2017 IEEE Conference on Computer Vision and Pattern Recognition Workshops (CVPRW), pp. 1110–1121 (2017)

25. Tong, T., Li, G., Liu, X., Gao, Q.: Image super-resolution using dense skip connections. In: 2017 IEEE International Conference on Computer Vision (ICCV), pp. 4809–4817 (2017)

26. Xie, T., Yang, X., Jia, Y., Zhu, C., Li, X.: Adaptive densely connected single image super-resolution. In: 2019 IEEE/CVF International Conference on Computer Vision Workshop (ICCVW), pp. 3432–3440 (2019)

27. Zhang, X., Ng, R., Chen, Q.: Single image reflection separation with perceptual losses. In: 2018 IEEE/CVF Conference on Computer Vision and Pattern Recognition, pp. 4786–4794 (2018)

28. Zhang, Y., Li, K., Li, K., Wang, L., Zhong, B., Fu, Y.: Image super-resolution using very deep residual channel attention networks. In: Proceedings of the European Conference on Computer Vision (ECCV), pp. 294–310 (2018)

29. Zhang, Y., Li, K., Li, K., Zhong, B., Fu, Y.: Residual non-local attention networks for image restoration. In: International Conference on Learning Representations (2019)
30. Zhang, Y., Tian, Y., Kong, Y., Zhong, B., Fu, Y.: Residual dense network for image super-resolution. In: 2018 IEEE/CVF Conference on Computer Vision and Pattern Recognition, pp. 2472–2481 (2018)

Attention Residual Convolution Neural Network Based on U-Net for COVID-19 Lung Infection Segmentation

Qiang Zuo, Songyu Chen, and Zhifang Wang[(✉)]

Department of Electronic Engineering, Heilongjiang University, Harbin 150080, China
wangzhifang@hlju.edu.cn

Abstract. In 2021, the COVID-19 is still widespread around the world, which has a great impact on people's daily lives. However, there is still a lack of research on the fast segmentation of lung infections caused by COVID-19. The segmentation of the COVID-19- infected region from the lung CT is of great significance for the diagnosis and care of patients. In this paper, attention gate residual U-Net (AGRU-Net) based on residual network and attention gates is proposed for the segmentation. As COVID-19- infected regions varies greatly from one to another, the deeper network is needed to extract segmentation features. The residual unit is an effective solution to the degradation problem of deeper network. The addition of attention gates to U-Net suppresses irrelevant areas in the image for more significant segmentation characteristics. In this paper, the experiments on a public COVID-19CT dataset show that AGRU-Net has good performance in the segmentation of COVID-19- infected region.

Keywords: Medical image segmentation · AGRU-Net · Residual · Attention gate

1 Introduction

An unknown virus was first discovered in Wuhan in 2019. Later, the virus appeared in many parts of China and the world. In February 2020, the WHO officially named it COVID-19. As of February 2021, the number of confirmed COVID-19 cases has exceeded 100 million, and the number of death cases has exceeded 2 million. Quick and accurate diagnosis of COVID-19 is the key to preventing the spread of the virus and treating patients in a timely manner. Currently, Real-time Reverse Transcription-Polymerase Chain Reaction (RT-PCR) is the standard method for detecting COVID-19. However, RT-PCR detection is time-consuming and there is a possibility of false negatives. In addition, many countries in the world still lack kits [1]. Infection with COVID-19 will cause symptoms like inflammation of the alveoli and fluid accumulation, which make patients have difficulty in breathing [2]. In recent years, imaging technology has played a huge role in the quantification and diagnosis of various diseases. As a routine diagnostic tool for pneumonia, lung CT can supplement the limitations of RT-PCR analysis [3]. The research of Fang [4] et al. showed that when the patient's RT-PCR

© Springer Nature Singapore Pte Ltd. 2021
J. Zeng et al. (Eds.): ICPCSEE 2021, CCIS 1451, pp. 350–359, 2021.
https://doi.org/10.1007/978-981-16-5940-9_26

test result is negative, lung CT can further screen COVID-19 patients and achieve better identification and accurate diagnosis of COVID-19.

Manual annotation of the infected region of the patient's lung CT image is labor-intensive and time-consuming. The accuracy of annotation relies heavily on the professional knowledge of doctors. The spread of COVID-19 has brought a very large burden to radiologists. Therefore, automatic segmentation algorithm for the infected region is urgently needed. The position and shape of the pulmonary infected regions of different patients are quite different, and the contrast of infected regions is low. Compared with traditional methods, deep learning can be faster and more accurate for image processing. In the COVID-19 image analysis based on deep learning, the segmentation of the ROI area of the lung CT image is the key to predict the development of disease, such as lung, lobes, bronchopulmonary segments, and infected regions [5]. The application of various deep learning models in image segmentation has made great progress, which has further improved the work efficiency of clinicians.

At present, there are many researches on COVID-19. For example, Fan et al. [6] proposed Inf-Net, which uses parallel partial decoder to aggregate high-level features and generate a global map, and uses implicit reverse attention as well as explicit edge-attention to model boundaries and enhance representations. Zhou et al. [18] proposed that the spatial and channel attention mechanism can be added to U-Net to re-weight the representation in terms of space and channel. Wang et al. [17] proposed an artificial intelligence system that can quickly diagnose COVID-19. U-Net [16] is currently the most widely used encoder-decoder segmentation network. The encoder extracts features, and the decoder combines semantic features to restore the original image size. At present, U-Net has been successfully applied to the segmentation of CT images, including heart segmentation [7], liver segmentation [8], or multi-organ segmentation [9]. As the network deepens, U-Net will face the problem of degradation [10]. Additionally, not all the features extracted by the encoder are useful. Therefore, AGRU-Net is proposed in this paper, which is a new deep learning model based on U-Net framework. The residual unit can train deeper network models and solve the problem of network degradation. The attention gate can use decoder feature as a gating signal to modify the encoder feature, thereby weakening the features of irrelevant background regions. The network model proposed in this paper uses both the residual network and the attention gate to help U-Net segment the infected region of COVID-19 more quickly and accurately, and conduct experiments on a public lung CT dataset of COVID-19 to evaluate ARU- Net network.

The paper is structured as follows: The second section describes the deep learning model in detail, including the structure of AGRU-Net, the improvement method of the encoder and decoder, and the attention gate. The third section introduces the dataset, evaluation metrics, experimental results, and the analysis of the results. The fourth section discusses the proposed model and concludes the paper.

2   AGRU-Net

2.1   The Proposed Network Architecture

This section will introduce Attention Gate Residual U-Net (AGRU-Net) model used for the lung CT image segmentation of COVID-19 in detail. The network proposed in this paper integrates the residual network and attention gate on U-Net framework. First, the residual unit is added to the encoder and decoder convolution block to solve the problem of network degradation. Then, the image features obtained by the encoder are input into the attention gate, and the deep features obtained by the decoder are used as the gating signal to modify the encoder features. Finally, the decoder outputs the final segmentation result through a 1 × 1 convolution and Sigmoid. The main components of the network model will be described below, including encoder, decoder, residual block and attention gate. The AGRU-Net architecture scheme is described in Fig. 1. The left ResNet blocks constitute the encoder, which is used to extract features from the original image. The right blocks constitute the decoder. Each block of the decoder receives the feature transmitted from the encoder and splices it with the feature transmitted from the previous layer of the decoder.

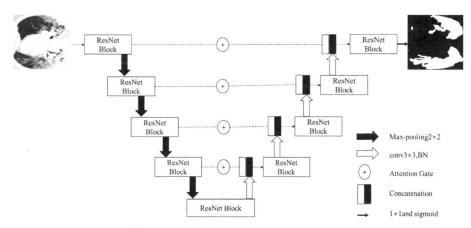

Fig. 1. The architecture of the proposed AGRU-Net.

2.2   Encoder-Decoder

U-Net network was first proposed by Ronneberger [16]. It is a variant of Fully Convolutional Neural Network (FCN) [11]. The AGRU-Net structure is based on to U-Net. The encoding path of AGRU-Net is composed of five residual blocks. Each time downsampling is performed, the size of the image is reduced by half and the number of filters is doubled. The initial number of filters in this paper is 32, and finally increased to 512. Each step in the decoding path up-samples the output of the previous layer. In a standard U-Net, the output of the encoding path transfers image features to the decoding path

through skip connections. And it is directly spliced with the features of the corresponding decoding layer through cropping and copying. AGRU-Net uses attention gates to
deal with the two features. The last layer of the decoder restores the size of the original
image, and outputs the final segmentation result through a 1 × 1 convolution and Sigmoid function. ReLU and Sigmoid are defined as formula (1) and formula (2), as shown
below.

$$\text{ReLU} : f(x) = \max\{0, x\} \tag{1}$$

$$\text{Sigmoid} : f(x) = \frac{1}{1 + \exp(-x)} \tag{2}$$

2.3 Residual Block

Traditional U-Net structure can easily segment the area of lung in the CT image. But for
more complex medical images, such as COVID-19, the infected regions varies greatly.
Traditional U-Net structure has limitation in segmentation in this case, so a deeper
network is needed to extract more complex features of the image. When the depth of the
network deepens, the accuracy will gradually increase, and then quickly decline. This

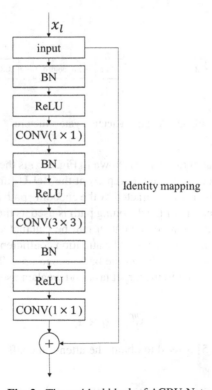

Fig. 2. The residual block of AGRU-Net.

situation is defined as degradation problem. ResNet can solve the problem very well. The definition of the residual block is as follows.

$$y_l = F(x_l, w_l) + x_l \tag{3}$$

x_l and y_l is the input and output mapping of the L_{th} layer. $F(x_l, w_l)$ represents the residual mapping to be learned. The shortcut connections of the residual network make network easier to be optimized. In this work, the residual block contains three convolutional layers and three Batch Normalization (BN) layer, followed by the BN layer is a ReLU activation function, as shown in Fig. 2.

2.4 Attention Gate

In U-Net, not all the features obtained by the encoder are beneficial to segmentation. In this paper, the skip connection is replaced by the attention gate in U-Net. The attention gate is related to the human visual attention mechanism. The human visual attention mechanism automatically focuses on the target area, which is a brain signal processing mechanism unique to human vision. The main function of attention gates is to obtain the target area that needs attention, so as to obtain more detailed information of the target that needs attention, thereby suppressing the feature response of irrelevant areas, highlighting the features benefited to the segmentation tasks and enhance network performance [13].

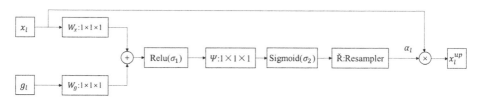

Fig. 3. Additive attention gate structure.

The structure of the attention gate is shown in Fig. 3. x_l is the encoding path feature. g_l is the gating signal vector. σ_1 and σ_2 represent the ReLU function and the Sigmoid function respectively. The feature extracted by the encoding path is x_l, which is input into the attention gate. The feature of the decoding path is used as the gating signal to select the focus region. The attention coefficient α_l can be calculated through the attention gate, and the value is in the range [0, 1]. The attention coefficient will get a larger value in the target area and a smaller value in the background area. Therefore, it can derive characteristic response related to the target task and weaken useless characteristic. The formula is as follows.

$$x_l^{up} = x_l \times \alpha_l \tag{4}$$

Additive attention [15] is used to obtain the attention coefficient, and the following is the formula.

$$Q_l = \Psi\left(\sigma_1\left(W_x x_l + W_g g_l + b_g\right)\right) + b_\psi \tag{5}$$

$$\alpha_l = \sigma_2\left(\check{R}(Q_l)\right) \tag{6}$$

σ_1 and σ_2 represent the ReLU function and the Sigmoid function respectively. W_x and W_g are weights, b_g and b_ψ are bias terms. Vector concatenation-based attention is used in this paper [9]. The linear transformation is calculated using the channel direction $1 \times 1 \times 1$ convolution of the tensors. The Grid resampling of attention coefficients is completed using trilinear interpolation. The update of attention gate parameters is trained in accordance with the standard backpropagation.

3 Experiments and Analysis

3.1 Evaluation Metrics

There are four metrics to measure the performance of AGRU-Net, namely Accuracy (AC), Sensitivity (SE), Specificity (SP), F1-Score (F1) and Area Under ROC Curve(AUC).

(1) AC: It is used to evaluate the accuracy of the model's classification of pixels.

$$AC = \frac{TP + TN}{TP + TN + FP + FN} \tag{7}$$

(2) SE (also called the recall): It measures the true positive rate correctly identified by the model.

$$SE = \frac{TP}{TP + FP} \tag{8}$$

(3) SP: It measures the true negative rate correctly identified by the model.

$$SP = \frac{TN}{TP + FP} \tag{9}$$

(4) F1: It is a metric used to measure the accuracy of the binary classification model, while taking into account the accuracy rate and recall rate of the classification model.

$$F_1 = \frac{precision \times recall}{precision \times recall} \tag{10}$$

(5) AUC: It is a metric to evaluate the quality of classification models. Class imbalance problem of samples makes traditional evaluation metrics like accuracy unable to reflect the performance of classification models properly.

$$AUC = S_{\text{Area under ROC curve}} \tag{11}$$

TP is the number of pixels in the CT image correctly identified as a COVID-19 infected area. TN is the number of pixels in the CT image correctly identified as background area. *FP* is the number of pixels in the CT image where the background area is wrongly identified as COVID-19-infected area. *FN* is the number of pixels in the CT image where the COVID-19-infected area is wrongly identified as the background area. In this paper, the receiver operating characteristics (ROC) curve and precision recall (PR) curve of AGRU-Net are provided. In addition, the performance of networks are compared through the area under the curve (AUC) of ROC.

3.2 Dataset and Implementation Details

The CT images used in the experiment comes from the COVID-19 CT image dataset collected by the Italian Society of Medical and Interventional Radiology, which is the first open dataset for the CT segmentation of lung infected by COVID-19. It consists of 100 axial two-dimensional CT images of the lung from different patients. The initial sizes of the images are different. The sizes are unified into 256×256 size in this paper. The experiments are conducted by using the tensor flow with NVIDIA GeForce RTX 2080Ti and the ADAM optimizer. The initial learning rate is 1e-4 and the learning rate factor is 0.1. The number of training times is 100, and the dataset is randomly divided into 80% as the training set and 20% as the test set. The first line of Fig. 4 shows the CT image of lungs infected by COVID-19, and the second line describes the corresponding ground truth images of lungs infected by COVID-19 annotated by the doctor.

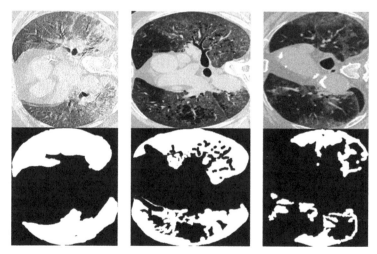

Fig. 4. The example CT segmentation images of lung infected by COVID-19.

3.3 Experiment Results

Table 1 shows the results of AGRU-Net and other networks. Attention U-Net represents the improved U-net network that only uses the attention gate. Compared with U-Net, Attention U-Net has improved in all parameters. This indicates that the attention gate can weaken the background area of lung CT images irrelevant to COVID-19 and improve the accuracy of segmentation. Compared with the other four networks, AGRU-Net shows good performance. This shows that the replacement of original convolution block with residual block can help to train a deeper network, improve the segmentation accuracy of U-Net, and avoid the problem of network degradation to a certain extent. Therefore, the attention gate and the residual block play an important role in AGRU-Net and improve the segmentation performance of network. The visualized results of the CT images of COVID-19-infected lungs are shown in Fig. 5. Each column represents the segmentation results of a different network. It can be seen from Fig. 5 that FCN and SegNet can hardly generate the CT segmentation image of lung infected by COVID-19. The segmentation images of U-Net are not accurate enough. The segmentation results generated by AGRU-Net are similar to the ground truth images annotated by doctors. By contrast, AGRU-Net has good segmentation effect. Figure 6 provides the ROC curve and PR curve of AGRU-Net.

Table 1. Comparison of the segmentation performance among AGRU-Net and other networks

Methods	F1-Score	Sensitivity	Specificity	Accuracy	AUC
FCN8s [11]	0.551	0.672	0.821	0.792	0.746
SegNet [19]	0.443	0.314	0.976	0.857	0.645
U-Net [16]	0.642	0.562	0.958	0.887	0.761
Attention U-Net [20]	0.652	0.565	0.962	0.891	0.764
AGRU-Net	0.728	0.642	0.973	0.913	0.808

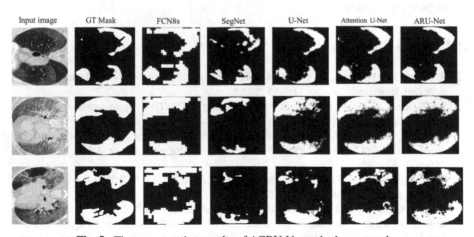

Input image   GT Mask   FCN8s   SegNet   U-Net   Attention U-Net   ARU-Net

Fig. 5. The segmentation results of AGRU-Net and other networks.

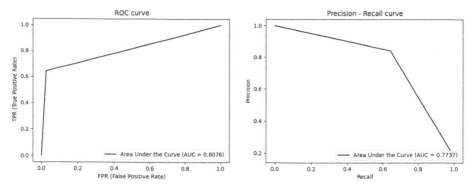

Fig. 6. The ROC curve and PR curve of AGRU-Net.

4 Conclusion

In this paper, AGRU-Net is proposed for the lung CT segmentation of COVID-19-infected region. AGRU-Net takes U-Net as the main framework and enhances its performance through residual network and attention gate. The residual connection can be used to alleviate the problem of network degradation. The attention gate can be used to suppress the characteristic response of areas irrelevant to COVID-19 and highlight the characteristic response of COVID-19-infected areas. The experiments are conducted on the COVID-19 CT dataset. Compared with U-Net, the metrics of AGRU-Net increased by 8.6%, 8%, 1.5%, 2.6% and 4.7% in respects of F1-Score, Sensitivity, Specificity, Accuracy, and AUC respectively. AGRU-Net has good performance on the lung CT segmentation of COVID-19-infected area, which can help doctors reduce the workload in practical application.

References

1. Shan, F., Gao, Y., Wang, J., et al.: Lung infection quantification of COVID-19 in CT images with deep learning. arXiv:2003.04655 (2020)
2. Shi, F., Xia, L., Shan, F., et al.: Large-scale screening of COVID-19 from community acquired pneumonia using infection size-aware classification. arXiv:2003.09860 (2020)
3. Cheng, Z., Lu, Y., Cao, Q., et al.: Clinical features and chest CT manifestations of coronavirus disease 2019 (COVID-19) in a single-center study in Shanghai, China. Am. J. Roentgenol. **215**(1), 1–6 (2020)
4. Fang, Y., Zhang, H., Xie, J., et al.: Sensitivity of chest CT for COVID-19: comparison to RT-PCR. Radiology **296**(2), 200432 (2020)
5. Shi, F., Wang, J., Shi, J., et al.: Review of artificial intelligence techniques in imaging data acquisition, segmentation and diagnosis for COVID-19. IEEE Rev. Biomed. Eng. **99**, 1 (2020)
6. Fan, D.-P., Zhou, T., et al.: Inf-Net: automatic COVID-19 lung infection segmentation from CT images. IEEE Trans. Med. Imag. **39**(8), 2626–2637 (2020)
7. Ye, C., Wang, W., Zhang, S., et al.: Multi-depth fusion network for whole-heart CT image segmentation. IEEE Access **7**, 23421–23429 (2019)
8. Liu, Z., Song, Y.-Q., et al.: Liver CT sequence segmentation based with improved U-Net and graph cut. Expert Syst. Appl. **126**(JUL), 54–63 (2019)

9. Dong, X., Lei, Y., Wang, T., et al.: Automatic multi-organ segmentation in thorax CT images using U-Net-GAN. Med. Phys. **46**(5), 2157–2168 (2019)
10. He, K., Sun, J.: Convolutional neural networks at constrained time cost. In: 2015 IEEE Conference on Computer Vision and Pattern Recognition (CVPR), pp. 5353–5360 (2015)
11. Long, J., Shelhamer, E., Darrell, T.: Fully convolutional networks for semantic segmentation. IEEE Trans. Pattern Anal. Mach. Intell. **39**(4), 640–651 (2015)
12. Ioffe, S., Szegedy, C.: Batch normalization: accelerating deep network training by reducing internal covariate shift. In: Proceedings of the 32nd International Conference on Machine Learning, pp. 448–456 (2015)
13. Wang, F., Jiang, M., Qian, C., et al.: Residual attention network for image classification. In: 2017 IEEE Conference on Computer Vision and Pattern Recognition (CVPR), pp. 6450–6458 (2017)
14. Abraham, N., Khan, N.M.: A novel focal tversky loss function with improved attention U-Net for lesion segmentation. In: 2019 IEEE 16th International Symposium on Biomedical Imaging (ISBI 2019), pp. 683–687 (2018)
15. Bahdanau, D., Cho, K., Bengio, Y.: Neural machine translation by jointly learning to align and translate. Comp. Sci. arXiv:1409.0473 (2014)
16. Ronneberger, O., Fischer, P., Brox, T.: U-Net: convolutional networks for biomedical image segmentation. In: Navab, N., Hornegger, J., Wells, W.M., Frangi, A.F. (eds.) MICCAI 2015. LNCS, vol. 9351, pp. 234–241. Springer, Cham (2015). https://doi.org/10.1007/978-3-319-24574-4_28
17. Wang, B., Jin, S., Yan, Q., et al.: AI-assisted CT imaging analysis for COVID-19 screening: building and deploying a medical AI system. Appl. Soft Comp. (2020)
18. Zhou, T., Canu, S., Ruan, S.: An automatic COVID-19 CT segmentation network using spatial and channel attention mechanism. arXiv:2004.06673 (2020)
19. Badrinarayanan, V., Kendall, A., Cipolla, R.: SegNet: a deep convolutional encoder-decoder architecture for image segmentation. IEEE Trans. Pattern Anal. Mach. Intell. **39**(12), 2481–2495 (2017)
20. Oktay, O., Schlemper, J., Folgoc, L.L., et al.: Attention U-Net: learning where to look for the pancreas. arXiv:1804.03999 (2018)

Fast Target Tracking Based on Improved Deep Sort and YOLOv3 Fusion Algorithm

Yanqing Wang$^{(\boxtimes)}$, Zhiyong Liang, and Xinyu Cheng

Nanjing Xiaozhuang University, Nanjing 211171, China

Abstract. Aiming at fast moving targets, such as ships, high-speed vehicles and athletes, this paper discusses a series of target detection algorithms based on neural network, YOLOv3 and background modeling. Compared KCF tracking with SSD tracking, Gaussian filter was applied to remove noise from pictures, and edge preserving filter was used to preserve edge features. Moreover, the algorithm combining deepsort tracking algorithm with YOLOv3 detection algorithm can improve the accuracy of YOLOv3 target detection, solve the problem of target loss during target tracking, adjust the frame size in real time, and improve the fit with the target position. Experiments show that the proposed algorithm based on detection before tracking has strong learning ability and robustness to unknown environment.

Keywords: Target detection · Target tracking · YOLOv3 · Kalman filter

1 Introduction

Recognition and detection of moving objects is one of the key research directions in the field of computer vision. The learning, cognition and description of unknown environment are far greater than those of structured environment, which makes target extraction and tracking particularly difficult. However, unstructured environment is the main research topics, which is also a hot spot in the field of image processing at present.

Using the deep learning model, because the size of bounding boxes is fixed, the tracked targets are easy to be lost, and the problems of missing detection and false detection are endless. In view of the situation in the experimental process, this paper has done a series of classic target detection and tracking algorithm experiments, and compared the advantages and disadvantages of different algorithms and their adaptability to unknown environment. This paper presents a tracking algorithm which detects first and tracks later. firstly, Gaussian blur is used to denoise, and edge features are retained by edge preserving filtering. then, the transformed image information is input to YOLOv3 for detection, and the features are enhanced and the detection accuracy is improved by denoising and filtering images. Finally, the detection results are input into the Kalman filter tracking algorithm. Due to the particularity of unstructured environment, the target may have some non-rigid motion, so Gaussian blur is introduced to preprocess the image, remove noise and fill the data. The edge preserving filtering is realized by Gaussian bilateral filtering and mean shift filtering, While processing pictures, the features of

© Springer Nature Singapore Pte Ltd. 2021
J. Zeng et al. (Eds.): ICPCSEE 2021, CCIS 1451, pp. 360–369, 2021.
https://doi.org/10.1007/978-981-16-5940-9_27

image edges are preserved. The algorithm of Deep Sort is improved based on the simple online and real time tracking (sort) algorithm [7]. The basic process is roughly divided into four steps. Firstly, reading the position and target characteristics of the target detection frame of the current frame; Secondly, screening the detection frames according to the confidence level, and deleting the redundant detection frames with low confidence level; And third, Suppress the non-maximum value of the detection frame to eliminate the multi-frame situation; Finally, Kalman filter is used to predict the target position.

2 Relevant Knowledge

2.1 Deep Sort Target Tracking

Deep Sort can be divided into three major problems. The first is the allocation problem, which includes motion matching and appearance matching. This algorithm uses Mahalanobis distance to solve the allocation problem, and uses two appropriate measures to integrate motion information and appearance information. When the Mahalanobis distance does not match the appearance, the minimum cosine distance can be used. The second is trajectory processing and state estimation, Trajectory processing and state estimation is about how to match tracker with motion trajectory and state representation of prediction results. The third is cascade matching, which deals with the tracker's preference for tracks with long occlusion time but smaller Mahalanobis distance.

Appearance Assignment
The traditional algorithm to solve the relationship between predicted Kalman filter tracking state and newly arrived measurement is based on the assignment problem solved by Hungarian algorithm. In the formula of this assignment problem, two appropriate measures are used to integrate motion and appearance information. Mahalanobis distance between predicted Kalman state and newly arrived measurement, hereinafter referred to as Mahalanobis distance [8]:

$$d^{(1)}(i,j) = (d_j - y_i)^T S_i^{-1} (d_j - y_i) \tag{1}$$

In the Formula (1), (y_i, S_i) is used to represent the prediction of the distribution of the ith orbit to metric space, and d_j indicates the detection state of the jth Bounding Boxes.

In real life, many objective factors will lead to the failure of Mahalanobis distance matching, such as the influence of environment, camera jitter, etc. Therefore, the algorithm introduces the minimum cosine distance [8], which is different from Mahalanobis distance. Cosine distance refers to the minimum cosine distance of the I-th tracking and the J-th detection. This allocation has a significant effect on recovery tracking after a certain period of occlusion, and the formula is as follows:

$$d^{(2)}(i,j) = \min\left\{1 - r_j^T r_k^{(i)} \mid r_{(k)}^{(i)} \in R_i\right\} \tag{2}$$

If both indexes meet the threshold condition, the Mahalanobis distance of two distances is obtained and cosine distance, followed by weighted fusion, the formula of correlation measure [8] is as follows:

$$c_{i,j} = \lambda d^{(1)}(i,j) + (1 - \lambda)d^{(2)}(i,j) \tag{3}$$

Among them, λ is a superparameter, different weights are adjusted, and different data sets adopt different superparameters. In this algorithm, there are two superparameters, cosine and Euclid, which can be calculated by these two methods. Distance measurement deals with the prediction and tracking of short-term occlusion, and appearance allocation deals with the prediction and tracking of long-term occlusion.

Trajectory Processing and State Estimation

Trajectory Processing

For each tracker, each track has a threshold value A. In the algorithm, a counter is used to record the difference between the frame and the frame that was successfully matched last time. The counter is incremented during the prediction period of Kalman filter. When the counter value is greater than the original set threshold value A, the track stops and the track is removed from the set [8]. In the tracking process,If there are no targets that can be successfully matched with the tracker, it is judged that there are new targets, and the trajectories of these new targets are determined as temporary in the first three frames. If continuous matching is successful, a new target is generated; otherwise, it is a false track and is deleted.

State Estimation

Tracking scenes generally use 8-dimensional space to describe the state of the track at a certain time: $(u, v, r, h, \dot{x}, \dot{y}, \dot{r}, \dot{h})$, (u,v) is the center coordinate of Bounding Boxes, r is the aspect ratio, and h is the height. The last four variables refer to the corresponding velocity information (derivative) in the image coordinates. In this method, a standard Kalman filter based on constant velocity model and linear observation model is used to predict the target state, and the prediction result is (u,v,r,h).

Cascade Matching

Cascade matching algorithm [8] appears in order to connect the detected target with a relatively short moving track. In the process of tracking and matching, two or even more tracks may compete for the same tracker. Because the continuous prediction of Kalman filter is not updated, the covariance matrix is dispersed.However, the longer the disappearing time, the smaller Mahalanobis distance will be obtained, which leads to the choice of the trajectory with longer time interval. The target with longer disappearing time may have higher uncertainty and less detection possibility than the target with shorter disappearing time. From the experimental results, what the tracker should get is a track with a shorter distance from disappearing time, In this paper, cascade matching is adopted, which gives priority to the targets with higher frequency and improves the robustness of the algorithm.

2.2   YOLOv3 Target Detection

Backbone Network

YOLOv1 target detection algorithm was proposed in 2016, and YOLOv3 was published in 2018. From the first generation to the third generation, the progress of each generation is inseparable from the replacement of backbone networks, among which Darknet-19 is the backbone network of YOLOv2, and Darknet-53 is the backbone network of YOLOv3. In many papers, there are comparative diagrams of backbone networks of YOLOv2 and YOLOv3, and the feature diagram is reduced to one when input. In the process of forward propagation, YOLOv3 realizes the dimension transformation of tensor through the step transformation of convolution kernel, and maximizes the detection effect [9] on the basis of ensuring real-time performance, so YOLOv3 is no longer blindly pursuing speed. The following table shows the different performances of YOLOv3 under different networks. The experimental data is quoted from other literatures, based on two backbone networks,Darknet and ResNet (Table 1).

Table 1. YOLOv3 performance on different backbone networks

Backbone	Top-1	Top-5	BnOps	BFLOP/s	FPS
Darknet-19	74.10	91.80	7.29	1246	171
ResNet-101	77.10	93.70	19.70	1039	53
ResNet-152	77.60	93.80	29.40	1090	37
Darknet-53	77.20	93.80	18.70	1457	78

Darknet-19 still leads Darknet-53 in speed, but Darknet-53 performs well in other aspects. BFLOP/s in the table means the number of convolution floating-point number operations every time, and the calculation complexity of the model is the sum of BFLOP/s of each layer, which can reflect the calculation speed of the model. YOLOv3 official website also has some lightweight weights and network configuration files.

Border Prediction

YOLOv3's prediction frame is obtained by clustering in data set, which has 9 prior frames, 3 different scales and 10, 467 candidate frames, while tiny-YOLOv3 has 6 prior frames. The prior frames depend on the data set, and each prediction frame has two parameters—height and width. The three different scales are $13 \times 13 \times 255$, $26 \times 26 \times 255$ and $52 \times 52 \times 255$. These three different output scales can detect targets with different scales respectively, thus optimizing the effect of YOLOv3 on small target detection. YOLOv3 used the idea of logistic regression to predict the boundary box, and finally got (x,y,w,h, confidence). The formula is as follows:

$$b_x = \sigma(t_x) + C_x \tag{4}$$

$$b_y = \sigma(y) + C_y \tag{5}$$

$$b_w = p_w e^{t_w} \tag{6}$$

$$b_h = p_h e^{t_h} \tag{7}$$

$$\text{Pr}(object) \ * \ IOU(b, object) = \sigma(t_0) \tag{8}$$

Logistic regression will score the target for the content of each prediction frame. This target score is how likely it is to predict the target in this frame. According to this score, the prior frame is selected for prediction, and it does not predict all prediction frames, thus reducing the amount of calculation. Moreover, YOLOv3 will only produce a prior frame with the highest score. If it is not matched successfully in the end, Then it will affect the confidence of the target, but fortunately it will not affect the prediction of other parameters or even other layers.

Multiscale Prediction

In YOLOv3, three pre selection boxes with different scales are set in each grid cell, so each pre selection box has five basic parameters - (x, y, w, h, confidence). The COCO dataset has 80 categories, so the depth of these three output tensors is 255, that is 3 * (5 + 80).

Then YOLOv3 first obtains two scale feature maps from the convolution layer of two low-scale convolution blocks, then samples the two scale feature maps twice, and then concatenates the two feature maps from the convolution blocks earlier than the two convolution blocks and two deeper scale feature maps respectively, so as to obtain more semantic information, And get fine-grained information from low-scale feature map.Then, the convolution layer is used to process the feature map to predict twice the original similar tensor [10].

Using the same network to design the scale prediction frame of prediction (8 * 8), it is helpful to classify prediction, such prediction can integrate operation and fine-grained features.

Through these evaluation indicators, we can roughly evaluate the accuracy and speed of the algorithm, and provide a reference for the applicable scenarios of the algorithm.

YOLOv3 mainly introduces the same three points. First, the progress of YOLOv3 depends on the update of the backbone network. YOLOv3 is based on the darknet-53 network, which improves the accuracy. The second point is to use the frame prediction of logical regression and prior detection to calculate the target score for each prediction box, and compete for the object with the highest score to match the labeled object, so as to reduce the calculation and increase the robustness. The third point is to use multiscale prediction, multiscale prediction uses three scale prediction box and convolution operation to obtain feature map and fine-grained feature, and predict tensor.

3   The Algorithm in this Paper

The algorithm core of Deep Sort adds the appearance information of the moving target, and puts the appearance information into the matching calculation, so that even if the target is occluded in a short time, the detection box can match the original target ID in

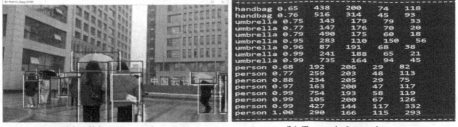

handbag 0.65	438	200	74	118
handbag 0.76	516	314	45	93
umbrella 0.75	143	179	79	33
umbrella 0.77	147	176	70	20
umbrella 0.79	490	175	60	18
umbrella 0.95	283	110	150	56
umbrella 0.96	87	191	68	38
umbrella 0.99	241	188	65	21
umbrella 0.99	735	164	94	45
person 0.68	192	206	29	82
person 0.77	259	203	48	113
person 0.88	234	205	29	75
person 0.97	163	200	47	117
person 0.99	754	193	58	119
person 0.99	105	200	67	126
person 0.99	427	144	117	332
person 1.00	290	166	115	293

(a) All items tested   (b) Target information

Fig. 1. Deep_Sort test results and data analysis

the case that the target appears again in the future, thus reducing the problem of frequent ID transformation in the detection.

Figure 1 shows Deep Sort YOLOv3 experiment, a frame detection screenshot and detection results are given.Fig. 1-a shows that under the detection of YOLOv3, for the extraction effect of the target, the blue box is the tracking box, and the white box is the detection box. These two frames are constantly updated, and the tracking effect is better than that in a single experiment. The tracking box is no longer a fixed size, and can be detected again even if the target is lost. Figure 1-b shows the detection results corresponding to the current frame, and prints out the coordinates (x, y, w, h) of the moving object, confidence level and frame for detection and recognition. The general structure of the algorithm is as follows (Fig. 2):

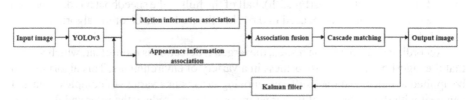

Fig. 2. Algorithm flow chart

YOLOv3 is implemented based on OpenCV, VideoCapture() function is used to obtain video source, parameters are initialized, cosine distance is set to 0.5, non maximum suppression is set to 1, Kalman filter needs Pb file, so network configuration file and weight of YOLOv3 need to be transformed into Pb file. Using Freeze_Model.py file to convert. After successful conversion, Create_box_The encoder() function loads the model file.

Then, the nearest neighbor distance returned by each target is measured as the nearest distance [8] observed so far from any sample, and the oldest sample reaching the target value is removed each time. Two parameters: cosine and Euclidean, can be selected to calculate the nearest neighbor distance. After the weight file and threshold setting are completed, the method can be used to calculate the nearest neighbor distance,Each frame of the video is read circularly, and the read image mode is changed from BGR to RGB through the model. After filtering the bounding boxes, the boxes with low confidence scores are removed, and the non maximum suppression algorithm is performed for the remaining bounding boxes to remove overlapping bounding boxes.Two for loops are used to match the trajectories with the trackers to remove the false trajectories, and update the position of the target to draw a rectangular box. Draw through the Rectangle() method of OpenCV, and use To_The tlbr() function gets the latest location of the target.

Finally, each frame image with box is stored in the form of video through the class Videowriter in OpenCV, through Video_capture.get() method gets the width and height of video frame. Parameter value 3 is the frame width and parameter value 4 indicates the frame height.

4   Deep_Fast Moving Target Detection Based On Sort And YOLOv3

In the detection results of the above experiments, roadside pedestrians are detected. For fast moving targets, such as vehicles and ships, such as Fig. 3. In this scene, there are three speedboats on the lake. The left side is the final detection result, and the right side is the parameter information of the detected target.

At the beginning of the detection, the target of the speedboat is too small, and the waves generated by the water surface also have serious interference on the detection and tracking. In Fig. 3-a, the images of frames 4–5 are shown. The detection results are shown in Fig. 3-b. because the target is too small, "person" is detected; At about frame 45, as shown in Fig. 3-c, "boat" is detected, but the smaller speedboat behind is still identified as "person"; Around frame 130,The volume of the speedboat is obviously much larger, and the confidence level of ship identification is as high as 0.98, as shown in Fig. 3-f; By about 150 frames(Fig. 3-h), half of the hull of the speedboat has disappeared, and the confidence level is reduced to 0.61, but the performance of the algorithm is still considerable.

According to the detection results, the targeted detection is more clear, which verifies that the algorithm has good robustness in a variety of environments. This algorithm can be applied to detect and track different moving targets in a variety of complex scenes. It can deal with the occlusion of the target in a short time, reduce the jitter and floating of the tracking frame, and the recognition frame tends to be stable.

On the basis of YOLOv3, the recognition effect of the target is relatively considerable. As shown in Fig. 4-a, a moving target is blocked, but it is still detected. The confidence level is 0.53, and other athletes are basically in the range of 0.90 to 1.00; In Fig. 4-c, the detection effect of the algorithm for vehicles is also very significant. After a short period of occlusion, it can be connected, and the confidence level is relatively high. This algorithm can solve the problem that the target appears in the middle of the way, and the detection and tracking frame will change with the change of the target size. The algorithm can deal with the occlusion in a short time.

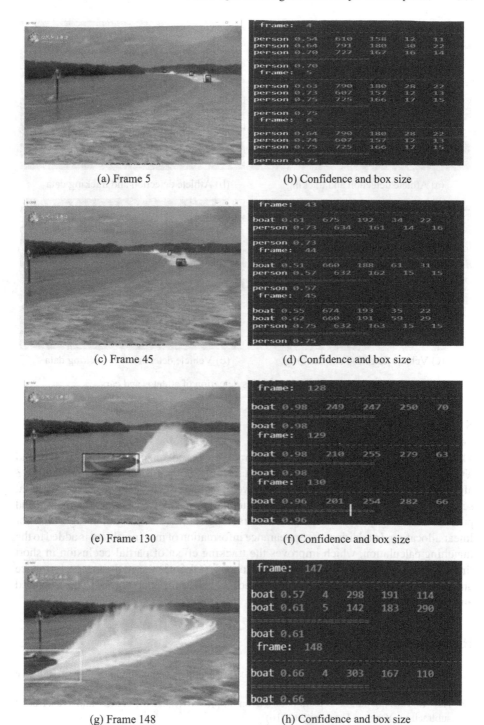

(a) Frame 5

(b) Confidence and box size

(c) Frame 45

(d) Confidence and box size

(e) Frame 130

(f) Confidence and box size

(g) Frame 148

(h) Confidence and box size

Fig. 3. Inspection effect and data analysis of ships

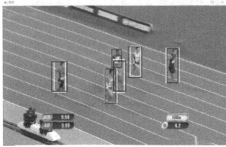

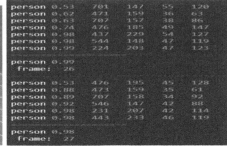

(a) Athlete detection and tracking   (b) Athlete detection and tracking data

(c) Vehicle detection and tracking   (d) Vehicle detection and tracking data

Fig. 4. Deep_Sort_YOLOv3's detection of athletes and cars

5 Summary

This paper mainly describes fast target tracking based on improved deep sort and YOLOv3 fusion algorithm. The experimental results of the fusion of sort and YOLOv3 algorithm are used to detect and track ships, vehicles and athletes in multiple unstructured scenes. Deep Sort uses recursive Kalman filter and frame by frame data association, and puts the matching of detection frame and tracking frame into Hungarian algorithm for linear allocation. In addition, the appearance information of moving object is added to the matching calculation, which improves the tracking effect of partial occlusion in short time. Combined with the real-time detection of YOLOv3, the real-time performance and robustness of the algorithm are improved, as well as the adaptability to unstructured environment.

References

1. Wang, C.: Overview of moving object detection methods in intelligent video surveillance system. Autom. Instrum. **03**, 1–3 (2017)
2. He, Y.: Moving target detection based on improved frame difference and background subtraction. Yanshan University (2016)
3. Zhang, J., Xu, C., Tang, M., Lu, W., Bian, Z.: Research on improved target detection method based on SSD [J]. Laser and Infrared, 2019, 49(08):1019-1025

4. Redmon, J., Farhadi, A.: YOLOv3: An Incremental Improvement (2018)
5. Jiang, X., Xiang, C., Liang, D.: An improved Mean Shift moving target tracking algorithm. Inf. Technol. **01**, 127–130 (2017)
6. Henriques, J.F., Caseiro, R., Martins, P., et al.: High-speed tracking with kernelized correlation filters. IEEE Trans. Pattern Anal. Mach. Intell. **37**(3), 583–596 (2015)
7. Bewley, A., Ge, Z., Ott, L., et al.: Simple Online and Realtime Tracking (2016)
8. Wojke, N., Bewley, A., Paulus, D.: Simple online and realtime tracking with a deep association metric. In: IEEE International Conference on Image Processing (ICIP). IEEE (2017)
9. Liu, C., Song, P., Di, K.: Multi-target tracking based on deep learning and information fusion. Ind. Control Comput. **32**(10), 108–109+112 (2019)
10. Josefina, T., et al.: Do children with overweight respond faster to food-related words? Appetite **161**, 105134 (prepublish) (2021)
11. Bisht, M., Gupta, R.: Offline handwritten Devanagari modified character recognition using convolutional neural network. Sādhanā **46**(1), 20 (2021)
12. Cai, J., et al.: To explore the changes and differences of microstructure of vocal fold in vocal fold paralysis and cricoarytenoid joint dislocation by diffusion tensor imaging. J. Voice (prepublish) (2020)
13. Fonseca, E., Santos, J.F., Paisana, F., DaSilva, L.A.: Radio access technology characterisation through object detection. Comput. Commun. **168**, 12–19 (prepublish) (2020)
14. Fan, L., Zhang, T., Du, W.: Optical-flow-based framework to boost video object detection performance with object enhancement. Expert Syst. Appl. **170**, (prepublish) (2020)
15. Ding, L., Xu, X., Cao, Y., Zhai, G., Yang, F., Qian, L.: Detection and tracking of infrared small target by jointly using SSD and pipeline filter. Digital Sig. Process. **110**, 102949 (prepublish) (2020)
16. Zhao, D., Yuan, B., Shi, Z., Jiang, Z.: Selective focus saliency model driven by object class-awareness. IET Image Process. **15**(6) (2020)

Intelligent Express Delivery System Based on Video Processing

Chuang Xu, Yanqing Wang(✉), Ruyu Sheng, and Wenjun Lu

College of Information Engineering, Nanjing Xiaozhuang University, Nanjing 211171, China

Abstract. Although the scale of the express industry is large, it is difficult to achieve the function of fully intelligent receiving and sending express. In this paper, the intelligent express delivery system is proposed based on the image and video processing technology of OpenCV, the Faster R-CNN object detection algorithm and other technologies. Through the depth camera and electronic scale, it can identify the object category, volume and weight of the items placed on the scale by the sender and store the video of the objects packed into the cabinet. The overall framework of the system was constructed; key technologies were applied to realize the system; the function of the system was tested. The experimental results show that it achieves the intelligent automation of delivery and delivery through the integrated express delivery system of intelligent identification and information traceability, which promotes the development of express delivery industry.

Keywords: Video processing · Object recognition · Intelligent express delivery system

1 Introduction

By the end of January 30, 2021, data released by the State Post Bureau showed that in 2020, the national postal industry is expected to achieve business revenue of 1.1 trillion yuan, including express delivery business volume and business revenue of 83 billion yuan and 875 billion yuan respectively, year-on-year growth of 30.8% and 16.7%. This shows that the express industry market demand is huge [1].

But now the express delivery is still staying in the most primitive way, there are certain problems. For example, in terms of logistics monitoring, it cannot realize automatic identification on the cloud and timely and accurately inform users of the logistics situation; in terms of the security of goods, it lacks the information record and traceability of the whole process of logistics; and in terms of the convenience of operation, it cannot realize the complete self-help of receiving and sending items in the express cabinet. In the aspect of human output, the delivery of the express cabinet still needs the Courier of each express company to manually review the delivery, which cannot realize the optimization of human resources. This paper will start from these problems, show an intelligent express delivery system based on video processing technology [2].

© Springer Nature Singapore Pte Ltd. 2021
J. Zeng et al. (Eds.): ICPCSEE 2021, CCIS 1451, pp. 370–379, 2021.
https://doi.org/10.1007/978-981-16-5940-9_28

2 Overall Framework of the System

2.1 Software Part

The intelligent express system is mainly composed of five parts: express cabinet terminal, bank server, Web client, background database server, and system management, as shown in Fig. 1. Each part is independent of each other but connected with each other through the wireless network, which has good expansibility, operability and application.

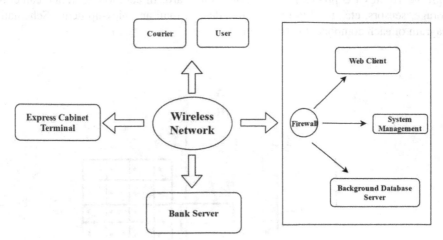

Fig. 1. General framework of the system

The "Express Cabinet Terminal" module runs the Windows system and connects to the server program through 4G or 5G network to conduct inquiry and submission operations. In each operation, the case of network disconnection is considered. The express cabinet terminal includes "intelligent mail and pick up" module and "self-service printing" module, "intelligent mail and pick up" also includes "object intelligent identification" module, "logistics information traceability and management" module, "user identity verification" module. The "Bank Server" mainly realizes the user's bank card payment and the data exchange with Alipay and other platforms. In addition to the display of information when operating on the counter, "Web Client" also has an independent website for senders, couriers and logistics companies to search information on the Internet at ordinary times. The "Background Database Server" first uses an independent desktop server, and later can be expanded into a cluster server network. It mainly realizes the communication with numerous express cabinet systems in various places, monitors the online status of each express cabinet terminal, pushes instructions and information to the express cabinet, and processes the requests and reports from the terminal. The "System Management" module mainly manages the personal user information, the terminal data management of the express cabinet, the log inquiry and so on [3].

2.2 Hardware Part

The hardware part is mainly undertaken by the intelligent terminal machine. The architecture composition of the intelligent terminal integrated machine includes: Display (3), host (1), RFID reader (5), printer (2), depth camera (4) (to complete the measurement of object volume), weighing table (8) and object cabinet (7), (6) are the packages to be sent by the customer [4]. The display, the RFID reader, the printer, the GPS/LBS module, the temperature and humidity monitor and the depth camera are respectively connected with the 16-bit CPU processing chip extension board. In addition, there are cameras, storage, sensors, etc., used to record video data of mail and pick-up items. Schematic diagram of each component is shown in Fig. 2.

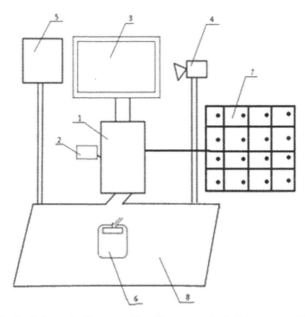

Fig. 2. Schematic diagram of intelligent terminal all in one machine

Acquired independent intellectual property rights:
Patent name: integrated intelligent terminal based on postal service.
Inventor (designer): Yanqing Wang; Kezheng Lin; Chaoxia Shi.

The intelligent terminal all-in-one machine is different from the present express cabinet. It adds a camera to record the delivery process of the consignor and uses image recognition technology to judge whether the goods can be delivered. The depth camera can identify the volume of the object, directly judge the size of the object, and pop open the corresponding cabinet door to facilitate customers to put the express. The intelligent terminal all-in-one machine is also equipped with weighing equipment, which can directly measure the weight of items and calculate the express fee, so that there is no need for the Courier to weigh them and save manpower and material resources [5].

3 Key Technologies and Implementation Process

3.1 Key Technologies

At the core of the system is a depth camera located in a smart terminal integrated machine, which will record video of the entire package process, and then upload it to the terminal for analysis, and finally identify the item [6]. Video contains a lot more information than pictures, and processing and analyzing video is becoming more and more mainstream in computer vision, and video is essentially made up of frames of image, so video processing ultimately comes down to image processing, but in video processing, there's a lot more temporal information available. The following will mainly introduce some OpenCV video processing methods used in this system [7].

For reading video frames, OpenCV provides a class VideoCapture for reading video frames. First, when defining the class, use the open () method to open a video or the default camera: "capture.open("../video. avi")". If you change the file name to Settings ID, the camera can be turned on. The default camera is 0. Then get the video frame: "capture.read(frame)". Then get the parameters of the video: "double rate = capture.get (CV_CAP_PROP_FPS)". A video has many parameters, such as frame rate, total number of frames, size, format, etc. VideoCapture's "get" method can capture many of these parameters. The set method of the VideoCapture class "set" allows us to fetch a frame at a certain location in the video. It has parameters, either by time, or by frame number, or in proportion to the length of the video. For example, get frame 100: "double position = 100.0; capture set (CV_CAP_PROP_POS_FRAMES, position)". Of course, the "set" method is only used to get the position of the video frame. You can also set the frame rate and brightness of the video.

Writing to video are similar to reading, OpenCV is implemented using the VideoWriter class. This class has several methods, which are very simple. In addition to the constructor, the "open", "IsOpen", "write", and overloaded operators " < <" are provided [8]. "VideoWriter::VideoWriter (const string & filename, int fourcc, double fps, Size frame Size, bool is Color = true);" The constructor takes the same parameters as the "open" method. The first parameter specifies the filename, the second is the encoding format, the third is the frame rate, and the fourth is the video size. Then segment the read video, intercept a picture every certain frame number, and set the frame number C. Split video into frames of images and save them in the specified location: "cv2.imwrite('out/'+STR ('%06d'%c)+'.jpg', frame)". Then, the Faster R-CNN algorithm is used for detection. Finally, the object image obtained is compared with the established model to obtain the name of the recognized object [9].

3.2 Implementation Process

The system first records the whole process of getting the package through the camera, and then decomposes the video from the intelligent terminal integrated machine into pictures frame by frame according to the requirements, and finally integrates all the pictures to identify what the package is [10]. This saves labor costs. The whole identification process is as follows (Fig. 3):

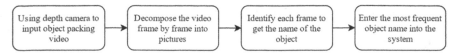

Fig. 3. Object identification process

The details of how each part is implemented are described below. First, the video recorded by the depth camera on the intelligent terminal integrated machine was decomposed frame by frame. Then we use OpenCV to provide a VideoCapture class for reading video. It then reads in the parameters of the video, such as: number of frames, total frames, size, format, etc. It then iterates through each frame from the beginning and breaks down the video. Each frame of decomposed video is treated as a picture. First, the Faster R-CNN object detection was used to determine the precise location of the express, and then the image content was identified [11] (Fig. 4).

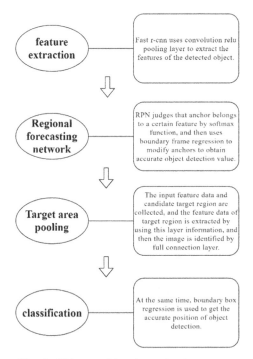

Fig. 4. Object position determination process

In the identification process, the model and model parameters are first imported. In the initial stage, faster_rcnn_inception_v2 and faster_rcnn_resnet50 can be used. The two models are the data trained on different training sets. After being widely used, you can import your own trained models. Then, OpenCV is used to build a neural network to obtain the parameters of the length and width of the picture, adjust the length and width of the picture to 300 * 300 to calculate the picture BLOB, and then the BLOB is passed into the neural network to calculate, and the forward propagation is used to predict the picture. Finally, the predicted value is iterated to extract the name of the object that set the threshold [12] (Fig. 5).

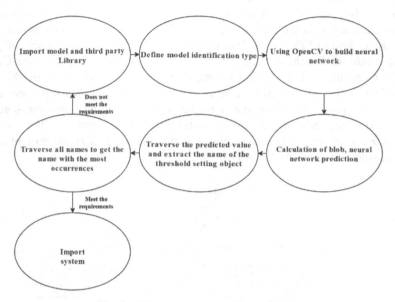

Fig. 5. Object name recognition and entry

For each frame of the picture to get the name of the object statistics, and then traversal the entire name directory, for the most occurrence of the name into the system. At the same time, for the video with too long packaging time, you can choose the image recognition every ten frames to speed up the input time. If the occurrence frequency of the name with the most frequency does not reach 90%, and the identification object name conflicts too much, it can be re-recognized, or re-recorded the video, and then a new result can be obtained and input into the system. This completes the process of automatic object recognition [13].

4   Use and Test the System

4.1   System Usage

The specific delivery process of the intelligent terminal all-in-one machine is as follows: in the postal cabinet, add a delivery button. Before putting the goods into the postal cabinet, the shipper records the video of the goods into the local storage system one by one through the camera in the all-in-one machine of the postal cabinet. The system can automatically identify the contents of the goods mailed. If the identification is wrong, it can be manually corrected, which ensures the verification process of the goods mailed and the information can be traced back. After that, the shipper will pack and weigh and record the weight through the scale in the all-in-one machine. At the same time, the all-in-one machine can record the shipping and shipping address corresponding to the mobile phone number. When you use it again, you can directly choose the same address without having to input it again. Then, the intelligent postal terminal all-in-one machine will print the mailing information, which will be pasted on the goods by the shipper and put into the container. After the Courier picks up the goods, the all-in-one machine will automatically send text messages to the mobile phone to remind the business and record the quantity of the goods. Postal personnel can collect the quantity of goods and other information through the network. When receiving the goods, the Courier will also place the goods in the postal storage cabinet, and the all-in-one machine will send SMS to inform the goods have been sent to the container. When picking up the goods, the consignee can choose to pick up the goods by sending a short message. The following process, and then the existing postal information system seamless docking. Real realization of sending goods, receiving goods unmanned intervention of intelligent express system [14] (Fig. 6) (Table 1).

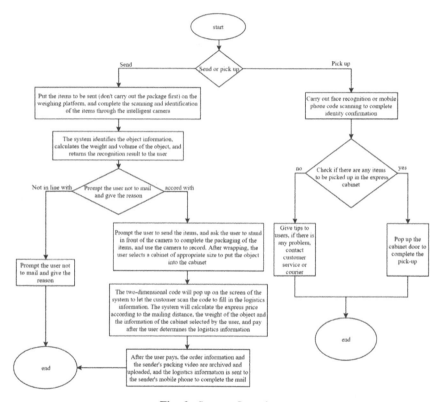

Fig. 6. System flow chart

Table 1. Common object identification

Types of goods	Gram (g)	Accurate identification rate (%)	Time needed (ms)	Can the customer send it
Cup	420.39	95.62	120	Yes
Book	240.20	99.87	90	Yes
Clothes	200.97	80.26	150	Yes
Shoe	796.64	97.63	85	Yes
Flour	120.57	92.26	95	No
Tool	530.89	85.39	130	No
Dumbbell	19998.87	82.31	113	No

4.2 Test of the System

Some items as shown in the above table were tested, such as water cups, books, clothes, shoes and other common postal items, which passed the test. Except for clothes, the recognition accuracy of other categories was higher, because there were some patterns on clothes, which might affect the recognition result. As for such items as flour, the system directly gives a warning and does not send it by mail because the eigenvalue extraction analysis only through the camera cannot simply determine whether they are drugs. Knives are dangerous, and they are not mailed directly. 10 kg dumbbells will not be mailed because the weight exceeds the upper limit [15].

5 Functional Analysis

The intelligent express delivery system is based on the image and video processing technology of OpenCV to transform the postal service and express delivery service. Using the recognition scale of video processing technology, information traceability, cloud identification, electronic receipt and intelligent escort can be realized. Get through the last step of express delivery towards intelligence. The logistics can save manpower, material resources, simplify the process of sending and receiving goods [16]. The following is a brief description of the project's functionality:

5.1 Intelligent Object Identification

This system adopts weighing table, depth camera, temperature and humidity detector and other sensors to carry out intelligent recognition of type, volume, weight and contraband of customers' objects through computer image recognition.

5.2 Information Traceability

Sensors and equipment used in this system to parcel delivery process in the whole process of records, including the video, will sign face to face process into local video, simply state qualified or not can be remote, convenient and quick, make the package complete logistics information traceability links, trace information can be used to deal with such as accountability for late problem [17].

5.3 Electronic Signature

Using the mobile phone number of both parties as the transmission carrier of identity verification, through the setting of a password or must personally sign (face recognition) conditions to achieve the purpose of not repudiation sign.

5.4 Butler Service

When the package cannot be picked up in time, the item can be stored online (for a storage period charge), or sent back to the original address or to another address (for a transfer charge).

6 System Summary

Intelligent delivery system based on OpenCV video processing technology and implementation of intelligent terminal machine through the depth camera and electronic scale for items on the scales, the sender is used to identify the object category, volume, weight and the item packaged into the ark of the video store to realize intelligent identification, traceability information integration of intelligent delivery system [18]. This article through the software and hardware two parts, the specific introduction of the system functions, and emphatically introduced the video processing technology in the application of the system. Then through testing and use to show the convenience and practicability of the system. The system will be based on the weight of the object, the required size of the cabinet, mail distance three information integrated intelligent calculation reasonable freight, to complete the intelligent automation of mail and pickup. In line with the current social contactless distribution trend at the same time, the system saves manpower, material resources, thereby increasing economic and social benefits, promote the whole express industry to the development of intelligence.

7 Thanks

This article is supported by the 2020 Innovation and Entrepreneurship Training Program for College Students in Jiangsu Province (Project name: Traceable multi-functional intelligent express cabinet, No. 201911460090P, No. 202011460090T).

This article is supported by the National Natural Science Foundation of China Youth Science Foundation project (Project name: Research on Deep Discriminant Spares Representation Learning Method for Feature Extraction, No. 61806098).

This article is supported by Scientific Research Project of Nanjing XiaoZhuang University (Project name: Multi-robot collaborative system, No. 2017NXY16).

References

1. Dong, P.: The development path of logistics express industry under the background of artificial intelligence. Mod. Econ. Inf. **05**, 407 (2019)
2. Tu, Y.: Practice and exploration of video post processing based on artificial intelligence. Inf. Record. Mater. **21**(11), 74–75 (2020)
3. Jiang, D., Shi, R., Fang, Q.: research on the construction method of 720° panoramic scene based on OpenCV and OpenGL. Survey. Mapping Spatial Geograph. Inf. **44**(02), 68–72 (2021)
4. Han, J.: Design of intelligent express container based on MCU. Electron. Technol. Softw. Eng. **05**, 84–85 (2021)
5. Zeng, W., Zhang, S., Ma, Y., et al.: Survey of image detection and recognition algorithm based on convolution neural network. J. Hebei Acad. Sci. **37**(04), 1–8 (2020)
6. Yanping, T.: Practical exploration of artificial intelligence for video post-processing. Mater. Inf. Record. **21**(11), 74–75 (2020)
7. Cheng, Y.: Teaching software of wood defect image segmentation based on OpenCV Python. Forestry Mach. Amp; Woodworking Equipment **49**(01), 36–39 (2021)
8. Zhang, Y., Bai, J., Meng, F., et al.: Application of convolution neural network in image recognition. New Technol. New Process (01), 52–55 (2021)
9. Liu, W., Liu, Y.: Moving object tracking and detection based on OpenCV. Electron. World **01**, 156–157 (2021)
10. Wang, Y., Wang, C., Shishui, Y.: Panoramic Video processing technology analysis. Res. Propag. Force **3**(22), 287 (2019)
11. Mingwei, X., Wang, D., Ye, J.: Design of port fire prevention system based on convolution neural network image recognition. J. Tianjin Polytech. Normal Univ. **29**(04), 21–24 (2019)
12. Hau, N.-X., Tran-Thai, S., Van Thinh, L., et al.: Learning spatio-temporal features to detect manipulated facial videos created by the deepfake techniques. Forensic Sci. Int. Digital Invest. **36**, 301108 (2021)
13. Duan, L., Ding, W., Shao, W., et al.: Design of intelligent monitoring system based on OpenCV. Autom. Technol. Appl. **39**(12), 34–39 (2020)
14. Zhang, X., Wu, F., Li, Z.: Application of convolutional neural network to traditional data. Expert Syst. Appl. **168**, 114185 (2021)
15. Yang, Y., He, P., Deng, L., et al.: Object recognition and weighing system based on internet of things. Commun. World **27**(01), 200 (2020)
16. Deng, X., Huang, R., He, J., et al.: Artificial intelligence: a new way of innovation and development of express industry. Netw. Secur. Technol. Appl. (11), 101–102 (2019)
17. Xie, H.: Implementation of face recognition and delivery system based on SSM framework. Inf. Commun. **02**, 83–84 (2020)
18. Jin, H., Chen, C.: Systematic analysis of express industry. Mod. Econ. Inf. (08), 387 (2019)

Speech Dictation System Based on Character Recognition

Wenjun Lu, Yanqing Wang[✉], and Longfei Huang

College of Information Engineering, Nanjing Xiaozhuang University, Nanjing 211171, China

Abstract. To solve students' dictation problems, a speech dictation system based on character recognition is proposed in this paper. The system applied off-line handwritten Chinese character recognition technology, denoised the image through Gaussian filter, segmented the text through projection method, and converted the image to text through OCR technology. The straight line mark in the picture was detected by Hough transform technology, and then SKB-FSS algorithm and WST algorithm were used for speech synthesis. Experiments show that the system can effectively assist students in dictation.

Keywords: Character recognition · Speech synthesis · Hough transform · Feature extraction · Image preprocessing

1 Introduction

In the current teaching, most teachers will leave homework related to dictation, and the parents of students are middle-aged, their work is often busy. There is no time for parents to help students with dictation. Some students' grandparent's pronunciation is not standard, it is difficult for them to be competent. Therefore, this kind of dictation system is a research hotspot.

In order to solve students' dictation problems and exercise their dictation ability, this paper presents an intelligent speech dictation system based on character recognition. The system realizes the transformation from picture text to text through OCR technology. Firstly, the weighted average method is used for image graying, and a Gaussian filter is used for noise reduction. The paper contour in the image is determined by contour detection, and then the tilt is corrected by perspective transformation. Finally, the midpoint of the gray image is traversed, and the image information is binarized to complete the image preprocessing. Through the combination of horizontal projection and vertical projection, the text in the image is segmented. HMM is used to build a language model for character recognition. Hough transform is used to detect whether there are specially marked words in the picture to meet the requirement that only part of the text information needs dictation. In speech synthesis technology, the acoustic model is established by HMM, and the recognized text information is synthesized by the straight synthesizer.

J. Zeng et al. (Eds.): ICPCSEE 2021, CCIS 1451, pp. 380–392, 2021.
https://doi.org/10.1007/978-981-16-5940-9_29

2 Application of Character Recognition and Speech Synthesis in Pedagogy

Chinese character recognition technology can be divided into printed Chinese character recognition and handwritten Chinese character recognition. Handwritten Chinese character recognition can be divided into on-line handwritten Chinese character recognition and off-line handwritten Chinese character recognition [1], as shown in Fig. 1.

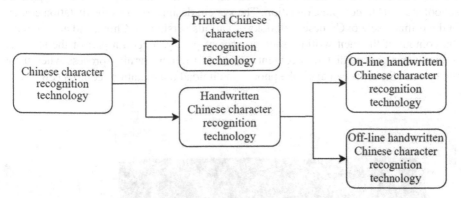

Fig. 1. Classification of Chinese character recognition

Printed Chinese character recognition technology and online handwritten Chinese character recognition technology have become mature, while off-line handwritten Chinese character recognition still has room for development. The system mainly uses off-line handwriting recognition technology. Off-line handwritten Chinese character recognition is the recognition of handwritten Chinese characters by two-dimensional images collected by scanners or cameras. This technology is widely used in the fields of automatic input of Chinese electronic data processing, Chinese text compression, office automation and computer-aided teaching, which has brought huge economic and social benefits, but it is also one of the difficulties in the field of pattern recognition [2].

Speech synthesis technology is a technology that converts text information into speech signals. Due to the unique pronunciation characteristics of Chinese, the Chinese speech synthesis system has not achieved satisfactory results [3]. Speech synthesis plays an important role in human-computer interaction. With the development of deep learning and the improvement of people's needs, speech synthesis has entered a new stage of development [4, 5].

With the continuous development of science and technology, artificial intelligence has become a key means of auxiliary teaching in the teaching classroom. It can accurately determine the quality of teachers in class, and put forward the direction of teaching and learning for the quality of the classroom, so as to provide a certain guarantee for the quality of classroom teaching. On this basis, artificial intelligence can also put forward corresponding teaching strategies, so that students can more easily find their own problems in classroom learning, and generate teaching strategies and methods suitable for students [6]. Fully integrating artificial intelligence into teaching and providing important support for education and teaching will have a more profound impact on the development of the education industry [7].

3   System Design

3.1   System Function

Content Selection Function. The system aims to meet the requirements of primary school students to complete dictation. The system design contains the dictation content in the primary school Chinese textbooks of various versions in China, and also contains the content of the tacit writing, such as ancient poetry, a certain part of the text, etc. Students can choose dictation content according to their learning process when using the system. Students can also take photos and upload the dictation required (Fig. 2).

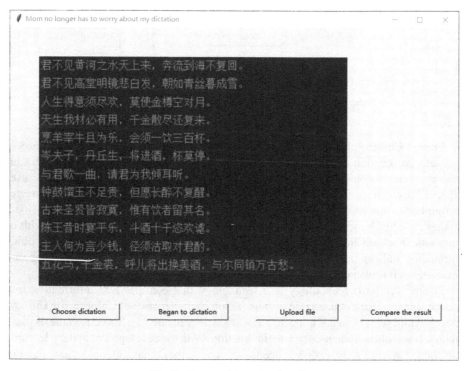

Fig. 2. Demo of dictation interface

Voice Dictation Function. The system uses standard voice to play the dictation content selected by students for dictation. Dictation speed can be adjusted in the system.

Operation Approval and Modification Function. There are two ways to dictation, one is to write on the mobile phone, the other is to write on the paper homework. If it is written on mobile phone, it will be submitted directly; if it is written on the working paper, it shall be submitted after taking photos and uploading. After students submit their dictation assignments, the system will compare the written content with the read dictation content, and rate the dictation grade according to the degree of similarity of the characters (Fig. 3).

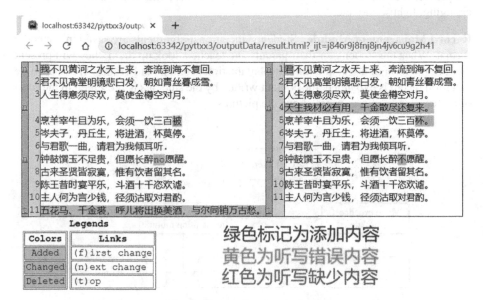

Fig. 3. Comparison result chart

Learning Record Management Function. The system can save the records and results of each dictation, which is easy to access and delete. The system can generate a text description for each dictation assignment, which can be quickly shared with parents and teachers on their mobile phones.

3.2 Main Modules of the System

Resource Library Module. During the development of the system, the dictation content in the primary school Chinese textbooks used all over the country is input in the form of words and made into resource files. According to the structure of resources, the method of selecting function or object of dictation content is developed.

Character Recognition Module. In dictation, students can choose to write on their mobile phones or on paper exercise books according to their own conditions. If you choose to write on the mobile phone, the system will turn off the fuzzy matching function of the input method to prevent cheating. When correcting the homework, the module is called for font recognition. The writing area of the mobile phone is at the bottom of the interface. Each word written is reduced to the middle of the interface and arranged in the order of broadcasting.

Voice Broadcast Module. For dictation, the system uses the function module of reading Chinese characters, not playing the recording module. In this way, the storage space of resources is greatly reduced and the operating efficiency of the system is improved. But the problems of some spelling rules of Chinese Pinyin are still unsolvable, such as tone changing, light sound, etc.

Record Management Module. Students' dictation assignments are saved in a specific format. When the students or their parents and teachers look up them, they will present them in the form of a list, and only display the time, content and grades. When a single dictation record is expanded, the content written by the students is presented in the form of pictures, and the scores are above the pictures.

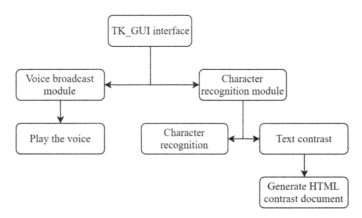

Fig. 4. System module call diagram

After decomposing the functions of each part of the system, it mainly includes two modules: character recognition of pictures and character to speech broadcast. When users are required to upload pictures, the system will recognize the pictures as text, and then output the text in the form of sound, that is, the user starts dictation, uploads the content of the user's dictation again, and uses the character recognition module to compare the two texts. The output format of the comparison result is an HTML document, as shown in Fig. 4.

The specific process is: first select the content to be dictated to upload the picture, and then convert the array read into image form through the image preprocessing operation, then extract the text information of the picture. Generally, the image is binary first, and then the picture is digitally morphological (mainly on operation). Because of the pictures identified by the built-in function of pytesseract, it is image form, not multidimensional array form in OpenCV. So before recognition, we need to use the image function in pil to transform picture format, and then identify it through functions in pytesserar.

After successful recognition, save the text, and then start dictation and voice broadcast. The program calls pyttsx3. init() factory function to get the reference to pyttsx3.

Engine instance. During the build, the engine initializes a pyttsx3 driver. Object, which is responsible for loading the voice engine driver implementation from the pyttsx3. Drivers module. After the build is complete, the application uses engine objects to register and unregister event callbacks.

After dictation, the content will be photographed and uploaded, then the text recognition will be carried out again, and the text will be saved. The difference between the two texts is compared by using Difflib of Python standard library. The HTML document marking error content will be generated from the comparison results for users to view. The specific system flow is shown in Fig. 5, and the main technical flow is shown in Fig. 6.

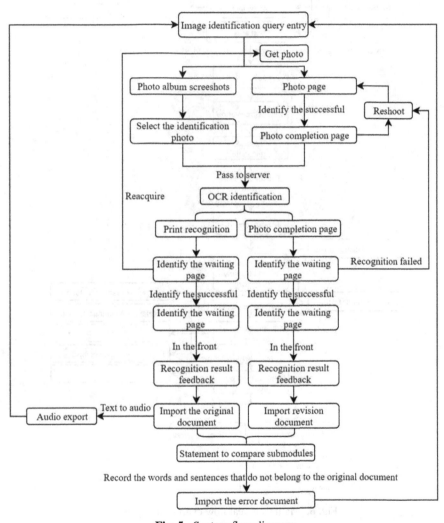

Fig. 5. System flow diagram

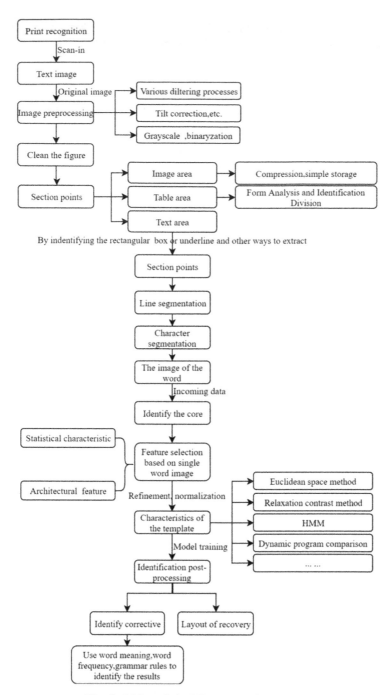

Fig. 6. Main technical flow chart of the system

4 System Key Technologies

4.1 Off-Line Handwritten Chinese Character Recognition Technology

In recent years, handwriting recognition technology has developed rapidly and become mature. Handwriting recognition technology is mainly divided into four stages: image preprocessing, feature extraction, classification recognition and post-processing [1, 2, 8].

Due to the complexity of character segmentation and feature extraction in offline handwritten recognition, the first thing to do is to preprocess the image in order to improve the accuracy of font recognition. Image preprocessing mainly includes the removal of paper background, graying, denoising and binarization of handwriting image, tilt correction, punctuation removal, Chinese character segmentation, etc. The commonly used methods for image graying are the maximum value method, average value method and weighted average method. The system uses the weighted-average method for image graying, the formula is

$$GRAY = 0.114 * B + 0.587 * G + 0.299 * R$$

Gaussian filter is used to denoise the image, and a threshold based on OpenCV is used to binarize the image. The image binarization effect is shown in Fig. 7 (a) (b). After graying and binarization, the horizontal projection is used to obtain the horizontal segmentation position, and then the vertical projection is used to obtain the vertical segmentation position, so as to determine the character segmentation position. The image segmentation effect is shown in Fig. 7 (c) (d).

By detecting the inner border of the image or extracting the special color, we can process the special mark of the text in the image. After image graying, the rectangle's kernel detection level line and kernel detection vertical line is defined according to the length and width of the image. After defining the kernel, the rectangle's kernel detection level line and kernel detection vertical line are detected according to the shape control operation. After adding the horizontal line and vertical line, the inner frame of the image is detected. Special color extraction is realized by OpenCV. Firstly, the image is transferred from BGR space to HSV space, the special color to be extracted is defined in HSV space, the background in the image is removed by setting threshold, then the original image and mask are bitwise combined to separate the color to be extracted, and finally, the BGR space image is converted into RGB image. The effect of special tag extraction is shown in Fig. 7 (e) (f).

The main features used in feature extraction can be divided into two categories: transform coefficient features and structural features. The method of feature extraction is mainly based on the statistical features of structural features, which combines structural features and statistical features [1, 2]. The classifiers used in classification and recognition can be divided into the single classifier and multi classifier ensemble, among which the most used one is multi classifier ensemble. At present, the post-processing technology mainly uses the hidden Markov method (HMM) for post-processing. First, the hidden Markov method (HMM) is used to establish the language model, then the text to be recognized is passed through the recognizer to get the candidate word array, and finally, the candidate words are sent to the language model for post-processing [2].

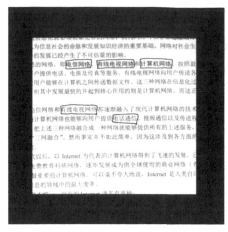

(a) Before image binarization   (b) After image binarization

test001.png   test002.png   test003.png   test004.png

(c) Image before segmentation   (d) Results after segmentation

(e) Before text mark processing

(f) After text mark processing

Fig. 7. Image preprocessing rendering

The system uses OCR (optical character recognition), an off-line handwritten character recognition technology, and deep learning technology based on CNN neural network, which makes the character recognition more efficient.

4.2 Hough Transform Technique

Hough transform is an effective method for line detection of binary images in the field of pattern recognition [12, 13]. The simplest Hough transform is to recognize the straight line in the image. The system mainly uses the detection of the straight line in the Hough transform. The basic idea of the Hough transform is [13, 14]: a point in the original image coordinate system corresponds to a straight line in the parameter coordinate system, and a straight line in the same parameter coordinate system corresponds to a point in the original coordinate system. All the points of the straight line in the original coordinate system have the same slope and intercept, so they correspond to the same point in the parameter coordinate system. In this way, after each point in the original coordinate system is projected to the parameter coordinate system, it is judged whether there are aggregation points in the parameter coordinate system, and such aggregation points correspond to the lines in the original coordinate system.

A straight line can be expressed by $y = kx + b$ in the rectangular coordinate system. The main idea of Hough transform is to exchange the parameters and variables of the equation, that is, x, y as the known quantity; k, b is the variable coordinate, so the line $y = kx + b$ in the rectangular coordinate system is represented as a point (k, b), and a point (x_1, y_1) is represented as a straight line $y_1 = x_1 \cdot k + b$ in the rectangular coordinate system, where (k, b) Is any point on the line. In order to calculate conveniently, the coordinate of parameter space is expressed as ρ and θ [14, 15] in polar coordinates. Since the points on the same line correspond to the same (ρ, θ), the edge detection of the picture can be performed first, and then every non-zero pixel point on the image can be transformed into a straight line in the parameter coordinate. Then, the points belonging to the same line in the rectangular coordinate form multiple lines in the parameter space and intersect at one point.

As shown in Fig. 8, any point (x, y) in the original graph can form a straight line in the parameter space. Taking a straight line in the graph as an example, there is a parameter $(\rho, \theta) = (69.641, 30°)$. All the points belonging to the same line will intersect at a point in the parameter space, which is the parameter of the corresponding line. A series of corresponding curves are obtained in the parameter space from (ρ, θ) obtained from all the lines in the graph, as shown in Fig. 9. The system uses the Hough transform technology to detect the lines in the image, and checks whether the adjacent lines can be

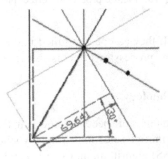

Fig. 8. Results of parameter space transformation

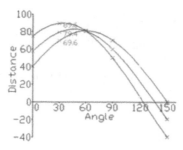

Fig. 9. Results of Hough statistical transformation

(a) original image

(b) picture processing process

(c) picture processing results

Fig. 10. Picture processing effect picture

combined into a rectangle. If they can be combined into a rectangle, the rectangle text can be segmented. The image processing effect is shown in Fig. 10 (a) (b) (c).

4.3 Speech Synthesis Technology

The main application technologies of speech synthesis in this system are feature extraction, acoustic model and primitive selection, frame synchronous search based on statistical knowledge (SKB-FSS) algorithm, and word search tree (WST) constrained by morphology.

In the aspect of feature extraction, acoustic model and primitive selection, it is assumed that the autoregressive coefficient and cepstrum coefficient of speech are independent of each other. The modified hidden Markov model [16] center distance continuous probability model is used to model the speech, and the straight synthesizer is used to synthesize the speech. In terms of primitive selection, the smaller the recognition unit, the greater the flexibility, but the more difficult the database calibration, the worse the recognition performance; and the larger the recognition unit, the smaller the flexibility, but the better the recognition effect. The system chooses syllables as a compromise choice of recognition unit, and the effect is good [17]. The frame synchronous search algorithm of statistical knowledge [17] uses the search algorithm combining the information of state dwell distribution (SDD) and the information of differential state dwell distribution (DSDD) between adjacent states to improve the search accuracy and search efficiency. Word search tree constrained by morphology is an algorithm for searching across primitives. Compared with linear search, this algorithm has higher search efficiency and space efficiency [17].

4.4 System Development Environment and Deployment

Development and debugging tools include Pycharm, Anaconda3, Python 3.6 and related dependency packages.

The system is currently deployed in the Lenovo XiaoXin Air15IEL notebook. The operating system uses Windows 10 version, Intel Core i5 CPU, 8GB memory and 256GB solid-state drive.

5 Conclusion

This paper mainly introduces an intelligent speech dictation system based on character recognition, which solves the dictation problem of primary school students through Chinese character recognition technology and speech synthesis technology. Off-line handwritten Chinese character recognition technology realizes the conversion from picture to text through Gaussian filter and OCR technology, and then uses SKB-FSS algorithm and WST algorithm to synthesize the extracted text, so as to improve students' dictation ability and make learning efficient and intelligent. At present, the off-line handwritten Chinese character recognition technology and speech synthesis technology are developing rapidly, and the system will also make continuous progress to bring better service to users.

Acknowledgements. This article is supported by the 2020 Innovation and Entrepreneurship Training Program for College Students in Jiangsu Province (Project name: Mom doesn't have to worry about my dictation any more—dictation software based on character recognition, No. 202011460104T).

This article is supported by the National Natural Science Foundation of China Youth Science Foundation project (Project name: Research on Deep Discriminant Spares Representation Learning Method for Feature Extraction, No. 61806098).

This article is supported by Scientific Research Project of Nanjing Xiaozhuang University (Project name: Multi-robot collaborative system, No. 2017NXY16).

References

1. Zhao, J., Zheng, R., Wu, B., et al.: A survey of off-line handwritten Chinese character recognition. Acta Electronica Sinca **38**(02), 405–415 (2010)
2. He, Z., Cao, Y.: A survey of off-line handwritten Chinese character recognition. Comput. Eng. **15**, 201–204 (2008)
3. Jing, X., Luo, F., Wang, Y.: A survey of Chinese speech synthesis technology. Comput. Sci. **39**(S3), 386–390 (2012)
4. Xhang, X., Xie, J., Luo, J., et al.: Research on deep learning speech synthesis technology. Comput. Age **9**, 24–28 (2020)
5. Zhang, B., Quan, C., Ren, F.: Speech synthesis methods and development. J. Chin. Comput. Syst. **37**(1), 186–192 (2016)
6. Li, Y.: Application analysis of artificial intelligence technology in the field of education. Theoret. Res. Pract. Innov. Entrep. **3**(06), 150–151 (2020)
7. Sun, L., Ma, Y.: A review on the application of deep learning education. Inf. Technol. Educ. China **17**, 98–101 (2019)
8. Ibrahim, M., Zhang, H., Liu, C., et al.: On-line handwritten Uyghur letter recognition method. Pattern Recogn. Artif. Intell. **25**(06), 979–986 (2012)
9. Gao, Y., Yang, Y.: A survey of off-line handwritten Chinese character recognition. Comput. Eng. Appl. **7**, 74–77 (2004)
10. Bai, B., Zhang, W.: Research on preprocessing algorithm of Chinese handwriting identification. Comput. Eng. Des. **30**(22), 5189–5191 (2009)
11. Zhang, Y., Li, E.: Research on preprocessing algorithm in off-line handwritten Chinese character recognition. J. Shenyang Univ. Technol. **06**, 534–537 (1999)
12. Chen, Z., Gao, M., Yang, S.: Line tracking method based on Hough transform. Comput. Appl. **23**(10), 30–32 (2003)
13. Sun, F., Liu, J.: Fast Hough transform algorithm. J. Comput. Sci. **10**, 1102–1109 (2001)
14. Qiu, S., Xia, Y.: A fast Hough transform algorithm. Comput. Eng. **02**, 148–150 (2004)
15. Tang, J., Wang, Z., Zhang, X.: The technology of line detection based on Hough transform. Sci. Technol. Inf. **14**, 33–35 (2011)
16. Yijian, W., Wang, R.: HMM based trainable Chinese speech synthesis. J. Chin. Inf. Processing **04**, 75–81 (2006)
17. Zheng, F., Mou, X., Xu, M., et al.: Research and implementation of Chinese speech dictator technology. J. Softw. **4**, 101–109 (1999)

Human Body Pose Recognition System Based on Teaching Interaction

Kaiyan Zhou, Yanqing Wang(✉), and Yongquan Li

School of Information Engineering, Nanjing Xiaozhuang University, NanJing 211171, China

Abstract. In view of the problems of the high time cost and low accuracy of manual supervision in traditional classroom teaching, this paper proposes a human body pose recognition system based on teaching interaction. The enhanced basic network (ResNext-101 + FPN) was used in Mask R-CNN to extract the features of the input images. Then based on the behavior analysis algorithm and face detection data, the behavior data of each student in the classroom were obtained. Moreover, the behavior data were applied to support multi-dimensional visualization. The experimental results show that the system can timely and effectively reflect the learning status of students, and help teachers accurately grasp the classroom learning state of students, so as to adjust teaching strategies in a targeted way and help improve the quality of teaching.

Keywords: Deep learning · Mask R-CNN · Body pose recognition · Face recognition · Visual presentation

1 Introduction

Classroom behavior is an important reference factor for evaluating students' classroom concentration and teachers' teaching quality [1]. Students' classroom behavior indirectly reflects students' learning state and learning efficiency.

In 2017, the State Council issued the Notice on the Development Plan of the New Generation of Artificial Intelligence, which clearly stated: "Build a new education model system that includes intelligent learning and interactive learning, and promote the application of artificial intelligence in teaching, management, resource construction and other whole processes." In 2019, the Beijing Municipal Education Commission issued the Beijing Action Plan to Promote the Integrated Development of Artificial Intelligence and Education, which clarified the main tasks of the integrated development of artificial intelligence and education in basic education, vocational education and higher education. On the one hand, studying students' classroom behaviors in the context of informatization and intelligentization of education can help teachers accurately grasp students' classroom learning, reduce the burden of class evaluation, and help teachers improve teaching methods and adjust teaching strategies. On the other hand, it provides a channel for students to understand their learning behavior and learning state during class, and urges students to reflect on it and adjust their own state [2].

© Springer Nature Singapore Pte Ltd. 2021
J. Zeng et al. (Eds.): ICPCSEE 2021, CCIS 1451, pp. 393–405, 2021.
https://doi.org/10.1007/978-981-16-5940-9_30

To this, this paper proposes a human body pose recognition system based on teaching interaction, which combines artificial intelligence with education and teaching research. Through combing the academic history of classroom behavior research at home and abroad, as well as the mainstream literature database at home and abroad related research information dynamic that in video images for human to human body gesture recognition and supervision, the human body target accurate segmentation and recognition is the premise and basis [3]. When the camera captures the classroom image, there may be situations in which the students are in close contact with each other and the students have a similar appearance. Such factors for the most part increased the difficulty of segmentation and recognition of target the human body, and directly affects the accuracy and efficiency of target body segmentation and feature extraction. On the other hand, most of the data obtained from the current classroom behavior analysis research only stay on the results of pose recognition, and there is almost no further analysis and guidance for education and teaching by using the multi-dimensional visualization analysis of the data.

With the improvement of new technology, more and more researchers begin to use computer vision, deep learning and other technologies to study body pose recognition. Zhang Hongyu, [4] through the scenarios used to access equipment in the classroom, realize the human bone the extraction of feature vector, and with the help of support vector machine classifier feature vector classification and recognition, but this method need to set up special equipment in the classroom and the use of traditional machine learning method for classification of attitude and behavior of the students, Makes the operation more complex and the accuracy is not high; Cao Z et al. [5] proposed a method to obtain human body pose based on the confidence and affinity of human skeleton key points. Although human body pose of single or multiple people in vedio images can be obtained quickly and accurately, these postures are not given specific guidance from the educational level. In addition, this method is limited to the training data set, resulting the results can not be directly used to identify students' classroom behavior.

Therefore, in order to improve the accuracy of the information of human body pose and try to provide more comprehensive and objective reference guidance for education and teaching, this paper adopts the technology of multi-feature fusion and deep learning, through the Mask R-CNN (Mask Region – based Convolutional Neural Network) [6] algorithm for each frame of video image feature extraction, mainly by extracting skeleton and the direction of the key information to establish position vector Angle, and combining the face information were collected on students' classroom behavior to make decision. Then, the Python based data analysis tool - Pandas, to achieve a multi-dimensional visual display of classroom behavior data, for classroom teaching evaluation and management to provide a higher reference basis, so as to better serve the field of education.

2   Overall Design of Human Body Pose Recognition System

The overall structural framework of the body posture recognition system proposed in this paper is shown in Fig. 1:

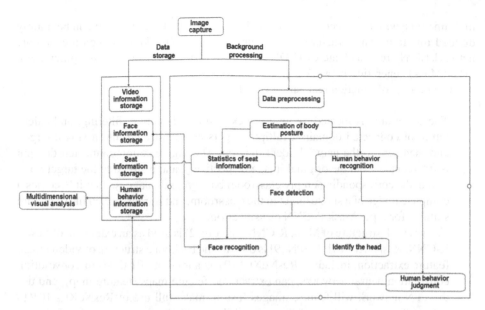

Fig. 1. General design of human body pose recognition system

The human body pose recognition system is mainly composed of data storage and background processing, in which the background processing part is composed of two main modules: human behavior recognition and face detection. The system is trained and tested on the deep learning development framework, and through the human body key point detection algorithm of Mask R-CNN and face feature points extraction (ASM) [7], the human body information and face image in the space are transformed into quantifiable data information through the camera. Combined with the above algorithm is used to identify the corresponding, the background system can integrate and classify the identified behavior posture, and then give it a weight and count the number of statistical action produced. Based on the above basis behavior data, diversified data presentation can be realized (such as: personal attention can be obtained by analyzing the number of times students make the "head up" movement, using part of the attitude determination results for determining the students' level of fatigue, using identification to each student classroom behavior attitude get class data statistics, and then come to the class overall behavior trend, etc.). The standardized, scientific and institutionalized teaching quality monitoring and evaluation system can be established to ensure the improvement of teaching quality and provide effective reference for the classroom teaching quality evaluation system, so as to provide more possibilites for making up for the existing teaching evaluation problems.

3   Human Body Pose Recognition

The body pose recognition function of this system is realized by using Mask R-CNN, which is a deep learning framework that can segment and recognize multiple targets at the same time with high accuracy by using labeled images for training. It can be mainly divided into four main structures: backbone architecture(Backbone), region candidate network (RPN), region of interest (ROI) classifier and border regressor, and segmentation mask(Mask) generation network [8].

The process of function realization is as follows:

(1) The target area is located by image background subtraction, filtering, and calculation of connected domain. Multiple targets are segmented into a single target, and then sent to the network identification. The segmentation connected domain is intercepted from the original image respectively, and restored to the target position in the corresponding empty classroom background image, so that it becomes a complete image of a single target in the classroom, and is transformed into a picture suitable for input Mask R-CNN classification.

(2) The network structure of Mask R-CNN (see Fig. 2) is used as an extension of Faster R-CNN. ResNeXt - 101 + FPN [9] is used as the backbone structure of video image feature extraction, including ResNeXt - 101 is a total of 101 times of convolution convolution neural networks, can extract and feature maps (feature map), and use the FPN network will feature images fusion, make full use of ResNeXt - 101 to extract the characteristics of the network layers, the main architecture of feature extracting figure input to regional candidate networks (Region Proposal Network, RPN) [10]. The regional candidate network is divided into two categories: target body and background. A region containing target body can be obtained through the regional candidate network, which will be pooled into a fixed-size feature map in Roi Align, and then input into two branches respectively. One branch network uses ROI classifier and Bounding-box regression for target body recognition. Another branch network is constructed by convolution (fully convolutional networks, FCN) consisting of segmentation Mask to generate network, the network will produce

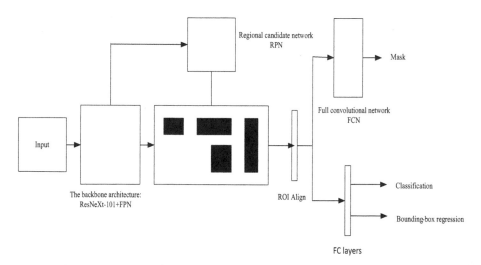

Fig. 2. Structure diagram of Mask R-CNN

consistent with the target body posture of the Mask of target image segmentation, will eventually identification combined with the results, to get to a target of human body and human body posture consistent segmentation Mask images.

(3) Extract the coordinates of key points: extract the information of human key points predicted by the network from the function. If it is the ideal state, 11 groups of key point information $\{\overrightarrow{AA'}, \overrightarrow{AB}, \overrightarrow{AB'}, \overrightarrow{BC}, \overrightarrow{B'C'}, \overrightarrow{CD}, \overrightarrow{C'D'}, \overrightarrow{EF}, \overrightarrow{E'F'}, \overrightarrow{FG}, \overrightarrow{F'G'}\}$, will be obtained successively by the system (see Fig. 3). Then, neck vectors $\overrightarrow{AA'}$, left shoulder vectors $\overrightarrow{AB}$ right shoulder vectors $\overrightarrow{AB'}$, left arm vectors $\overrightarrow{BC}$ right arm vectors $\overrightarrow{B'C'}$, left hand vectors $\overrightarrow{CD}$, right hand vectors $\overrightarrow{C'D'}$, left leg vectors $\overrightarrow{EF}$, right leg vectors $\overrightarrow{E'F'}$, left foot vectors $\overrightarrow{FG}$ and right foot vectors $\overrightarrow{F'G'}$ were established respectively.

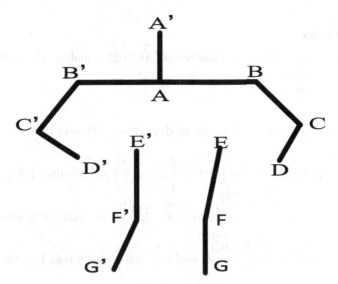

Fig. 3. Key points labeling of human body

Finally, according to the extracted Angle and direction and other information, design conditions to determine human movements, and take the determination process of "Head Up" movement as an example to illustrate (see Fig. 4):
The sample data of this action is:

[[601, 95], [601, 382], [601, 382], [786, 382], [416, 382], [832, 620], [323, 573], [740, 620], [508, 620], [832, 429], [416, 382]]

[[601, 382], [786, 382], [416, 382], [832, 620], [323, 573], [832, 620], [323, 620], [832, 429], [416, 382], [786, 620], [323, 620]]

Fig. 4. "Head Up" movement detection

1) Parameters involved

   $\overrightarrow{AA'}$ (neck vector), $\overrightarrow{AB}$ (left shoulder vector), $\overrightarrow{AB'}$ (right shoulder vector), Human_face
   (number of faces)

2) Implementation method

When face information can be collected, And when Human face < 1, if $\left|\overrightarrow{AA'}\right| >$ $\dfrac{2\left(\left|\overrightarrow{AB}\right|+\left|\overrightarrow{AB'}\right|\right)}{3}$, and $\cos\left\langle\overrightarrow{AA'},\ \overrightarrow{AB}\right\rangle$ and $\cos\left\langle\overrightarrow{AA'},\ \overrightarrow{AB'}\right\rangle$ are less than 1/2, judged to be looked up; If $\left|\overrightarrow{AA'}\right| > \dfrac{2\left(\left|\overrightarrow{AB}\right|+\left|\overrightarrow{AB'}\right|\right)}{3}$, $\cos\left\langle\overrightarrow{AA'},\ \overrightarrow{AB}\right\rangle$ greater than 1/2, it is considered to be left head. If $\left|\overrightarrow{AA'}\right| > \dfrac{2\left(\left|\overrightarrow{AB}\right|+\left|\overrightarrow{AB'}\right|\right)}{3}$, $\cos\left\langle\overrightarrow{AA'},\ \overrightarrow{AB'}\right\rangle$ greater than 1/2, it is considered to be right head;

When the face information can not be collected, the key point coordinates of the shoulders and neck of the human body are extracted, the corresponding vector is established, and the length of the shoulder and neck segment is calculated. If the length of the neck is greater than two-thirds of the sum of the shoulders, and the Angle of the neck vector and shoulders is greater than 70°, the head can be determined; If the Angle between the neck and the left shoulder is less than 60°, it is judged to look up in the left direction. If the Angle between the neck and the right shoulder is less than 60°, the head is judged to be headed to the right.

3.1 Face Detection

The study fatigue degree of students will have a direct impact on the class efficiency of students, so this system will also include the fatigue degree of students in the partial posture judgment of students. Under normal circumstances, fatigue states are generally reflected in facial states such as "Yawn" and "Blink", and these states will be maintained for a certain period of time. This system will be analyzed from the opening and closing degree of students' eyes and mouth.

For the detection of "Blink" posture, the system will use the face detection method of 68 key points to extract the key points of left and right eyes in the dictionary, and use the six-point detection method of human eyes to analyze the distance between upper and lower eyelids. When less than a certain threshold, continuous frame judgment is carried out;

For the detection of "Yawn" posture, this system will mainly adopt the double threshold detection method, which combines the mouth opening degree and mouth opening time to detect the internal contour.

Next, this paper takes the detection process of "Yawn" pose as an example to explain the relevant functions of the system in detail:

Firstly, the system uses ASM technology to locate the mouth area, and takes the range within the mouth as the input image. At the same time, the input RGB image is transformed into grayscale image, so as to reduce the image processing time. Then the image is enhanced and the gray value of the image is adjusted at the same time until the gray value between the mouth area and the background exceeds the threshold. Then binarization processing was performed on the image of the mouth area. Morphological reconstruction technology [11] was used to eliminate the influence of noise and fill other blank areas. Last mouth area is extracted height and black and white pixels than near the mouth, is used to evaluate the extent of the mouth, the mouth of the black and white pixels in the vicinity of open distance and the ratio of the must meet the following three basic conditions are short of one cannot at the same time, system will process to determine the attitude "Yawn", and at the back of the frame in the video repeat the following steps:

(1) The ratio of the number of black pixels in the current frame to the number of black pixels in the reference frame is greater than the set threshold A;
(2) The ratio of the number of black pixels in the mouth area of the current frame to the number of white pixels near the mouth is greater than the set threshold B;
(3) The height of the mouth should be greater than the set threshold value C.

In short, the main work of this system is by calling the Dlib library for face recognition and key point detection, using Dlib existing training model for 68 feature points extraction and localization of human face [12], mouth and facial signs index, through OpenCv to grayscale video streaming, detect the location information of the mouth, and set up "the aspect ratio of the mouth" "Yawn" gesture recognition, using threshold method to realize state judgment of "Yawn" (see Fig. 5).

Fig. 5. Detection of "Yawn" movement

Mouth Aspect Ratio (MAR) of this facial action sample was 1.31.

(1) In involved parameters: MAR
(2) Implementation method: When MAR value in facial recognition is greater than the threshold value and is maintained at more than 5 frames, it can be determined as yawning; Then when MAR value is less than the threshold value, the judgment ends and the number of yawning increases. To sum up, after obtaining the feature points of students' mouths, the system will judge the opening and closing state and opening time of students' mouths in combination, and use a large number of test sets to obtain the accuracy under different methods and compare them. If the number of "Yawn" gesture is more than 15 times, the student is considered to be sleepy. Finally, the system will generate a visual analysis and display of learning fatigue degree of class students, which can be used by the school and teachers to analyze the teaching situation and adjust the teaching plan.

4 Analysis of Experimental Results

The experimental environment of the class student behavior analysis algorithm based on Mask R-CNN is: video surveillance camera is used in the hardware environment; The software environment uses the deep learning experimental platform, the CPU is I7-5830K, the memory is 128G, the GPU is GTX1080, the framework is Tensorflow1.4, and the language used is Python3.6.

4.1 Production of Data Sets

The experimental video of a primary school classroom was collected with a total of 30 students. Each student was individually filmed for 5–6 min, and the students were asked to send out some common actions in class respectively. Each action lasted for about 60 s. 25 frames of images were randomly selected from each behavior pattern of each shooting Angle, totaling about 9000 images, to form the training set. Similarly, we get 360 verification sets and 360 test sets. In order to avoid over-fitting and large generalization errors, data enhancement was conducted in two ways:

(1) 30 students' various classroom behavior postures were collected in the experiment to increase the information contained in the experimental data;
(2) Horizontal flip to generate images, making the database 2 times larger. In order to reduce the correlation between training samples, the images of each student's three behavior patterns were randomly scrambled and emitted, and labels were generated to form the training data set.

4.2 Experimental Results

Every 5 s or so random collection of students in class teaching video a frame of image behavior recognition and face detection, for the main parts of the human body, the system to achieve the "Bow", "Head Up", "Chin", "Lie On the Table", "Yawn", "Blink" and other common movements in the classroom process detection. Part the result of the experiment ("Chin") as shown in Fig. 6, implementation method is: remove the two rival and neck point coordinates, the vertical distance calculation of hand ends to the neck line, if the distance is less than a quarter of the shoulder length, and keep more than 3 s, are judged to be "Chin" hand gesture.

Fig. 6. Detection results of "Chin" movement

Then, the system carries out multi-dimensional visualization analysis of the data based on the above basic attitude data. In this paper, the generation of "the overall behavior trend chart of the class" is taken as an example (see Fig. 7).

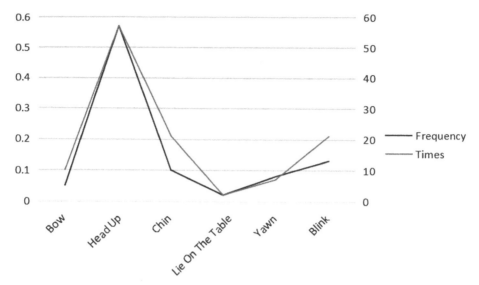

Fig. 7. The overall behavior trend of the class

Through many experiments, the recognition accuracy of the system for "Lie On The Table", "Head Up" and "Bow" reaches 100%; The recognition rate of "Yawn" and "Blink" was 98%. The recognition rate of "Chin" movement was 95%. The analysis of the data shows that the error results are caused by the fact that the subjects' working range is too small, resulting in the posture not reaching the specified threshold value. Through training, such errors can be avoided and the correct rate can reach 99.99%. The above results show that the human body pose recognition method based on Mask R-CNN adopted in this paper has a high recognition rate. And the multi-dimensional visual data analysis provided by this system is a channel of self-supervision and detection for students themselves. For teachers, for its teaching quality evaluation can provide a quantitative standard, so as to better help the students in the school and teachers to learn the classroom focus and improve and adjust the teaching plan, so as to realize the process of a two-way supervision, has a two-way promoting role to teaching, the experimental results verify the feasibility and reliability of the system, presented in this paper, indicates that the system has the very good application value in the class.

5   Improvement Methods

In the introduction part of this paper, some shortcomings of traditional attitude recognition algorithms are mentioned, such as low recognition accuracy, limited data set application, slow network convergence and so on. Therefore, this paper proposes corresponding improved methods to solve these problems. Deep learning algorithm based on Mask R-CNN is used to extract key points, form corresponding skeleton structure, establish corresponding vector sequence, and realize human behavior detection, which further improves the accuracy and speed of testing.

Mask R-CNN is a multi-task neural network based on Faster R-CNN. Compared with the latter, Mask R-CNN adds a segmentation Mask used for candidate box prediction in the final frame prediction and classification, which eliminates the phenomenon of inter-class competition. The original RoI Pooling algorithm for feature extraction of candidate boxes is upgraded to RoI Align algorithm, that is to say, the two quantization operations of border regression are canceled, instead, the bilinear interpolation method is used to obtain the pixel coordinates with floating point values, so as to reduce the deviation and avoid the occurrence of mismatching.

In addition, the experimental results show that the training condition using Mask R-CNN network is better, and the accuracy of the test is also at a high level. In order to directly represent the advantages and disadvantages of Mask R-CNN algorithm and other algorithms, this paper lists their performance and characteristics under different circumstances(see in Table 1).

Table 1. Algorithm comparison table

Algorithm	R-CNN	Faster R-CNN	Mask R-CNN
Attitude recognition test	The effect of feature extraction is good	The classification effect is very good with high accuracy	The pixel level detection results can be obtained with higher accuracy
Training time and space	Training and testing are slow and take up a lot of disk space	Test speed is still no qualitative leap, and the training space required	Fast training time
Network training effect	Real-time detection cannot be achieved	Can not meet the real-time application, no real implementation of end-to-end training testing	It can realize real-time detection, and the training is simple and convenient to extend to the human pose

As can be seen from Table 1, Mask R-CNN network is more suitable for human posture recognition module. When video and image information is obtained, Mask R-CNN algorithm can realize real-time detection of targets. Compared with other algorithms, the recognition accuracy is greatly improved, which further improves the system's smoothness and stability.

6 Conclusion

This paper proposes and implements a monitoring in place, high accuracy, easy operation based on the teaching interactive human gesture recognition system, the system adopts the Mask R-CNN points of human body detection algorithm and face the key location to obtain the key body and facial information, compared with the traditional target detection algorithm, this method effectively solves the interference factors in a wide range of application scenarios and instability, and the traditional detection algorithm is limited to a single test, etc.; This system realizes objective and accurate statistics, analysis and evaluation of classroom student behavior through multi-dimensional visualization analysis of basic data, which is beneficial to standardize the classroom behavior of teachers and students, create a good classroom learning atmosphere, and also provides a high reference basis for teaching evaluation and management.

Acknowledgement. This parper is supported by the 2019 Innovation and Entrepreneurship Training Program for College Students in Jiangsu Province (Project name: Human posture recognition based on teaching interaction, No. 201911460042Y).

This parper is supported by the National Natural Science Foundation of China Youth Science Foundation project (Project name: Research on Deep Discriminant Spares Representation Learning Method for Feature Extraction, No. 61806098).

This parper is supported by Scientific Research Project of Nanjing Xiaozhuang University (Project name: Multi-robot collaborative system, No. 2017NXY16).

References

1. Jun, Z., Bingjiang, G.: Research on classroom behavior detection based on SSD algorithm. Comput. Knowl. Technol. **16**(34), 212–214 (2020)
2. Xiuling, H., Fan, Y., Zengzhao, C., Jing, F., Yangyang, L.: Modern Educational Technology **30**(11), 105–112 (2020)
3. Yingna, D., Hong, Z., Wei, L., Huifang, Q.: A crowd target segmentation method based on attitude model. Comput. Eng. **36**(07), 195–197 (2010)
4. Zhang, H.Y.: Design and Implementation of Classroom Learning Behavior Measurement System, pp. 6–39. Huazhong University of Science and Technology, Wuhan (2016)
5. Zhe, C., Simon, T., Wei, S.E., et al.: Realtime multi-person 2D pose estimation using part affinity fields. In: 2017 IEEE Conference on Computer Vision and Pattern Recognition (CVPR). IEEE (2017)
6. Yaole, W., Linyan, L., Xinru, S., Fuyuan, H.: An improved mask-RCNN feature fusion instance segmentation method. Comput. Appl. Softw. **36**(10), 130–133 (2019)
7. Wang, S., Zhao, L., He, W., Li, F.: Detection of train driver's eye state based on improved ASM algorithm. Transsens. Microsyst. **38**(05), 129–132 (2019)
8. Zhong, W., Liu, X., Yang, K., Li, F.: Multi-objective blade segmentation and recognition based on mask-RCNN in complex background. Acta Agriculturae Zhejiangensis **32**(11), 2059–2066 (2020)
9. Lin, T.Y., Dollar, P., Girshick, R., et al.: Feature pyramid networks for object detection. In: 2017 IEEE Conference on Computer Vision and Pattern Recognition (CVPR). IEEE Computer Society (2017)

10. Fan, Z., Hongyuan, W., Ji, Z.: An improved pedestrian fine-grained detection algorithm based on mask R-CNN. Comput. Appl. **39**(11), 3210–3215 (2019)
11. Fenghui, L.: Image noise processing based on mathematical morphology. Inf. Technol.**30**(6), 45–46,142 (2006)
12. Kazemi, V., Sullivan, J.: One millisecond face alignment with an ensemble of regression trees. In: 2014 IEEE Conference on Computer Vision and Pattern Recognition (CVPR). IEEE, pp. 1867–1874 (2014)

Camera Calibration Method Based on Self-made 3D Target

Yanyu Liu[1(✉)] and Zhibo Chen[2]

[1] Hebei University of Economics and Business, Shijiazhuang 050061, China
[2] Beijing Forestry University, Beijing 100083, China

Abstract. A camera calibration algorithm based on self-made target is proposed in this paper, which can solve the difficulty of making high precision 3D target. The self-made target consists of two intersecting chess board. With the classic scale method, the 3D coordinates of selected points in the target were derived from the distance matrix. The element in distance matrix is the distance between every two points, which can be obtained by measurement. The spatial location precision of points in the target was ensured by measurement instead of manufacturing, which reduced the production cost and the requirements for the production accuracy greatly. Camera calibration was completed using 3D target based method. It can be further extended to the applications where the target cannot be produced. The experimental results show the validity of this method.

Keywords: Computer vision · Image processing · Camera calibration · 3D target

1 Introduction

1.1 A Subsection Sample

The purpose of camera calibration is to recover intrinsic and extrinsic parameters, which is a preliminary step for further developing a computer vision system [1]. Using calibration objects (target) is a common approach to camera calibration. Based on the geometry of the camera model and the object, intrinsic and extrinsic parameters can be recovered. According to the dimension of calibration objects, camera calibration can be classified into four categories: 3D object based methods, 2D object based methods, 1D object based methods and Self-calibration methods [2].

3D object based methods [3] require the use of some precisely made calibration patterns. This will involve the design and use of some highly accurate tailor-made calibration patterns, which are often difficult and expensive to manufacture. To overcome these difficulties, Scholars proposed many modifications. Three coordinates measuring machine was used as target for generating the spatial points in [4], which saved the cost of target manufacturing. However, the price of machine limited its application. Some symmetrical objects (such as spheres and cylinders) were introduced for calibration in [5–9], which need lower manufacturing precision due to their shape feature.

© Springer Nature Singapore Pte Ltd. 2021
J. Zeng et al. (Eds.): ICPCSEE 2021, CCIS 1451, pp. 406–416, 2021.
https://doi.org/10.1007/978-981-16-5940-9_31

Compared with the 3D based methods, 2D object based methods are more flexible, which use a planar point pattern shown at a few different orientations [10–13]. A confocal conics planner pattern is proposed in [14] to encode the metric information, which obtains high accuracies for both intrinsic and extrinsic camera parameters. This kind of method reduced the cost of target, but increased the calibration steps.

Many 1D based methods are proposed to solve the problem of shelter in multi-camera calibration [15–18]. The calibration object usually consists of three or more collinear points with known relative position. It needs to move around a fixed endpoint to complete the camera calibration.

Self-calibration does not use a calibration object. Their aim is to provide further flexibility by not requiring a prior knowledge of the 3D to 2D correspondences. Various methods [19–21] have been proposed in this category that rely on scene or motion constraints, most of which require good initialization, and multiple views.

3D target based methods have the advantages of high accuracy and fast calibration but the disadvantage of high cost. While the other methods are less cost but more complicate calibration steps. Here, we propose a self-made 3D target based method for camera calibration. The accuracy of target is ensured by measurement not by manufacturing, which can lower the cost greatly with less loss in accuracy.

This paper is organized as follows. Section 2 gives some preliminaries such as camera model, intrinsic matrix, and foundation matrix. Calibration principle based on 3D target is also discussed in Sect. 2. And Sect. 3 gives the parameter computing of self-made 3D target. Experimental example is shown in Sect. 4. And Sect. 5 is some concluding remarks.

2 Projection Matrix Based Camera Calibration

2.1 Camera Model

In this paper, camera is modeled as pinhole camera. The pin-hole model is composed with intrinsic and extrinsic parameters. Intrinsic parameters contain the geometry of the camera and the optical features of the sensor. Extrinsic parameters map the 3D points in the scene to the coordinate system of the camera.

Let $A = (X, Y, Z)^T$ be the Euclidean 3D coordinates of a point in the world frame, $\tilde{A} = (X, Y, Z, 1)^T$ be its 3D homogeneous coordinates, and $\tilde{a} = (u, v, 1)^T$ be the 2D homogeneous coordinates of its projection in the image plane. The camera model can be described as follows

$$s\tilde{a} = M[R\,t]\tilde{A} \tag{1}$$

Where s is scale factor; $[R\,t]$ is a matrix formed by the extrinsic camera parameters that is the rotation and translation from the world frame to the camera frame; M is a matrix formed by the intrinsic camera parameters. And

$$M = \begin{bmatrix} f_u & \gamma & u_o \\ 0 & f_v & v_o \\ 0 & 0 & 1 \end{bmatrix} \tag{2}$$

Where (f_u, f_v) is the focal length; (u_o, v_o) is the principal point; γ is the incline factor of image plane. $P = M[R\ t]$ is referred as the camera projection matrix, which can be written as follows

$$P = \begin{bmatrix} P_{11} & P_{12} & P_{13} & P_{14} \\ P_{21} & P_{22} & P_{23} & P_{24} \\ P_{31} & P_{32} & P_{33} & P_{34} \end{bmatrix} \tag{3}$$

2.2 Solution of Camera Parameters

Eliminate s in (1). And (1) can be rewritten as follows

$$\begin{bmatrix} \tilde{A}^T & 0_{1\times 4} & -u\tilde{A}^T \\ 0_{1\times 4} & \tilde{A}^T & -v\tilde{A}^T \end{bmatrix} H = \begin{bmatrix} 0 \\ 0 \end{bmatrix} \tag{4}$$

Where $H = [P_{11}, P_{12}, P_{13}, P_{14}, P_{21}, P_{22}, P_{23}, P_{24}, P_{31}, P_{32}, P_{33}, P_{34}]$. Suppose that there are n known points in the 3D target. We can get an equation set contains $2n$ equations. When $n \geq 6$, there is a single least square solution of H. And P is obtained too.

Let $P = [Q\ p_4]$, and $Q = [p_1\ p_2\ p_3]^T$, we have

$$\rho P = \rho [Q\ p_4] = M[R\ t] \tag{5}$$

Where ρ is a scale factor which ensure that $|\rho p_i| = 1$. Let $R = [r_1\ r_2\ r_3]^T$, we have

$$\rho Q = \begin{bmatrix} f_u r_1^T + \gamma r_2^T + u_o r_3^T \\ f_v r_2^T + v_o r_3^T \\ r_3^T \end{bmatrix} \tag{6}$$

Consider that R is an orthogonal matrix, and its row vectors are unit vectors. We can get further

$$\begin{cases} \rho = \frac{\pm 1}{|p_3|} \\ r_3 = \rho p_3 \\ u_o = \rho^2 (p_1 \cdot p_3) \\ v_o = \rho^2 (p_2 \cdot p_3) \end{cases} \tag{7}$$

$$\begin{cases} \rho^2 (p_1 \times p_3) = -f_u r_2 - \gamma r_1 \\ \rho^2 (p_2 \times p_3) = f_v r_1 \end{cases} \tag{8}$$

then

$$\begin{cases} f_v = \rho^2 (p_2 \times p_3) \\ \gamma = -\dfrac{\rho^2 (p_1 \times p_3) \cdot (p_2 \times p_3)}{|p_2 \times p_3|} \\ f_u = \sqrt{\rho^2 |p_1 \times p_3|^2 - \gamma^2} \\ r_1 = \dfrac{(p_2 \times p_3)}{|p_2 \times p_3|} \\ r_2 = r_3 \times r_1 \end{cases} \tag{9}$$

From (5), we get

$$t = \rho M^{-1} p_4 \tag{10}$$

Here, all camera parameters are obtained as shown in (7), (9), and (10).

The pin-hole model is not sufficient and further parameters should be estimated to take lens distortion and data errors into account [22].

Lens distortion will cause the radial and tangential distortion. And radial distortion plays a major role, which can be described by polynomial expression. Ignore the high order terms and the tangential distortion, then

$$\begin{cases} u = u'\left(1 + k_1\left(u'^2 + v'^2\right)\right) \\ v = v'\left(1 + k_1\left(u'^2 + v'^2\right)\right) \end{cases} \tag{11}$$

Where k_1 is a factor of correction, (u,v) is a corrected coordinate, and (u',v') is an original coordinate in (11). Data errors may lead to equations has no solution, which can be solved by method of optimum estimate. We don't want to discuss lens distortion and data errors further in this paper. For details, see reference [3].

3 Self-made 3D Target

3.1 3D Coordinates Calculation of the Target

From Sect. 2, we know that camera parameters can be calculated if the 3D coordinates of some points in the target are given. Here, we made a 3D target by ourselves, and calculate the 3D coordinates of some selected points from the distance matrix whose elements are the distances between every two points. And the distances can be obtained by measurement.

Let n be the number of points in the 3D target, $X = [x_1, x_2, \dots, x_n]^T$ be the set consists of the n points. Where $x_i = [x_{1i}, x_{2i}, x_{3i}]$ is a 3D point. Define $D^{(2)}$ as the distance matrix whose elements are the square of distances between the n points each other.

$$D^{(2)}(i,j) = \left\{d_{ij}^2\right\}_{n\times n} = |x_i|^2 + |x_j|^2 - 2x_i x_j^T \tag{12}$$

Where d_{ij}^2 is the square of distances between point i and point j. Comparing $D^{(2)}$ with X, We have

$$D^{(2)} = c1^T + 1c^T - 2XX^T \tag{13}$$

where $c = [|x_1|^2, |x_2|^2, \dots, |x_n|^2]^T$; $1 = [1,1,\dots,1]^T$. $D^{(2)}$ has nothing to do with the location of the coordinate origin. Without losing generality, we suppose that

$$\sum_{i=1}^{n} x_i = 0 \tag{14}$$

Define a centralized matrix

$$J = I - \frac{1}{n}11^T \tag{15}$$

Where I is an unit matrix. Multiplied both sides of (13) by J, we can get

$$JD^{(2)}J = J\left(c\mathbf{1}^T + \mathbf{1}c^T + 2XX^T\right)J$$
$$= -2XX^T \tag{16}$$

$$XX^T = -\frac{1}{2}JD^{(2)}J \tag{17}$$

$-(1/2)JD^{(2)}J$ in (17) is a symmetric matrix which can be decomposed using SVD method. That is

$$XX^T = U\Lambda U^T \tag{18}$$

$$X = \frac{1}{2}U\Lambda^{1/2} \tag{19}$$

Where U is an orthogonal matrix consists of eigenvectors of $-(1/2)JD^{(2)}J$; Λ is a diagonal matrix consists of Eigen value of $-(1/2)JD^{(2)}J$. And (19) is the expression for 3D coordinates calculation.

3.2 Error Analysis

There must be some errors in distance measurement, which can cause errors in 3D coordinates of points in the target. Suppose that $x_{ij} = x_{ij}{}^* + \Delta x_{ij}$, and $d_{ij} = d_{ij}{}^* + \Delta d_{ij}$, where x_{ij} and d_{ij} are measured values, $x_{ij}{}^*$ and $d_{ij}{}^*$ are actual values, and Δx_{ij} and Δd_{ij} are errors. We have

$$\sqrt{\sum_{k=1}^{3}\left(\left(x_{ik}^* + \Delta x_{ik}\right) - \left(x_{jk}^* + \Delta x_{jk}\right)\right)^2} = d_{ij}^* + \Delta d_{ij} \tag{20}$$

In order to complete the error analysis, we designed two simulation experiments as follows.

1) Distance errors caused by coordinate errors.
In this simulation experiment, we calculate the distance (d_{ij}) using (12). Then add errors (Δx_i) to x_i. And calculate the distance (d'_{ij}) again. Finally, calculate the maximum value of the difference (max Δd_{ij}).

We take 8 points to simulate the distance errors caused by coordinate errors using MATLAB. The coordinates of points are $(-10, -10, 0)$, $(0, -10, 10)$, $(10, -10, 0)$, $(0, -10, -10)$, $(-10, 10, 0)$, $(0, 10, 10)$, $(10, 10, 0)$, and $(0, 10, -10)$ respectively. We add errors governed by a Gaussian distribution to the coordinates of the eight points to see the variation of distance errors. The experiment was repeated 50,000 times. The results are shown as Fig. 1. Where (a), (b), and (c) in Fig. 1 are the maximum distance errors caused by coordinate errors of 0.1 mm, 0.01 mm, and 0.001 mm respectively.

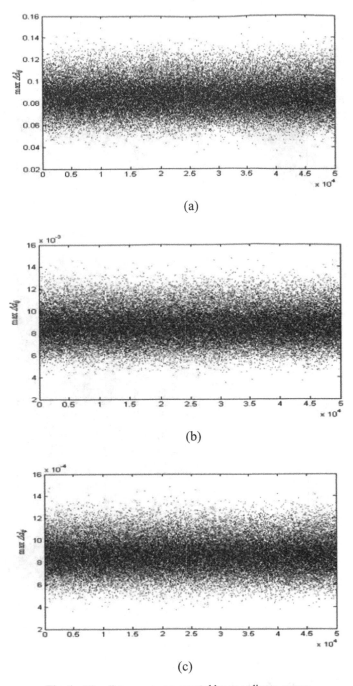

Fig. 1. The distance errors caused by coordinate errors

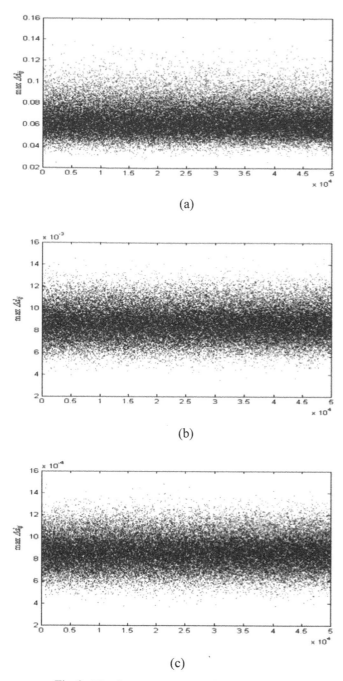

(a)

(b)

(c)

Fig. 2. The distance errors caused by our method

2) the distance errors caused by our method.

In this simulation experiment, we calculate the distance (d_{ij}) using (12). Then add errors (Δd_{ij}) to d_{ij}. And calculate the coordinates $(x_i + \Delta x_i)$ using (19). Then calculate the distance (d'_{ij}) using (12) again. Finally, calculate the maximum value of the difference (max Δd_{ij}).

We also take the same eight points to simulate the distance errors caused by our method using MATLAB. We calculate the actual distances using the coordinates given in 1). Then add errors governed by a Gaussian distribution to them to simulate the measurement errors. And calculate the coordinates of points using (19). Finally calculate the distances using these coordinates. The differences between the calculated distances and the actual distance are the distance errors. The experiment was repeated 50,000 times. The results are shown as Fig. 2. Where (a), (b), and (c) in Fig. 2 are the maximum distance errors caused by our method with measurement errors of 0.1 mm, 0.01 mm, and 0.001 mm respectively.

Compare 2) with 1), we can see that distance error in 1) and distance error in 2) have the same ranges but different average values. Average distance error in 2) is less than the one in 1). Take (a) in Fig. 1 and (a) in Fig. 2 as an example. Distance errors caused by our method with 0.1mm measurement error are less than the one caused by 0.1mm coordinates error. It shows that when the coordinates of points in the self-made target are obtained using our method, the coordinate errors are less than the measurement errors of distance.

4   Experimental Example

In order to validate the method discussed in this paper, we calibrated a given camera using two methods: method in reference [10] and the proposed method in this paper

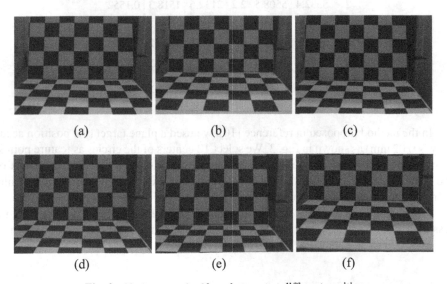

(a)   (b)   (c)

(d)   (e)   (f)

Fig. 3. Six images of self-made target at different position

respectively. The camera to be calibrated adopts the Elanus-UC1400C of the CatchBEST with 8mm lens of M0814-MP2. The effective pixel is 4608 × 3288. And the pixel size is 1.4 um × 1.4 um.

According to our method, we made a 3D target as shown in Fig. 2. It consisted of two intersecting planes with printed chequer on them. We select 12 corners of the chequer as feature points. And obtain the distances between any two points ($D^{(2)}$) by measurement with precision of 20 um. Then calculate the 3D coordinates of every point according to (19). At last, calibrate the camera according to (7), (9) and (10). We calibrated the camera six times with the image shown in Fig. 2. (a)–(f) respectively. After calibration, we calculated the distance matrix ($D^{(2)}$') of the selected points in the target using the measured camera parameters in turn. Then we compared with the original distance matrix ($D^{(2)}$). We defined average residual error

$$\Delta_{avr} = \frac{\sum\limits_{i=1}^{n}\sum\limits_{j=1}^{n}\left(d'_{ij} - d_{ij}\right)}{n(n-1)} \tag{21}$$

Where d_{ij}' is the calculated distance between point i and j using the calibrated camera parameters, and d_{ij} is the original distance between point i and j obtained by measurement. And Δ_{avr} can be used for the verification of the method. The calibration results are shown in Table 1 (Fig. 4).

Table 1. Calibration results using the proposed method

No	f_x	f_y	γ	u_0	v_{osss}	Δ_{avr}/mm
1	5676.8	5645.6	1.2	2212.1	1588.7	0.1187
2	5532.4	5508.5	2.2	2132.5	1518.3	0.1352
3	5562.2	5557.3	3.5	2316.3	1621.7	0.1302
4	5913.8	5875.1	2.7	2197.5	1689.2	0.1056
5	5871.1	5848.6	2.9	2398.7	1711.2	0.1268
6	5577.7	5549.3	5.6	2209.2	1553.5	0.1128

In the method proposed in reference [10], we used a plane target (the position accuracy is 0.02 mm) as shown in Fig. 3. We select 12 centers of the circles as feature points. In order to consistent with the experiment above, we calculated the distance matrix of the selected points and repeated the experiment six times too. The calibration results using the two methods are shown in Table 2.

From Table 2, we can see that the calibration differences between the two methods are less than 1%, which show that the two methods are with the same calibration accuracy.

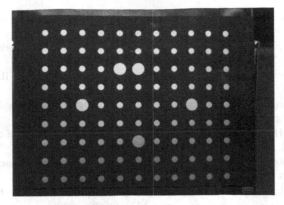

Fig. 4. Standard calibration board

Table 2. Comparison between two calibration results

Camera parameters	Our method (average value)	Method in reference [10] (average value)
fx	5689.0	5678.5
fy	5664.1	5660.2
γ	3.0	3.1
uo	2244.4	2233.7
vo	1613.8	1608.5
Δ_{avr}/mm	0.1216	0.1208

5 Conclusion

In this paper, we propose a camera calibration method based on self-made 3D target instead of high precision 3D target. The 3D coordinates of self-made target can be calculated by classic scale method. And the obtained coordinate's errors of points in the target are less than the measurement errors. So, the accuracy of coordinates of the feature points in target is ensured by measurement instead of manufacturing. It reduces the target cost greatly without losing the advantages of 3D target based method. The experiment example shows the validity of this method.

References

1. Yan, L., Payandeh, S.: On the sensitivity analysis of camera calibration from images of spheres. Comput. Vis. Image Underst. **114**, 8–20 (2010)
2. Zhu H, Li S, 2D pattern vs. surrounding 3D pattern for fisheye camera calibration. Chinese J. Sci. Instr. **29**(7), 1512–1516 (2008)
3. Jie, Z.: Li X, Dai X, Camera calibration method based on 3D board. J. Southeast Univ. (Natural Science Edition) **41**(3), 543–548 (2011)

4. Huang, F., Qian, H.: Camera calibration technology driven by three coordinate measuring machine. Optics Precis. Engi. **18**(4), 952–957 (2010)

5. Sun, J., Cheng, X., Fan, Q.: Camera calibration based on two-cylinder target. Optics Express **27**(20), 29319–29331 (2019)

6. Cao, X., Foroosh, H.: Camera calibration using symmetric objects. IEEE Trans. Image Process. **15**(11), 3614–3619 (2006)

7. Zhang, H., Wong, K.-Y., Zhang, G.: Camera calibration from images of spheres. IEEE Trans. Pattern Anal. Mach. Intell. **29**(3), 499–503 (2007)

8. Gu, F., Zhao, H., Bu, P.: Analysis and correction of projection error of camera calibration ball. Acta Optica Sinica **32**(12), 12150011–12150017 (2012)

9. Shen, E., Hornsey, R.: Multi-camera network calibration with a non-planar target. IEEE Sens. J. **11**(10), 2356–2364 (2011)

10. Luo, H., Zhu, L., Ding, H.: Camera calibration with coplanar calibration board near parallel to the imaging plane. Sensors Actuators A **132**, 480–486 (2006)

11. Kruger, L., Wohler, C.: Accurate chequerboard corner localization for camera calibration. Pattern Recogn. Lett. **32**, 1428–1435 (2011)

12. Yang, M., Chen, X., Yu, C.: Camera calibration using a planar target with pure translation. Appl. Optics. **58**(31), 8362–8370 (2019)

13. Yu, L., Han, Y., Nie, H., et al.: A calibration method based on virtual large planar target for cameras with large FOV. Optics Lasers Eng. **101**, 67–77 (2018)

14. Kim, J.-K., Gurdjos, P., Kweon, I.S.: Euclidean structure from confocal conics: Theory and application to camera calibration. Comput. Vis. Image Understanding **114**, 803–812 (2010)

15. Wei, Z., Li, J., Ji, X., Bo, Y.: A Calibration Method Based on Multi-Lin-ear Structured Light. In: 2010 Symposium on Security Detection and Information. Procedia Eng.**7**(2010), 345–351 (2010)

16. Xue, J., Su, X.: Camera calibration with single image based on two orthogonal one-dimensional objects. ACTA OPTICA SINICA **32**(1), 01150011–01150017 (2012)

17. Lv, Y., Liu, W., Xu, X.: Methods based on 1D homography for camera calibration with 1D objects. Appl. Optics. **57**(9), 2155–2164 (2018)

18. Manuel, E.: Loaiza, AlbertoB. Raposon, MarceloGattass, Multi-camera calibration based on an invariant pattern. Comput. Graph. **35**, 198–207 (2011)

19. Dornaika, F., Chung, R.: An algebraic approach to camera self-calibration. Comput. Vis. Image Underst. **83**, 195–215 (2001)

20. Brito, D.N., Pádua, F.L.C., Lopes, A.P.C.: Using geometric interval algebra modeling for improved three-dimensional camera calibration. J. Math. Imaging Vis. **61**(9), 1342–1369 (2019). https://doi.org/10.1007/s10851-019-00907-x

21. Feng, W., Zhang, S., Liu, H., et al.: Unmanned aerial vehicle-aided stereo camera calibration for outdoor applications. Optical Eng. **59**(1), 014110 (2020)

22. Ricolfe-Viala, C., Sanchez-Salmeron, A.-J.: Using the camera pin-hole model restrictions to calibrate the lens distortion model. Opt. Laser Technol. **43**, 996–1005 (2011)

A2Str: Molecular Graph Generation Based on Given Atoms

Xilong Wang[✉]

Heilongjiang University, Harbin 150080, Heilongjiang, China

Abstract. Molecular graph generation is an emerging area of research with numerous applications. Most existing approaches generate new molecules by learning a generative model from a collection of known molecules. However, how to obtain the molecular structure from its atomic composition is a problem to be solved. In this paper, it proposes an autoencoder-based model called A2Str which takes the atomic composition of a molecular as input. Its molecular structure was generated and model structures were designed. The effects were verified one by one. In addition, experiments were implemented with isomer generation and determined that the ratio of more than 3% of molecules with different structures can be obtained from the same combination of atoms.

Keywords: Molecular generation · Atomic composition · Autoencoder

1 Introduction

Deep learning has been successfully applied in various fields including molecular generation tasks with remarkable results. Molecular generation task is to discover the appropriate molecule from the vast molecular space. Molecular data is usually stored as simplified molecular input entry linear specification (SMILES) strings, thus LSTM that is suitable for processing sequence data such as strings are usually used for molecular generation. After VAE and other generative models were proposed, they are also used in the field of molecular generation. These models usually take a known molecule as an input, and output some new molecules.

However, in practical applications, we often face such a problem: we only know the atomic composition of the molecule, but not the structure of the molecule. But it is necessary to obtain the molecular structure. As mentioned in [1], it is very important for high school students and university students to determine the structural formula of a molecule in the teaching of organic chemistry. However, to infer the structural formula from the molecular formula also needs to know a series of complex physical and chemical properties related to the molecule, and it is difficult to infer the structural formula only knowing the molecular formula. But as far as we know, this problem has never been toughed. To address the above problem, we propose a model called A2Str from atomic composition to molecular structure, which takes the atomic features of a molecule graph as an input and obtains the adjacency matrix of the molecule graph. In this way, we can infer its structural formula from the molecular formula of a compound.

© Springer Nature Singapore Pte Ltd. 2021
J. Zeng et al. (Eds.): ICPCSEE 2021, CCIS 1451, pp. 417–428, 2021.
https://doi.org/10.1007/978-981-16-5940-9_32

Taking epoxyethane as an example, the inference process of the A2Str model is shown in Fig. 1.

Assuming that our molecule has at most three atoms and the number of atomic types is 2 (carbon and oxygen), each row of atomic features is the one-hot representation of the corresponding atom. Considering that not all molecules have the maximum of atoms, the number of columns should plus 1 to indicate that there is no atom at this position. We know from the atomic features that our molecules contain two carbon atoms and one oxygen atom. The adjacency matrix that we can infer tells us how the atoms are connected to each other, and by combining these two pieces of information, we will know which atoms are connected to which bonds, and determine the structure of a molecule.

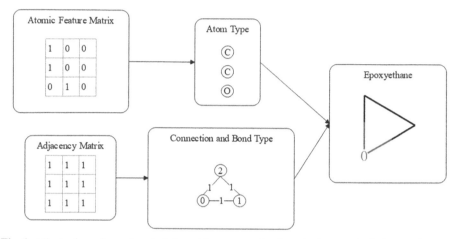

Fig. 1. Illustration of a molecule inferred from a matrix. Atomic features include atomic types, and the adjacency matrix includes atomic connection and chemical bond types. All information of a molecular graph can be obtained by combining the two matrixes.

Overall, our contributions are as follows:

We propose a new method to generate molecules, and the relationship between atomic features and chemical bond features is established.

The structural formula can be inferred from the molecular formula, and the isomers are considered at the same time, which can provide a certain reference for the teaching task of organic chemistry.

Our model performs well on the ZINC dataset, reaching almost 100.00% validity.

2   Related Work

2.1   Molecular Generative Model Based on Recurrent Neural Network

SMILES is a common way for molecules to be stored in computers. Recurrent Neural Network (RNN) and Long Short Term Memory Network (LSTM) are often used for string generation as excellent models for sequence data processing. Segler et al. [2] used

LSTM for molecular generation. Bjerrum et al. [3] and Gupta et al. [4] also adopted this approach, and added *temperature* parameter to the generation of molecules in the model to avoid singleness of the generated molecules. Ertl et al. [5] replaced Cl, Br, [nH] and other multi-character atoms with a single character, so as to generate characters one by one and not easy to lead to errors. However, stereo molecules and charged molecules were not considered during training. Zheng et al. [6] further considered multi-character atoms in stereo molecules and charged molecules, such as [C@@H], [C@H], [K+], etc., and regarded them as one sign. The character-by-character generation was developed into sign-by-sign generation.

2.2 Molecular Generative Model Based on Variational Autoencoder

Variational AutoEncoder [7] (VAE) is a common generative model, which approximates the probability distribution of data sets through continuous learning, and then finds new samples from the distribution. Lim et al. [8] integrated molecular property vectors into encoders and decoders to generate molecules with desired properties. Kang et al. [9] used data set x to train a property prediction network whose output is y, and then used x and y as the inputs of VAE to get the output x', in which the property prediction network and VAE are different parts of the same model for joint training. Gomez-Bombarelli et al. [10], similar to Kang et al., whose idea contained the nature of the prediction network, but the difference was that the former property prediction network was connected in VAE after the hidden space, Kusner et al. [11] made VAE learn the semantic nature of SMILES sequences by integrating context-free grammar into VAE.

2.3 Molecular Generative Model Based on Generative Adversarial Network

Generative Adversarial Network [12] (GAN) is a generative model which uses generators to generate samples and allows discriminators to determine whether the samples they receive come from the data set or the generator to improve their accuracy. Guimaraes et al. [13] trained the discriminator as a classifier and trained the generator by reinforcement learning. De Cao et al. [14] integrated reinforcement learning, graph neural network [15] and GAN, and sent the two matrices into the discriminant by sampling an adjacency matrix and node type matrix from the generator. In addition, a reward network was used to make it easier for GAN to generate molecules with desired properties.

2.4 Other Generative Models

In addition to the commonly used generative models mentioned in Sects. 2.1–2.3, there are several generative models also applied to molecular generation. For example, Adversarial autoencoders combining VAE and GAN (AAE), Kadurin et al. [16] compared the hidden space of VAE as a discriminator with the prior distribution. Graph neural networks, mentioned in Sect. 2.3, are also used for molecular generation. Jin et al. [17] regarded the two atoms connected by chemical bond the ring structure in the molecule as a cluster, and made the neighbour clusters join into a tree structure, which reduced the error rate of molecule generation. You et al. [18] combined graph neural network with reinforcement learning to generate atom-by-bond (that is, node-by-edge). Zang et al. [19] used flow model for molecular generation, which was different from VAE and GAN to approach the real probability distribution of data set by optimizing upper bound or adversarial training. It directly found the mapping between prior distribution and data set distribution through Jacobian determinant.

3 Method

3.1 Problem Definition

If the molecule is represented as a graph, the atoms and bonds can be the nodes and edges in the graph, respectively. We define a molecular graph as $\mathbb{G} = (\mathbf{F}, \mathbf{A})$, where $\mathbf{F} \in \{0, 1, 2, 3\}^{N*(d_a+1)}$ is atoms feature matrix, N is the maximum number of atoms in a molecule, d_a is the number of atomic types. It is important to note that not all molecules have N atoms, so we need to make $\mathbf{F}$ own one more column to represents non-atom. $\mathbf{A} \in \{0, 1, 2, 3\}^{N*N}$ is the adjacency matrix, the elements in $\mathbf{A}$ represent the connections between atoms, namely, no connection, single bond, double bond and triple bond.

3.2 A2Str Model

Our model takes atomic features as input and obtains two outputs, one is the reconstruction of atomic features and the other is the association matrix. The purpose of atomic features reconstruction is to enable the model to learn the implied expression of atomic features better and thereby infer a more accurate adjacency matrix. The model training framework is shown in Fig. 2:

The model takes the known atomic feature matrix $\mathbf{F}$ as the input, obtains the latent vector z through the encoder, and then obtains the reconstruction $\mathbf{F}_j$ of the input through the decoding of z, and predicts the corresponding adjacency matrix $\mathbf{A}_j$ according to z.

Our loss function is defined as follows. The loss is divided into two parts. The first part $\mathcal{L}_{recon}$ is the reconstruction loss of the atomic feature matrix, and the second part $\mathcal{L}_{pred}$ is the prediction loss of the adjacency matrix, where D is the amount of data in the data set, $p_i(\cdot)$ is the real value of the corresponding element in the i th data, $q_i(\cdot)$ is

$$\mathcal{L} = \mathcal{L}_{recon} + \mathcal{L}_{pred} \tag{1}$$

$$\mathcal{L}_{recon} = -\frac{1}{D} \sum_{i=1}^{D} \sum_{j=1}^{N} \sum_{k=1}^{d_a+1} p_i(\mathbf{F}_{jk}) \log\left(q_i\left(\mathbf{F}'_{jk}\right)\right) \tag{2}$$

$$\mathcal{L}_{pred} = -\frac{1}{D} \sum_{i=1}^{D} \sum_{j=1}^{N} \sum_{k=1}^{N} p_i(\mathbf{A}_{jk}/3) \log\left(q_i\left(\mathbf{A}'_{jk}/3\right)\right) \tag{3}$$

the predicted value of the corresponding element in the ith data, and $\mathbf{F}_{jk}$ is the element in the jth row and kth column of the matrix. The element in the adjacency matrix $\mathbf{A}$ may get values in $\{0,1,2,3\}$. We need to normalize it to the interval of $[0,1]$, so we divide it by 3.

3.3 Model Inference

In the model inference process, we still take the atomic feature matrix as the input, but we no longer perform the reconstruction steps, and directly infer the adjacency matrix. The inference model is shown in Fig. 3.

Our actual purpose is to generate new molecules, and the model is to infer the adjacency matrix of the graph through the atomic features. In this case, the atomic features and the adjacency matrix can be combined to obtain the complete graph information. The atomic feature matrix contains the information of atomic types, and the adjacency matrix contains the connection between atoms and chemical bonds and the types of chemical bonds.

Since the molecular graph is a connected graph and does not contain self-loop nodes, and we specify that the elements of the adjacency matrix have four choices, it can be regarded as the adjacency matrix of the weighted graph. In view of these points, our adjacency matrix has these two properties: (1) each column has only two non-zero elements, and the value of these two non-zero elements is the same, (2) the sum of elements in each row represents the weight of the corresponding node. These two properties can help us to correct the prediction error. The adjacency matrix of property (1) can correct prediction adjacency matrix of a column element errors, ensure that the edge of the connection is correct, and the corresponding node weights in the molecular graph of property (2) is the number of shared electrons pairs. We already know the atomic feature matrix, so we know the corresponding maximum number of electron pairs that the atoms can share, which can correct the connection error of the atoms.

3.4 Isomers

The same molecular formula may have different structures, including tectonic isomerism and stereoisomerism. The treatment of stereoisomerism is relatively easy, for it is possible to specify the type of isomerism of an atom or chemical bond when forming an atom. The difficulty of tectonic isomerism lies in the need to change the position of non-zero elements of the adjacency matrix. Since the model parameters are fixed, the same input will inevitably only get the same output. In order to obtain different structure of the same molecular formula, our strategy is to change the order of the input atomic features, so that different output can be obtained without changing the atomic types. It is important to note that what we call isomers do not contain hydrogen atoms, only heavy atoms such as carbon, oxygen and nitrogen.

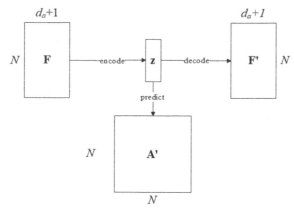

Fig. 2. Illustration of A2Str model. The latent vector is obtained in the encoding stage, and then reconstructed by decoding. Meanwhile, the adjacency matrix is predicted.

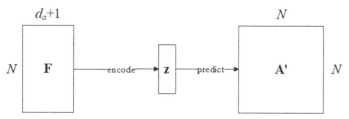

Fig. 3. Illustration inference model. The difference between Fig. 2 and this is that the reconstruction of atomic feature matrix is omitted.

4   Experiment

4.1   Data

The data comes from the open-source dataset ZINC [20], from which we extract 422,357 drug molecules. Each unit in the dataset is made up of a SMILES string, combinations of different characters containing information about a molecule's atoms, chemical bonds, branched chains, ring structure and isomerism properties, etc.

4.2   Experimental Environment and Parameter Configuration

This experiment is run in the development environment of Anaconda under Windows 10 system. The Python version is 3.7, the deep learning framework is Keras-2.4.3, the RDKit version of the open-source chemical informatics library is 2020.03.2, and the GPU is NVIDIA GeForce GTX 2080 Ti.

The choices embedding size are 2/4/8/16. The encoder is constructed as a two-layer convolutional neural network. The number of filters in the first layer is 300 with the kernel size (1, embedding size), and the number of filters in the second layer is 300 with the kernel size (N, embedding size). That is, the first convolution layer is to capture the features of a single atom, and the second convolution layer is to capture the features of all atoms. In addition, we will pool the results of the first convolution layer to obtain the deep representation of the input. The hidden space dimension is 16. The decoder structure is multilayer perceptron. The dimension is $N * (d_a + 1)$, and then the atomic feature matrix is reconstructed by shape transformation. In the prediction part, we use two structures. The first structure is three-layer transpose convolution, whose filter sizes are 32, 128 and 512, and whose kernel sizes are (4,4), (8,8) and $(N-7, N-7)$. The second structure is a three-layer perceptron with dimensions of 512, 1024 and $N * N$, respectively.

4.3 Evaluation Indicators

We first use the evaluation indicators frequently used in previous work:

Validity: if the generated string is a legitimate SMILES molecule, is as a valid molecule. This is the percentage of the generated legal SMILES molecules to all generated strings.

Uniqueness: if SMILES is not generated before, then regard it as a unique molecule. This is the percentage of all the different SMILES to all the generated SMILES.

Novelty: if SMILES molecule generated is not in the training set, is as a novel molecule. This is the percentage of all SMILES generated that are not present in the training set.

In addition, Quantitative Estimation of drug-likeness [21] (QED) and Synthetic Accessibility [22] (SA) of molecules are investigated. The value of the former is located in the interval (0, 1), the higher the value represents the molecular drug similarity degree is higher, it investigates the molecular eight physical and chemical indicators: (1) the molecular weight, (2) octanol-water partition coefficient, (3) the hydrogen bond donors number, (4) the hydrogen bond receptors number, (5) topological polar surface area, (6) rotatable bonds number, (7) aromatic ring number and (8) alarm structure number, every physical and chemical indicators can assign different weights, we choice to maximize the information content of the first 1,000 the average of the weights, the information content can be calculated by Shannon entropy. The value of the latter is located in the interval (0, 10), the higher the value of representative molecules are more difficult to synthesis, the calculation of SA is divided into two parts, the first part is weighted based on the frequency of known ECFP4 fingerprint of compounds database, we get the number of contributed fragment of which we think that exist in the existing compounds the higher frequency, then the part is easy to synthesize. The second part is to calculate the molecular complexity, including ring complexity and molecular weight. The difference between the first part and the second part is calculated, usually within the interval (−4,2.5), and then the value is multiplied by −1 and normalized to the interval (0,10).

4.4 Molecular Generation

Direct Generation. We use the trained model to generate molecules, randomly selected 20,000 different atomic feature matrices as inputs, and obtain the predicted adjacency matrices through the model, and get the molecular graph from the two. We introduced an embedding layer after the model input layer and verified the effects of different embedding dimensions on the experimental results. The experimental results are shown in Table 1:

Table 1. Indicators in different embedding dimensions

Embedding dimension	Validity	Uniqueness	Novelty
2	0.50%	100.00%	100.00%
4	0.94%	100.00%	100.00%
8	2.17%	100.00%	100.00%
16	1.22%	100.00%	100.00%

Isomer Generation. It can be seen that when the embedding dimension is set to 8, the effect is better than other cases. It is worth noting that there are 9 kinds of atoms in our data, so compressing the representations of the 9 kinds of atoms to the appropriate dimension is helpful to improve the model effect. In addition, we also verified the generation of isomers. We randomly selected 100 non-repeated atomic features, shuffled the order for 2000 times, and calculated the average situation of evaluation indexes, as shown in Table 2:

Table 2. Indicators of isomers

Validity	Uniqueness	Novelty
3.80%	83.21%	100.00%

As can be seen, by shuffling the atoms out of order, our model can also generate corresponding molecules. The reason why the uniqueness is not 100.00% is that there are too few atomic species in some molecules. Even if the order is randomly shuffled, it cannot guarantee that the species of atoms changed after the position change.

We also tried to change the partial structure of the model. As described in Sect. 4.2, we changed the neural network of the part predicting the adjacency matrix to a multilayer perceptron. As we know from Table 1, the effectiveness is the highest when the embedding dimension is 8, so we directly used this dimension to verify the model. The results are shown in Table 3.

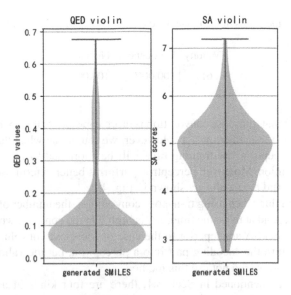

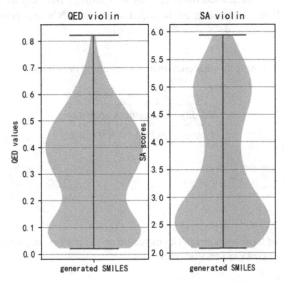

Fig. 4. Violin plots of QED and SA distribution of output from different prediction networks. The upper half is the case where the prediction network is transpose convolution, and the lower half is the case where the prediction network is perceptron.

Table 3. Indicators of the model whose prediction part consists of perceptron

Validity	Uniqueness	Novelty
1.61%	100.00%	100.00%

It can be seen that transpose convolution is more appropriate than perceptron in terms of effectiveness in the prediction step. However, we found that when the prediction part is changed to perceptron, the training time will be shortened to 1/13 of that of using transpose convolution. Moreover, perceptron performs better in terms of the distribution of QED and SA, and the results are shown in Fig. 4.

It can be seen that when using transpose convolution, the number of molecules with low QED is large, and at the same time, SA is high. On the contrary, when perceptron is used, both QED and SA are superior to the case of transpose convolution. In addition, we also tried to turn the encoding part into a perceptron, but the validity of the result was almost zero, so we did not use this method.

In addition, as mentioned in Sect. 3.1, there are four kinds of adjacency matrix elements. Although we can predict the probability of three types of chemical bonds or non-connection, but it increased the difficulty of training. We tried to use the adjacency matrix with only 0 and 1 elements as the input and used CNN to predict the output. The results are shown in Table 4.

Table 4. Indicators of using an adjacency matrix containing only 0 and 1 as input to predict the outputs

Validity	Uniqueness	Novelty
7.84%	99.36%	99.74%

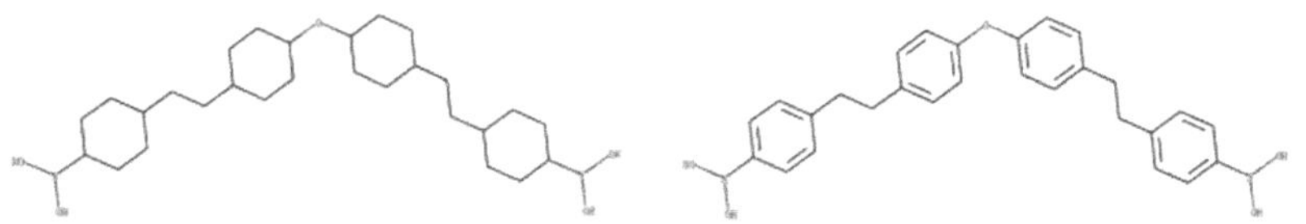

Fig. 5. Illustration of local retouching. The molecule on the left has only single bonds, while the molecule on the right contains benzene rings.

Obviously, compared with Table 1, all indicators have been improved, which indicates that reducing the types of elements in the adjacency matrix can reduce the prediction difficulty of the model. Since the chemical bonds in the molecules predicted by this method are only single bonds, we can adopt the local modification method to replace the six membered carbon rings with the benzene in accordance with the chemical rules. Such local modification method can enrich the diversity of generated molecules, as shown in Fig. 5.

5 Conclusion

We propose an A2Str model, which takes molecules as graph structures and atomic features as inputs, reconstructs them, uses latent vectors to predict their adjacency matrices, and then combine atomic features with adjacency matrices to obtain new molecules. A2Str model can also generate isomers, and the average ratio is not less than 3%.

References

1. Giac, C.C., Thanh, P.H., Ninh, N.X., Hang, N.T.D.: Designing exercises to determine the structural formula of organic compounds based on the experimental data. World J. Chem. Educ. **5**(2), 23–28 (2017)
2. Segler, M.H., Kogej, T., Tyrchan, C., Waller, M.P.: Generating focused molecule libraries for drug discovery with recurrent neural networks. ACS Central Sci. **4**(1), 120–131 (2018)
3. Bjerrum, E.J., Threlfall, R.: Molecular generation with recurrent neural networks (RNNs). arXiv preprint arXiv:1705.04612 (2017)
4. Gupta, A., Müller, A.T., Huisman, B.J.H., Fuchs, J.A., Schneider, P., Schneider, G.: Generative recurrent networks for de novo drug design. Mol. Inform. **37**(1–2), 1700111 (2018)
5. Ertl, P., Lewis, R., Martin, E., Polyakov, V.: In silico generation of novel, drug-like chemical matter using the lstm neural network. arXiv preprint arXiv:1712.07449 (2017)
6. Zheng, S., et al.: QBMG: quasi-biogenic molecule generator with deep recurrent neural network. J. Cheminform. **11**(1), 1–12 (2019)
7. Kingma, D.P., Welling, M.: Auto-encoding variational bayes. arXiv preprint arXiv:1312.6114 (2013)
8. Lim, J., Ryu, S., Kim, J.W., Kim, W.Y.: Molecular generative model based on conditional variational autoencoder for de novo molecular design. J. Cheminform. **10**(1), 1–9 (2018)
9. Kang, S., Cho, K.: Conditional molecular design with deep generative models. J. Chem. Inf. Model. **59**(1), 43–52 (2018)
10. Gomez-Bombarelli, R., et al.: Automatic chemical design using a data-driven continuous representation of molecules. ACS Central Sci. **4**(2), 268–276 (2018)
11. Kusner, M.J., Paige, B., Hern´andez-Lobato, J.M.: Grammar variational autoencoder. In: International Conference on Machine Learning. PMLR, pp. 1945–1954 (2017)
12. Goodfellow, I.J., et al.: Generative adversarial networks. arXiv preprint arXiv:1406.2661 (2014)
13. Guimaraes, G.L., Sanchez-Lengeling, B., Outeiral, C., Farias, P.L.C., Aspuru-Guzik, A.: Objective-reinforced generative adversarial networks (organ) for sequence generation models. arXiv preprint arXiv:1705.10843 (2017)
14. De Cao, N., Molgan, T.K.: An implicit generative model for small molecular graphs. arXiv preprint arXiv:1805.11973 (2018)
15. Scarselli, F., Gori, M., Tsoi, A.C., Hagenbuchner, M., Monfardini, G.: The graph neural network model. IEEE Trans. Neural Netw. **20**(1), 61–80 (2008)
16. Kadurin, A., et al.: The cornucopia of meaningful leads: applying deep adversarial autoencoders for new molecule development in oncology. Oncotarget **8**(7), 10883 (2017)
17. Jin, W., Barzilay, R., Jaakkola, T.: Junction tree variational autoencoder for molecular graph generation. In: International Conference on Machine Learning. PMLR, pp. 2323–2332 (2018)
18. You, J., Liu, B., Ying, R., Pande, V., Leskovec, J.: Graph convolutional policy network for goal-directed molecular graph generation. arXiv preprint arXiv:1806.02473 (2018)
19. Zang, C., Wang, F.: MoFlow: an invertible flow model for generating molecular graphs. In: Proceedings of the 26th ACM SIGKDD International Conference on Knowledge Discovery & Data Mining, pp. 617–626 (2020)

20. Irwin, J.J., Sterling, T., Mysinger, M.M., Bolstad, E.S., Coleman, R.G.: ZINC: a free tool to discover chemistry for biology. J. Chem. Inform. Model. **52**(7), 1757–1768 (2012)
21. Bickerton, G.R., Paolini, G.V., Besnard, J., Muresan, S., Hopkins, A.L.: Quantifying the chemical beauty of drugs. Nat. Chem. **4**(2), 90–98 (2012)
22. Ertl, P., Schuffenhauer, A.: Estimation of synthetic accessibility score of drug-like molecules based on molecular complexity and fragment contributions. J. Cheminform. **1**(1), 1–11 (2009)

WSN Data Compression Model Based on K-SVD Dictionary and Compressed Sensing

Liguo Duan, Xinyu Yang, and Aiping Li$^{(\boxtimes)}$

Taiyuan University of Technology, Jinzhong 030600, China

Abstract. Aiming at the problems of different monitoring data characteristics, limited energy consumption of nodes, and low data compression efficiency in wireless sensor networks, a data compression model based on K-SVD dictionary and compressed sensing is proposed. The model used the K-SVD dictionary learning algorithm to train the sparse base, transferred the sparse transformation from the sensing nodes to the base station, and reduced the energy consumption of the sensing nodes. Compared with the existing OEGMP algorithm and the CS compression algorithm based on DCT sparse basis on the same data set, the experimental results show that the model in this paper has a significant improvement in data compression rate and recovery accuracy.

Keywords: Wireless sensor network (WSN) · Compressed sensing · K-SVD dictionary · Data compression · Spatial-temporal correlation

1 Introduction

With the rapid development of various supporting technologies, as an important part of the Internet of Things (IoT) [1], Wireless sensor networks (WSN) [2] has been widely used in various fields such as environmental monitoring, military defense and smart agriculture. The amount of data generated by the popularization of applications is increasing day by day, and the resulting demand for storage, transmission, and processing is also increasing. The research on WSN data compression has been one of the most critical technologies for IoT application promotion.

In recent years, there have been many methods to compress WSN data [3], but the compression effect still has room for further improvement. LUO et al. [4] first proposed the application of Compressed Sensing (CS) [5] theory to large-scale WSN data collection. This method can greatly and effectively improve the data compression ratio and provide new ideas for WSN data collection research. In view of the differences in the characteristics of the monitoring data in different application scenarios, when a fixed sparse basis operation is used, the sparse representation results are inaccurate and the data reconstruction accuracy is low. In addition, the processing capability of the sensing node is limited, the classic CS-based WSN data compression methods directly perform a large number of operations such as sparse transformation and matrix measurement on the sensing nodes, which not only shortens the network life but also delays the transmission time.

© Springer Nature Singapore Pte Ltd. 2021
J. Zeng et al. (Eds.): ICPCSEE 2021, CCIS 1451, pp. 429–442, 2021.
https://doi.org/10.1007/978-981-16-5940-9_33

To address the above issues, in this paper, we propose a WSN data compression model based on K-SVD dictionary and CS. The model improves the K-SVD initial dictionary, and uses the historical data to train the K-SVD sparse dictionary that adapts to the data characteristics, to ensure that the data transmission volume is reduced while improving the data recovery accuracy. And the CS-based WSN data collection model is optimized, the perception layer is only performed data observation, transferring the complex sparse transformation to the base station and extends the life of the network. The results of the experiment on the temperature data set [6] collected in Berkeley Laboratory show that the compression and recovery effect of the monitoring data with strong spatial-temporal correlation is significantly improved than the existing OEGMP and CS compression based on DCT sparse basis.

2 Related Work

2.1 Composition and Characteristics of WSN

WSN [2] consists of many sensing nodes deployed in the monitoring area and powerful fusion center (base station or sink node). The sensing nodes transmit the monitoring data to the fusion center in the way of multi hop and self-organization. The computing capacity, storage, communication bandwidth and power of WSN sensing nodes are very limited, so we need to save energy as much as possible. Therefore, it is necessary to compress a large amount of data collected by the sensing layer before transmission, and then decompress it in the application layer before use.

General WSN data has the characteristics of large amount of data, strong real-time performance, spatial-temporal correlation, and the characteristics of monitoring data generated by WSN in different scenarios are also different.

2.2 Compressed Sensing Theory

Since the compressed sensing (CS) theory was formally proposed by Donoho et al. [5] in 2006, it has been widely used in wireless communication, image processing, data acquisition and other fields. Its main content includes three parts: signal sparse transformation, matrix measurement and signal reconstruction, as shown in Fig. 1.

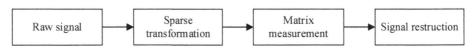

Fig. 1. CS theory components

Suppose the N-dimensional signal X is sparse, or sparse decomposition can be performed on a certain transform domain to obtain Eq. (1).

$$X = \Psi \Theta \tag{1}$$

Where Θ is a $N \times 1$ dimensional column vector with sparsity K ($K \ll N$), that is, there are only K non-zero items in Θ. Then the sparse coefficient is projected onto another observation matrix Φ with $M \times N$ ($M \ll N$) dimension that is not related to the transformation basis Ψ, and the M $\times$ 1 dimensional observation set y is obtained, as shown in Eq. (2), where A^{CS} is called the sensing matrix.

$$y = \Phi\Theta = \Phi\Psi^T X = A^{CS}X \tag{2}$$

Candès, Tao et al. Give a sufficient and necessary condition for the existence of definite solution of Eq. (2) is that Φ and Ψ are not related to each other [7], then we can use these observations to solve Eq. (3) and get the accurate recovery of signal X.

$$min\left\|\Psi^T X\right\|_0 \ s.t. \ \Phi\Psi^T X = y \tag{3}$$

The sparsity or compressibility of signal is the basis and premise of CS application, and it is also the key of data reconstruction. Therefore, finding a sparse basis that can effectively represent the signal sparsely becomes one of the important contents of CS research.

2.3 Existing WSN Data Compression Methods

With the continuous expansion of the scale of WSN applications, the amount of data is constantly multiplying. Many staff have carried out in-depth researches on improving the efficiency of data compression and reducing the amount of data transmission [3, 8].

Most of the WSN monitoring data are correlated in time and space. One of the classic algorithms to eliminate time redundancy is linear data compression algorithm, including prediction coding [9], linear regression algorithm [10], etc. Generally, the data with spatial correlation is processed by routing structure. LEACH (low energy adaptive clustering hierarchy) algorithm [11] has been well applied in the field of WSN data compression. The algorithm randomly selects cluster head nodes in a circular way and then clusters them. Sensing nodes transmit their data to the cluster head, and then cluster heads transmit the data which are processed to the base station.

The CDG (compressed data gathering) method proposed by Luo et al. [4] not only has strong load balancing characteristics and can extend the lifetime of the whole sensor network, but also reduces the communication cost without introducing intensive computing. However, the data transmission volume is still large without considering the data connection between nodes. Qiao comprehensively elaborated the compression data collection algorithm of WSN based on CS, showed the applicability of CS theory in WSN, and summarized the problems of small compression ratio, unsuitable sparse basis, inaccurate data recovery, etc. [8]. Gao compress the sparse transform coefficients of the original data to improve the data compression ratio [12]. However, the sparse transform in the sensing layer consumes energy of nodes and increases the complexity of reconstruction. Xie et al. used the difference matrix as the sparse basis, combined with the improved LEACH algorithm to study the compression of WSN data [13]. The experimental results show that this method effectively reduces the amount of data transmission, but the compression step is more complicated. Therefore, the use of routing algorithms is one of the effective methods to improve the data compression rate, meanwhile, transferring the sparse transform to the base station can effectively reduce the compression complexity and save energy consumption of nodes.

Whether the data can be accurately reconstructed depends not only on the reconstruction algorithm, but also on the sparse representation of the data. DCT (Discrete Cosine Transform) has a fixed structure and a strong "energy concentration" property, and most of the signal energy is concentrated in the low frequency part [14]. K-Singular Value Decomposition (K-SVD) dictionary learning algorithm [15] was proposed by Aharon et al. It can train a learning dictionary suitable for data characteristics and show good performance when the data is sparsely represented. The combination of K-SVD dictionary learning algorithm and compressed sensing theory has been studied in many fields. In references [16, 17], it is used for image denoising and image compression reconstruction respectively. Also, it is used for target location without equipment to improve the accuracy of position estimation [18]. Gong use it to seismic data denoising. Under the framework of compressed sensing, K-SVD dictionary shows good sparse representation effect for all kinds of data [19]. Therefore, this paper trains K-SVD dictionary for sparse representation of WSN data, and researches WSN data compression combined with compressed sensing.

3 K-SVD Dictionary Learning

Signal sparsity is the premise of using CS. Although most of the signals collected in the physical environment are not sparse, because of the strong spatial-temporal correlation and compressibility of WSN data, it is necessary to perform s parse transformation on the signal. The fixed sparse basis can't be used for sparse representation of all kinds of signals. The proposed dictionary learning solves this problem. It does not need to fix the form of dictionary in advance, but through continuous iteration and updating, which can train the appropriate dictionary according to the existing signals.

The main idea of K-SVD dictionary learning [15]: train the complete dictionary matrix $D \in R^{n \times p}$ according to the original sample data Y, D contains p atoms, $d_k \in \{d_1, d_2, ..., d_p\}$, and S is the sparse representation of Y on dictionary D, such as Eq. (4).

$$Y = DS \tag{4}$$

To find the best dictionary D that can represent the sample Y as sparsely as possible, then the sparse representation coefficient S will be obtained accordingly. At this time, S has the highest sparseness and contains the least non-zero elements. The dictionary solving problem can be expressed as Eq. (5), T_0 represents the maximum threshold of the number of non-zero elements in the vector s_i.

$$\min\|Y - DS\|^2 \ s.t. \ \|s_i\|_0 \leq T_0 \tag{5}$$

Which contains two independent variables: the dictionary D and the sparse coefficient S. The value of the independent variable when the objective function takes the minimum value is required. Generally, one of the variables is fixed, the other is solved, and so on. The dictionary D is updated column by column by SVD decomposition or least square method, and the sparse coefficient S can be solved by the existing classic methods, such as Orthogonal matching pursuit (OMP) [20], basis pursuit and other algorithms. We intends to use OMP algorithm to solve S in this paper.

4 WSN Data Compression Based on K-SVD Dictionary and Compressed Sensing

A WSN data compression model based on K-SVD dictionary and compressed sensing is proposed in this paper, as shown in Fig. 2, which is mainly divided into three parts: (1) training K-SVD sparse basis suitable for this data type in the base station using historical data sets; (2) sensing nodes collect data to cluster heads for CS compression; (3) reconstructing compressed data in the base station.

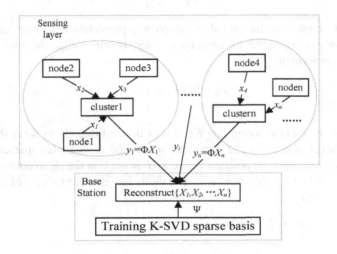

Fig. 2. WSN data compression model based on K-SVD dictionary and CS

4.1 Improved K-SVD Dictionary Training

This algorithm uses historical data to train sparse basis in base station, which not only reduces the burden of sensing nodes, but also provides more reliable and powerful computing power for sparse transformation.

WSN monitoring data usually has large spatial and temporal redundancy, and the variation range between values is small. When the original data is set as the initial dictionary for direct dictionary learning training, because the values are too similar, there is a certain linear correlation in the dictionary, which will affect the solution effect of the sparse coefficient, resulting in the situation that the dictionary is stagnant in iterative updating. To solve this problem, the DCT sparse basis [14] is set as the initial dictionary. For highly correlated data, DCT has very good energy compactness. As an initial dictionary, DCT can effectively improve the convergence speed of sparse dictionary.

When the dictionary [15] is updated column by column, the k-th column d_k of dictionary D and the k-th row sk T of sparse coefficient S, that is, Eq. (6) holds.

$$Y - DX = Y - \sum_{j=1}^{K} d_j s_T^j$$
$$= (Y - \sum_{j \neq k} d_j s_T^j) - d_k s_T^k$$
$$= E_k - d_k s_T^k \tag{6}$$

Where E_k represents the error after the sparse representation of other columns when the k-th atom is removed. At this time, the problem is converted to the solution of d_k, sk T, as shown in formula (7).

$$\arg\min_{d_k, s_T^k} ||E_k - d_k s_T^k||_F^2 \tag{7}$$

However, it is worth noting that E_k cannot be used directly to solve the problem, otherwise the calculated new sk T element diverges and no longer has sparseness. Therefore, we need to extract the position where sk T is not 0 in the E_k to obtain a new E_K' and perform SVD decomposition on it. The dictionary training process of this article is shown in Algorithm 1.

Algorithm 1. Improved K-SVD dictionary learning

K-SVD dictionary learning

Input: Y, D_0, S
Output: D
Begin
1. $D_0 \leftarrow$ DCT, j = 1;
2. Sparse coding;
3. Dictionary updating:
   ① $E_k = Y - \sum_{j \neq k} d_k s_T^j$
   ② $\omega_k = \{i \,|\, 1 \leq i \leq n, s_T^k(i) \neq 0\}, ,$
      $s_T'^k = \{s_T^k(i) \,|\, 1 \leq i \leq n, s_T^k(i) \neq 0\}$
   ③ $E_k' = U\Sigma V^T$, 记 $d_k = U(\cdot, 1)$, $s_T^k = \Sigma(1,1)V(\cdot,1)^T$
   ④ $j = j+1$
4. Determine whether the error reaches the specified threshold after the sparse representation of the sample, if it reaches the output dictionary, otherwise continue to step 2-3
End

4.2 Sensor Data Compression

In order to deal with the spatial-temporal correlation of data, this paper uses LEACH [11] routing algorithm to cluster sensing nodes, and adopts CS in the cluster to achieve WSN data compression. In view of the limited computing and processing capabilities of sensing nodes in WSN, we assumes that WSN monitoring data are all compressible, and the collected data set can be directly subjected to matrix observation, eliminating the need for sparse transformation and reducing the complexity of data compression. WSN data compression is divided into four steps: (1) Sensing nodes compete for cluster head nodes according to their remaining energy, and divide clusters; (2) Nodes in the clusters transmit monitoring data for a period of time to their cluster head nodes; (3) CS observations are Completed at the cluster head nodes to achieve data compression; (4) The compressed information is transmitted to the base station in multi-hops.

The data collected by node i in a period of time is as in Eq. (8), and the data collected by n nodes in the cluster during this period is as in Eq. (9).

$$x_i = [x_{i1}, x_{i2}, \cdots, x_{im}]^T \tag{8}$$

$$X = \begin{bmatrix} x_{11}, x_{21}, \cdots, x_{n1} \\ x_{12}, x_{22}, \cdots, x_{n2} \\ \vdots \\ x_{1m}x_{2m}, \cdots, x_{nm} \end{bmatrix} \tag{9}$$

In Eq. (9), the column vector of the matrix represents the time series data of each node in the cluster, and the row vector represents the data sensed by each node at the same time. In this way, the collected data has strong spatial-temporal correlation, and the completion of data compression can eliminate the spatial-temporal redundancy at the same time. In order to facilitate CS observation, we expand the elements of the matrix in the order of columns to obtain the $N \times 1$ ($N = m * n$) dimensional column vector, as shown in Eq. (10).

$$\text{vec}(X) = [x_{11}, x_{12}, \cdots, x_{1m}, x_{21}, x_{22}, \cdots, x_{nm}]^T \tag{10}$$

We can see in Fig. 2, at each cluster head node, the $M \times N$ ($M \ll N$) dimensional observation matrix Φ is multiplied by the cluster head data $vec(X)$ to obtain the $M \times 1$-dimensional compressed data y, as shown in Eq. (11). The compressed data arrives at the base station through multi hop transmission. In order to ensure that the data can be reconstructed, Gaussian matrix [5] is used as the observation matrix, which is not related to most orthogonal basis.

$$y = \Phi \cdot \text{vec}(X) \tag{11}$$

4.3 Data Reconstruction in Base Station

Data reconstruction is the process of decompressing compressed data by base station. According to the above analysis, the base station knows the compressed data y from the sensing layer, the observation matrix Φ and the sparse basis Ψ suitable for the environmental data (dictionary D). Due to the original data X can be expressed as Eq. (12), X can be accurately recovered by solving Eq. (13).

$$X = \Psi\Theta \tag{12}$$

$$\min\|\Theta\|_0 \quad s.t. \quad y = \Phi\Psi\Theta \tag{13}$$

Since the number of equations is much smaller than the number of unknowns, there is no deterministic solution. The l_0 norm problem of Eq. (13) is NP-hard problem and difficult to solve. Usually, the l_0 norm problem is regarded as l_1 norm or l_2 norm to solve. Candès gave that the necessary and sufficient condition for Eq. (13) to have a definite solution, that is: Φ and Ψ are not correlated with each other [7], then the accurate recovery of signal X can be obtained by solving Eq. (13) with these compressed data. In this paper, the classic OMP [20] algorithm is used to reconstruct the compressed data, and the threshold control algorithm process is added. When the data is restored to the desired effect and the error is less than the set threshold, the iteration can end directly regardless of whether the number of iterations reaches the total number of initial settings. The specific reconstruction process is shown in Algorithm 2.

Algorithm 2. Reconstruction steps of OMP algorithm

OMP

Input: $r_0,\ t,\ \Lambda_0,\ A,\ U_0 = \varnothing,\ ,k,\ \varepsilon$

Output: $\hat{X}$

Begin

1) $r_0 \leftarrow y$, $t \leftarrow 1$, $\Lambda_0 \leftarrow \varnothing$, $U_0 \leftarrow \varnothing$;

2) $\lambda_t = \arg\max |< r_{t-1}, a_j >|$, $j = 1, 2, \cdots, \lambda, \cdots, n$;

3) $\Lambda_t = \Lambda_{t-1} \cup \lambda_t$, $U_t = U_{t-1} \cup a_\lambda$;

4) $\hat{\theta}_t = \arg\min \| y - U_t\theta_t \|_2$;

5) $r_t = y - U_t\hat{\theta}_t$;

6) $t = t+1$, if $r_t < \varepsilon$ to 8), else to 7) ;

7) if $t > k$ to 8), else return 2) ;

8) $\hat{X} = \Psi\hat{\theta}_t$.

End

Where r_0 is the residual, t is the number of iterations, Λ_0 is the index set, U_0 is the matrix selected by Λ_0, $A = \Phi\Psi$, k is the number of iterations required, and ε is the threshold. Here, θ is k-sparse, that is, only k terms in the vector are nonzero, and the number of reconstruction iterations is k. It is noticed that the more iterations during reconstruction, the more time it takes and the more accurate the recovered data is. But in practical application, it is necessary to choose the appropriate number of iterations according to the recovery accuracy requirements.

5   Experiment and Result Analysis

5.1   Data Set and Experimental Environment

In this paper, the temperature data set [6] collected by Berkeley laboratory was used for experiments. The algorithms involved are implemented by Python programming language. The experimental environment is Intel Core i7-8550U CPU@1.8 GHz, running memory 8 GB, 64-bit Windows 10 operating system, and PyCharm development platform.

5.2   Performance Evaluation Index

This paper chooses three evaluation indexes to test the effectiveness of the model in this paper. Compression ration (CR) refers to the ratio of the amount of data transmitted to the sink node after the cluster head node is compressed to the amount of original data. Root mean square error (RMSE) is used to weigh the reconstructed value and the actual. The smaller the RMSE is, the smaller the deviation is and the more accurate the recovery data is. Relative reconstruction error (RRE) reflects the reliability of data recovery. Respectively as shown in Eq. (14), Eq. (15) and Eq. (16).

$$CR = \frac{\text{The amount of data transferred}}{\text{The amount of raw data}} \tag{14}$$

$$RMSE = \sqrt{\frac{\sum_{i=1}^{mn} (\hat{X}_i - X_i)^2}{N}} \tag{15}$$

$$RRE = \frac{\sum_{i=1}^{mn} \left| \hat{X}_i - X_i \right|}{\sum_{i=1}^{mn} X_i} \tag{16}$$

Among them, N is the number of data collected by the cluster heads, $\hat{X}_i$ represents the reconstructed data after decompression, and X_i represents the true value of data.

5.3   Analysis of Results

Dictionary Training. In the experimental data set [6], the temperature data generated by each node every minute is taken as the sample training set, DCT is selected as the initial dictionary, the threshold is set, and the algorithm in Table 2 is used to train the K-SVD dictionary. The experimental parameter configuration is shown in Table 1, and the sparse transformation dictionary suitable for the characteristics of the data set is obtained.

Table 1. Experimental parameter configuration

Parameters	Value
The size of training set	[1024, 8]
The number of iteration	20
Error threshold	1e−4

The sample training set is as follows:

$$
\begin{bmatrix}
19.988 & 19.862 & 19.567 & 19.43 & & 19.596 & 19.393 \\
19.097 & 19.26 & 19.44 & 19.361 & & 16.362 & 16.484 \\
18.969 & 18.99 & 19.342 & 19.293 & \cdots & 16.068 & 15.999 \\
& \vdots & & & \ddots & \vdots & \\
24.457 & 24.241 & 24.016 & 17.999 & & 19.988 & 19.878 \\
24.359 & 24.241 & 23.869 & 17.989 & \cdots & 20.233 & 20.363
\end{bmatrix}
$$

A 1024 × 1024 dimensional dictionary is generated as follows:

$$
\begin{bmatrix}
-0.067 & -0.247 & -0.545 & -0.534 & & -0.028 & -0.038 & -0.068 \\
-0.067 & -0.19 & -0.516 & 0.552 & & 0.074 & 0.096 & 0.108 \\
-0.066 & -0.277 & -0.498 & 0.043 & \cdots & -0.053 & -0.063 & -0.052 \\
& \vdots & & & \ddots & \vdots & \\
-0.061 & -0.047 & 0.023 & -0.099 & & 0.944 & -0.062 & -0.075 \\
-0.061 & -0.045 & 0.02 & -0.14 & & -0.062 & 0.927 & -0.084 \\
-0.061 & -0.041 & 0.014 & -0.167 & \cdots & -0.072 & -0.082 & 0.896
\end{bmatrix}
$$

Effect of Cluster Size on Data Reconstruction. In order to choose the optimal cluster size for WSN data compression using this model, we need to analyze the impact of different cluster sizes on the results of data reconstruction. The amount of data in a cluster is 32, 64, 128, 256, 512 and 1024 respectively, and the RMSE values under different compression ratios are compared. When the amount of data in a cluster is 32 and the compression ratio is 0.9, RMSE is 0.74, so the deviation between the recovered data and the real data is large. The experimental results show that the model is not suitable for the case of the data quantity in a cluster is less than 32 because of the small amount of data collected. The experimental results of other values are shown in Fig. 3.

Analysis of the results in Fig. 3 shows that RMSE tends to be stable with the growth of data. When the amount of data in a cluster is 128 and 256, the compression effect of this model is better.

Influence of the Number of Reconstruction Iterations on the Data Reconstruction. It can be seen from Fig. 3 that when the amount of original data in a cluster is 512, the RMSE between the reconstructed data and the original data is about 0.14, and the

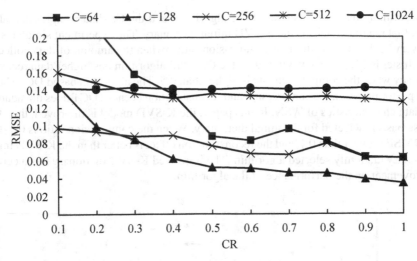

Fig. 3. The influence of different cluster sizes on the results of data reconstruction

convergence rate of the model is shown in Fig. 4. When RMSE is 0.14 for different reconstruction iterations, due to the adaptive characteristics of K-SVD dictionary, OMP is used to reconstruct only for 5 times, the data compression rate can reach 0.2, which shows that the model can converge to the optimal solution quickly.

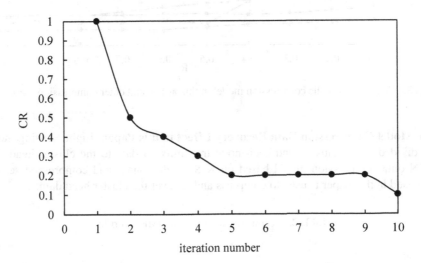

Fig. 4. Relation between reconstruction iteration number and data compression rate

Comparison Experiments with Other WSN Compression Algorithms. In order to compare the performance of the model in this paper with other WSN data compression algorithms, the compression model proposed in this paper is compared with the OEGMP

[9] algorithm, the fixed dictionary DCT algorithm, and the CS data compression algorithm that randomly selects the K-SVD initial dictionary. The experiment simulates the recovery of data under different compression ratios when the amount of data collected in a cluster is 512. As shown in Fig. 5, the OEGMP algorithm has higher data recovery accuracy when the compression rate is higher than 0.5. The algorithm based DCT shows poor performance on the compression and reconstruction because DCT does not adapt to the data characteristics of WSN. In this paper, the K-SVD model is improved, the DCT sparse basis is selected for the initial dictionary. When the compression ratio is 0.1–0.2, the RMSE is lower than 0.1, and the reconstruction effect is better than the K-SVD initial dictionary randomly selected algorithm. The improved K-SVD dictionary has a certain improvement on the performance of the algorithm.

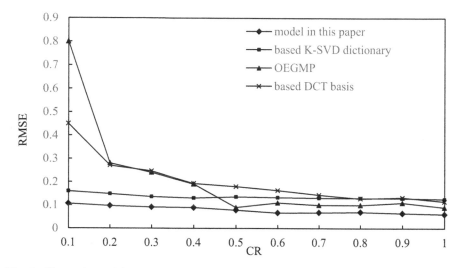

Fig. 5. Comparison of the compression model in this article and other compression algorithms

The Model Compression Data Recovery Effect in this Paper. Eight sensing nodes are divided into a cluster, and each node transmits 64 data to the cluster head. The WSN data compression model based on K-SVD dictionary and compressed sensing proposed in this paper is used to compress and recover the cluster head data.

Table 2. Reconstruction steps of OMP algorithm

Indexes	Values									
CR	0.1	0.2	0.3	0.4	0.5	0.6	0.7	0.8	0.9	1
RRE	0.073	0.061	0.057	0.054	0.046	0.044	0.043	0.041	0.042	0.036

The paper calculates RRE between the reconstructed data and the original data under different compression rates, as shown in Table 2 (the significant number takes three decimal places). With the increase of compression rate, the smaller the error of data reconstruction is. When the compression ratio is 0.2 and the number of reconstruction iterations of OMP algorithm is 5, the data reconstruction only takes 0.0691 s, the comparison effect between the reconstructed data and the original data is shown in Fig. 6. With the increase of compression rate, the smaller the error of data reconstruction is. When the compression ratio is 0.2 and the number of reconstruction iterations of OMP algorithm is 5, the data reconstruction only takes 0.0691 s, the comparison effect between the reconstructed data and the original data is shown in Fig. 6.

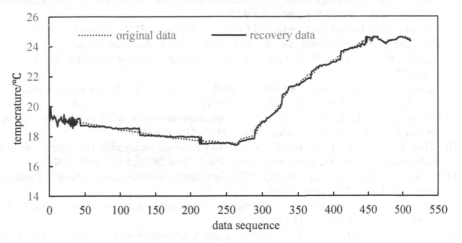

Fig. 6. Comparison chart of temperature recovery data and original data

It can be seen from Fig. 6 that the model in this paper can effectively restore the original data when the compression rate is 0.2, and RMSE of the reconstructed data and the original data is 0.097, which can meet most practical application requirements.

6 Summary

Aiming at the data collection problem of WSN monitoring environment, we proposes a WSN data compression model based on improved K-SVD dictionary and CS. This model uses the self-adaptability of K-SVD dictionary learning to train a sparse basis suitable for the monitoring data. At the same time, in order to reduce the burden of sensing nodes, the sparse transformation in CS is transferred to the base station. Which provides a new idea for WSN data compression.

Combined with theoretical analysis and experimental verification, the use of this model in WSN data compression collection not only improves compression efficiency, reduces network transmission energy consumption, but also restores the original data with higher accuracy. However, the Gaussian observation matrix used in the article requires large storage space and high computational complexity, which is not considered in this stage of work. In the future, we will study the observation matrix with small memory and simple structure that is more suitable for WSN.

References

1. Jeretta, H.N., Alex, K., Joanna, P.: The Internet of Things: review and theoretical framework. Exp. Syst. Appl. **133**(1), 97–108 (2019)
2. BenSaleh, M.S., Saida, R., Kacem, Y.H., Abid, M.: Wireless sensor network design methodologies: a survey. J. Sens. **2020**(1), 1–13 (2020)
3. Tuama, A.Y., Mohamed, M.A., Muhammed, A.: Recent advances of data compression in Wireless Sensor Network. J. Eng. Appl. Sci. **13**(21), 9002–9015 (2018)
4. Luo, C., Wu, F., Sun, J., et al.: Compressive data gathering for large-scale wireless sensor networks. In: Proceedings of the 15th Annual International Conference on Mobile Computing and Networking, 20–25 September 2009 (2009)
5. Donoho, D.L.: Compressed sensing. IEEE Trans. Inf. Theor. **52**(4), 1289–1306 (2006)
6. Intel Lab Data. http://db.csail.mit.edu/labdata/labdata.html. Accessed 12 May 2021
7. Candès, E.J., Wakin, M.B.: An introduction to compressive sampling. IEEE Sig. Process. Mag. **25**(2), 21–30 (2008)
8. Qiao, J., Zhang, X.: Compressed sensing based data gathering in wireless sensor networks: a survey. J. Comput. Appl. **11**(11), 229–237 (2017)
9. Duan, L., Zhu, L., Li, X., Li, A.: Data compression method of WSN used improved grey model. J. Beijing Univ. Posts Telecommun. **41**(2), 119–124 (2018)
10. Chen, C., Zhang, L., Tiong, R.L.K.: A new lossy compression algorithm for wireless sensor networks using Bayesian predictive coding. Wirel. Netw. **26**(8), 5981–5995 (2020)
11. Li, D., Xu, D.M.: Improvement of LEACH algorithm for wireless sensor networks. Comput. Eng. Des. **41**(7), 1852–1857 (2020)
12. Gao, J.F.: Research on application of compressed sensing in wireless sensor networks. Wirel. Internet Technol. **16**(8), 13–15 (2019)
13. Xie, X., Wang, J., Hu, F., et al.: An improved spatial-temporal correlation algorithm of Wsns based on compressed sensing. In: 2017 IEEE International Conference on Computational Science and Engineering (CSE) and IEEE International Conference on Embedded and Ubiquitous Computing (EUC), pp. 159–164 (2017)
14. Kuang, X.H., Gao, X.F., Wang, L.F., Zhao, G., et al.: A discrete cosine transform-based query efficient attack on black-box object detectors. Inf. Sci. **546**(3), 596–607 (2021)
15. Aharon, M., Elad, M., Bruckstein, A.: K-SVD: An algorithm for designing overcomplete dictionaries for sparse representation. IEEE Trans. Sig. Process. **54**(11), 4311–4322 (2006)
16. Zhu, L., Liu, S., Cao, S.N., et al.: Nonparametric Bayesian dictionary learning in sparse gradient domain for image denoising. Comput. Eng. Des. **41**(3), 802–807 (2020)
17. Tang, X.R., Liu, Y.T., Zhang, Y., et al.: Compressed sensing reconstruction of core image based on K-SVD dictionary learning. J. Jilin Univ. (Inf. Sci. Edn.) **38**(3), 108–114 (2020)
18. Jin, J., Ke, W., Lu, J.: Device-free localization based on link selection learning algorithm. Chin. J. Radio Sci. **33**(5), 583–590 (2018)
19. Gong, Z., Song, W.Q., Wang, C., et al.: Seismic data denoising based on K-SVD dictionary learning method, June 2020. https://kns.cnki.net/kcms/detail/11.2982.P.20200608.1134.066. html
20. Li, J., Chow, P., Peng, Y., Jiang, T.: FPGA implementation of an improved OMP for compressive sensing reconstruction. IEEE Trans. Very Large Scale Integr. (VLSI) Syst. **29**(2), 259–272 (2021)

Channel Drop Out: A Simple Way to Prevent CNN from Overfitting in Motor Imagery Based BCI

Jing Luo[1]([✉])(iD), Yaojie Wang[1], Rong Xu[2], Guangming Liu[1], Xiaofan Wang[1], and Yijing Gong[1]

[1] Shaanxi Key Laboratory for Network Computing and Security Technology, School of Computer Science and Engineering, Xi'an University of Technology, Xi'an, China
`luojing@xaut.edu.cn`
[2] Shaanxi Province Institute of Water Resources and Electric Power Investigation and Design, Xi'an, China

Abstract. With the development of deep learning, many motor imagery brain-computer interfaces based on convolutional neural networks (CNNs) show outstanding performances. However, the trial number of EEG in the training set is usually limited, and redundancy extensively exists in multiple channel EEG. Thus, overfitting often appears in CNN based motor imagery recognition model and greatly affects the performances of model. In this paper, channel drop out is proposed to address this problem by data augmentation and ensemble learning. Specifically, one of all EEG channels will be dropped and replaced by the mean signal of all EEG channels. In this way, the trial number in the training set was enlarged by channel drop out. And at the testing stage, all the EEG trials processed by channel drop out were fed to the CNN model and the average output probabilities of them were applied to determine the prediction. The experiments were conducted on two popular CNN models applied in motor imagery BCI and BCI Competition IV datasets 2a to verify the performances of the proposed channel drop out approach. The results show that average improvements provided by channel drop out in two-category or four-category motor imagery classification are 2.83% and 2.65% compared with the original CNN model. So the channel drop out approach significantly improves the performances of motor imagery based BCI.

Keywords: Brain-computer interface (BCI) · Channel drop out · Motor imagery (MI) · Convolutional neural network (CNN)

1 Introduction

Electroencephalography (EEG) is a kind of simple, flexible, non-invasive brain monitoring technology. Motor imagery based EEG recognition is an important

Supported by the National Natural Science Foundation of China, grants 61906152 and 61976177 and Key Research and Development Program of Shaanxi (Program No. 2021GY-080).

branch of brain-computer interface (BCI). The motor imagery recognition based BCI system collects EEG signals when the subject performs specific motor imagery, recognizes the motor imagery content according to the EEG signals, and then converts the recognition results into control commands of peripheral devices, such as a robotic arm [1,2]. EEG signals have the characteristics of low signal-to-noise ratio and low spatial resolution. Thus, how to extract effective features from EEG signals is the key to success of the motor imagery BCI system [3].

Based on the features applied in motor imagery BCI, existing approaches mainly consist of three classes: common spatial pattern (CSP)-based approaches, frequency feature-based approaches and neural network-based approaches. Firstly, common spatial pattern (CSP) has achieved great success in motor imagery BCI. The main idea of CSP is searching for the optimal spatial filters to minimize the variance of EEG in one class and maximize the variance of EEG in another class simultaneously [4]. Many variants of the CSP have been proposed. The filter bank CSP algorithm used 9 sub-band band-pass filters to process EEG signals, and then extracted CSP features. The discriminant CSP features were selected and fed to naïve Bayesian Parzen window classifier [5]. Luo proposed an ensemble support vector learning based approach to combine the advantages of the event-related desynchronization and event-related synchronization (ERD/ERS)-based CSP features and the event-related potential-based features in motor imagery-based EEG classification [6]. Secondly, spectrum-based features is extracted and applied in motor imagery BCI based on detection of ERD/ERS. Shahid and Prasad utilized an advanced and robust high-order statistical method called bispectrum analysis to extract discriminant feature from EEG signals [7]. Three common spectral analysis methods, i.e. wavelet packet decomposition, discrete wavelet transform and empirical mode decomposition, are used by Morioka to extract features from EEG signals for motor imagery based BCI. The high-order wavelet packet decomposition method combined with multi-scale principal component analysis noise reduction method obtained the best recognition accuracy in the experiment results [8].

With the fast development of deep learning in computer vision and natural language processing, many neural network models have appeared and provided significant performances in motor imagery based BCI, especially the convolutional neural network (CNN) model is widely used in the field of motor imagery based BCI. Schirrmeister investigated CNNs with different architectures and design choices for decoding executed motion or motor imagery tasks from raw EEG signals (EEGDecoding) [12]. The experiment results showed that the proposed "Deep ConvNet" and "Shallow ConvNet" performed best among the compared CNN structures. Recent advances in neural network improved the result significantly, such as batch normalization [9], dropout [10] and ELUs activation function [11]. EEGNet is a compact convolutional network presented robust performances for four BCI paradigms, i.e. movement-related cortical potentials, sensory motor rhythms, P300 visual-evoked potentials and error-related negativity responses [13]. Ren proposed a convolutional deep belief network to learn feature from EEG and used them in motor imagery classification task [14]. Sakhavi proposed a scheme combining static energy network and CNN to learn the dynamic energy representation of EEG signal for motor imagery EEG classification [15]. Tabar proposed a new input form of CNN combining time, frequency and location information of EEG. Then, a shallow CNN consisting one convolution layer and one max-pooling layer, and another deep CNN combining stacked autoencoders were proposed for motor imagery classification [16]. Tang built a 5-layer CNN structure to classify single-trial motor imagery EEG based on the spatio-temporal characteristics of EEG [17]. A hybrid convolution scale network and a data augmentation method were proposed for motor imagery BCI. And the experiment result outperformed the state-of-the-art methods [18].

However, as the collection of motor imagery based EEG is difficult, the number of EEG trials is usually limited. In addition, redundancy extensively exists in multiple channel EEG. For this small sample task, overfitting often appears in CNNs and greatly affects the performances. To prevent overfitting, the channel drop out approach is proposed in this paper. Specifically, one of all EEG channels is dropped and replaced by the mean signal of all EEG channel. Thus, the trial number in the training set can be enlarged by channel drop out. And at the testing stage, all the EEG processed by channel drop out are fed to the CNN classification model and the average output probabilities of them are applied to determined the prediction. This procedure is an probability ensemble method. The proposed channel drop out approach is universal and can be applied to any motor imagery recognition based CNN model.

The rest of this paper is organized as follows. In Sect. 2, we describe the proposed channel drop out approach in details. The data and the experiment is introduced in Sect. 3. The Experiment results is given in Sect. 4 and the conclusion is drawn in Sect. 5.

2 Methodology

For motor imagery based EEG classification task, overfitting extensively exists in CNN based model. Firstly, motor imagery based EEG collection is more complex and expensive compared with natural image capturing and many other forms of data, and the model training is usually based on samples of individual subject. So the sample number in the training set is usually limited. Secondly, for motor imagery BCI, multiple channels EEG with high sampling rate are collected. For example, in BCI Competition IV dataset 2a, EEG signals from 22 channels are monitored 250 Hz sampling rate. The positions of the electrodes (International 10–20 system) used are shown in Fig. 1. Thus, redundant information extensively exists in EEG because of multiple channel collecting and high sampling rate. As a result, overfitting greatly affects the performances of motor imagery based BCI. To prevent overfitting, enlarging the sample number in the training set is a simple but efficient method. In this paper, channel drop out is applied at the training stage to reasonably expand the number of training sample. In addition, model combination is usually applied to improve the classification performance.

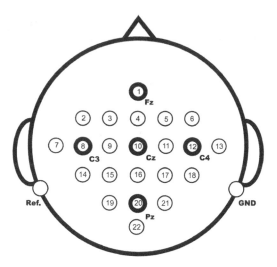

Fig. 1. The locations of the electrodes in the International 10–20 system.

But model combination requiring different model architectures or different training sets is usually expensive. On the contrary, channel drop out can be easily applied to achieve probability ensemble at the testing stage in single CNN without model parameter increasing.

2.1   Channel Drop Out at the Training Stage

At the training stage, artificial EEG trials manufactured by channel drop out are applied in model training. Firstly, the average EEG signal of all channel in one EEG trial is calculated:

$$\bar{E} = \frac{1}{C} \sum_{i=1}^{C} E_i \qquad (1)$$

where E_i is the EEG signal of channel i and C is the total channel number. Secondly, One of all channels is replaced by the average EEG signal.

$$E_i^d = \{E_1, E_2, ..., E_{i-1}, \bar{E}, E_{i+1}, ..., E_C\} \qquad (2)$$

where the ith channel of ith artificial EEG signal is replaced by the average EEG signal. If one channel of EEG is replaced by the average EEG signal, the modified EEG become a new different sample (E_i^d). In this way, C more modified EEG trials are manufactured by channel drop out. The original EEG trial and all C modified EEG trials are combined to expand sample number of the training set. Then, the model training is executed as same as the original CNN model.

Similar to the unit drop out approach [10], training set augmented by channel drop out will make the classification model more robust. This augmented training set drives the model to learn useful features from the all EEG channel, instead of depending on one specific channel, because one specific channel is dropped and replaced by the mean EEG signal. This also can be regarded as a data augmentation method.

2.2 Channel Drop Out at the Testing Stage

At the testing stage, the channel drop out approach is also applied to manufacture artificial EEG trials. The original EEG trials combined with the artificial EEG trials are both fed to the trained CNN model. As the cross entropy loss is applied in CNN training, the output of CNN means the classification probability of each EEG trial. Then the CNN model outputs of original EEG trial and artificial EEG trials are combined to determine the prediction of motor imagery, the average probability is calculated as:

$$P = \frac{\sum_{i=0}^{C} P_i}{C+1} \tag{3}$$

where P_i is the output probabilities of ith artificial EEG trials and P_0 is the original EEG trial. This operation acts as a probability ensemble method and the class with the largest average probability is the final predicted label. Through this probability ensemble method, we can improve the performance by ensemble learning method in single CNN model without increasing of parameters number.

3 Data and Experiment

BCI Competition IV dataset 2a [19] is employed to verify the efficiency and universality of our proposed channel drop out approach.

3.1 BCI Competition IV Dataset 2a

BCI Competition IV dataset 2a provides a four-category motor imagery recognition task [19]. Motor imagery based EEG signals from 9 subjects are collected. For each subject, there are 288 trials (72 trials per class) in training session and testing session recorded on different dates. According to the prompts displayed on the screen, subjects perform four different types of motor imagery tasks (left hand, right hand, tongue and feet), and EEG signals from 22 channels are monitored. The positions of the electrodes used are shown in Fig. 1. The sampling rate of the EEG signal 250 Hz, and the resolution of the amplifier is set to 100 μV. In each trial, the prompt appears on the screen at 2 s, and the execution time of the motor imagery task is from 3 s to 6 s. 0.5–100 Hz band pass filter 50 Hz notch filter are used to preprocess the collected EEG signal. After recording, experts mark abnormal trials as artifacts. More details can be found in [19].

Table 1. The two-category motor imagery classification precision of channel drop out approach is compared with the original CNN.

	Accuracy			MID AVG accuracy			Last AVG accuracy			Max accuracy		
	Original	CDO	Improve	Original	CDO	Improve	Original	CDO	Improve	Original	CDO	Improve
EEGNet-2a-0 Hz	81.59%	82.85%	1.27%	80.27%	82.25%	1.98%	81.39%	82.95%	1.56%	83.80%	86.70%	2.90%
Shallow-2a-0 Hz	83.15%	**88.80%**	5.65%	82.20%	**87.70%**	5.51%	82.74%	**88.47%**	5.73%	85.94%	91.10%	5.15%
EEGNet-2a-4 Hz	77.25%	76.42%	−0.83%	74.62%	75.83%	1.21%	76.15%	75.90%	−0.25%	79.37%	80.83%	1.47%
Shallow-2a-4 Hz	87.87%	88.26%	0.39%	86.45%	86.84%	0.39%	87.44%	88.20%	0.76%	89.81%	**91.62%**	1.81%
Average	82.47%	84.08%	1.62%	80.88%	83.16%	2.27%	81.93%	83.88%	1.95%	84.73%	87.56%	2.83%

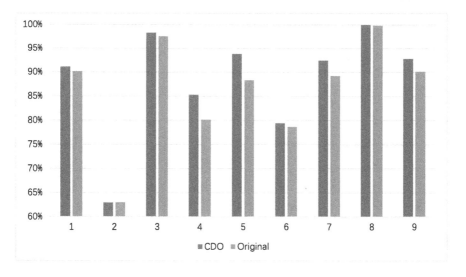

Fig. 2. The two-category motor imagery classification accuracy comparison for each subject with Shallow ConvNet.

3.2 Experiment

The proposed channel drop out approach is independent to CNN structure and can be utilized on any motor imagery recognition CNN model. To verify the efficiency and universality of channel drop out approach, two popular CNN models in motor imagery recognition, including EEGNet [13], Shallow ConvNet from EEGDecoding [12], are employed in the experiment comparison. EEGNet consists of 3 convolution layers and 1 full connect layer. Batch normalization, dropout, maxpooling and exponential linear unit (ELU) activation function [11] are applied in the EEGNet. Convolution kernels with sizes of [2, 32] and [8, 4] are employed in the experiment as it performed best in sensory motor rhythm data set in [18]. The Shallow ConvNet inspired by the FBCSP pipeline consists of 2 convolution layers and 1 full connect layer. Square function is set as activation function in Shallow Covnet. In EEGDecoding model, maxpooling, dropout and batch normalization are applied. Only Shallow ConvNet from EEGDecoding is employed in this paper, because they performed best in experiment comparison reported in [12]. In this paper, the channel drop out approach is evaluated on a two-category classification task (left hand and right hand motor imagery) and a four-category classification task separately.

To make full use of feature learning ability of CNN model, the EEG trials are only minimally preprocessed. A 4.5 s EEG segment from 0.5 s before the appearance of motor imagery prompt to 4 s after it is employed in the experiment. The EEG signal is filtered by a 0–38 Hz or 4–38 Hz band pass filter as the configuration in [12], and then fed to the CNN model. The parameters of each model are set according to the published paper of EEGNet and EEGDecoding. The adam optimization method [20] and cross entropy loss are used in model training. The experiments in this paper are implemented by Python and Pytorch deep learning framework [21].

4 Experiment Results

At first, the two-category motor imagery classification precisions of CNN models with and without channel drop out are compared in this section. The maximum epochs number is set as 1000 to ensure model convergence. The accuracy of epoch with the least training loss is presented to evaluate the classification precision (Accuracy in the Table 1 and 2). In order to reduce the influence of randomness on the experimental results, the average testing accuracy of 501–600 epochs (MID AVG Accuracy), the average testing accuracy of 100 epochs before the least training loss epoch (Last AVG Accuracy) and the maximum testing accuracy (Max Accuracy) are also presented. In addition, these measurements of precision can help readers to observe overfitting of the model. The average results of five runs and all 9 subjects are showed in Table 1. The improvement is measured by the accuracy difference. The first column indicates the original CNN model name and the band pass filter applied in the preprocessing. The highest accuracy is marked in bold. We also present accuracy comparison (the average testing accuracy of 100 epochs) for each subject with Shallow ConvNet and 0–38 Hz band pass filter in Fig. 2.

Some conclusions can be drawn from the left/right hand two-category experimental results: (1) the channel drop out approach improves the precision of motor imagery recognition at 1.62%, 1.95% and 2.83% on average compared with the original CNN model; (2) the Shallow Convnet with channel drop out performs best; (3) the performances improvement of channel drop out is universality, which is independent to the original CNN model and the band pass filter applied in the preprocessing; (4) no obvious overfitting appears in CDO models.

In addition, the four-category motor imagery classification precision of CNN model with and without channel drop out is also compared. Because of the outstanding performance of Shallow ConvNet, it is employed in four-category classification. The other configurations of experiments are set as same as two-category motor imagery classification and the results are presented in Table 2. We also present accuracy comparison (the average testing accuracy of 100 epochs) for each subject with Shallow ConvNet and 0–38 Hz band pass filter in Fig. 3. We can conclude from the experimental comparison that: (1) the channel drop out

Table 2. The four-category motor imagery classification precision of channel drop out approach is compared with the original CNN.

	Accuracy			MID AVG accuracy			Last AVG accuracy			Max accuracy		
	Original	CDO	Improve	Original	CDO	Improve	Original	CDO	Improve	Original	CDO	Improve
Shallow-2a-0 Hz	75.70%	78.32%	2.62%	74.51%	77.84%	3.33%	75.00%	78.00%	3.01%	78.67%	81.40%	2.73%
Shallow-2a-4 Hz	76.41%	**78.82%**	2.41%	75.43%	**78.34%**	2.91%	75.71%	**78.53%**	2.82%	79.44%	**82.00%**	2.57%
Average	76.06%	78.57%	2.51%	74.97%	78.09%	3.12%	75.35%	78.27%	2.91%	79.05%	81.70%	2.65%

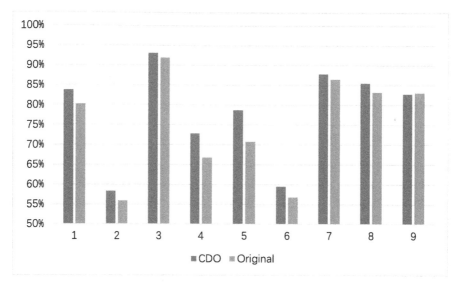

Fig. 3. The four-category motor imagery classification accuracy comparison for each subject with Shallow ConvNet and 0–38 Hz band pass filter.

approach improves the precision of motor imagery recognition at 2.51%, 2.91%, 2.65% on average compared with the original Shallow Convnet model; (2) the performances improvement of channel drop out is independent to the band pass filter applied in the preprocessing; (3) no obvious overfitting appears in CDO models.

5   Conclusion

In this paper, to prevent the overfitting problem existing in the motor imagery recognition based CNN model, a channel drop out approach is proposed by the idea of data augmentation and probability ensemble. Specifically, one channel from EEG signal is dropped and replaced by the mean EEG signal of all channel. Thus, the trial number in the training set is enlarged by channel drop out. And at the testing stage, all the EEG processed by channel drop out are fed to the CNN classification model and the output probabilities of them are combined to determine the prediction as a probability ensemble method. The experiment results showed that the performances improvement of channel drop out is universality, which is independent to the original CNN model and the band pass filter applied in preprocessing. However, this study is focusing on subject-independent model. In future work, we plan to study the regularization method improving the subject-to-subject generalization ability applied in multi-subject motor imagery recognition based CNN model.

References

1. Edelman, B.J., et al.: Noninvasive neuroimaging enhances continuous neural tracking for robotic device control. Sci. Robot. **4**(31), eaaw6844 (2019)
2. Penaloza, C.I., Nishio, S.: BMI control of a third arm for multitasking. Sci. Robot. **3**(20), eaat1228 (2018)
3. Luo, J., Wang, J., Xu, R., Xu, K.: Class discrepancy-guided sub-band filter-based common spatial pattern for motor imagery classification. J. Neurosci. Methods **323**, 98–107 (2019)
4. Lemm, S., Blankertz, B., Curio, G., Muller, K.: Spatio-spectral filters for improving the classification of single trial EEG. IEEE Trans. Biomed. Eng. **52**(9), 1541–1548 (2005)
5. Ang, K.K., Chin, Z.Y., Wang, C., Guan, C., Zhang, H.: Filter bank common spatial pattern algorithm on BCI competition IV datasets 2a and 2b. Front. Neurosci. **6**, 39 (2012)
6. Luo, J., Gao, X., Zhu, X., Wang, B., Lu, N., Wang, J.: Motor imagery EEG classification based on ensemble support vector learning. Comput. Methods Programs Biomed. **193**, 105464 (2020)
7. Shahid, S., Prasad, G.: Bispectrum-based feature extraction technique for devising a practical brain-computer interface. J. Neural Eng. **8**(2), 025014 (2011)
8. Morioka, H., et al.: Learning a common dictionary for subject-transfer decoding with resting calibration. Neuroimage **111**, 167–178 (2015)
9. Ioffe, S., Szegedy, C.: Batch normalization: accelerating deep network training by reducing internal covariate shift. arXiv preprint arXiv:1502.03167 (2015)
10. Srivastava, N., Hinton, G., Krizhevsky, A., Sutskever, I., Salakhutdinov, R.: Dropout: a simple way to prevent neural networks from overfitting. J. Mach. Learn. Res. **15**(1), 1929–1958 (2014)
11. Clevert, D.-A., Unterthiner, T., Hochreiter, S.: Fast and accurate deep network learning by exponential linear units (ELUS). arXiv preprint arXiv:1511.07289 (2015)
12. Schirrmeister, R.T., et al.: Deep learning with convolutional neural networks for EEG decoding and visualization. Hum. Brain Mapp. **38**(11), 5391–5420 (2017)
13. Lawhern, V.J., Solon, A.J., Waytowich, N.R., Gordon, S.M., Hung, C.P., Lance, B.J.: EEGNet: a compact convolutional neural network for EEG-based brain-computer interfaces. J. Neural Eng. **15**(5), 056013 (2018)
14. Ren, Y., Wu, Y.: Convolutional deep belief networks for feature extraction of EEG signal. In: 2014 International Joint Conference on Neural Networks (IJCNN), pp. 2850–2853. IEEE (2014)
15. Sakhavi, S., Guan, C., Yan, S.: Parallel convolutional-linear neural network for motor imagery classification. In: 2015 23rd European Signal Processing Conference (EUSIPCO), pp. 2736–2740. IEEE (2015)

16. Tabar, Y.R., Halici, U.: A novel deep learning approach for classification of EEG motor imagery signals. J. Neural Eng. **14**(1), 016003 (2016)
17. Tang, Z., Li, C., Sun, S.: Single-trial EEG classification of motor imagery using deep convolutional neural networks. Optik **130**, 11–18 (2017)
18. Dai, G., Zhou, J., Huang, J., Wang, N.: HS-CNN: a CNN with hybrid convolution scale for EEG motor imagery classification. J. Neural Eng. **17**(1), 016025 (2020)
19. Tangermann, M., et al.: Review of the BCI competition IV. Front. Neurosci. **6**, 55 (2012)
20. Kingma, D.P., Ba, J.: Adam: a method for stochastic optimization. arXiv preprint arXiv:1412.6980 (2014)
21. Paszke, A., et al.: Pytorch: an imperative style, high-performance deep learning library. In: Advances in Neural Information Processing Systems, pp. 8026–8037 (2019)

Quantum Color Image Scaling on QIRHSI Model

Guanglong Chen and Xianhua Song[✉]

School of Science, Harbin University of Science and Technology, Harbin, China
songxianhua@hrbust.edu.cn

Abstract. Scaling operations are widely used in traditional image processing. Therefore, in this paper, an improved quantum image representation based on HSI color space (IQIRHSI) is proposed, which extends the original $2^n \times 2^n$ size to general $2^{n_1} \times 2^{n_2}$ size. Then, the quantum algorithms and circuits were designed to implement quantum image scaling. Interpolation was introduced to recover the lost information in the scaled image. The nearest neighbor interpolation method was researched on scaled IQIRHSI to make the interpolation method easy to implement. Finally, the complexity of the quantum circuit for image scaling was analyzed and the process of quantum image scaling was described in detail by examples.

Keywords: Quantum image processing · Quantum computation · QIRHSI representation · Image scaling · Nearest neighbor interpolation

1 Introduction

Quantum computation [1] is the product of a new discipline formed after the mutual integration of quantum mechanics and computer science. The earliest idea of quantum computation was proposed by Feynman in 1982, considering that a new type of computer designed using the principles of quantum mechanics could solve certain problems more efficiently than traditional computers [2]. In 1985, Deutsch first introduced the concept of quantum Turing machines, i.e., general-purpose quantum computers, by drawing on the properties of quantum mechanics and modeling them after the machines defined by Turing [3]. In 1994, Shor proposed a quantum computer-based large number factorization algorithm that could decompose the large number factorization from NP problem into P problem [4], further challenging the security basis of the RSA public key cryptosystem. In 1996, Grover proposed the quantum search algorithm [5], a method that could achieve accelerated search computations by an order of magnitude of the square root of the original speed. The emergence of these two quantum algorithms posed a fundamental threat to the current cryptosystem, and this particular potential, in turn, accelerated the development of quantum computers. Quantum computation is a new model of computation based on the principles of quantum mechanics [1], such as superposition and entanglement of quantum states to store, process and transmit information.

© Springer Nature Singapore Pte Ltd. 2021
J. Zeng et al. (Eds.): ICPCSEE 2021, CCIS 1451, pp. 453–467, 2021.
https://doi.org/10.1007/978-981-16-5940-9_35

Quantum image processing (QIP) is an extremely significant part of quantum information processing. Quantum image representation is the cornerstone of QIP, and the main model sets of quantum image representation in QIP are Qubit Lattice [6], Real Ket [7], Entangled Image [8], FRQI [9], NEQR [10], QSMC&QSNC [11], QIRHSI (Submission in progress [12]) and so on. Qubit Lattice arranges the qubits required for the image in the form of a quantum matrix [6]. Real Ket model maps each pixel to a ground state in a 4-dimensional sequence of qubits to store the image [7]. Entangled Image model gives the relationship between image pixels through entangled states [8]. The FRQI model represents the color and location information of the grayscale image as a quantum superposition state of $2n + 1$ qubits [9]. The NEQR model stores the color and location information of the grayscale image with the help of $2n + q$ qubits [10]. QSMC&QSNC are color image representation models extended from the Qubit Lattice model using two sets of quantum states representing m colors and n coordinates respectively [11].

The QIRHSI method on which our designed quantum circuit is based is constructed from FRQI and NEQR, and QIRHSI has the advantages of both of them making it suitable for image scaling in QIP.

- QIRHSI uses two qubits to represent hue and saturation information and q qubits binary sequence to encode intensity information, which not only reduces the number of qubits needed to represent the image but also makes it easier to perform various transform operations on the intensity.
- Both intensity information and position information are stored in a binary quantum sequences using q qubits and $2n$ qubits, respectively, so it is easier to port certain algorithms from conventional computer to quantum computer.
- Intensity and position information of any part of the image can be easily processed with the help of Controlled-NOT and Toffoli gates.

On the basis of the increasing maturity of quantum image representation, research on related quantum computation methods is also flourishing. Examples include geometric transformations [13], image processing [14], feature extraction [15], image encryption [16], image segmentation [17], image scrambling [18], digital watermarking [19, 20], quantum movies [21, 22], etc.

In classical image processing, image scaling is a common image transformation operation, and so far it is still in its infancy. In 2014, Jiang and Wang firstly used the scaling operation for quantum intensity images, based on INEQR [23], using nearest neighbor interpolation method for scaling $2^{n_1} \times 2^{n_2}$ quantum images, followed by the study of quantum intensity image scaling based on GQIR [24] for $H \times W$. Next year, Sang, Wang and Niu constructed the halving operation to scaled quantum images using nearest neighbor interpolation. Both FRQI and NEQR models were studied, and the quantum circuits were improved with the help of Controlled-NOT gate [25]. In 2017, Zhou studied quantum multidimensional color image scaling based on FRQI with the help of nearest neighbor interpolation [26]. In 2018, Li and Liu studied bilinear interpolation of quantum Fourier transform based on NEQR model. Next, the bilinear interpolation and nearest neighbor scaling effects of interpolation were compared [27].

Image scaling is very different from other geometric transformations, and the reasons for this phenomenon are:

- The image needs to be resized.
- New pixels need to be added or redundant pixels need to be removed using interpolation methods.

Since image scaling involves interpolation methods, different interpolation methods strike a balance between accuracy and computational complexity. The essence of interpolation is the process of estimating pixels at unknown locations from known data. Nearest neighbor interpolation, bilinear interpolation and bicubic interpolation are three common grayscale interpolation methods [28, 29]. The effect of the three interpolation methods on the color image "Lena" after scaling is shown in Fig. 1, where the size of the scaled image (**b**) (**c**) (**d**) is two times the size of the image before scaling (**a**).

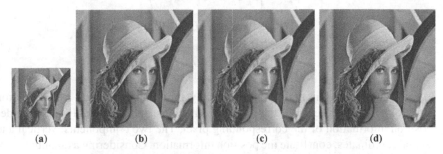

(a)   (b)   (c)   (d)

Fig. 1. The effect after scaling with different interpolation methods. (**a**) Original "Lena" image (**b**) Nearest neighbor interpolation (**c**) Bilinear interpolation (**d**) Bicubic interpolation

The main contributions of this paper are presented here: Firstly, QIRHSI is extended to general size; Secondly, QIRHSI color images are scaled with the help of nearest neighbor interpolation method; Thirdly, the scaling up and scaling down operations of QIRHSI color images are investigated and the corresponding quantum circuits are designed; Finally, the complexity analysis of quantum circuits are presented.

The remaining work in this paper is distributed as follows: Sect. 2 describes the basic knowledge about quantum color image scaling; Sect. 3 describes the improved QIRHSI color image model; The algorithms for color image scaling and quantum circuits and their complexity analysis are presented in Sect. 4; Sect. 5 gives some specific examples; The last section summarizes the content of the paper and indicates the future direction of the work.

2 Basic Knowledge

2.1 Novel Quantum Image Representation Based on HSI Color Space (QIRHSI)

Assume that the range of values of color image intensity is $[0, 2^q - 1]$, binary sequence $C_k^0 C_k^1 \cdots C_k^{q-1}$ encodes the intensity value I_k of the corresponding position k as in (1):

$$|I(\theta)\rangle = \frac{1}{2^n} \sum_{k=0}^{2^{2n}-1} |C_k\rangle \otimes |k\rangle = \frac{1}{2^n} \sum_{k=0}^{2^{2n}-1} |H_k\rangle |S_k\rangle |I_k\rangle \otimes |k\rangle \qquad (1)$$

Where

$$\begin{aligned}
|H_k\rangle &= \cos\theta_{hk}|0\rangle + \sin\theta_{hk}|1\rangle \\
|S_k\rangle &= \cos\theta_{sk}|0\rangle + \sin\theta_{sk}|1\rangle \\
|I_k\rangle &= \left| C_k^0 C_k^1 \cdots C_k^{q-2} C_k^{q-1} \right\rangle
\end{aligned} \qquad (2)$$

$$\begin{aligned}
\theta_{hk}, \theta_{sk} &\in [0, \pi/2], \quad C_k^m \in \{0, 1\} \\
m &= 0, 1, \cdots, q-1 \\
k &= 0, 1, \cdots, 2^{2n} - 1
\end{aligned} \qquad (3)$$

θ_{hk}, θ_{sk} constitute the main phase encoding information for hue and saturation vectors. $|C_k\rangle$ is used to encode the color information, and accordingly, $|k\rangle$ encodes the position information of the corresponding pixel. The two components, vertical and horizontal coordinates, constitute the position information. Considering a quantum color image in $2n$-qubits system,

$$\begin{aligned}
|k\rangle = |y\rangle |x\rangle &= |y_0 y_1 \cdots y_{n-2} y_{n-1}\rangle |x_0 x_1 \cdots x_{n-2} x_{n-1}\rangle \\
y, x &\in \{0, 1, 2, \ldots, 2^n - 1\} \\
y_j, x_j &\in \{0, 1\}, j = 0, 1, \ldots, n-1
\end{aligned} \qquad (4)$$

Here $|y\rangle = |y_0 y_1 \cdots y_{n-1}\rangle$ encodes the first n-qubits along the vertical direction and $|x\rangle = |x_0 x_1 \cdots x_{n-1}\rangle$ encodes the second n-qubits along the horizontal direction. Figure 2 shows a simplified 2×2 color image and its QIRHSI representation, respectively.

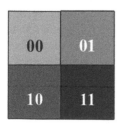

$$\begin{aligned}
|I(\theta)\rangle = \frac{1}{2} \Bigg[&\left(\cos\frac{39\pi}{100}|0\rangle + \sin\frac{39\pi}{100}|1\rangle \right)\left(\cos\frac{21\pi}{100}|0\rangle + \sin\frac{21\pi}{100}|1\rangle \right)|11111111\rangle \otimes |00\rangle \\
+ &\left(\cos\frac{17\pi}{100}|0\rangle + \sin\frac{17\pi}{100}|1\rangle \right)\left(\cos\frac{7\pi}{20}|0\rangle + \sin\frac{7\pi}{20}|1\rangle \right)|01111110\rangle \otimes |01\rangle \\
+ &\left(\cos\frac{11\pi}{25}|0\rangle + \sin\frac{11\pi}{25}|1\rangle \right)\left(\cos\frac{3\pi}{10}|0\rangle + \sin\frac{3\pi}{10}|1\rangle \right)|01011000\rangle \otimes |10\rangle \\
+ &\left(\cos\frac{27\pi}{100}|0\rangle + \sin\frac{27\pi}{100}|1\rangle \right)\left(\cos\frac{23\pi}{100}|0\rangle + \sin\frac{23\pi}{100}|1\rangle \right)|01001011\rangle \otimes |11\rangle \Bigg]
\end{aligned}$$

Fig. 2. A 2×2 color image and its QIRHSI representation

2.2 Nearest Neighbor Interpolation for Classical Image Scaling

Image scaling is one of the more widely used image processing operations, the main purpose of which is to resize the image. It usually adds new pixels or removes remaining pixels with the help of interpolation methods. Nearest neighbor, bilinear, and bicubic interpolation [28, 29] are three common interpolation methods. In comparison, the easiest to implement and the most efficient interpolation method is the nearest neighbor.

The nearest neighbor interpolation procedure can usually be decomposed into two directions: horizontal and vertical. As shown in Eq. (5).

$$I' = S(I, r_x, r_y) = S_x(S_y(I, r_y), r_x) = S_y(S_x(I, r_x), r_y) \tag{5}$$

among them the scaling function is S, the initial image is I, the image after scaling is I', the scaling in the x-axis and y-axis directions are r_x and r_y, respectively. S_x and S_y are compounded into a scaling function S, and where S_x denotes the horizontal scaling function and S_y indicates the vertical scaling function. The order of action of horizontal and vertical scaling functions does not affect the scaled image, hence S_y and S_x are interchangeable. In other words, the scaling of the whole image is a composite of r_y and r_x. Namely, the scaling of the whole color image is compounded by the horizontal and vertical scaling. Figure 3 demonstrates us the scaling as shown in Eq. (5), where $r_x = 4$ and $r_y = 2$. Therefore, we first study the scaling of the quantum color image in one direction, and then extend to the scaling of the entire quantum color image, and finally design the complete quantum circuits through some detailed examples.

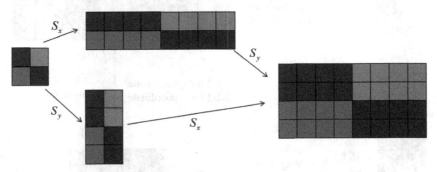

Fig. 3. The decomposability and the exchangeability of S

Suppose the original image size is $2^{n_1} \times 2^{n_2}$, the scaled image size is $2^{m_1} \times 2^{m_2}$, and the scaling ratios are $r_y = 2^{m_1-n_1}$ and $r_x = 2^{m_2-n_2}$, where n_1, n_2, m_1, m_2 are non-negative positive integers. Under this premise, we will take the horizontal direction as an example to further expand the scaling of quantum color images [28–30].

- Scaling up: Repeat the pixel value r_x times. For instance, "AB" is the original image, the scaling ratio $r_x = 4$, and "AAAABBBB" is the scaled image.
- Scaling down: Each $1/r_x$ pixel points in the initial image will be reduced to one pixel after scaling. The color value of the new pixel location x' is equal to the color value of

the first set of $1/2r_x$ pixels, i.e., pixel $x'/r_x + 1/2r_x = x' \cdot 2^{n_2-m_2} + 2^{n_2-m_2-1}$ in the initial image, as presented in Fig. 4. For instance, "ABCDEFGH" is the initial image, $r_x = 1/4$ is the scaling ratio, and "BDFH" is the scaled image; if the scaling ratio $r_x = 1/2$, the scaled image is "CG". Figure 5 describes a simple example of image scaling.

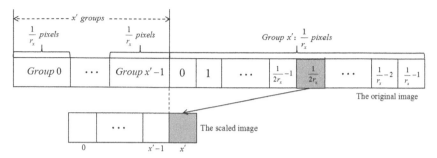

Fig. 4. The principles of scaling down (Fig. 3 in Ref. [23])

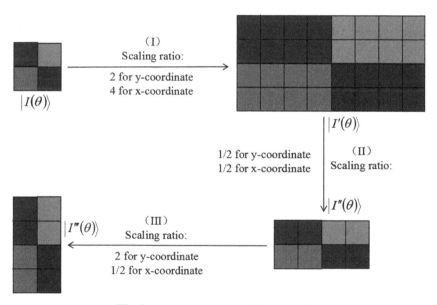

Fig. 5. A simple example of image scaling

3   The Improved QIRHSI (IQIRHSI)

The QIRHSI model can only deal with color images of size $2^n \times 2^n$. Nevertheless, if the image is scaled to $r_x \neq r_y$, it will not continue to apply to color images of size $2^n \times 2^n$.

For the sake of solving this problem, in this section, with the help of the method in Ref. [23], the improved QIRHSI model (IQIRHSI) can represent the size $2^{n_1} \times 2^{n_2}$ is presented, as shown in Eq. (6).

$$
|I(\theta)\rangle = \frac{1}{\sqrt{2}^{n_1+n_2}} \sum_{y=0}^{2^{n_1}-1} \sum_{x=0}^{2^{n_2}-1} |C_{yx}\rangle \otimes |yx\rangle = \frac{1}{\sqrt{2}^{n_1+n_2}} \sum_{y=0}^{2^{n_1}-1} \sum_{x=0}^{2^{n_2}-1} |H_{yx}\rangle|S_{yx}\rangle|I_{yx}\rangle \otimes |yx\rangle
$$

$$
|yx\rangle = |y\rangle|x\rangle = |y_0 y_1 \cdots y_{n_1-2} y_{n_1-1}\rangle|x_0 x_1 \cdots x_{n_2-2} x_{n_2-1}\rangle, y_j, x_j \in \{0, 1\}
$$

(6)

Therefore, IQIRHSI uses $q + 2 + n_1 + n_2$ qubits to represent a $2^{n_1} \times 2^{n_2}$ color image with intensity range 2^q. Figure 6 shows a simple 4×2 IQIRHSI image and its representation, respectively.

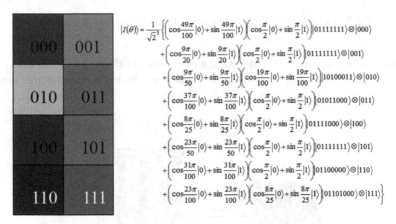

Fig. 6. A 4×2 IQIRHSI quantum color image and its representation

4 Quantum Color Image Resizing

Imagine that the dimensions of the primal color image are $2^{n_1} \times 2^{n_2}$, and the dimensions of the color image after scaled are $2^{m_1} \times 2^{m_2}$. The scale ratio is $2^{m_1-n_1} \times 2^{m_2-n_2}$. Based on the IQIRHSI, the color and position information of the primary color image and the scaled color image are C_{yx}, (y, x) and $C_{y'x'}$, (y', x'), respectively.

According to Eq. (5), the scaling of the image as a whole is compounded by the scaling in both directions. Therefore, in Sect. 4, we only take the quantum color image scaling in the y-axis direction as an example. For more convenience, the subscripts 1 in n_1 and m_1 are omitted here [23].

4.1 Quantum Circuit for Color Image Zoom Up Operation

A quantum circuit that uses the quantum module operation UP(n, m) for color image zooming, where n with m denote the color image size from 2^n to 2^m, respectively. Since the grayscale image is targeted in Ref. [23], the quantum circuit needs to be modified. The structure of the improved UP(n, m) is seen in Fig. 7, for which $m > n$.

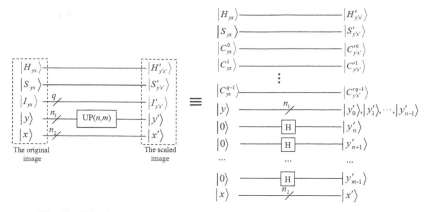

Fig. 7. Color image of the quantum circuits scaled up the y-axis direction

- Need to set up the $(m - n)|0\rangle$ qubits to a new y-axis position.
- To enlarge the axis, $(m - n)$ Hadamard gates are added to expand the new axis $y'_n, y'_{n+1}, \cdots, y'_m$.

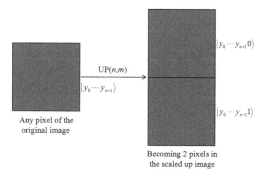

Fig. 8. A brief example of UP(n, m)

Obviously, the addition of $(m - n)$ Hadamard gates acts on the initial state, which in turn presents $|0\rangle$ and $|1\rangle$ with equal probability, with all possible values of $\{|00\cdots00\rangle, |00\cdots01\rangle, \cdots, |11\cdots10\rangle, |11\cdots11\rangle\}$. That is, each pixel $|y_0y_1\cdots y_{n-1}\rangle$ of the original image is enlarged by a factor of 2^{m-n}, and corresponding the coordinates become

$$|y_0y_1\cdots y_{n-1}0\cdots0\rangle, \cdots, |y_0y_1\cdots y_{n-1}1\cdots1\rangle$$

while keeping the color values unchanged. Following the above method, r_y repeated for each pixel in the initial color image. Figure 8 shows an illustration of image scaled up where $m - n = 1$, i.e., $r_y = 2$.

4.2 Quantum Circuit for Color Image Zoom Down Operation

For the purpose of designing circuits for quantum color image scaling down, using the principle of scaling down, further analysis is required with the help of Sect. 2.2, especially Fig. 4. The location information of the pixel in the original color image is $|y_0 y_1 \cdots y_{m-1} y_m y_{m+1} \cdots y_{n-2} y_{n-1}\rangle$. The position information was divided into two parts: $|y_0 y_1 \cdots y_{m-2} y_{m-1}\rangle$ and $|y_m y_{m+1} \cdots y_{n-2} y_{n-1}\rangle$ and it has the following characteristics:

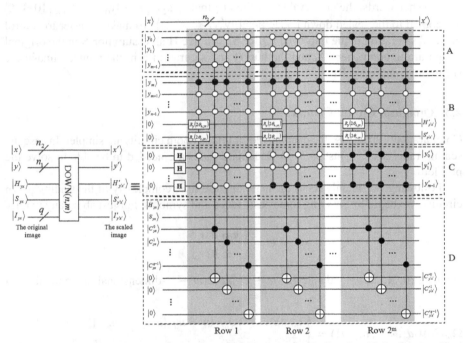

Fig. 9. Color image of the quantum circuits scaled down the y-axis direction (Color figure online)

(I) All pixels belonging to the same group have the same $|y_0 y_1 \cdots y_{m-1}\rangle$ and are equal to the group's label, i.e., the location information of the pixels in the resulting scaled-down color image of the group.

(II) The value of the color of the pixel y' in the zoomed-down image is equal to the value of the color of the pixel $|y'_0 y'_1 \cdots y'_{m-1} 1 0 \cdots 0\rangle$ in the original image, i.e., $|y_m y_{m+1} \cdots y_{n-1}\rangle = |1 0 \cdots 0\rangle$.

The proofs of the above two features are given in Ref. [23]. The quantum circuit scaling down in the same direction is represented by the modulo DOWN(n, m), where n with m denotes the image zoomed-down from 2^n to 2^m. The structure of DOWN(n, m) seen in Fig. 9 with $m < n$ is listed in the following:

- Add $(q + 2)$ $|0\rangle$ qubits as a new color qubits and m $|0\rangle$ qubits as a new position y of the scaled down image.
- Add m Hadamard gates to generate $|y'_0\rangle, |y'_1\rangle, \cdots, |y'_{m-1}\rangle$.

- When $|y_m y_{m+1} \cdots y_{n-1}\rangle = |10 \cdots 0\rangle$, with the help of the Controlled-NOT gates, copy $|C_{yx}\rangle$ to $|C_{y'x'}\rangle$.

  - Based on the characteristic (I) in Sect. 4.2, the same control values for components A and C.
  - Based on the characteristic (II) in Sect. 4.2, the control value that can be derived for component B is $|10 \cdots 0\rangle$.

In other words, the color values of the original image pixel $|y_0' y_1' \cdots y_{m-1}' 10 \cdots 0\rangle$ are copied to the scaled down image pixel $|y_0' y_1' \cdots y_{m-1}'\rangle$. To make it easier to control hue and saturation in the improved IQIRHSI, the hue H and saturation S in the original image are in component A, and the hue H and saturation S in the reduced image are shifted to component B.

4.3 Quantum Circuit Complexity

Comparatively speaking, the quantum color image zoom circuits are simpler. In case the scaling ratio is 2^{m-n}, just $(m - n)$ Hadamard gates are needed to be able to solve the problem.

Following, we mainly focus on analyzing the time complexity of the scaling down circuits. We begin by covering some basic ideas and symbols. For just about any unitary

$$U = \begin{bmatrix} u_{00} & u_{01} \\ u_{10} & u_{11} \end{bmatrix}$$

and $m \in \{0, 1, 2, \cdots\}$, define the $(m + 1)$-bit $(2(m + 1)$-dimensional) operator $\Omega_m(U)$ as

$$\Omega_m(U)(|a_1, \cdots, a_m, d\rangle) = \begin{cases} u_{d0}|a_1, \cdots, a_m, 0\rangle + u_{d1}|a_1, \cdots, a_m, 1\rangle, & \text{if } \wedge_{j=1}^m a_j = 1 \\ |a_1, \cdots, a_m, d\rangle, & \text{if } \wedge_{j=1}^m a_j = 0 \end{cases}$$

For all $a_1, a_2, \cdots, a_m, d \in \{0, 1\}$, $\wedge_{j=1}^m a_j = 1$ denotes the AND of the Boolean variables $\{a_k\}$. The operator $\Omega_m(U)$ is known as generalized Toffoli gate, and extended family of all $\Omega_0(U)$ together with $\Omega_1(X)$, of which $X = \begin{bmatrix} 0 & 1 \\ 1 & 0 \end{bmatrix}$, is called whole family of basic operations or basic gates [31].

When the determinant of a 2×2 matrix U is 1, then it is called a special unitary matrix. The set of all 2×2 special matrices are denoted $SU(2)$.

Lemma 1. ([31], Corollary 5.3) For any unitary 2×2 matrix W, a $\Omega_1(U)$ gate can be simulated by at most six basic gates: four 1-bit gates (Ω_0), and two CNOT gates ($\Omega_1(X)$).

Lemma 2. ([31], Corollary 6.2) For any unitary 2×2 matrix U, a $\Omega_2(U)$ gate can be simulated by at most sixteen basic gates: eight 1-bit gates (Ω_0) and eight CNOT gates ($\Omega_1(X)$).

Next, the complexity of $\Omega_n(U)$ is given, whereby U is a unitary 2×2 matrix with $m \geq 3$.

Lemma 3. ([32], Lemma 3) Given $n \in \{3, 4, \cdots\}$ and a unitary 2×2 matrix U, a $\Omega_n(U)$ gate can be simulated by $4n - 10$ $\Omega_2(X)$ gates and two $\Omega_2(U)$ gates, assume the existence of $n - 2$ auxiliary qubits.

Lemma 4. ([31], Lemma 7.2) If $n \geq 5$ and $m \in \{3, \cdots, \lceil n/2 \rceil\}$ then a $\Omega_m(X)$ gate can be simulated by a network consisting of $4(m - 2)$ $\Omega_2(X)$ gates that is of the form.

Next, we start to investigate the complexity of scaled-down circuit for quantum color images. The number of basic quantum gates plays a decisive role in the complexity of the quantum circuit. In this paper, controlled-not gates and controlled rotation gates are chosen as the basic units.

Theorem 1. The gates complexity of preparing a quantum image scaling down circuit of IQIRHSI state of size $2^{n_1} \times 2^{n_2}$ is $O\left(3 \times 2^{m+3} \times (q+2)(n+m-1)\right)$.

Proof. In Fig. 9, for the quantum circuit has 2^m thin gray shadow regions, a region with q layers multi-control-qubit-NOT gate and one layer multi-control-qubit-rotation gate, i.e., it has $q \times 2^m$ layers multi-control-qubit-NOT gates and 2^m layer multi-control-qubit-rotation gates.

For each layer, it is a multi-control-qubit-NOT gate. The number of controlled qubits is

$$
\begin{aligned}
&m \text{ (from Component } A) \; + \; (n - m) \text{ (from Component } B) \\
&+ m \text{ (from Component } C) \; + \; 1 \text{ (from Component } D) \\
&= n + m + 1
\end{aligned}
\tag{7}
$$

With the help of Lemma 4, a $\Omega_{n+m+1}(X)$ gate is equivalent to

$$
4(n + m - 1) \; \Omega_2(X) \text{ gate}
\tag{8}
$$

In the same way, with regard to each row, it is a multi-control-qubit-rotation gate. The number of controlled qubits is

$$
\begin{aligned}
&m \text{ (from Component } A) \; + \; (n - m) \text{ (from Component } B) \; + \; m \text{ (from Component } C) \\
&= n + m
\end{aligned}
\tag{9}
$$

From Lemma 3, a $\Omega_{n+m}(U)$ gate is equivalent to

$$
4(n + m) - 10 \; \Omega_2(X) \text{ gates} \; + \; 2 \; \Omega_2(U) \text{ gates}
\tag{10}
$$

Therefore, depending on Eqs. (7) and (9), the quantum image scaling down complexity of one row in Fig. 9 is

$$
8(n + m) - 20 \; \Omega_2(X) \text{ gates} \; + \; 4 \; \Omega_2(U) \text{ gates} + q \times 4(n + m - 1) \; \Omega_2(X) \text{ gates}
$$

Consequently, the quantum module DOWN(n, m) can be simulated by

$$2^m \times \left[8(n+m) - 20\ \Omega_2(X) \text{ gates } + 4\ \Omega_2(U) \text{ gates } + q \times 4(n+m-1)\ \Omega_2(X) \text{ gates}\right]$$
$$+ (n+m-1) \text{ NOT gates} \tag{11}$$

[1] has pointed out one $\Omega_2(X)$ gate can be simulated by 6 $\Omega_1(X)$ gates. By Lemma 2, Eq. (11) can be reduced to

$$3 \times 2^{m+3} \times \left((q+2)\cdot(n+m-1) - \frac{5}{3}\right) \Omega_1(X) \text{ gates } + 2^{m+5}\ \Omega_0 \text{ gates } + (n+m-1) \text{ NOT gates}$$

In other words, the complexity of the gate in the quantum circuit is

$$O\left(3 \times 2^{m+3} \times (q+2)(n+m-1)\right)$$

5 Quantum Color Image Scaling Example

Here we choose the image zooming transform in Fig. 5 a example to explain the quantum scaling circuit in some details.

The initial image is

$$|I(\theta)\rangle = \frac{1}{2}\sum_{y=0}^{1}\sum_{x=0}^{1}|H_{yx}\rangle|S_{yx}\rangle|I_{yx}\rangle \otimes |y\rangle|x\rangle$$

$$= \frac{1}{2}\left\{|H_{00}\rangle|S_{00}\rangle\left|C_{00}^0 C_{00}^1 \cdots C_{00}^7\right\rangle \otimes |0\rangle|0\rangle + |H_{01}\rangle|S_{01}\rangle\left|C_{01}^0 C_{01}^1 \cdots C_{01}^7\right\rangle \otimes |0\rangle|1\rangle\right.$$

$$\left. + |H_{10}\rangle|S_{10}\rangle\left|C_{10}^0 C_{10}^1 \cdots C_{10}^7\right\rangle \otimes |1\rangle|0\rangle + |H_{11}\rangle|S_{11}\rangle\left|C_{11}^0 C_{11}^1 \cdots C_{11}^7\right\rangle \otimes |1\rangle|1\rangle\right\}$$

Where $n_1 = n_2 = 1$ with q equals 8.

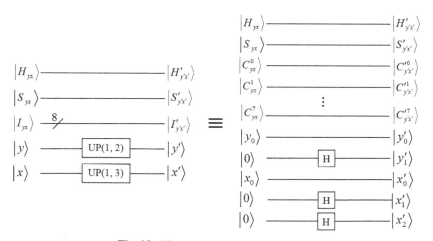

Fig. 10. Figure 5 circuit design in step (I)

(I) $|I(\theta)\rangle \rightarrow |I'(\theta)\rangle$, here $r_y = 2$ and $r_x = 4$, $m_1 = 2$ and $m_2 = 3$. Figure 10 depicts the scaled circuits.

(II) $|I'(\theta)\rangle \rightarrow |I''(\theta)\rangle$, where $r_y = r_x = \frac{1}{2}$, $m_1' = 1$ with m_2' takes the value 2. Figure 11 depicts the scaled circuits.

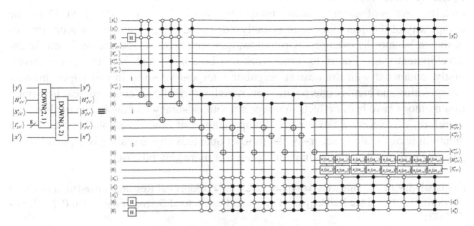

Fig. 11. Figure 5 circuit design in step (II)

(III) $|I''(\theta)\rangle \rightarrow |I'''(\theta)\rangle$, where $r_y = 2$ and $r_x = \frac{1}{2}$, $m_1'' = 2$ and $m_2'' = 1$. The scaling circuit is presented in Fig. 12.

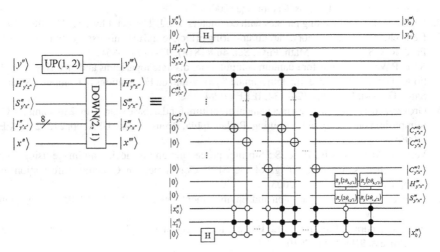

Fig. 12. Figure 5 circuit design in step (III)

6 Conclusion

In this paper, we succinctly discuss the quantum color image representation and algorithms concerning quantum image processing. First, the improved of QIRHSI to IQIRHSI makes it possible to store quantum images from $2^n \times 2^n$ to $2^{n_1} \times 2^{n_2}$. Second, the circuits of scaling quantum images based on IQIRHSI are designed. Third, the complexity of the quantum circuit is analyzed. Finally, the scaling of quantum images is shown by specific examples. Up to now, quantum color image scaling is still in the development stage. Future research work can be approached from two perspectives: firstly, compared with the nearest neighbor interpolation used in this paper, bilinear interpolation and bicubic interpolation even higher order linear and nonlinear interpolation methods have better effects in image smooth, accuracy and the image quality loss, therefore, it is necessary to explore the application of other interpolation methods in quantum image scaling; secondly, the ratio of quantum image scaling is 2^r, and how to perform scaling when the scaling multiplier is a more general case.

Acknowledgement. This work is supported by the Postdoctoral Research Foundation of China (2018M631914) and the Heilongjiang Provincial Postdoctoral Science Foundation (CN) (LBH-Z17042).

References

1. Nielson, M.A., Chuang, I.L.: Quantum Computation and Quantum Information, 2nd edn. Cambridge University Press, Cambridge (2000)
2. Feynman, R.P.: Simulating physics with computers. Int. J. Theoret. Phys. **21**, 467–488 (1982)
3. Deutsch, D.: Quantum theory, the church-turing principle and the universal quantum computer. Proc. Roy. Soc. Lond. A Math. Phys. Sci. **400**(1818), 97–117 (1985)
4. Shor, P.W.: Algorithms for quantum computation: discrete logarithms and factoring. foundations of computer science. In: 35th Annual Symposium on Foundations of Computer Science, Santa Fe, NM, USA, pp. 124–134. IEEE (1994)
5. Grover, L.K.: A fast quantum mechanical algorithm for database search. In: 28th annual ACM Symposium on Theory of Computing, Philadelphia, Pennsylvania, USA, pp. 212–219. IEEE (1996)
6. Venegas-Andraca, S.E., Bose, S.: Storing, processing and retrieving an image using quantum mechanics. In: Proceedings of the SPIE Conference on Quantum Information and Computation, pp. 137–147 (2003)
7. Latorre, J.I.: Image Compression and Entanglement. Quantum Physics (2005). arXiv:quant-ph/0510031v1
8. Venegas-Andraca, S.E., Ball, J.L.: Processing images in entangled quantum systems. Quant. Inf. Process. **9**(1), 1–11 (2010)
9. Le, P.Q., Dong, F.: A flexible representation of quantum images for polynomial preparation, image compression and processing operations. Quant. Inf. Process **10**(1), 63–84 (2011)
10. Zhang, Y., Lu, K.: NEQR: a novel enhanced quantum representation of digital images. Quant. Inf. Process **12**(12), 2833–2860 (2013)
11. Li, H.S., Zhu, Q.: Image storage, retrieval, compression and segmentation in a quantum system. Quant. Inf. Process **12**(9), 2269–2290 (2013)

12. Chen, G.L., Song, X.H., Venegas-Andraca, S.E.: QIRHSI: novel quantum image representation based on HSI color space model. Quant. Inf. Process (submitted)
13. Le, P.Q., Iliyasu, A.M.: Fast geometric transformations on quantum images. Int. J. Appl. Math. 40(3), 113–123 (2010)
14. Wang, J., Jiang, N., Wang, L.: Quantum image translation. Quant. Inf. Process. 14(5), 1589–1604 (2014)
15. Zhang, Y., Kai, L., Kai, X., Gao, Y., Wilson, R.: Local feature point extraction for quantum images. Quant. Inf. Process. 14(5), 1573–1588 (2014)
16. Jiang, N., Wang, L., Wen-Ya, W.: Quantum Hilbert image scrambling. Int. J. Theoret. Phys. 53(7), 2463–2484 (2014)
17. Caraiman, S., Manta, V.I.: Image segmentation on a quantum computer. Quant. Inf. Process. 14(5), 1693–1715 (2015)
18. Zhou, R.-G., Sun, Y.-J., Fan, P.: Quantum image gray-code and bit-plane scrambling. Quant. Inf. Process. 14(5), 1717–1734 (2015)
19. Iliyasu, A.M., Le, P.Q., Dong, F., Hirota, K.: Watermarking and authentication of quantum images based on restricted geometric transformations. Inf. Sci. 186(1), 126–149 (2012)
20. Song, X.-H., Wang, S., Liu, S., Abd, A.A., El-Latif, X.-M.: A dynamic watermarking scheme for quantum images using quantum wavelet transform. Quant. Inf. Process. 12(12), 3689–3706 (2013)
21. Iliyasu, A.M.: A framework for representing and producing movies on quantum computers. Int. J. Quant. Inf. 09(06), 1459–1497 (2011)
22. Yan, F., Jiao, S.: Chromatic framework for quantum movies and applications in creating montages. Front. Comput. Sci. 12(4), 736–748 (2018)
23. Jiang, N., Wang, L.: Quantum image scaling using nearest neighbor interpolation. Quant. Inf. Process. 14, 1559–1571 (2015)
24. Jiang, N., Xiaowei, L., Hao, H., Dang, Y., Cai, Y.: A novel quantum image compression method based on JPEG. Int. J. Theoret. Phys. 57(3), 611–636 (2017)
25. Sang, J., Wang, S., Niu, X.: Quantum realization of the nearest-neighbor interpolation method for FRQI and NEQR. Quant. Inf. Process. 15(1), 37–64 (2015)
26. Zhou, R.-G., Tan, C., Fan, P.: Quantum multidimensional color image scaling using nearest-neighbor interpolation based on the extension of FRQI. Mod. Phys. Lett. B 31(17), 1750184 (2017)
27. Li, P., Liu, X.: Bilinear interpolation method for quantum images based on Quantum Fourier Transform. Int. J. Quant. Inf. 16(04), 1850031 (2018)
28. Gonzalez, R., Woods, R.: Digital Image Processing, 3rd edn. Prentice Hall, New Jersey (2007)
29. Anthony Parker, J., Kenyon, R.V., Troxel, D.E.: Comparison of interpolating methods for image resampling. IEEE Trans. Med. Imaging 2(1), 31–39 (1983)
30. https://ww2.mathworks.cn/help/images/ref/imresize.html (2021)
31. Barenco, A., et al.: Elementary gates for quantum computation. Phys. Rev. A 52(5), 3457–3467 (1995)
32. Khan, R.A.: An Improved Flexible Representation of Quantum Images. Quant. Inf. Process 18(7), 201 (2019)

Modeling and Analysis of EEG Brain Network in High Altitude Task State

Yao Ma, Lei Wang, Yu Yang, Xuepeng Li, Zeng Xu, and Haifang Li[✉]

Taiyuan University of Technology, Taiyuan 030024, China
lihaifang@tyut.edu.cn

Abstract. Long-term exposure to high altitude, low pressure and low oxygen will seriously threaten people's cognitive function. To explore the changes in whole-brain network dynamics during brain activity in long-term high-altitude migrants, EEG signals from three subjects of 75 different altitudes were analyzed using the Stroop experimental paradigm and the network recombination prediction model. The sliding window method was used to explore the dynamic change process of the brain network. At the same time, the time period with significant difference between the brain networks of the altitude group was selected as the real response network to measure the model prediction accuracy. Then, according to different network prediction model rules, the weights of brain network 200 ms before stimulation were updated for each subject. Finally, the prediction model with the least difference between the prediction network and the real response network was selected for each subject. The experimental results showed that the prediction accuracy of the model reach 98.95%, and there is a significant difference in model selection between the elevation groups. It helps to understand the brain dynamics of healthy people, and reveals the abnormal changes in the brain networks of those who have stayed at high altitude for a long time, providing an important reference for the cognitive rehabilitation training of victims ex-posed at high altitude.

Keywords: High altitude · EEG · Dynamic brain network · Task state · Prediction model

1 Introduction

Today, the number of people traveling to work, study, or live at high altitude continues to grow [1]. People exposed to high altitude hypoxic environments often exhibit deficits in perceptual and cognitive domains such as working memory, vision, attention, and executive function [2, 3]. Notably, most studies on changes in cognitive function due to high altitude exposure are from acute altitude [4], but the changes in brain networks of cognitive function in long-term high altitude migrants are not well studied, and the dynamic evolutionary characteristics of their brain networks are poorly understood.

In network research, characterizing the topological relationships of complex networks through graph-theoretic methods is an important tool for studying network properties. Using functional connectivity to study the characteristics of brain networks when

J. Zeng et al. (Eds.): ICPCSEE 2021, CCIS 1451, pp. 468–480, 2021.
https://doi.org/10.1007/978-981-16-5940-9_36

the brain performs cognitive activities is a very effective way. Studying the topological properties of brain networks helps us to better examine properties such as functional separation and integration of the brain when performing cognitive activities, which are closely related to brain function. Moreover, functional connectivity can be reorganized by later cognitive training [5, 6] and remodeled back to a healthy brain. Therefore, it is important to determine the effects of plateau exposure on cognitive function and thus elucidate the underlying neural mechanisms.

The current dynamical state of the brain is determined by a specific brain coding state [7], and the coding of a specific state will have predictive power for subsequent states [8], and the changes of altitude on the structure of human brain networks can be explored by predicting abnormalities in the coding dynamics mechanism. Based on this basis, a brain network reorganization prediction model is proposed, which can simulate the change process of brain network of the subject's cognitive task and be used to reveal abnormalities in brain network dynamics.

In this study, EEG data were collected from three different elevation task states, with channels as nodes and Phase Locking Value PLV between channels as edges, and the sliding window technique was applied to construct a dynamic PLV brain network over time, calculate the topological parameters of each subject's dynamic network, analyze the dynamic change process of the brain network, and select the elevation group between The accuracy of the prediction model was evaluated by selecting the time windows with significant differences between altitude groups. Six different probability models for updating weights were designed and experimentally validated in order to eliminate the influence of individual weights and to fit the brain network change process to the maximum extent. α-wave activity is an important manifestation of hypoxic environment [9], and the network reorganization prediction model iteratively predicted the brain network before α-band stimulation, and the best-fitting network model was selected for each subject of the network model. By analyzing the variability of topological parameters and the variability of network model selection for each altitude dynamic network, abnormal manifestations of network impairment in highland sedentary subjects were identified, which in turn provided an important reference for cognitive rehabilitation training for highland injury.

2 Subjects and Experiments

2.1 Subjects

Seventy-five healthy male youths, 25 in each group, with a mean age of 20.27 ± 2.56 years, born in the plains and never having been on the plateau before entering the plateau, were recruited and moved to Weinan (400m), Lhasa (3700m) and Nagqu (4500m). The mean length of residence on the plateau for all subjects was 2.12 ± 0.31 years. All subjects were right-handed and had no mental impairment. Visual acuity or corrected visual acuity was normal, and there were no disorders and no medications. No coffee or alcohol was consumed within 24 h before the experiment, no psychostimulant drugs were taken, and sleep was normal. All subjects had completed an informed consent form.

2.2 Stroop Tasks

The stimulus material consisted of two Chinese color words (red and green) and two colors (red and green) paired in two pairs to form two stimuli. There were 38 consistent stimuli and 102 inconsistent stimuli in each task, presented in pseudo-random order. The color words were in 48-point font and were presented in the center of a 14-in. computer monitor screen with a viewing angle of 2.2°. The experiment consisted of a color judgment task and a word sense judgment task. Subjects were allowed to make judgments by keystrokes, and were allowed to practice until they fully understood them before the formal experiment. The accuracy rate and average hit response time were recorded for analysis.

1) Color task: Subjects were asked to overcome word-sense interference by pressing "z" for red font color and "m" for green font color. The order of the tasks was balanced among the subjects. For each group of experiments, a "+" was first presented in the center of the screen as a gaze point, and the presentation time was fixed at 300 ms. The presentation time of the color stimulus was fixed at 200 ms, and the subject had 2000 ms to make a keystroke judgment, followed by the next color word waiting for the subject's keystroke response, and if the subject made a keystroke error. If the subject pressed a wrong key, the next round of the same task would be performed after 2200 ms. In the task, the subject was required to respond as quickly and accurately as possible.

2) Word meaning task: Similar to the color task, the subject is required to overcome the interference of font color and judge the meaning of the presented stimulus (Fig. 1).

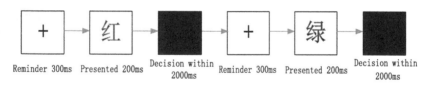

Fig. 1. Schematic diagram of stroop paradigm task

3 Methods

3.1 EEG Data Acquisition and Pre-processing

The data were acquired using Neuroscan's 36-conductor EEG system with electrode distribution conforming to the international 10–20 system standard, with reference electrodes placed on the earlobe as A1 and A2, and four recording electrodes placed at the upper and lower part of the eye and at the outer bared eye for the acquisition of vertical and horizontal EEG. The sampling rate used for EEG data acquisition was 1000 Hz, and the impedance between the scalp and the electrodes was guaranteed to be less than 5 kΩ.

All data analysis for the experiments was performed using matlab2017b and the open source toolbox EEGLAB for preprocessing to remove the four electrodes from the electrooculography. The EEG data were routinely preprocessed for noise removal, extracting all inconsistencies from 200 ms before stimulus onset to 600 ms after stimulus onset for one trial, ensuring no overlap with subsequent trials, and correcting the baseline for each period.

3.2 Construction of Brain Network

Neuroscience studies have concluded that effective information communication can occur between populations of neurons that undergo phase locking [10], and PLV quantifies the degree of synchronization of two neural signals entering a phase locking state in a specific frequency band and time region [11]. Therefore, in this study, a weighted network was constructed using scalp electrodes as nodes and PLV values between channels as connected edges. PLV can be obtained by the following equation.

$$PLV(t) = \frac{1}{N} \left| \sum_{n=1}^{N} e^{i(\Delta \varphi_n(t))} \right| \tag{1}$$

where $\Delta \varphi n = \varphi x(t) - \varphi y(t)$ represents the instantaneous phase difference between signal x and signal y at time t, and N indicates the total number of time points of the data. the PLV value is between 0 and 1, where a value of 1 indicates that the two signals are perfectly synchronized in phase; a value of 0 indicates that the two signals are not synchronized in phase and there is no phase locking.

To assess changes in brain networks over time during a cognitive task, a sliding window was used in this study to assess changes in brain networks over time during a cognitive task. A sliding window with an overlap of 90% and a window size of 200 ms was used to construct the brain network for each subject's data from 200 m before to 600 ms after each stimulus, and to eliminate specificity, all stimulus networks were averaged for each subject and the network properties were extracted for each subject [12].

3.3 Network Properties

The topological nature of brain functional networks can be characterized by some topological structure parameters, and in this paper, the characteristic path length, clustering coefficient [5], network density, and global efficiency [13] are used to evaluate brain networks.

The characteristic path length characterizes the degree of integration of the network [13], which is defined as the average shortest path length between all pairs of nodes in the network:

$$PL = \frac{1}{N} \sum_{i \in n} \frac{\sum_{j \in N, j \neq i} d_{ij}}{n - 1} \tag{2}$$

dij is the length of the shortest path between node i and node j. N is the number of network nodes.

The average clustering coefficient quantifies the degree of separation of the network. In the case of weighted networks, the average clustering coefficient can be generalized as follows in order to avoid the influence of the primary connection weights [14]:

$$CLC = \frac{1}{N}\left[\binom{N}{3}\sum_{jh\in n}\left(W_{ij}W_{ih}W_{jh}\right)^{1/3}\right] \tag{3}$$

Wij denotes the weight between electrodes i and j. N is the number of network nodes.

Network density evaluates the strength of connectivity of a network and is calculated as follows.

$$D = \frac{2L}{N(N-1)} \tag{4}$$

L is the sum of network ownership values and N is the number of network nodes.

Global efficiency measures how fast or slow the information flows in the network [13], and the global efficiency is defined by Eq:

$$GF = \frac{1}{N}\sum_{i\in N}\frac{\sum_{j\in N, j\neq i}(d_{ij})^{-1}}{N-1} \tag{5}$$

dij denotes the random distance from node i to node j. N is the number of network nodes.

3.4 Technical Route

The overall experimental technical route is shown in Fig. 2. First, EEG data from 200 ms before to 600 ms after stimulation were selected, and 31 consecutive dynamic PLV brain networks were constructed for each subject using a sliding window with a window size of 200 ms and an overlap rate of 90%, and then network properties were extracted for each brain network separately, and the one with the greatest variability in network properties among the three altitude groups after stimulation was selected window as a measure of the accuracy of the model prediction.

Finally, the PLV brain network was constructed for each subject's EEG data 200 ms before stimulation separately, and in order to determine the brain network reorganization prediction model for each subject, the brain network weights before stimulation were probabilistically modified for each subject according to different network prediction model rules, and the model type with the smallest mean squared error (MSE) between simulated and real network attributes was selected for each subject.

3.5 Network Prediction Model

The six network prediction models are shown below:

1. Primary connection enhancement: this model assumes that the primary connections of the stimulated forebrain network (i.e., connections with larger network weights) are more likely to increase in weight during the cognitive task with a probability of increase of p = P.

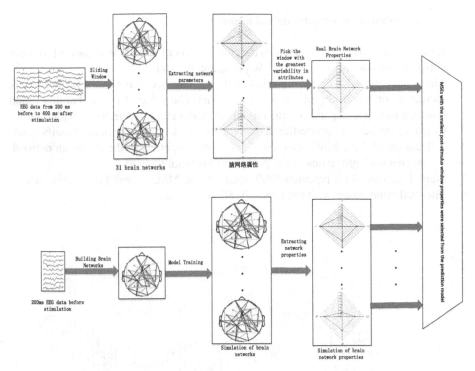

Fig. 2. Technology roadmap

2. Secondary connection enhancement: The model assumes that secondary connections in the stimulated forebrain network (i.e., connections with lower network weights) are more likely to increase in weight during a cognitive task with a probability of increase of p = (1−P).

3. Random enhancement: The model assumes that the brain network connections are randomly increasing before stimulation. That is, the probability of increase in the weights of the selected connections p = 1.

4. Main connection weakening: the model assumes that the primary connections of the stimulus forebrain network (i.e., connections with larger network weights) are more likely to have reduced weights during the cognitive task, with j probability p = P.

5. Secondary connections weakening: The model assumes that secondary connections in the stimulated forebrain network (i.e., connections with lower network weights) are more likely to have reduced weights during the cognitive task, with j probabilities of p = (1−P)

6. Random connection weakening: The model assumes that the stimulus forebrain network connections are randomly decreasing. That is, the probability of reduction in the weights of the selected connections p = 1.

The algorithm steps are summarized as follows:

1. Random selection of a connection in the connectivity matrix of the stimulus forebrain network (commonly referred to as W_{ij} in graph theory).
2. The connection value Wij is modified by the weights according to the probability p, increasing or decreasing in size by (1% * Wij). The value of the probability p and how it is modified depends on the particular model currently under consideration.
3. Calculate the network properties of the brain network after each model modification.
4. Calculation of 3 the MSE between the network properties of the midbrain network and the network properties of the post-stimulus brain network.
5. Steps 1, 2, 3 and 4 are repeated 5000 times and the MSE is stored for each iteration.
6. The prediction model with the lowest MSE is selected.

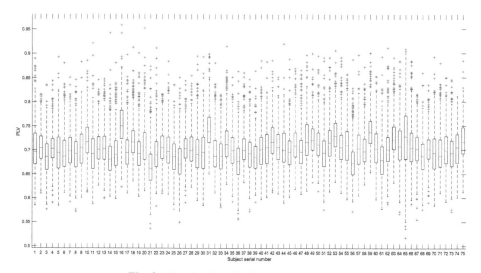

Fig. 3. Distribution of PLV weights of subjects

As shown in Fig. 3, the network weights of high-altitude subjects are normally distributed, and the distribution of weights is relatively concentrated, and there are differences in the centers of network weights distribution among subjects. Therefore, in order to eliminate the problem of concentrated distribution of individual network weights, highlight the priority of network weights, and reflect the probabilistic problem of updating weights, the design of the update probability P of the network reorganization prediction model is particularly important. The model is set to P = num/N, num is the ranking position of the selected connections, and N is the total number of connections in the network. This probability setting eliminates the network weight distribution problem and highlights the variability among models.

4   Experimental Results and Analysis

4.1   Behavioral Data Results and Analysis

The correct rate and reaction time of the Stroop task were investigated, and the Information Translate Rate (ITR) was used to quantify the brain information processing rate of the subjects, so as to assess the variability of the cognitive processing rate in different altitude groups. Assuming that the Stroop task provides N choices (or categories) on each trial, and assuming that the probability P that the subject's response to each experiment is consistent with the correct outcome of the actual trial is always the same (P is the experimental correctness rate), then the probability that each of the other categories is chosen, i.e., $(1-P)/(N-1)$, T denotes the subject's response time, and the ITR (or bitrate) formula is shown in Eq. (6).

$$ITR = \left\{ \log_2 N + P \log_2 P + (1 - P) \log_2 \frac{1 - P}{N - 1} \right\} \tag{6}$$

From Table 1, it can be seen that the ITR values for each elevation word sense judgment task are higher than the ITR values for the color judgment task, this is because in the Stroop task reading words is automatic processing and color naming is controlled processing, automatic processing does not require attentional involvement, while controlled processing requires intentional control, and the color judgment task is more complex intensity compared to the word sense task [10].

Statistical analysis revealed significant differences in reaction time between groups in the color and word-meaning tasks (P color = 0.01, F color = 4.912; P word = 0.01, F word = 5.375), and reaction time increased with increasing altitude. After Bonferroni correction, in the color task, ITR was significantly different between the Weinan and Nagqu groups (P = 0.045). And the correct rate did not differ significantly between altitudes. Long-term residence in a high-altitude, low-pressure, low-oxygen environment results in relatively slower human responses and relatively balanced correct rates [11], which may be due to the brain being chronically under-oxygenated and exhausted in a high-altitude environment, resulting in neuronal damage to cognitive abilities [5, 12], compensating for cognitive responses with time consumption.

4.2   Network Dynamics Results and Analysis

To evaluate the time-varying nature of the network properties, the network proper-ties were calculated for each dynamic window. Figure 4 illustrates the kinetic variation of network properties under each altitude for the α-band of the color task. By means of feature path length, clustering coefficients, network density and global efficiency it can be observed that the high altitude group exhibits lower integration, lower separation, lower connection strength and lower propagation efficiency relative to the low altitude group brain network. Previous studies have shown significant differences in EEG properties between high and low altitude groups around 300 ms [3, 19, 20]. In the present study, we performed variance tests on the brain network network at the 20th window (200–400 ms) with characteristic path length (f = 5.187, P < 0.01), clustering coefficient (f = 5.189,

Table 1. Statistical analysis of Stroop experiment results

Group			Weinan	Lhasa	Nagqu
Color task	Inconsistent stimulation	Correct rate (%)	72.68	79.92	74.81
		Reaction time (ms)	402.98	481.60	493.55
		ITR	10.73	10.50	9.21
	Consistent stimulation	Correct rate (%)	74.28	80.44	83.72
		Reaction time (ms)	396.46	456.32	487.99
		ITR	8.05	8.59	7.89
Word task	Inconsistent stimulation	Correct rate (%)	75.92	81.25	80.39
		Reaction time (ms)	415.53	472.45	502.62
		ITR	10.77	10.92	9.85
	Consistent stimulation	Correct rate (%)	78.16	83.65	85.07
		Reaction time (ms)	405.32	468.83	501.75
		ITR	8.24	8.82	7.95

$P < 0.01$), network density ($f = 4.505$, $p < 0.05$), and global efficiency ($f = 5.021$, $p < 0.01$) were significantly different between groups. Elevation under this window was also correlated with feature path length ($r = 0.336093$, $p = 0.009$), clustering coefficient ($r = -0.336065$, $p = 0.009$), network density ($r = -0.338507$, $p = 0.008$), and global efficiency ($r = -0.332180$, $p = 0.009514$). The reduced integration and connection strength of functional brain networks indicates a reduced ability to communicate between brain networks, which is consistent with previous studies that have examined the phase results of generally reduced neural activity [2]. Reduced functional brain network separability indicates weaker localized brain communication or less separated neural processing [8]. Reduced global efficiency of the network indicates reduced information processing capacity of the network. Reduced functional brain network separability indicates weaker localized brain communication or less separated neural processing [8]. Reduced global efficiency of the network indicates reduced information processing capacity of the network. Prolonged plateau exposure leads to reduced neural activity in humans, which in turn responds to slower cognitive responses, but the cause of the reduced neural activity is uncertain whether it is due to neural damage or to insufficient oxygen supply. The EEG signals of repatriated individuals can be collected for further analysis in future studies.

4.3 Network Prediction Model Analysis

Based on the results of network dynamics, the network properties of the 200–400 ms (20th window) post-stimulus data were selected as the judgment criteria for the prediction model. The brain network of the 200-ms pre-stimulus data for each subject was placed into the prediction model for iterative prediction, and the cognitive response model that best fitted the pre-stimulus network changes was selected for each subject. The mean

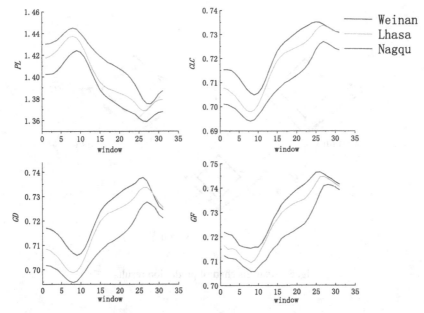

Fig. 4. Network properties change over time

square error (MSE) between the prediction model and the cognitive response as shown in Fig. 5 reached 98.9512%, which demonstrated the accuracy of the improved model.

The distribution of model selection for each group under the color task is shown in Fig. 6a, with enhancement of sub-connectivity (i.e., low-phase synchronous connectivity between brain regions prior to stimulus perception) as the dominant brain network dynamics model for both high- and low-altitude subjects during cognition; however, the distribution of model selection between altitude groups was statistically significant ($\chi 2(3, N = 75) = 11.04$, $p < 0.0 5$; chi-square test). There was a statistically significant relationship between the amount of change in network parameters from pre-stimulus to post-stimulus and the choice of network reorganization model (($F_{weinan} = 5.876$, $p_{weinan} = 0.009$), ($F_{lhasa} = 10.251$, $p_{lhasa} = 0.001$), ($F_{nagqu} = 85.903$, $p_{nagqu} < 0.001$); ANOVA). Figure 6b illustrates that the sub-connectivity enhancement model is associated with higher network reconfiguration in the cognitive task, which can be considered as "normal behavior". However, in the high-altitude group, some subjects showed less variability in network parameters and weaker network reconfiguration before and after stimulation, suggesting that long-term high-altitude exposure may cause abnormalities in cognitive brain function networks.

Hypobaric hypoxia induces alterations in gray matter volume and white matter fiber bundle function associated with abnormal activity in functional brain areas, and hippocampal damage induced by high altitude has a tendency to worsen with increasing altitude and exposure time [20]. Thus more subjects in the high-altitude group deviated from the model of enhanced subconnectivity, which may be due to reduced damage and activation of functional brain areas, making whole-brain network differences diminished [20].

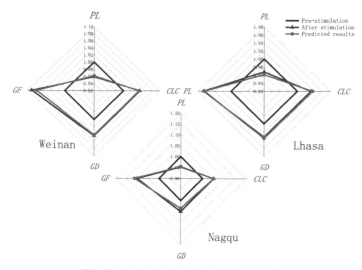

Fig. 5. Analysis chart of prediction results

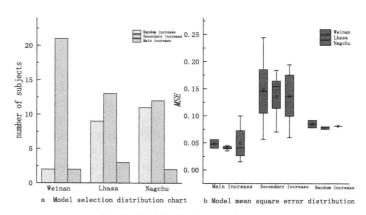

Fig. 6. Model selection result graph

5 Conclusion

This paper constructs brain functional networks from the perspective of sliding windows of complex networks, analyzes the dynamic change process of brain network properties, finds out the time windows with significant differences, and uses a brain network prediction model based on edge probability evolution to simulate the brain network change process. The sliding window reveals that the effect of altitude on human brain cognition mainly occurs around 300 ms, and the significance becomes more obvious with the increase of task intensity; the results of brain network prediction model indicate that long-term high altitude exposure will cause the weakening of brain network dynamics and damage the reconfiguration ability of brain functional network, and the degree of brain damage will be aggravated with the increase of altitude.

This study analyzes the effects of long-term high altitude exposure on the human brain in terms of dynamic brain networks and predictive modeling, reveals the basis of human brain dynamics, and provides experimental evidence for the neurodynamics of abnormal cognitive levels in high altitude exposed populations. Although the model proposed in this study largely fits the brain network evolution process, brain functions are complex and diverse, and the process of brain network dynamics may be more complex than the proposed model with more diverse intermediate change processes. Therefore, several models can be combined in the future to add more types of edge-increasing probabilities, but this may reduce the generalization ability of the model.

Acknowledgments. This work was supported by the Next Generation Internet Technology Innovation Project of Celtic Network (No. NGII20181206), the National Natural Science Foundation of China (No. 61976150), the Key R & D Projects of Shanxi Province (No. 201803D31038).

References

1. Weng, T.B., Pierce, G.L., Darling, W.G., et al.: Differential effects of acute exercise on distinct aspects of executive function. Med. Sci. Sports Exerc. **47**(7), 1460–1469 (2015)
2. Yan, X., Zhang, J., Gong, Q., et al.: Prolonged high-altitude residence impacts verbal working memory: an fMRI study. Exp. Brain Res. **208**(3), 437–445 (2011)
3. Ma, H., Huang, X., Liu, M., et al.: Aging of stimulus-driven and goal-directed attentional processes in young immigrants with long-term high altitude exposure in Tibet: an ERP study. Sci. Rep. **8**(1), 17417 (2018)
4. Zhang, D., Zhang, X., Ma, H., Wang, Y., Ma, H., Liu, M.: Competition among the attentional networks due to resource reduction in Tibetan indigenous residents: evidence from event-related potentials. Sci. Rep. **8**(1), 610 (2018)
5. Watts, D.J., Strogatz, S.H.: Collectivedynamics of 'small-world' networks. Nature **393**(6684), 440–442 (1998)
6. Bamidis, P.D., Vivas, A.B., Styliadis, C., et al.: A review of physical and cognitive interventions in aging. Neurosci. Biobehav. Rev. 44(Sp. Iss. SI), 206–220 (2014)
7. Friston, K., Kiebel, S.: Predictive coding under the free-energy principle. Philos. Trans. R. Soc. B Biol. Sci. **364**(1521), 1211–1221 (2009)
8. Gomez-Pilar, J., Poza, J., Gómez, C., et al.: Altered predictive capability of the brain network EEG model in schizophrenia during cognition. Schizophrenia Res. **201**, 120–129 (2018)
9. Bing, L., Buxin, H., Xin, A., et al.: Research progress on the effect of acute altitude hypoxia on resting EEG Power (2017)
10. Gao, J.F., Yang, Y., Huang, W.T., et al.: Exploring time- and frequency- dependent functional connectivity and brain net-works during deception with single-trial event-related potentials. Sci. Rep. **6**, 37065 (2016)
11. Hardmeier, M., Hatz, F., Bousleiman, H., et al.: Re-producibility of functional connectivity and graph measures based on the phase lag index (PLI) and weighted phase lag index (wPLI) derived from high resolution EEG 9(10), e108648 (2017)
12. Luyao, W., Wenhui, W., Tianyi, Y., et al.: Beta-band functional connectivity influences audiovisual integration in older age: an EEG study. Front. Aging Neurosci. **9**, 239 (2017)
13. Munarini, G.: Presentazione articolo: complex network measures of brain connectivity uses and interpretations. In: Corso Psicologia matematica (2013)
14. Newman, M.E.J.: The structure and function of complex networks. Siam Rev. **45**, 167–256 (2003)

15. Haiyan, L., Jun, C., Shaobei, X.: Research progress of stroop effect. J. Hainan Normal Univ. (Nat. Sci. Edn.) **22**(001), 100–103 (2009)

16. Lefferts, W.K., et al.: Changes in cognitive function and latent processes of decision-making during incremental ascent to high altitude. Physiol. Behav. **201**, 139–145 (2019)

17. Martin, K., Mcleod, E., Périard, J., et al.: The impact of environmental stress on cognitive performance: a systematic review. Human Fact. J. Human Fact. Ergon. Soc. 61(8), 1205–1246 (2019)

18. Feddersen, B., Neupane, P., et al.: Regional differences in the cerebral blood flow velocity response to hypobaric hypoxia at high altitudes. J. Cerebral Blood Flow Metab. Off. J. Int. Soc. Cerebral Blood Flow Metab. **35**(11), 1846–1851 (2015)

19. Ma, H., Wang, Y., Wu, J., et al.: Long-term exposure to high altitude affects response inhibition in the conflict-monitoring stage. Sci Rep. 5, 13701 (2015)

20. Altbäcker, A., Takács, E., Barkaszi, I., et al.: Differential impact of acute hypoxia on event related potentials: impaired task-irrelevant, but preserved task-relevant processing and response inhibition. Physiol. Behav. **206**, 28–36 (2019)

ECG-Based Arrhythmia Detection Using Attention-Based Convolutional Neural Network

Renxing Zhao[1]([envelope]) [iD] and Runnan He[2] [iD]

[1] Beijing University of Posts and Telecommunications, Beijing 100876, China
[2] Peng Cheng Laboratory, Shenzhen 518055, Guangdong Province, China
gorgebest@126.com

Abstract. This study constructs a multi-classification model of arrhythmia based on the dual-channel convolutional neural network with attention mechanism, in order to make automatic detection of arrhythmias. Firstly, the public arrhythmia data set and the previous data preprocessing work were introduced. Secondly, the processed data was input into the deep learning model constructed with convolutional neural network, to automatically extract features of arrhythmia from electrocardiogram signals. Thirdly, the designed deep learning model was used for classifications and diagnosis of arrhythmias. Then the performance of the proposed model and that from other research work were compared. The validation of the method is proved with five cross-validation strategy.

Keywords: Arrhythmia · Convolutional neural network · Attention

1 Introduction

1.1 Research Background

Cardiovascular disease (CVD) accounts for over 31% of worldwide death [1]. Over 80% of sudden cardiac deaths, were closely related to cardiac arrhythmias, and responsible for half of deaths caused by heart diseases [2]. Long-term analysis of heart signals of arrhythmia diagnosis is usually conducted by an experienced cardiologist. Therefore, it is necessary to automatically analyze the ECG (Electrocardiogram) signal to determine the type of arrhythmia, in order to assess the risk of cardiac disease such as stroke [3], heart attack [4], and ischemia [5]. Holter monitoring is often used to record data that may exceed 24 h or more, including thousands of heart signals.

For cardiologists, analyzing such a large amount of data or long-term records is time-consuming and laborious, with long-term large-scale analysis being prone to misdiagnosis. Automatic ECG examination with aid of computer, is an effective method that enable evaluating long-term heartbeat records to analyze rare occurrences of arrhythmia, helping doctors provide patients with the necessary cures and medical care. This automatic computer-aided diagnostic method can provide patients with automatic, efficient and fast treatment plans to increase the cure rate of heart disease.

The original version of this chapter was revised: the citation "Table 5" in the reference on Page 498 was changed to "Table 3". The correction to this chapter is available at https://doi.org/10.1007/978-981-16-5940-9_41

© Springer Nature Singapore Pte Ltd. 2021, corrected publication 2021
J. Zeng et al. (Eds.): ICPCSEE 2021, CCIS 1451, pp. 481–504, 2021.
https://doi.org/10.1007/978-981-16-5940-9_37

In the past few decades, a few researchers developed algorithms for arrhythmia detection and classification, which was based on three main steps including preprocessing, feature extraction and classification [6–13].

However, traditional machine learning algorithms cannot perform feature extraction and classification of several arrhythmias very well, with the effect of automatic diagnosis not accurate enough. With the application of deep learning in various fields, a lot of attempts have been made to apply deep learning in computer-aided medical diagnosis. Deep learning algorithm has following two major advantages. It automatically learns to perform appropriate feature analysis and representation on the ECG signals, which changes the complicated and cumbersome situation of manually extracting features. In addition, the use of deep learning technology can reduce noise interference and build a model which is more robust for different big data analysis, improving the accuracy of arrhythmia diagnosis [14–26].

This article constructs a multi-classification model of arrhythmia based on the dual-channel convolutional neural network, in order to make automatic diagnosis of a few arrhythmias. The procedure is as follows: Firstly, introduce the public arrhythmia data set and the previous data preprocessing work; Secondly, input the processed data into the deep learning model constructed with convolutional neural network, to extract features of arrhythmia automatically; Thirdly, use the designed deep learning model for classifications and diagnosis of arrhythmias, which is probably to play a key role in clinical decision-making for treatment of arrhythmias.

1.2 Medical Background

In order to clarify the classification research, the medical background is introduced in details to make sense of clinical manifestation of arrhythmia diseases.

Electrocardiogram Introduction. An electrocardiogram is a picture of the electrical conduction of the heart. Clinicians can identify a multitude of cardiac disease processes by examining changes from normal on the ECG. Most people learn a combination of both pattern recognition (the most common) and understanding the exact electrical vectors recorded by an ECG as they relate to cardiac electrophysiology.

The standard ECG has 12 leads. Six of the leads are considered "limb leads" which are placed on the arms and/or legs of the individual. The other six leads are considered "precordial leads" which are placed on the torso (precordium). The six limb leads are called lead I, II, III, aVL, aVR and aVF. The letter "a" stands for "augmented," as these leads are calculated as a combination of leads I, II and III. The six precordial leads are called leads V1, V2, V3, V4, V5 and V6.

The Normal ECG. The normal electrocardiogram (ECG) is composed of several different waveforms that represent electrical events during each cardiac cycle in various parts of the heart as Fig. 1.

As in Fig. 1, ECG waves are labeled alphabetically starting with the P wave, followed by the QRS complex, and the ST-T complex (ST segment and T wave). The PR interval is measured from the beginning of the P wave to the first part of the QRS complex. The QT interval consists of the QRS complex which represents only a brief part of the interval, and the ST segment and T wave which are of longer duration.

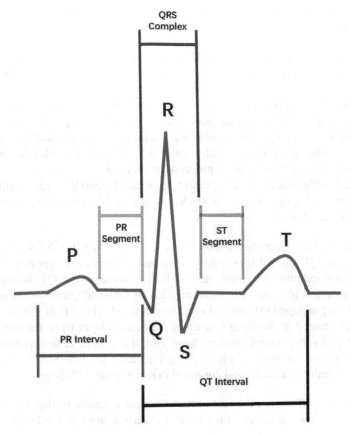

Fig. 1. The normal electrocardiogram

P Wave. The P wave represents atrial depolarization. The normal sinus P wave demonstrates depolarization from the right to left atrium and is an initial low amplitude deflection preceding the QRS complex that is positive in most leads.

PR Segment. The PR segment is the portion of the ECG from the end of the P wave to the beginning of the QRS complex. The PR segment is different from the PR interval, which is measured in units of time.

PR Interval. The PR interval includes the P wave as well as the PR segment. It is measured from the beginning of the P wave to the first part of the QRS complex (which may be a Q wave or R wave). It includes time for atrial depolarization (the P wave) and conduction through the AV node and the His-Purkinje system (which constitute the PR segment).

QRS Complex. A combination of the Q wave, R wave and S wave. The QRS complex represents the time for ventricular depolarization. If the initial deflection is negative, it is termed a Q wave. Small Q waves are often seen in leads I, aVL, and V4–V6 as a result of initial septal depolarization and are considered normal.

The first positive deflection of the QRS complex is called the R wave. It represents depolarization of the left ventricular myocardium. Right ventricular depolarization is obscured because the left ventricular myocardial mass is much greater than that of the right ventricle.

R Wave. The R wave is the first upward deflection after the P wave and part of the QRS complex. The R wave morphology itself is not of great clinical importance but can vary at times.

The R wave should be small in lead V1. Throughout the precordial leads (V1–V6), the R wave becomes larger—to the point that the R wave is larger than the S wave in lead V4. The S wave then becomes quite small in lead V6; this is called "normal R wave progression." When the R wave remains small in leads V3 to V4—that is, smaller than the S wave—the term "poor R wave progression" is used.

If a right bundle branch block is present, there may be two R waves, resulting in the classic "bunny ear" appearance of the QRS complex. In this setting, the second R wave is termed "R" or "R prime."

QT Interval. The QT interval consists of the QRS complex, the ST segment, and T wave. Thus, the QT interval is primarily a measure of ventricular repolarization. If the QRS complex duration is increased, this will lead to an increase in QT interval but does not reflect a change in ventricular repolarization. A widened QRS, therefore, must be considered if a prolonged QT interval is being evaluated. The QT interval is dependent upon the heart rate; it is shorter at faster heart rates and longer when the rate is slower. The QT interval is dependent upon the heart rate; it is shorter at faster heart rates and longer when the rate is slower. Thus, a QT interval that is corrected for heart rate (QTc) has been classically calculated and corrected in recent years [27–30].

ST Segment. The ST segment occurs after ventricular depolarization has ended and before repolarization has begun. It is a time of electrocardiographic silence.

T Wave. The T wave represents the period of ventricular repolarization. Since the rate of repolarization is slower than depolarization, the T wave is broad, has a slow upstroke, and a more rapid downslope to the isoelectric line following its peak. Thus, the T wave is asymmetric and the amplitude is variable.

Arrhythmia Introduction. The 12 leads of the ECG include 6 limb leads (I, II, III, aVR, aVL, aVF) and 6 chest leads (V1–V6). Limb leads include standard bipolar leads (I, II, and III) and compression leads (aVR, aVL, and aVF). The bipolar lead is named for recording the voltage difference between the two levels.

Left Bundle Branch Block. Left bundle branch block is abbreviated as bundle branch block, including left bundle branch main block and left anterior branch and left posterior branch double block. The incidence of left bundle branch block is far less than that of right bundle branch block. The incidence is low in patients under 30 years of age. The incidence of left bundle branch block in patients over 40 years of age is 3.6%, and the incidence of patients under 40 years of age is 0.9%. The incidence of right bundle branch block was 8–16 times higher than that of left bundle branch block. The characteristic of the left bundle branch block electrocardiogram is sinus, that is, the heart rhythm is regular. Only when the left bundle branch block, the R wave on the V5 and V6 leads of the electrocardiogram changes. If the QRS complex of ECG widens at the same time, it is called a complete left bundle branch block. If there is only a frustration on the R wave in leads V5 and V6, and the QRS waveform of the ECG does not exceed the normal range of 120 ms, it is called an incomplete left bundle branch block. The left bundle branch block waveform of a single heart beat in the dataset is shown in Fig. 2.

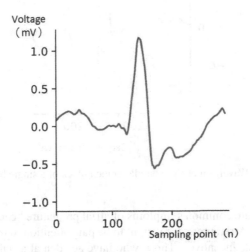

Fig. 2. Waveform of left bundle branch block of a single heartbeat

Right Bundle Branch Block. Right bundle branch block is a relatively common arrhythmia, which appears in normal people, such as children and young people. Incomplete right bundle branch block is relatively more common. About 1% of normal young people have incomplete right bundle branch block. Patients with right bundle branch block do not need to receive treatment if they are asymptomatic. The symptoms mostly cause the original right bundle branch block, such as symptoms of coronary heart disease, palpitations, chest tightness, shortness of breath, precordial pain, etc. The treatment mainly focuses on the treatment of primary diseases, with a generally good prognosis. When right bundle branch block occurs, it can be divided into complete right bundle branch block and incomplete right bundle branch block. The electrocardiogram characteristic of complete right bundle branch block is that the QRS complex time limit is no less than 0.12 s, with small R, small S, and large R patterns in leads V1 and V2, and thick and blunt R wave; leads V5 and V6 have the shape of small Q, big R, big S type, wide S wave, opposite T wave and QRS main wave. The graph of incomplete right bundle branch block is similar to the above, the difference is that the QRS time limit is less than 0.12 s. The right bundle branch block waveform of a single heart beat in the data set is shown in Fig. 3.

Atrial Premature. Atrial premature is atrial ectopic beats that occur before the basic heart rhythm. Abnormal heart structure and function are common causes of atrial premature beat. Some atrial premature beats appear in people with normal hearts. Palpitations and

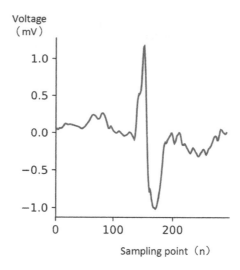

Fig. 3. Waveform of right bundle branch block of a single heartbeat

pauses in heartbeat are common symptoms of atrial premature beat, with some patients have no discomfort. Atrial premature ought to be paid attention to treatment of its cause and elimination of its incentives. Those who have accidental atrial premature beats or unobvious symptoms do not have to use anti-arrhythmic drugs. If the symptoms are obvious or induce tachycardia, antiarrhythmic drugs can be used for treatment.

Atrial premature is benign arrhythmia, with generally good prognosis. The p-wave morphology of atrial premature beat is different from the sinus p-wave, with the p-wave morphology of the upper right atrium heart rhythm being mostly similar to the sinus p-wave. The p waves are upright in lead I, II, aVL, aVF, and V3 to V6, with p wave inverted in lead aVR, indicating that it originates from the upper right pulmonary vein. In the lower right atrial rhythm, p wave is upright in lead I, aVL, and V1 to V6, with p wave inverted in lead II, III, and aVF. In the upper left atrial rhythm, the p-waves are inverted in lead I, aVL, V1 to V6, with p waves upright in lead II, III. The different forms of p waves indicate that the cardiac pacing site of the atrial rhythm is different. The atrial premature waveform of a single heart beat in the dataset is shown in Fig. 4.

Ventricular Premature Beat. Ventricular premature beat is the most common type of ventricular arrhythmia in clinical practice. It refers to the heart beat that occurs prematurely below the bifurcation of the His bundle and depolarizes the myocardium in advance. Its incidence rate increases with age. The symptoms of ventricular premature contraction vary greatly, as many patients are asymptomatic. Those with symptoms have clinical manifestations such as palpitations, cardiac arrest feelings, or feeling of overweightness or weightlessness similar to the rapid rise and fall of elevators. Patients with severe symptoms may have symptoms such as fatigue, syncope, angina pectoris, hypotension and shock.

The most significant feature of the electrocardiograph of ventricular premature beats is that it does not affect the normal heart rhythm, that is the rhythm of p-wave. This

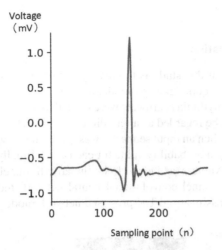

Fig. 4. Waveform of atrial premature of a single heartbeat

is also the reason why premature ventricular beats have complete compensatory pause. Premature ventricular beats are likely to fall within the relative refractory period of the previous heartbeat, with conduction delay leading to QRS complex lenient deformity, with the time limit greater than 0.12 s. Ventricular premature beat is a kind of premature beats, with QRS complexes appearing early. The electrocardiogram characteristics of ventricular premature beats are an early, wide deformed QRS complex, with complete compensatory pause. The waveform of ventricular premature of a single heartbeat in the dataset is shown in Fig. 5.

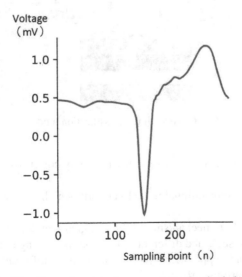

Fig. 5. Waveform of ventricular premature of a single heartbeat

2   Methods

2.1   Problem Formulation

The research problem in this study is to classify 5 types of arrhythmias. As the ECG signal uses one lead, the complexity of model training is reduced. In the preprocessing stage, the lengths of arrhythmia heartbeats were unified to length of 256 sampling points. The network input can be regarded as a one-dimensional vector sequence with a length 256 sampling points. When an input sequence was given, the network model was trained to classify it, outputting a probability value for each class. Finally, the class that obtains the largest probability value, was judged to be the arrhythmia class to which the input belongs. A new dual-channel convolutional neural network model was proposed for feature learning and classification. The proposed network model structure is as Fig. 6.

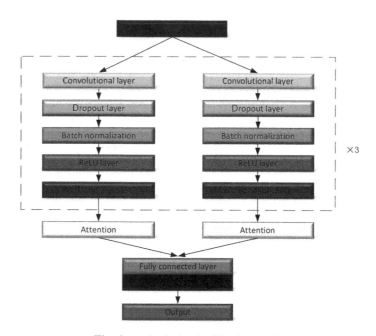

Fig. 6. Arrhythmia classification model

As shown in Fig. 6, the network structure is mainly composed of two parts:

1. The dual-channel convolutional neural network, which learns diverse features and merges them.
2. To apply the attention mechanism, the ECG signal features are automatically allocated with specific scale and different dimensions according to the weights, in order to discover relevant supplementary information from different scales, to assist the classification of various arrhythmia.

2.2 Arrhythmia Classification Model

Specifically, the dual-channel convolutional neural network, mainly uses a convolutional layer, connecting the dropout layer and the batch normalization layer as assistance. In order to make the features learned more diverse and refined, the convolutional neural network of each channel uses 3 identical convolution modules. The specific structure analysis of each channel is as follows.

For convolutional neural network of each channel, a convolutional layer with a stride of 1 to extract features was used, performing each convolution operation by moving the kernel once on the input vector. Then multiplication and addition of the superposition matrix were made. The convolutional layer can extract the spatial feature map to obtain the meaningful spatial information in the data.

Over-fitting is a common problem in deep learning. The model only learns to classify on the training set without the capability to achieve the corresponding effect on the test set. Many solutions to over-fitting problems have been proposed in recent years, in which dropout is simple and effective. From a theoretical perspective, dropout can be regarded as a kind of model averaging. For each sample input into the network (it may be a sample or a batch of samples), the corresponding network structure is different, while all these different network structures share the weight value of the hidden node at the same time.

As a result, different samples correspond to different models. During the training process of deep learning network, dropout refers to temporarily dropping neural network units out of the network according to a certain probability. For random gradient descent, the neural network unit was randomly discarded. Therefore, each mini-batch trains different network. This can randomly inactivate some neurons, excluding the connection weights from updating without participating in subsequent network training, so as to reduce model fitting and increase the generalization performance of the model. Therefore, in order to enhance the generalization performance of the model and reduce the overfitting, a dropout layer was used after the convolutional layer to operate on the extracted feature map.

The dropout layer was connected to a batch normalization layer. When the deep neural network performs nonlinear transformation, its activation input value deepens with the network depth during the training process, with its distribution gradually shifting or changing. The bad training effect of deep neural network generally comes from that the overall distribution gradually approaching the upper and lower limits of the value range of the nonlinear function, which causes the gradient disappearance of the lower neural network during backward propagation. It is the essential reason why convergence of the training deep neural network become slower and slower. The batch normalization layer uses a certain standardization method to force the distribution of the input value of any neuron in each layer of the neural network, back to a standard normal distribution with a mean of 0 and a variance of 1, which means that the increasingly skewed distribution is forced back to a standard distribution. The activation input value falls in the area where the nonlinear function is more sensitive to the input. A small change in the input will lead to a larger change in the loss function, making the gradient larger and avoiding the problem of gradient disappearance. When the gradient becomes larger, the convergence speed of the model learning will also become faster, which greatly accelerates the training speed of the model.

In each convolution module, a ReLU activation function layer was also included. The output of the standard sigmoid function does not satisfy sparseness. Some penalty factors were needed to train a large amount of redundant data close to 0, to generate sparse data, such as L1, L1/L2 or Student-t as penalty factors. Therefore, unsupervised pre-training was required. ReLU is a linear correction, which is a polyline version of purelin. Its function is to make the calculated value equal to 0 if it is less than 0, otherwise keep the original value unchanged. This is a method to force some data to be zero. It has been proved that the trained network has a moderate sparsity, with the visualization effect after training similar to that after traditional method. It shows that ReLU has the ability to guide moderate sparsity. Therefore, the use of ReLU enables the network to introduce sparsity, which greatly improves the training speed.

In addition, when a function like sigmoid was used as the activation function, the calculation work was huge as the derivation involves division when the gradient error was obtained by backpropagation. The use of ReLU activation function will save the calculation amount of the entire process. Therefore, for the deep convolutional network used in this study, the gradient disappears when the sigmoid function was backpropagated. When the sigmoid function was close to the saturation region, the transformation was too slow with the derivative, tending to be 0. This situation will cause information loss, being unable to complete the training of the deep convolutional network. At the same time, using the ReLU activation function will make the output of some neurons to be 0, resulting in the sparsity of the network. It also reduces the interdependence of parameters, and alleviates the occurrence of overfitting.

In order to reduce the size of the input representation by half, a max-pooling layer was used after each ReLU layer. There are many functions of adding the max-pooling layer to the convolutional network module. The first function was down-sampling. At the same time, it is able to reduce dimensionality, calculations, memory consumption,, the number of parameters, and to remove redundant information, compress features, simplify network complexity. Secondly, the max-pooling layer can achieve nonlinearity and expand the perceptual field of the convolutional neural network. Finally, it can achieve invariance, which includes translation invariance, rotation invariance and scale invariance. Therefore, at the end of the convolutional neural network module, a max-pooling layer was added. This layer can reduce the size of the feature map from the previous layer according to the area size, and generate a new feature map by obtaining the maximum value in the specified area obtained from the previous layer. Reducing the size of the feature map is an important way to reduce the computational cost of deep learning structure, which also reduces the complexity of the model and increases the generalization performance of the model.

For the convolution module of each channel, three identical modules were used to increase the depth of the model, so that it can extract features of different scales. In addition, in order to increase the diversity of learning features, a new dual-channel convolution module was used for construction of the model.

2.3 Attention Mechanism

In addition to the dual-channel convolutional neural network structure, the attention mechanism was added to the structure built in this study. The attention mechanism is a problem-solving method proposed by imitating human attention. In a word, it is to filter out high-value feature information from a large amount of information. This mechanism is mainly used to solve the problem that is difficult to obtain the final reasonable vector representation when the model input sequence is long. The method is to retain the intermediate results of the model, learn it with the new model, and associate it with the output, so as to achieve the purpose of information screening.

The attention mechanism can be understood in this way that when we are looking at something, what we are currently paying attention to, must be a certain part of the thing we focus. In other words, when we look elsewhere, our attention shift following the movement of eyes. This means that when people pay attention to a certain target or a certain scene, the attention distribution inside the target and each spatial position in the scene are different. This is also true in the following situations. When we try to describe something, the words and sentences we are talking about at the moment, are mostly relevant to a corresponding segment of the thing that is being described, while the relation between it and other parts are constantly changing as the description progresses.

Attention is actually a match between the current input and output. Though the model calculation amount will increase when we use the attention mechanism, its performance level can be improved. In addition, the attention mechanism is used to facilitate the understanding of how the information in the input sequence affects the final generated sequence during the output process. Specifically, the principle is that we selectively focus on the corresponding information in the input at the output step. The methods using attention mechanism have been widely used in various sequence prediction tasks, including text translation and speech recognition. In this paper, the attention layer in TensorFlow was used.

The attention mechanism, which learns the importance of each sampling point from the ECG signal sequence, and merges the sampling points according to their importance, is usually applied in the Encoder-Decoder model. Therefore, the attention mechanism, which provides the Decoder with information from the hidden state of each Encoder, can be regarded as an interface between Encoder and Decoder. With this setting, the model can selectively focus on the useful parts of the input ECG sequence, so as to learn the "alignment" between them. This indicates that when the Encoder encodes the input sequence elements, multiple semantic codes are obtained rather than a fixed semantic code, while different semantic codes are determined by different ECG sequence sampling points with combination of different weight parameters. A schematic diagram that simply reflects the operation of the attention mechanism is as Fig. 7.

Under the attention mechanism, the semantic coding C is not the direct coding of the input sequence X, but the weighted sum of each element according to its importance as formula (1):

$$C_i = \sum_{j=0}^{Tx} a_{ij} f\left(x_j\right) \tag{1}$$

In formula (1), the parameter i represents the time, j represents the j-th element in the sequence, T_x represents the length of the sequence, and $f(\cdot)$ represents the encoding of the element x. a_{ij} can be regarded as a probability, reflecting the importance of the

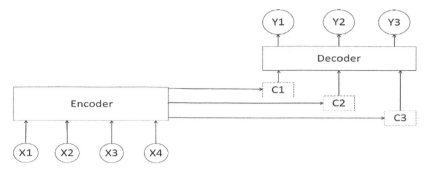

Fig. 7. Attention mechanism

element h_j to C_i which can be expressed by softmax as formula (2):

$$a_{ij} = \frac{\exp(e_{ij})}{\sum_{k=1}^{Tx} \exp(e_{ik})} \tag{2}$$

Here e_{ij} just reflects the matching degree between the element to be coded and other elements. When the matching degree is higher, it means that the element has a greater impact on it, with a_{ij} also bigger.

For the ECG signal, the manifestation of the ECG signal of each arrhythmia is different, and related to each other in time and space. For the learning and judgment of each arrhythmia disease, we focused on different positions of the ECG signal waveform. Inputting the features extracted by the dual-channel convolution module into the attention layer, can enable it to perform specific learning for different types of arrhythmias, and improve the classification performance of the model.

In the end, the learned features were input into a fully connected layer, then connected to a softmax layer to predict the categories of the input data. It can be seen from the subsequent classification results that the fusion of the established dual-channel convolution model with the attention mechanism, achieves a good classification performance, which proves the effectiveness of the constructed model.

3   Experiment

3.1   Environment

The proposed model was trained and tested on a server with an Intel Xeon E5-2637 CPU running at 3.5 GHz, 64 GB memory and an NVIDIA Quadro k6000 GPU. This server runs a Windows 10 system, and the model was implemented using the TensorFlow 2.3 framework.

3.2   Data Source

The article used the public standard MIT-BIH arrhythmia dataset for classification of arrhythmias [31]. The dataset contains 48 ECG records of 47 patients, including 110,109 beat labels. The ECG signal data is band-pass filtered with a cut-off frequency of 0.1–100 Hz. The ECG signals in the dataset are in the range of 10 mV, acquired with 11-bit resolution, with the sampling frequency of 360 Hz.

For the dataset used, multiple cardiologists labelled the rhythm type of each heartbeat in the dataset, and unified all the annotations through joint consultation. Each record in the dataset has two leads, with the coverage time of about 30 min. In this study, the corrected limb II lead, which is similar to the standard limb II lead, was selected for the classification of various arrhythmias. The difference is that it uses electrodes which is commonly used in Holter electrocardiographs, attaching to the torso to obtain signals. As certain types of heartbeat are very rare with small sample size, 5 common types of arrhythmia were selected in this study, including normal heartbeat (N), left bundle branch block (LBBB), right bundle branch block (RBBB), atrial premature beats (APB) and ventricular premature beats (PVC). The data volume of 5 arrhythmias is as Table 1.

Table 1. MIT-BIH arrhythmia dataset

Category	Quantity
Normal heartbeat (N)	75,906
Left bundle branch block (LBBB)	8058
Right bundle branch block (RBBB)	7176
Atrial premature beats (APB)	2516
Premature ventricular contraction (PVC)	7121
Total	99,777

3.3 Data Preprocessing

The preprocessing of the ECG signal is the key step in the detection and identification of arrhythmia. The original ECG signals contain various types of noise, which reduces the quality of data.

ECG noise includes baseline drift, power frequency interference, artifacts and electrode movement, which is caused by muscle contraction. Classifications of a variety of arrhythmias require efficient preprocessing operations in the early stage, to remove the noise associated with the ECG signal, and improve the signal-to-noise ratio.

This study applied average filter to extract the baseline drift signals from the ECG signals, and subtracted the baseline wandering signal from the original signal. There are many denoising methods for high-frequency noise, in which wavelet transform method has been verified to be efficient in denoising high-frequency noise. Wavelet transform is a new transform analysis method, which can perform localized analysis at the time (space) and frequency level, and gradually refine the signal (function) in multiple scales through expansion and translation operation, finally achieving time subdivision at high frequencies and frequency subdivision at low frequencies, which automatically adapt to the requirements of time-frequency signal analysis.

Analog filters and digital filters were always used to remove high frequency noise filters. In addition to the filter, the Fourier transform works on denoising, which is difficult to distinguish the high-frequency part of the useful signals from the high-frequency interference caused by noise, and not capable of keeping the high-frequency information of the signal which removes high-frequency noise.

In comparison, wavelet transform denoising can protect useful signal spikes and sudden changes. Therefore, wavelet transform is suitable for noise removal of transient signals. It suppresses the interference from high-frequency noise, and effectively distinguishes high-frequency information from high-frequency noise. Wavelet transform has good time-frequency localization characteristics, which can retain wavelet coefficients mainly controlled by signals, detecting and removing wavelet coefficients controlled by noise. The remaining wavelet coefficients were inversely transformed to obtain the denoised signal. The theoretical basis of wavelet transforms to remove signal high-frequency noise is that wavelet coefficient amplitude of the signal is greater than coefficient amplitude of the noise. The wavelet transform denoising method was used to remove the high-frequency noise (such as power frequency and EMG interference) in the ECG signal with baseline drift. The basic principle is as Fig. 8.

Fig. 8. The basic principle of wavelet transforms denoising.

After denoising the ECG signal, annotations marked by experts were used to cut the data set while a fixed-size window was set with the R point as the center. As the data sampling frequency was 360 Hz, the window size was set to be 256, in order to make the heartbeats of different arrhythmia types can contain a complete ECG signal waveform. Each heartbeat consists of only one type of arrhythmia taking 128 samples before and after the R peak, with the heartbeat length of each type around 0.71 s.

According to the human heart rate range, it is guaranteed that all the heartbeats cut in the dataset contain all the waveform information. The schematic diagram of the heartbeat segmentation processing is shown in Fig. 9. The example in Fig. 9 is a normal heartbeat. From the figure, the heartbeat segmentation was based on the R wave peak as the midpoint, using a sliding window with the size of 256 sampling points, with 128 sampling points intercepted before and after the R wave peak.

Afterwards, five types of arrhythmia data were randomly divided into five subsets after segmentation. The deep learning model was trained with the five-cross validation method, which not only increases the robustness of the model, but prevents the model from overfitting. Table 1 shows the data volume of 5 kinds of arrhythmia. In addition, the overall procedure of various arrhythmia classification in the study is shown in Fig. 10.

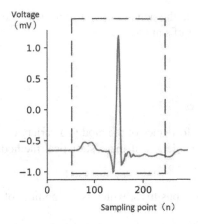

Fig. 9. Heartbeat segmentation diagram

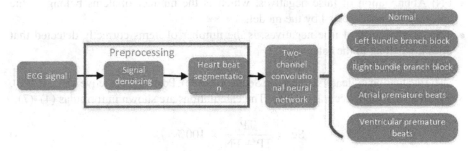

Fig. 10. The flowchart of arrhythmia classification algorithm

3.4 Training Setting

Five cross-validation strategies were used to prove the generalization performance of the model. The cross-validation value was selected as 5 to divide the data into 5 equal subsets. Each time, a new set of training and test data was used to train a new initialized model. In order to monitor the training process to prevent the model from overfitting, 20% of the training data was used at the end of each training period to verify the performance.

Then the backpropagation algorithm was used to train the model. After each training set was iterated, the validation set was used to test the model. When the loss value on the validation set no longer drops for 10 consecutive times, the training of each cross model was stopped. The test set was used to evaluate each crossover model. The learning rate of the model was set as 0.001. The Adam optimizer was used to accelerate the learning process of the model. In the process of updating the model, the cross-entropy loss function was used to evaluate the performance of the proposed network model. The loss function is as formula (3):

$$loss = -\frac{1}{n}\sum_i y_i ln a_i \qquad (3)$$

In the formula, y represents the actual label, a represents the predicted output, and n represents the total number of samples.

4 Result

4.1 Performance Metrics

In order to evaluate the performance of the model, a variety of standard statistical methods were adopted. The calculation of these statistical methods requires the following parameters:

- TP: Abbreviation for true positives, which is the number of correct detections in the category.
- FP: Abbreviation for false positives, which is falsely predicted as the number of other categories.
- FN: Abbreviation of false negatives, which is the number of items belong to the category but not detected by the model.
- TN: Abbreviation of true negatives, is the number of items correctly detected that does not belong to the category.

The evaluation indicators used were Sensitivity (Se), Positive Predictive Value (PPV), Specificity (Sp), and Accuracy (Acc). The calculations are shown in formulas (4)–(7):

$$Se = \frac{TP}{TP + FN} \times 100\% \tag{4}$$

$$PPV = \frac{TP}{TP + FP} \times 100\% \tag{5}$$

$$Sp = \frac{TN}{TN + FP} \times 100\% \tag{6}$$

$$Acc = \frac{TP + TN}{TP + FN + FP + FN} \times 100\% \tag{7}$$

Since we focused on the classification of a variety of arrhythmias, the total calculation index of all types was related to the ratio of data in each category to the total amount of data. The calculation indicators of each category were weighted by proportions, and summed to obtain the overall calculation indicators, which is reasonable and efficient. The calculation method for each indicator is shown in formula (8).

$$Z = \sum_{i=1}^{5} p_i x_i \tag{8}$$

Z represents all calculation indicators, p_i represents the proportion of data of each arrhythmia in total, and x_i represents calculation indicators of each type of arrhythmia.

4.2 Classification Performance

In order to make a fair performance comparison with other algorithms, the previous part briefly introduces the evaluation indicators of the model. Basing on this, the unified MIT-BIH arrhythmia dataset was used to evaluate indicators. In addition, in order to prevent overfitting of the constructed model and obtain a reliable and stable model, five-cross validation was used to train the arrhythmia classification model. The average results of five-cross validation of the model on the MIT-BIH arrhythmia dataset are shown in Table 2.

Table 2. Results of model on MIT-BIH arrhythmia dataset

		Predicted label					Se (%)	Sp (%)	PPV (%)	Acc (%)
		N	LBBB	RBBB	APB	PVC				
	N	74687	4	12	73	130	99.71	97.59	99.20	99.18
Label	LBBB	12	8024	1	1	20	99.58	99.96	99.48	99.93
	RBBB	13	1	7120	37	5	99.22	99.94	99.26	99.89
	APB	401	5	36	2070	4	82.27	99.88	94.56	99.43
	PVC	174	22	4	8	6913	97.08	99.83	97.75	99.63
Overall index							99.04	98.17	99.01	99.33

As in Table 2, classification results of all arrhythmia types have achieved good results except for atrial premature beats, whose sensitivity index achieved 82.27%. Compared with classification of other types of arrhythmia, its sensitivity index was 82.27%. The reason was mainly the small amount of data. In the training process of the model, extraction of the relevant feature was not comprehensive. Additionally, the data was not balanced in the pre-processing of the data. The balance of data has a great impact on deep learning algorithms, resulting in missed detections occurred in classifying atrial premature beats.

In addition, from the table, the positive predictive value of atrial premature beats was lower than other types of arrhythmia. When classifying atrial premature beats, many samples were judged as normal samples. The reason may be that the waveform shape was not significant different from that of normal samples leading to false detections. The model proposed has some limitations and needs improvement. In addition, it is required to focus on the separate detailed analysis of arrhythmia such as atrial premature beats, with more consideration of data balance problem.

5 Discussion

5.1 Performance Analysis

In order to verify the performance of the proposed model, the arrhythmia classification results of the proposed model with the results of other research works were compared.

Among the compared algorithms, there were traditional machine learning algorithms and deep learning methods for arrhythmia classification. From the comparison results in Table 3, the overall classification accuracy of this model was better than other algorithms.

Table 3. The performance of the algorithm proposed and its comparison with other work

Algorithm	Time	The data distribution					Se (%)	Sp (%)	PPV (%)	Acc (%)
		N	LBBB	RBBB	APB	PVC				
AMartis [32]	2013	10000	8069	7250	2544	7126	99.27	98.31	99.33	93.48
Li [33]	2017	2640	2973	3000	3000	1200	97.42	99.33	97.43	98.88
Oh [34]	2018	8245	344	660	1004	6246	97.52	98.41	97.49	98.55
Ji [35]	2019	5250	5037	4629	0	4564	98.06	99.45	98.04	99.21
Oh [36]	2019	71337	7890	7123	2123	6194	96.13	95.84	96.05	97.70
Proposed algorithm (No attention)	2021	74906	8058	7176	2516	7121	98.88	98.00	99.01	99.28
Proposed algorithm	2021	74906	8058	7176	2516	7121	99.04	98.17	99.01	99.33

The algorithm proposed by Martis includes traditional machine learning methods, while other researchers used methods based on deep learning. Many researchers have applied deep learning methods in the research of arrhythmia classification. Among the other studies listed in the table, some simply use convolutional neural networks for network construction, some use a faster R-CNN model. Others refer to the improved U-Net network in the field of image segmentation, and LSTM network to build the model (LSTM network has a greater advantage in processing time series signals). From the results in Table 3, the dual-channel convolutional neural network proposed combined with attention mechanism, could capture information at various scales more efficiently, focus on different parts of the ECG signal for different categories, and improve the overall classification performance.

Simply applying automatic feature extraction with convolutional neural network, has an inefficient classification performance, as it only contains the local spatial information of the ECG signal. In addition, the use of the attention mechanism in the model, was able to focus on spatio-temporal information, and improve the arrhythmia classification performance.

In order to prove the effectiveness of using the attention mechanism, ablation experiments (to remove the attention layer) were conducted with two-channel model for training and model testing. The classification results are shown in Table 3. From Table 3, when the attention layer is removed, indicators have declined except for the positive and positive indicators. Therefore, it is proved that the attention mechanism was able to perform more focused learning effectively on different types of arrhythmia in the process of model training.

Subsequent research may consider fusing traditional features and features automatically extracted by deep learning algorithms for more systematic research. For the classification of arrhythmia, there are still many researches that use different datasets to classify other categories. Table 3 does not list all other related studies. In addition, although all the research works mentioned in Table 3 use the same dataset with the same classification types, the amounts of data in test sets are not exactly the same.

Some research works test the performance of the algorithm only on a subset of the original dataset, some research works changed the ratio between the number of samples in each category, such as adding data with a small sample type, or randomly removing variety to achieve a relative balance of various types of data. These changes will undoubtedly have an impact on the classification results of arrhythmia, so the distribution of the test set in each research work was listed in the algorithm performance comparison, to make the comparison of research work more detailed and fairer. Our research on arrhythmia classification used all the data of five types of samples in the MIT-BIH arrhythmia dataset. The results showed that the performance of the proposed was still better than the performance of the other algorithms listed, which proved the validity of the model.

5.2 Model Parameter and Validation

In order to represent the parameters of each layer in the convolution module in details, Table 4 and Table 5 list the parameters of each layer in the dual channel convolution module (as the parameters of each layer in the dual channel convolution module are different, they are listed separately).

As in Table 4 and Table 5, the size and number of convolution kernels in the convolution layer in the dual channel are different, which can make the automatically extracted features more diversified improving the accuracy rate of multiple arrhythmia classification.

In order to obtain a relatively optimal arrhythmia classification model, it is necessary to analyze hyperparameters of different network structure. As the parameters of network structure and their combinations were very large, some hyperparameters were adjusted during the training process of the model, to enable the model to achieve a relatively optimal classification effect.

The selected parameters include learning rate, dropout rate, the size of the convolution kernel in the convolutional layer, the pool size of the Max pooling layer, and the number of neurons in the fully connected layer. The learning rate control the speed of weight updating during the training process. After using the selected combination of parameters for model training, the result of arrhythmia classification achieves the best performance. The values of each hyperparameter are shown in Table 6.

As in Table 6, learning rate is a critical hyperparameter that must be adjusted, which can make deep learning training more effective. Adam stochastic gradient descent optimizer was used for the training of the model. The learning rate controls the optimizer to move the weights in the opposite direction to the gradient of the minimum batch. When the learning rate was low, its convergence speed would slow down, affecting the speed

Table 4. Specific parameters of each layer of the first channel convolution module

Description of each layer of the first channel convolution module	Parameters of each layer and their descriptions
Convolutional layer	Convolution kernel size (16), number of convolution kernels (16)
Dropout layer	Dropout rate (0.2)
Batch normalization layer	Data processing without parameters
ReLU layer	Activation function without parameters
Max-pooling layer	Max-pooling layer core size (2)
Convolutional layer	Convolution kernel size (16), number of convolution kernels (16)
Dropout layer	Dropout rate (0.2)
Batch normalization layer	Data processing without parameters
ReLU layer	Activation function without parameters
Max-pooling layer	Max-pooling layer core size (2)
Convolutional layer	Convolution kernel size (8), number of convolution kernels (16)
Dropout layer	Dropout rate (0.2)
Batch normalization layer	Data processing without parameters
ReLU layer	Activation function without parameters
Max-pooling layer	Maxi-pooling layer core size (2)

of model training. An increase in the learning rate may lead to poor classification performance of arrhythmia, and reduce its classification accuracy. Therefore, if the learning rate was set to 0.001, the model would achieve relatively optimal performance. Subsequent research can also set the learning rate into segments and use varying learning rate to get a better performance. In addition, the selection of the pool size and dropout rate of the max pooling layer was also very important in simplifying the model, to reduce a large number of parameters to prevent the occurrence of model overfitting. The model which is constructed based on Keras, proposed a total of 10,741 trainable parameters.

Five-cross model was used for training, generally achieving convergence before about 40 iterations. When the five-cross model converges, the loss value of the training set was lower than the loss value of the validation set, so it was proved that the five-cross model was not over-fitting with the model and reaching a relatively stable state. Figure 10 is a graph of the loss function values of the training set and validation set of one of the cross models. As in Fig. 11, the loss function value of the arrhythmia training set is lower than the loss function value of the validation set.

Table 5. Specific parameters of each layer of the second channel convolution module

Description of each layer of the second channel convolution module	Parameters of each layer and their descriptions
Convolutional layer	Convolution kernel size (16), number of convolution kernels (8)
Dropout layer	Dropout rate (0.2)
Batch normalization layer	Data processing without parameters
ReLU layer	Activation function without parameters
Max-pooling layer	Maxi-pooling layer core size (2)
Convolutional layer	Convolution kernel size (16), number of convolution kernels (8)
Dropout layer	Dropout rate (0.2)
Batch normalization layer	Data processing without parameters
ReLU layer	Activation function without parameters
Maxi-pooling layer	Maxi-pooling layer core size (2)
Convolutional layer	Convolution kernel size (8), number of convolution kernels (8)
Dropout layer	Dropout rate (0.2)
Batch normalization layer	Data processing without parameters
ReLU layer	Activation function has no parameters
Maxi-pooling layer	Maxi-pooling layer core size (2)

Table 6. Model hyperparameters for arrhythmia classification

Parameter	The optimal value
Learning rate	0.001
Dropout rate	0.5
Convolution kernel size of the convolution layer	16
Pool size of the Max-pooling layer	2
Number of neurons in the fully connected layer	16

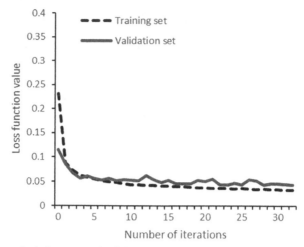

Fig. 11. The numerical change graph of the loss function of the arrhythmia training set and the validation set

6   Conclusion

This study focuses on the automatic classification of a variety of arrhythmias based on deep learning and obtain good results on most categories of arrhythmias, while the classification results for atrial premature beats was not perfect. The future work can focus on the following aspects:

1. Specific studies on solutions to classification for atrial premature beats.
2. For multiple arrhythmia classification problems, the size and balance of data has always been a core issue to be solved, which may cause learning deviations in the learning process of model. In the future study, the data ought to be supplemented to reduce data imbalance, without expanding the difference in data distribution.
3. Study can also be carried out on the use of a small number of samples for model training, to reduce data collection and improve the efficiency of model construction with its model performance guaranteed.
4. For the problem of data labeling, as a large amount of data labeling will cause a huge workload for heart disease experts, the future work could focus on unsupervised learning with unlabeled ECG signal data for model training, to make better use of big data for arrhythmia classification research and enhance the development of medical big data.
5. The arrhythmia classification study in this article used a single heartbeat as input, which contains relatively little information. Single-lead was used as input. In future, study on arrhythmia classification could focus on relatively long-term multi-lead ECG signals.

References

1. Mendis, S., Puska, P., Norrving, B., Organization, W.H., et al.: Global Atlas on Cardiovascular Disease Prevention and Control. World Health Organization, Geneva (2011)
2. Mehra, R.: Global public health problem of sudden cardiac death. J. Electrocardiol. **40**(6), 118–122 (2007)
3. Stamkopoulos, T., Diamantaras, K., Maglaveras, N., Strintzis, M.: ECG analysis using nonlinear PCA neural networks for is chemia detection. IEEE Trans. Signal Process. **46**(11), 3058–3067 (1998)
4. Leijdekkers, P., Gay, V.: A self-test to detect a heart attack using a mobile phone and wearable sensors. In: 2008 21st IEEE International Symposium on Computer-Based Medical Systems, pp. IEEE (2008)
5. Goldstein, D.S.: The electrocardiogram in stroke: relationship to pathophysiological type and comparison with prior tracings. Stroke **10**(3), 253–259 (1979)
6. Mathunjwa, B.M., Lin, Y.-T., Lin, C.-H., Abbod, M.F., Shieh, J.-S.: ECG arrhythmia classification by using a recurrence plot and convolutional neural network. Biomed. Signal Process. Control **64**, 102262 (2021)
7. Zhang, J., Liu, A., Gao, M., Chen, X., Zhang, X., Chen, X.: ECG-based multi-class arrhythmia detection using spatio-temporal attention-based convolutional recurrent neural network. Artif. Intell. Med. **106**, 101856 (2020)
8. Rajendra Acharya, U., Hamido Fujita, O., Lih, S., Hagiwara, Y., Tan, J.H., Adam, M.: Automated detection of arrhythmias using different intervals of tachycardia ECG segments with convolutional neural network. Inf. Sci. **405**, 81–90 (2017)
9. Sanamdikar, S.T., Hamde, S.T., Asutkar, V.G.: Classification and analysis of cardiac arrhythmia based on incremental support vector regression on IOT platform. Biomed. Signal Process. Control **64**, 102324 (2021)
10. Zhou, Z., Zhai, X., Tin, C.: Fully automatic electrocardiogram classification system based on generative adversarial network with auxiliary classifier. Expert Syst. Appl. **174**, 114809 (2021)
11. Chen, A., et al.: Multi-information fusion neural networks for arrhythmia automatic detection. Comput. Methods programs Biomed. **193**, 105479 (2020)
12. Xie, X., et al.: A multi-stage denoising framework for ambulatory ECG signal based on domain knowledge and motion artifact detection. Futur. Gener. Comput. Syst. **116**, 103–116 (2021)
13. Yao, Q., Wang, R., Fan, X., Liu, J., Li, Y.: Multi-class arrhythmia detection from 12-lead varied-length ECG using attention-based time-incremental convolutional neural network. Inf. Fusion **53**, 174–182 (2020)
14. Rajendra Acharya, U., Shu Lih, O., Hagiwara, Y., Tan, J.H., Adeli, H.: Deep convolutional neural network for the automated detection and diagnosis of seizure using EEG signals. Comput. Biol. Med. **100**, 270–278 (2018)
15. Srivastava, S., Soman, S., Rai, A., Srivastava, P.K.: Deep learning for health informatics: recent trends and future directions. In: 2017 International Conference on Advances in Computing, Communications and Informatics (ICACCI), pp. 1665–1670 (2017)
16. Dash, S., Acharya, B.R., Mittal, M., Abraham, A., Kelemen, A. (eds.): Deep Learning Techniques for Biomedical and Health Informatics. SBD, vol. 68. Springer, Cham (2020). https://doi.org/10.1007/978-3-030-33966-1
17. Kwak, G.H.-J., Hui, P.: DeepHealth: deep learning for health informatics reviews, challenges, and opportunities on medical imaging, electronic health records, genomics, sensing, and online communication health. arXiv preprint arXiv (2019)

18. Saha, J., Chowdhury, C., Biswas, S.: Review of machine learning and deep learning based recommender systems for health informatics. In: Dash, S., Acharya, B.R., Mittal, M., Abraham, A., Kelemen, A. (eds.) Deep Learning Techniques for Biomedical and Health Informatics. SBD, vol. 68, pp. 101–126. Springer, Cham (2020). https://doi.org/10.1007/978-3-030-339 66-1_6

19. David Naylor, C.: On the prospects for a (deep) learning health care system. JAMA **320**(11), 1099 (2018)

20. Beam, A.L., Kohane, I.S.: Big data and machine learning in health care. JAMA **319**(13), 1317–1318 (2018)

21. Mosavi, A., Ardabili, S., Varkonyi-Koczy, A.R.: List of deep learning models. In: International Conference on Global Research and Education, pp. 202–214 (2019)

22. Shickel, B., Tighe, P.J., Bihorac, A., Rashidi, P.: Deep EHR: a survey of recent advances in deep learning techniques for electronic health record (EHR) analysis. IEEE J. Biomed. Health Inform. **22**(5), 1589–1604 (2018)

23. Navamani, T.M.: Efficient deep learning approaches for health informatics. In: Deep Learning and Parallel Computing Environment for Bioengineering Systems, pp. 123–137 (2019)

24. Hinton, G.: Deep learning—a technology with the potential to transform health care. JAMA **320**(11), 1101–1102 (2018)

25. Martis, R.J., Lin, H., Javadi, B., Fernandes, S.L., Yasmin, M.: Editorial of the special issue DLHI: deep learning in medical imaging and health informatics. Pattern Recogn. Lett. **140**, 116–118 (2020)

26. Matsushita, H.: Innovation in health informatics. In: Matsushita, H. (ed.) Health Informatics. TSS, vol. 24, pp. 1–23. Springer, Singapore (2021). https://doi.org/10.1007/978-981-15-378 1-3_1

27. Malik, M., Färbom, P., Batchvarov, V., et al.: Relation between QT and RR intervals is highly individual among healthy subjects: implications for heart rate correction of the QT interval. Heart **87**, 220–228 (2002)

28. Bogossian, H., Frommeyer, G., Ninios, I., et al.: New formula for evaluation of the QT interval in patients with left bundle branch block. Heart Rhythm **11**(12), 2273–2277 (2014)

29. Rautaharju, P.M., Zhang, Z.M., Prineas, R., Heiss, G.: Assessment of prolonged QT and JT intervals in ventricular conduction defects. Am. J. Cardiol. **93**(8), 1017–1021 (2004)

30. Sriwattanakomen, R., Mukamal, K.J., Shvilkin, A.: A novel algorithm to predict the QT interval during intrinsic atrioventricular conduction from an electrocardiogram obtained during ventricular pacing. Heart Rhythm **13**(10), 2076–2082 (2016)

31. Moody, G.B., Mark, R.G.: The impact of the MIT-BIH arrhythmia database. IEEE Eng. Med. Biol. Mag. **20**(3), 45–50 (2001)

32. Martis, R.J., Rajendra Acharya, U., Mandana, K.M., Ray, A.K., Chakraborty, C.: Cardiac decision making using higher order spectra. Biomed. Signal Process. Control **8**(2), 193–203 (2013)

33. Li, D., Zhang, J., Zhang, Q., Wei, X.: Classification of ECG signals based on 1D convolution neural network. In: 2017 IEEE 19th International Conference on E-Health Networking, Applications and Services, Healthcom, pp. 1–16. IEEE (2017)

34. Shu Lih, O., Ng, E.Y.K., Tan, R.S., Rajendra Acharya, U.: Automated diagnosis of arrhythmia using combination of CNN and LSTM techniques with variable length heart beats. Comput. Biol. Med. **102**, 278–287 (2018)

35. Ji, Y., Zhang, S., Xiao, W.: Electrocardiogram classification based on faster regions with convolutional neural network. Sensors **19**(11), 2558 (2019)

36. Shu Lih, O., Ng, E.Y.K., Tan, R.S., Rajendra Acharya, U.: Automated beat-wise arrhythmia diagnosis using modified U-net on extended electrocardiographic recordings with heterogeneous arrhythmia types. Comput. Biol. Med. **105**, 92–101 (2019). https://doi.org/10.1016/j.compbiomed.2018.12.012

Study on Virtual Assembly for Satellite Based on Augmented Reality

Jie Du[1($\boxtimes$)], Yong Wang[1], and Haigen Yang[2]

[1] College of Automation and College of Artificial Intelligence, Nanjing University of Posts and Telecommunications, Nanjing, China
[2] Engineering Research Center of Wider and Wireless Communication Technology, Ministry of Education, Nanjing University of Posts and Telecommunications, Nanjing, China

Abstract. Assembly activities are a very important part of the production process of large and complex equipment. In order to solve the huge resource problem of large and complex equipment assembly process, this paper proposes a virtual assembly for satellite. Microsoft's latest holographic glasses HoloLens2 is used as a display facility, and Unity3D is applied to develop a satellite part assembly scene based on augmented reality (AR) technology. Voice input and gaze in the Mixed Reality Toolkit (MRTK), provided by Microsoft, are adopted in the development scenario to enable users to interact with virtual models in a more natural way. Finally, the development scenario has been tested in a small space. The results show its validity and good performance.

Keywords: AR · HoloLens2 · MRTK · Assembly · Satellite

1 Introduction

In manufacturing, assembly technology is designed to improve product design quality, shorten development cycle and save production costs by means of fast, flexible, reliable and low-cost assembly. The traditional assembly method is to create an equal-scale physical model of the part, although the part physical model can provide the operator's real assembly operation experience, but the production of a complete part model and perfect the details of the part is very time-consuming, energy and cost of a job, and in some of the more difficult assembly environment, the physical model may cause wear or complete damage due to various uncertainties, such a repeated process will increase part maintenance costs and cause a lot of time waste. By its aviation, aerospace products, such as the representative of large-scale complex assembly activities are the most obvious. Augmented Reality (AR) is a technique that accurately overlays computer-generated virtual images in real time and allows users to interact with virtual objects that are superimposed by real objects [1]. This technology enables a seamless interface between the virtual world and the real world, allowing assembly workers to quickly return to the real world after experiencing the virtual assembly. Makris [2] proposed an algorithm that automatically generates and inputs assembly sequence information into the system, establishes an AR instruction template for operational tasks, and finally provides

© Springer Nature Singapore Pte Ltd. 2021
J. Zeng et al. (Eds.): ICPCSEE 2021, CCIS 1451, pp. 505–514, 2021.
https://doi.org/10.1007/978-981-16-5940-9_38

an assembly simulation system for technicians through the AR application. Sabine [3] combined a vibrating bracelet with a tablet to create an AR-based mobile training platform. Ong [4] proposed an enhanced assembly system that uses bare hands for assembly between virtual components in an enhanced environment, and uses a constraint-based hybrid approach that enables the system to accurately identify the user's assembly intent without the need for additional CAD information, enhancing the interaction between the user and the virtual assembly object in a real-world environment. Khan [5] designed the generation method of high-quality ARToolkit labels. Fiala [6] developed the ARTag tag. Attard [7] designed the TangiBoard system, which uses cameras to detect black and white labels of physical objects on the whiteboard, and interactively projects digital content according to the semantics used by labels. Kurth [8] proposed a dynamic muti-projection mapping system composed of target objects, multiple projectors and depth-based tracking system to improve the detection and tracking algorithm efficiency when feature points are few, obtain relatively accurate internal and external parameters of the projector in the calibration stage, and complete the virtual/real registration on arbitrary surfaces. Although this method all mentioned that the physical object could be moved and the registration would be automatically optimized after moving, it did not mention that the object could be used for real-time interaction.

For large-scale equipment such as satellites, the human and financial resources required to use traditional assembly operations are undoubtedly enormous. This paper first built a complete satellite model on 3Dsmax software, then imported it into the Unity3D engine, decomposed the model, and then used MRTK to assist in the development, added behavior scripts to each model part. Considering the more natural way of interacting, in addition to basic gesture interaction, voice interaction and gaze interaction have been added throughout the development scenario.

2 Hardware Facilities and Development Tools

2.1 Hardware Facility-HoloLens2

HoloLens is a holographic computer that is completely free from any other facility [9]. Launched in 2019, the second-generation HoloLens are 10 g lighter than the previous generation, three times more comfortable than the previous generation, twice as comfortable, and have a larger field-of-view angle that gives users a greater perspective. HoloLens2 provides more powerful gesture recognition, compared to a generation, HoloLens2 can track 25 points in a single hand, both hands of the fingers can be well recognized, the corresponding gestures have increased, such as pick, grab, pinch this subtle gesture can be recognized. Due to the good wear comfort of the HoloLens2, the larger field of view angle, and the complete computer equipment nature of HoloLens itself, it is the optimal display device in a large complex product assembly environment.

2.2 Development Tools-Mixed Reality Toolkit

To lower the bar for developing AR applications, there are many open source AR SDKs on the market. Peter [10] designed an AR writing system that encapsulated a range of components such as display menus, rotating lists, and so on, allowing developers to create AR applications directly. Gimeno [11] has developed a tool that allows users with no programming experience to develop AR applications online and provides a rich API interface.

Mixed Reality Toolkit (MRTK) is the best choice for developing HoloLens applications. MRTK is an open source project developed by Microsoft to share the underlying components, providing a range of components and capabilities for the Augmented Reality scenario of the Unity project designed to accelerate mixed reality application development in the Unity platform.

3 Scenario Architecture Design

3.1 The Overall Architecture of the Scenario

The entire assembly scene consists of a satellite model dataset, a follow-up menu, a satellite part model management system, gaze function, and a voice input system. The entire scenario framework architecture is shown in Fig. 1.

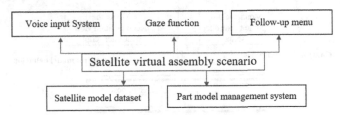

Fig. 1. The overall architecture of the satellite assembly scene

3.2 Satellite Model Dataset

Satellite model dataset is the most important part of the assembly scene, which contains three different forms of satellite data model. The original model, the decomposed model, and the model with the assembly part prompt. The three models are combined to form a satellite model dataset, as shown in Fig. 2.

Figure 3 shows the overall framework for the production of satellite model datasets. Where the decomposed satellite model is hidden in the final model set, its function is to set the world coordinates to be placed by each part after the satellite decomposition. By grabbing the detached model and placing it at the highlight prompt, the model is automatically glued to the missing parts of the satellite model. The auto-bonding function is controlled by the behavior script in the part model management system.

3.3 Part Model Management System

A part management system is a behavior script mounted on a satellite model dataset, which is a kind of instance object in the Unity3D engine that can add behavioral logic

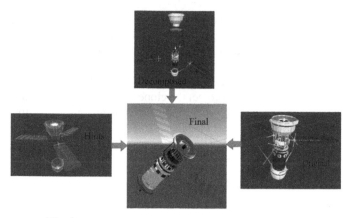

Fig. 2. The process of making satellite model datasets

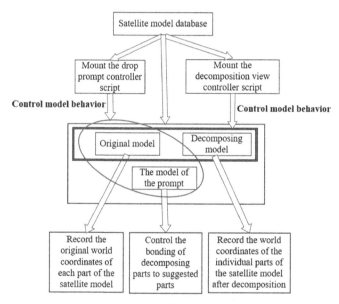

Fig. 3. The design idea of the satellite model dataset

to all objects in a scene. The part model management system implements two functions. First, the implementation of the satellite model decomposition visual effects of the script, the specific process of the script is shown in Fig. 4.

The second script of the part model management system implements the function of automatically bonding when the captured decomposed part model is close to the highlighted part model on the satellite model. This paper set that when the difference between the decomposed part model and the part model at the highlight prompt is greater than 0.1 cm and less than 10 cm, the bonding event is triggered, the values in the Transform assembly of the part model at the highlighted prompt are assigned to the

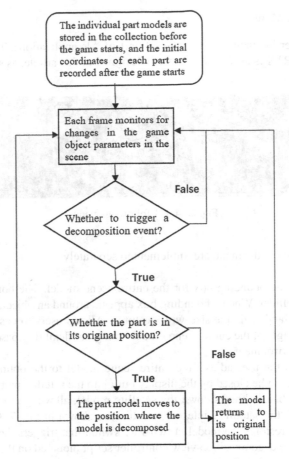

Fig. 4. Satellite model decomposition script flowchart

Transform assembly of the decomposed part model, and the sound effect is played back to the user feedback. This allows for visual and tactile bonding of the decomposition model, as shown in Fig. 5.

Fig. 5. The model is automatically glued when the specified distance is reached

3.4 Follow-Up Menu

Human-computer interaction is an integral part of AR applications. The 4 × 1 menu prefab in the MRTK is used in the scenario designed in this article, as shown in Fig. 6.

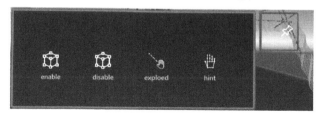

Fig. 6. 4 × 1 follow-up menu

The five twists in the menu are implemented separately:

- enable: Open the bounding box for the entire scene model. The bounding box is a special box collider. When a bounding box appears around an object, you can rotate, drag position, and scale the size with gestures. Allows users to customize the size and rotation angle of the entire scene model with great flexibility based on the actual application environment.
- disable: When the user adjusts the entire scene model to the optimal posture, the bounding box can be closed by the display button to the satellite part.
- exploed: This button breaks down the satellite model, allowing users to more intuitively observe the satellite's internal structure, as shown in Fig. 7. Clicking on the exploed again restores the model, a two-way switch that triggers the corresponding events set in the decomposition view controller script mounted on the satellite model dataset.
- hint: Clicking on the hint button can turn off the highlighted prompt model on the satellite dataset, click again can be restored, and is also a two-way button that triggers the corresponding event set in the placement prompt controller script mounted on the satellite model dataset.
- follow-up: This button is at the top right of Fig. 6, and when you click it, the menu follows the user's field of view based on the rotation of the user's head.

3.5 Voice Input System

This paper used the voice input function in the input system in the MRTK to add another kind of human-computer interaction to the scene. The SpeedInputHandler script provided in MRTK is used in the scenario to convert the functionality achieved by the four buttons in the follow menu in Sect. 3.4 into a voice trigger that triggers the corresponding event when the user says a keyword in the scenario. Speech recognition interaction enriches the means by which users manipulate scene objects, and in an assembly environment,

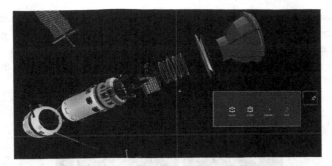

Fig. 7. The effect of the decomposed button

when the hands are occupied by other resources, corresponding events can be triggered by voice commands to achieve efficient work.

In MRTK's voice input system, the four buttons of the Sect. 3.4 were named as voice commands and then added an empty object to the Unity development scenario, which mounts the SpeedInputHandler script, added the voice commands set in voice input system to the component, and set the events triggered by the voice commands. Figure 8 shows.

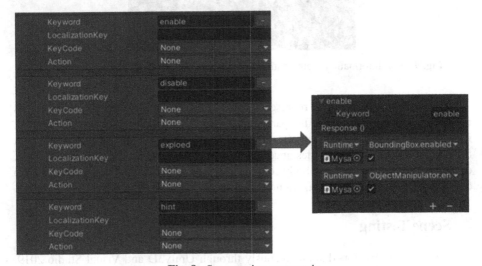

Fig. 8. Set up voice commands

3.6 Gaze Function

In the assembly scene designed in this paper, the process information of the part was displayed by gazing at the input. In an assembly scenario, if all part information is displayed directly with a panel assembly, the display panel will be very large, resulting in a very bloated application, and if the process information is materialized near each part model, the model will be directly obscured and cluttered, which will greatly affect the user's assembly operation. The best way is when the user looks at a part, the part automatically pops up process information, when the user's line of sight away, the text information is automatically hidden, the effect is shown in Fig. 9.

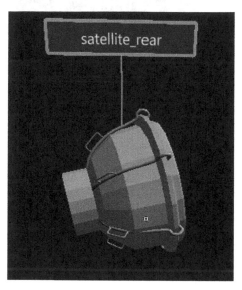

Fig. 9. Text information is automatically displayed when you gaze at the part model

The small circle in Fig. 9 represents that the ray emitted by the user's eyeball is blocked by the box collider of the part model. This is called ray detection in Unity3D, is the program when the scene application is turned on the camera with the camera emit ray detection scene, the scene designed in this paper all the parts model has added a box collider, when the ray and box-body collider intersect, the event that triggers the pop-up text information box, thus realizing the effect of gazing at the input.

4   Scene Testing

This article was developed simultaneously through Unity3D and Visual Studio 2019. First switched the Unity3D development platform to the Universal Windows Platform, packaged the scene to generate.sln files after the scene is developed, connected HoloLens2 to the computer via a USB cable, and finally deployed the application to HoloLens 2 in the.sln file.

The augmented reality scene is shown in Fig. 10. The scene showed a satellite model dataset, broken down model parts, and a dynamic follow menu. Figure 11 narrowed the entire assembly application by opening the bounding box, showing the scene of sitting at a table for assembly. This fully demonstrates the flexibility of the scenario. Figure 12 showed the decomposition of the satellite model in front of the user through voice input, allowing the user to observe the satellite's internal structure more closely.

Fig. 10. The original assembly scene

Fig. 11. The reduced assembly scene

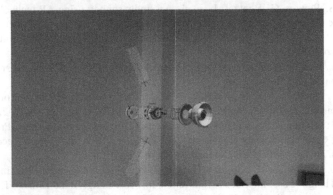

Fig. 12. Take a closer look at the satellite decomposition model

The test results showed that the model in this scene showed well, and the rich interactive means can meet the needs of users in different situations. The flexibility of the scene allows the user's hands, eyes, and mouth to maintain a very strong synergy, which greatly improves assembly efficiency, while the reusability of the software greatly reduces the cost of traditional assembly.

5 Conclusion

Based on augmented reality technology, this paper used Microsoft's latest HoloLens2 holograms, Unity3D and MRTK development kit to conduct research on virtual assembly of satellite parts. In the development scene of this paper, a wealth of human-computer interaction, satellite assembly model design and part model text information display results.

The internal structure of the satellite model is quite complex, and this paper provided a common virtual model assembly design, and used the satellite external parts for simple verification. If the internal structure of each satellite part is assembled in detail, one or more development scenarios need to be added, switched through a main page, and each scene is designed in much the same way as the individual scene proposed in this paper.

Acknowledgment. This work was supported by the Research Program of the Basic Scientific Research of National Defense of China under Grant JCKY2019210B005, JCKY2018204B025, and JCKY2017204B011, the Key Scientific Project Program of National Defense of China under Grant ZQ2019D20401, the Open Program of National Engineering Laboratory for Modeling and Emulation in E-Government, Item number MEL-20-02.

References

1. Zhou, F., Duh, B.L., Billinghurst, M.: Trends in augmented reality tracking, interaction and display: a review of ten years of ISMAR. In: IEEE ACM International Symposium on Mixed & Augmented Reality, Cambridge, UK, pp. 193–202. IEEE (2008)
2. Makris, S., Pintzos, G., Rentzos, L., et al.: Assembly support using AR technology based on automatic sequence generation. CIRP Ann. Manuf. Technol. **62**(1), 9–12 (2013)
3. Sabine, W., Uli, B., Timo, E., et al.: Augmented reality training for assembly and maintenance skills. Robot. Auton. Syst. **61**(4), 398–403 (2013)
4. Ong, S.K., Wang, Z.B.: Augmented assembly technologies based on 3D bare-hand interaction. CIRP Ann. Manuf. Technol. **60**(1), 1–4 (2011)
5. Khan, D., Ullah, S., Yan, D.M., et al.: Robust tracking through the design of high quality fiducial markers: an optimization tool for ARToolkit. IEEE Access **6**, 22421–22433 (2018)
6. Fiala, M.: ARTag, a fiducial marker system using digital techniques. In: Proceedings of the IEEE Computer Society Conference on Computer Vision and Pattern Recognition, Los Alamitos, pp. 590–596. IEEE Computer Society Press (2005)
7. Attard, G., de Raffaele, C., Smithm, S.: TangiBoard: a toolkit to reduce the implementation burden of tangible user interfaces in education. In: Proceedings of the 13th IEEE International Conference on Application of Information and Communication Technologies, Los Alamitos, pp. 1–7. IEEE Computer Society Press (2019)
8. Kurth, P., Lange, V., Siegl, C., et al.: Auto-calibration for dynamic multi-projection mapping on arbitrary surfaces. IEEE Trans. Visual Comput. Graphics **24**(11), 2886–2894 (2018)
9. https://docs.microsoft.com/en-us/hololens/hololens2-hardware
10. Ebbesmeyer, P., Krumm, H., et al.: Efficient creation of augmented reality content by using an intuitive authoring system. In: ASME 2004 Design Engineering Technical Conferences, Salt Lake City, Utah, USA, 53–60 (2004)
11. Gimeno, J., Morillo, P., Orduña, J.M., Fernández, M.: A new AR authoring tool using depth maps for industrial procedures. Comput. Industry **64**(9), 1263–1271 (2013)

Study on the Protection and Product Development of Intangible Cultural Heritage with Computer Virtual Reality Technology

Yimin Shen[⊠]

Jiaxing Nanyang Polytechnic Institute, Jiaxing, Zhejiang, China

Abstract. The rapid development of computer virtual reality technology has provided a brand-new idea for the inheritance, protection, product design and dissemination of China's intangible cultural heritage, which is of great practical significance. Based on the investigation of the status quo of computer virtual reality technology in the protection and development of intangible cultural heritage, this paper proposes an innovative method for the intangible cultural heritage protection mode integrated with computer virtual reality technology. Besides, this paper also attempts to obtain a development strategy for the inheritance, protection, integration, development of virtual reality of intangible cultural heritage by analyzing the Intangible Cultural Heritage Project of Zhejiang Province with Xiuzhou peasant paintings as a study case. This paper represents the achievement of the project titled Research on Collaborative Development of Design Majors under the Background of "Integration of Production and Education" (jg20190966), one of the second batch of teaching reform research projects of higher education in Zhejiang Province during the 13th Five-Year Plan.

Keywords: Virtual reality technology · Intangible cultural heritage · Protection · Product development

1 The Significance of Computer Virtual Reality Technology to the Protection and Development of Intangible Cultural Heritage

With digitization as an important carrier for information recording, presenting and spreading at present, virtual reality technology has provided a better reading experience of digital information. Given the popularity of virtual reality hardware and the efficiency of mobile Internet communication, presenting the intangible cultural heritage through virtual reality not only follows the Times featured by great development of science and technology, but also echoes the country's appealing of "culture export", so as to further demonstrate the cultural soft power with national and regional characteristics to the world under the background of China's "Belt and Road" Initiative [1].

© Springer Nature Singapore Pte Ltd. 2021
J. Zeng et al. (Eds.): ICPCSEE 2021, CCIS 1451, pp. 515–527, 2021.
https://doi.org/10.1007/978-981-16-5940-9_39

2 The Status Quo of Computer Virtual Reality Technology in the Protection and Development of Intangible Cultural Heritage

Thanks to the development of 5G communication technology, the immersive, interactive and imaginative virtual reality technology has conquered many technical difficulties including data storage and transmission. Therefore, as the next generation internet gateway and general information platform, virtual reality can exert many positive effects on the digital display of traditional culture, so that it is currently recognized as an effective technological means of traditional culture digitization.

In spite of the obvious trend and advantages of virtual reality technology in the digital protection and promotion of traditional culture, the utilization of virtue reality technology in China's traditional cultural digitalization remains not enough. Many problems still call for strategies. For instance, how to create a closer connection between the intangible cultural heritage and the daily life of the public in the era featured by great technological and scientific development for the inheritance and reservation of the excellent traditional culture? How to promote the growth of the virtual reality industry with the economic returns brought by developing virtual reality products while seeking to inherit and protect the intangible cultural heritage? How to realize the sustainable development of virtual reality technology in the protection and product development of intangible cultural heritage? [2].

3 Innovation and Case Study of the Intangible Cultural Heritage Protection Mode Integrated with Computer Virtual Reality Technology

The "technology + culture" integrative development mode can roughly be divided into two major types, namely, the "technology + culture" integrative internal growth and the "technology + culture" external support. Under these two major types, "technology + culture" integrative internal growth can be further subdivided into eight modes, that is, the creation-oriented mode, the technology-oriented mode, the platform innovation mode, the consumption-oriented mode, the industrial chain extension mode, the government-driven mode, the agglomeration and collaboration mode, and the industry-college-institute mode [3]. See Fig. 1 for the "technology + culture" integrative development mode.

Taking the Design and Information Branch of Jiaxing Nanyang Polytechnic Institute as an example and focusing on "technology + culture", the system flow that combines production, education, research and application, which connects the design major portfolio with the cultural and creative industries of intangible cultural heritage, is thus constructed.

All majors concerned in the portfolio shall be connected based on the characteristics of "technology + culture". The existing four majors of the branch, including Visual Communication Design and Production, Product Art Design, Industrial Design and Digital Media Application Technology, shall give full play to their own strengths to form the "technology + culture" results production flow highlighting "inheritance of intangible

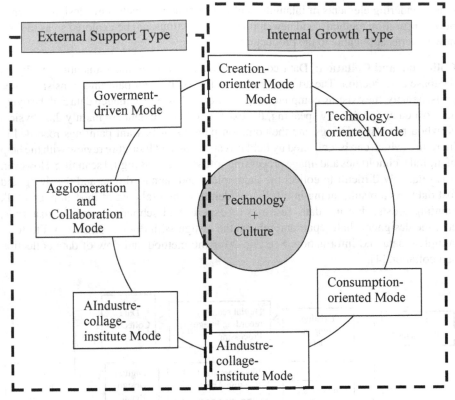

Fig. 1. "Technology + culture" integrative development mode

heritage design elements - art design and production of cultural and creative products - promotion of digital media cultural and creative products".

Visual Communication Design and Production, Product Art Design, Industrial Design and Digital Media Application Technology shall be linked by Zhejiang's intangible cultural heritage—Xiuzhou peasant paintings, to pool the advantages of the major portfolio, innovate the talent cultivation mode, deepen the integration of industry and education, and coordinate the innovation among schools, enterprises and industries. In addition, with the course project as the breakthrough, the studio operation mode shall be introduced to incorporate the needs of enterprise products and highlight the charm of traditional culture.

3.1 Digital Inheritance and Presentation of Intangible Heritage Design Culture

The major of Visual Communication Design and Production emphasizes the inheritance of research and has therefore established a basic resource database to form systematic digital resources by collecting, sorting and interpreting Xiuzhou peasant paintings, thus providing support for the cultural and creative design with effective protection and dissemination at the same time. Besides, based on the inheritance and research of the plane rendering elements of intangible cultural heritage projects, new design elements, new constitution methods, and new application directions can be explored so as to form new patterns applicable to the product design.

Collection and Collation: Data collection and collation is the foundation of digital database construction. The arts materials of Xiuzhou peasant paintings consist of three parts, namely, the existing complete data, the incomplete and damaged data and the dying and lost data. Specifically speaking, the existing complete data are mainly the physical Xiuzhou peasant paintings and the common patterns of peasant paintings recorded in literature, which can be collected by field investigation and literature review with the help of digital technologies and image processing software such as digital scanning. However, it is relatively difficult to collect the incomplete and damaged data and the dying and lost data. As a result, taking into full consideration the oral data of inheritors and folk painting artists, drafting data, research of experts and scholars, historical materials, etc., the designers shall repair and rebuild the design with digital technologies to form complete data and information. See Fig. 2 for the method and flow of data collection and collation [4].

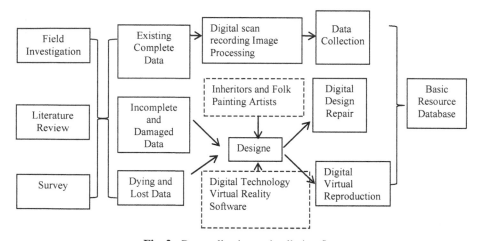

Fig. 2. Data collection and collation flow

Classification and Retrieval: Scientific and systematic classification of data not only functions as the premise of retrieval and reference, but also represents one of the core tasks of database construction. When constructing the database for the time-honored Xiuzhou peasant paintings featured by different categories and variegated artistic forms, digital technologies shall be applied to standardize the definition of intangible cultural heritage works and serve the database framework design on the basis of the needs, the characteristics of service objects and the opinions solicited from inheritors, experts and users [5].

The main content modules of the database are as follows:

Publicity Window: It covers the three columns of Introduction to Xiuzhou Peasant Paintings, Latest News, and Art Exhibition News.

Intangible heritage inheritance base of peasant paintings: It consists of the three columns of Artist Introduction, Art Village Introduction and Cultivation Action.

Artwork Appreciation: The works are divided into four parts of the 1970s, the 1980s, the 1990s and the 21st century according to the age of creation, with their distinctive styles of different ages highlighted by different page effects.

Product Management Workshop: It is composed of two columns: Product Recommendation and Product Distribution. Works for sale are displayed on the Recommendation page, presented with the dynamic mounting effect with the virtual reality technology of mounting for the customers' appreciation.

See Fig. 3 for the basic resource database structure of Xiuzhou peasant paintings.

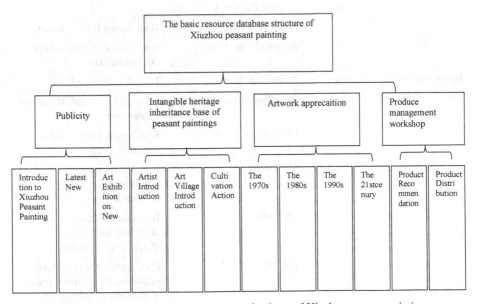

Fig. 3. Frame structure of basic resource database of Xiuzhou peasant paintings

Introduction and Interpretation: *Classification and Retrieval:* Intangible cultural heritage, as a kind of culture which incorporates life experience, historical tradition, collective memory and social practice of special groups of people, is a cultural heritage. The current digital protection tends to attach greater importance to technology than culture, encountered with a number of cultural problems such as ignorance of locality, difficulty in showing dynamism of intangible heritage and focusing on form instead of meaning. Apart from protecting the cultural heritage, the resource database of Xiuzhou peasant painting should also popularize and inherit culture. What's more, its design and construction shall cover not only the collection and collation of resource data, but also the systematic introduction and cultural interpretation of intangible heritage resources. Similarly, the introduction and interpretation of Xiuzhou peasant paintings is also beset by the numerous and complicated contents. To solve this problem, the key lies in the scientific and systematic frame structure, which can be realized through classification and grading. Adopting the deductive method, the systematic and comprehensive introduction and interpretation can be made in rich forms at three levels. See Table 1 for more details.

Table 1. The introduction and interpretation of the basic resource database presentation form of Xiuzhou peasant paintings

Type	Content	Presentation form
Publicity	Introduction to Xiuzhou peasant Paintings	Text, digital image
	Latest news	Text, digital image, video
	Art exhibition news	Text, digital image, virtual exhibition tour
Intangible heritage inheritance base of peasant paintings	Artist introduction	Text, digital image
	Art village introduction	Text, digital image, virtual art village tour
	Cultivation action	Text, digital image, video, virtual drafting
Artwork appreciation	The 1970s	Text, digital image, virtual exhibition tour, virtual scene of dynamic mounting
	The 1980s	Text, digital image, virtual exhibition tour, virtual scene of dynamic mounting
	The 1990s	Text, digital image, virtual exhibition tour, virtual scene of dynamic mounting
	The 21st century	Text, digital image, virtual exhibition tour, virtual scene of dynamic mounting
Product management workshop	Product recommendation	Text, digital image, virtual scene of dynamic mounting, virtual environment of dynamic mounting
	Product distribution	Text, digital image

3.2 Digital Design and Production of Intangible Cultural Heritage Design Elements

On the one hand, Xiuzhou peasant paintings take the product design as its main form of activation and inheritance and the major of Product Art Design stresses design and development. The inherited appearance and modeling elements shall be refined and used in the appearance and modeling design of cultural and creative products, daily supplies, and tourist souvenirs. On the other hand, the industrial design major is dedicated to craft production. To be specific, by properly organizing the product structure relations and using suitable carrier materials in accordance to the pattern design of product appearance and modeling, the industrial model shall be created for the industrial production of products [6].

In view of this, the resource database of cultural and creative products functions as both the resource regeneration bank and the extension of "Product Recommendation" under "Product Management Workshop" for the basic resource database. As one of the distinctive characteristics from other databases, the construction of "Cultural and Creative Products Recommendation Database" will include two parts, namely, the "Virtual Display of Products" and the "Virtual Display of Cultural and Creative Specialties", the former of which shall be an open resource database, where designers can upload works at will for display and communication. However, relevant works shall be basically examined and approved to avoid any violation or infringement. The latter is a semi-open resource database, where the works will only be available to the public on the paid basis or after a period of time. The works entering the database will be evaluated by experts according to the scientific evaluation system, mainly including the professional design works from inheritors, professional designers, universities and other design communities. Furthermore, the works in the "Virtual Display of Products" can be selected and incorporated by experts into the "Virtual Display of Cultural and Creative Specialties". Finally, the industries and enterprises shall select appropriate products for mass production and provide relevant services and remuneration. See Fig. 4 for the specific construction process of cultural and creative product resource database [7].

3.3 Promotion of Digital Display and Design Products of Intangible Cultural Heritage

The major of Digital Media Application Technology appreciates digital display and design product promotion. Digital media application technology, which is involved in the recording and image data sorting of the whole research and development process, established the virtual video and audio display environment and created the tour display in the virtual environment, the dynamic display in the virtual scene and the omnidirectional display of virtual products, so as to build a digital display and promotion platform for the inheritance and product results of intangible cultural heritage in terms of the channel of communication [8].

In addition to the role of assisting cultural communication and development of cultural industries, intangible cultural heritage should also be able to popularize specific cultural and creative design products. Besides, despite the static communication function, the database still lacks driving forces. Therefore, the promotion and communication

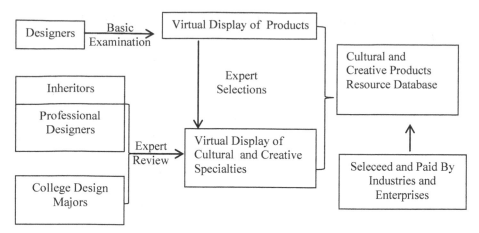

Fig. 4. Construction process of cultural and creative products of Xiuzhou peasant painting

platform, as the marketer of cultural communication and industrial development, shall give play to the dynamic interactive active communication function dominated by virtual reality technology and the role as a window by promoting exchanges and activities for higher public participation. See Fig. 5 for the structure and function framework of the promotion and communication platform.

4   Integrative Development Strategy for the Digital Products of Intangible Cultural Heritage

4.1   Break Down Industry Barriers and Continue the In-Depth Cross-Border Integration of "Technology + Culture"

With the arrival of the 5G era and the support of artificial intelligence, the Internet of Things and other emerging infrastructure, the cultural industry shall be further integrated with other industries. Traditional culture is more a carrier for contents, but cultural communication shall undergo revolutionary changes in both breadth and depth, given the support by technologies. For instance, technologies shall help present the traditional culture and history behind the intangible cultural heritage in a more diversified manner by holographic projection and virtual reality, thus enhancing both the cultural and creative industrial deposits and the competitiveness of the cultural and creative industry of intangible cultural heritage. The extensive integration of "technology + culture" can increase the economic value added of cultural products and nurture a new cultural format. What's more, the mergers and reorganizations across industries and regions during the integration of culture and technology can create conditions for related enterprises to integrate resources at a larger scale, extend the industrial chain and enhance their market competitiveness.

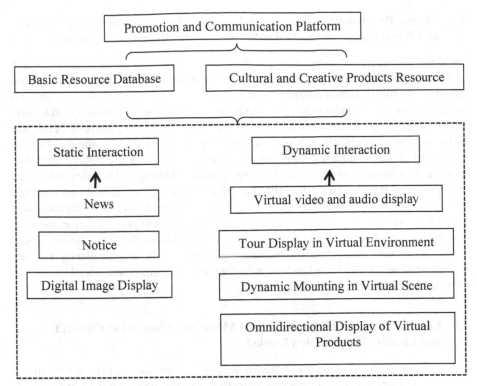

Fig. 5. Structure and function framework of the promotion and communication platform

4.2 Promote Resource Sharing and Enable "Technology + Culture" the Mainstream of Communication

Thanks to the deep integration of culture and technology, the form of sharing culture resources is increasingly diversified. For one thing, these cultural contents are reproduced with copyright by virtue of virtual reality technology to play a greater role in the cross-media communication. For another, cultural resources can enter the daily life of the public gradually with virtual reality technology. The in-depth integration of "technology + culture" has enriched the expression of cultural resources. In other words, the interaction with traditional culture is realized through the virtual culture world and vivid teaching produced with digitalization instead of cold showcases and books. Thus, with the application of virtual video and audio, virtual environment tour, virtual scene interaction, virtual product display and other new technologies, the public are able to interact highly with museum and library collections. In the context of deep integration of culture and technology, cultural resources will be further available and display more forms of expression. For example, present more excellent traditional culture step by step to the public in the vivid and interesting digital manner. Besides, in the future, the further improvement in science and technology signifies a growing number of more vivid forms of expression, among which digital culture shall be the mainstream for the opening of cultural resources [9].

4.3   Change the Structure of Traditional Cultural Industries and Drive Scientific and Technological Innovation with the Cultural Industrial Revolution

The deep integration of "technology + culture" will bring a new cultural format, which will impact the original cultural industry structure, accelerate the upgrading of cultural industry structure, and thus enhance its international competitiveness. The new cultural format caused by the integration of "technology + culture" will promote the extension of the industrial chain, eliminate the original traditional cultural subdivision fields, and add new cultural fields which is more advanced and attractive to consumers. First of all, virtual reality technology, with further expanded impact on the cultural field, has profoundly influenced and changed the development of the cultural industry, which not only promoted the integration of traditional culture industry and cultural undertakings, but also broadened the boundary of cultural industry, contributing to the thriving new cultural format. Secondly, due to the development, equipment popularization and content innovation of new immersive and super-immersive technologies including virtual reality, augmented reality, holographic imaging, naked eye 3D graphic display, interactive entertainment engine development and interactive video, the audio-visual interactive experience is upgraded in an all-round way.

4.4   Encourage the Emerging Industrial Mode and Change the Cultural and Creative Consumption Market

The integration and development of "technology + culture" has led to a host of emerging industrial models, as well as a new industrial model suitable for the further development of cultural industry under the background of digital economy. The emerging cultural industry, marked by film anime, network broadcast, music games, mobile terminal information retrieval, and VR/AR experience, has been operated successfully with a brand-new business model, revolutionarily changing the depth and breadth of people's participation in cultural life. Therefore, it has been accepted to a considerable extent that "technology + culture" shall be taken as the guiding strategy for the economic development to drive the growth of urban economy. The deep integration of "technology + culture" has changed the consumption pattern of cultural industry.

Given the further developed science and technology in the future, virtual reality technology will bring a fiercer storm to the traditional cultural industry and continuously enhance the international competitiveness of China's cultural industry by means of industrial revolution [10].

5   Reflection on Virtual Reality Technology in the Digitalization of Traditional Culture

Virtual reality technology has revitalized a lot of things close to extinction. Traditional culture and industry, seemingly independent of each other, can form new connections and crossover with each other through virtual reality technology to exploit many emerging formats of business. This kind of crossover and integration is exactly what helps intangible cultural heritage and other Chinese historical and cultural treasures find new ideas for inheritance and development. From the perspective of protection and product development of intangible cultural heritage, this paper constructs the database for Xiuzhou peasant paintings, so that this intangible cultural heritage project of Zhejiang Province can inherit the cultural spirit, carry forward the artistic charm and renew the artistic vitality [11].

In the process of digitizing intangible cultural heritage and developing cultural and creative products, virtual reality technology should also strengthen the cultivation of characteristic talents, who should be interdisciplinary talents instead of generalists, highlighting superior humanistic quality, general knowledge of basic science and technology, specialized technical skills, passion for traditional culture, awareness of cultural protection and communication and ability of innovation and creativity. According to the analysis of talents' ability, talents of virtual reality technology entail a solid science and engineering background, but science and engineering talents are featured by uneven perception and aesthetic appreciation of intangible cultural heritage value because of different emphasis in education. As a result, there is a gap between the academic backgrounds of cultural, artistic and scientific talents under the present education system of China. So that's why the reserve base of talents with such characteristics is weak [12]. Nevertheless, it is noteworthy that Visual Communication Design and Production, Product Art Design, Industrial Design, Digital Media Application Technology and other majors of general college and universities, which pay equal attention to technology and art, are devoted not only to the cultural and artistic cultivation, but also to the training of various modern media science skills in the talent cultivation. Moreover, given their considerable scale, these majors can be the backbone forces for the rapid supply of characteristic talents for digital intangible cultural heritage [13].

Despite the relevance of original records of intangible cultural heritage to the recreation of products based on the cultural elements of intangible cultural heritage, they are different in essence. Therefore, classified management and policy guidance should be implemented. Creators have to identify the historical materials that need to be taken seriously and the derivative products that can enhance the cultural vitality of the intangible cultural heritage. At present, virtual reality technology is still faced with certain limitations in the digitization of intangible cultural heritage. For example, the sense of touch cannot yet be simulated digitally, which might be a defect in the digital presentation of peasant paintings drawn on different media, and therefore has greatly impaired the detailed record and reproduction of the traditional cultural atmosphere. As time goes on, the communication form of intangible cultural heritage keeps developing and evolving. The online culture constructed by the large-scale network distributed virtual reality will also become an indispensable part in the traditional cultural chain.

For the time being, the application of virtual reality technology in the digitization of intangible cultural heritage in China is still unsatisfied, which can be solved in the following ways as far as the author considered [14].

Firstly, for the intangible cultural heritage, it can be protected in a permanent, safe and stable manner with the digital transformation strategy and its can be widely spread with the strategy of "virtual reality + effective internet". The closed loop of industrial model shall be formed by taking the content innovation as the original driving force, so as to help realize the continuance and value enhancement of intangible cultural heritage.

The strategy of "pan-creators" participating in crowd innovation shall be adopted to strengthen the connection between the public and traditional culture, which shall also promote the innovation of intangible cultural heritage, thus highlighting the sustainability of virtual reality technology in the digitalization of intangible cultural heritage.

Secondly, the implementation of strategies should also reflect the geographic advantages (industrial environment advantages, talent advantages, advantages of local policies). The realization of intangible cultural heritage protection and product promotion should feedback and drive the development of virtual reality content industry. In addition, the participation of social forces shall be increased by combining intangible cultural heritage more closely with the daily life of the public under the background of great scientific and technological development. By doing so, excellent traditional culture shall be inherited and reserved [15].

Thirdly, attention should also be paid to the protection of digital copyright related to virtual reality. Apart from gradually improving legal norms, we should also take advantage of the high compatibility of the digitalization of intangible cultural heritage to actively seek the immutable and traceable solutions built by block chain and other technologies under the concept of decentralization. More importantly, intangible cultural heritage, as an important resource for the development of human society, entails vigorous protection, development and promotion. Since China takes the lead in 5G deployment around the world, Chinese traditional culture should take full advantage of the 5G era to integrate virtual reality technology, and show the international visibility of China's excellent traditional culture under the background of the "Belt and Road" initiative, thus bringing an opportunity to the Chinese people to enhance their cultural confidence [16].

References

1. Jin, Y.P.: New changes in the integration of the fifth generation mobile technology and Chinese cultural and creative industries under global competition. J. Shandong Univ. Philos. Soc. Sci. 5, 74–85 (2020)
2. Kuang, Y.X., Wu, Q.: Innovative Design Strategy of Cultural Products for Protection and Development of Intangible Cultural Heritage, pp. 70–74. Old Liberated Area Built, Jiangxi, February 2019
3. China Bridge: Report on China's Culture + Technology Integrative Development 2020 (2020)
4. Xing, J.H., Wang, H.N., Wu, Z.J.: Construction of art database for nou mask in Hunan province under the guidance of cultural creativity. Pack. Eng. 77–81 (2019)
5. Kong, C.T.: Study on the construction scheme of the resource base for Huxiang cultural and creative product design. Master's Thesis, Hunan University, May 2017
6. Yu, J.A., Xu, L., Yin, K.: The intellectualization of traditional cultural products: integration of culture and modern technology, pp. 54–61. Forum Science and Technology in China, Beijing (2020)
7. Wang, Y.N., Yu, I..: Network Structure and Innovation Knowledge Flow in the Cultural and Creative Industries Cluster—Based on the Perspective of Social Network Analysis, pp. 158–163. Science and Technology Management Research, Guangdong (2017)
8. Yang, X., Xiang, Q.F.: The approach to creative industry: strategic orientation of the integrated development of culture, science and technology. J. Southwest Univ. National (2018)
9. Jia, J.: Advancement of digital dissemination of intangible cultural heritage under the background of artificial intelligence. Contem. Commun. 98–101 (2020)

10. Wang, Y.Y.: Digital Protection and Promotion Strategy of Traditional Culture from the Perspective of Virtual Reality Technology Integration, Creation and Design, Jiangsu, pp. 23–27, February 2020
11. Xu, M., Li, X.L.: Inheritance and innovation of non-heritage culture and creation in the era of digital economy -- comment on cultural and creative revolution in the digital economy. Contemp. Finan. Econ. 1–3 (2020)
12. Li, M.Y., Li, A.X.: Research on training mode of higher vocational cultural creative talents based on "one room, two bases and one platform". Mod. Vocat. Educ. 84–85 (2020)
13. Qin, F.: Study on cultural and creative product innovation based on digital technology. Cult. Ind. Res. Jiangsu 1, 234–246 (2015)
14. Mao, L.J., Hao, B.E.Z.N.: Research on the development path of cultural and creative industry in the era of digital economy. Heilongjiang Soc. Sci. 56–60 (2020)
15. Yi, N.: Research on the development of international cultural and creative industries under the influence of digital economy. J. Renmin Univ. China 50–60 (2020)
16. Wu, C.Z.: Cultural Industry Innovation in the 5G era. J. Shenzhen Univ. (Human. Soc. Sci. 51–60 (2019)

Channel Context and Dual-Domain Attention Based U-Net for Skin Lesion Attributes Segmentation

XueLian Mu, HaiWei Pan[✉], KeJia Zhang, Teng Teng, XiaoFei Bian, and ChunLing Chen

Harbin Engineering University, Harbin, People's Republic of China

Abstract. Skin melanoma is one of the most common malignant tumors originating from melanocytes, and the incidence of the Chinese population is showing a continuous increasing trend. Early and accurate diagnosis of melanoma has great significance for guiding clinical treatment. However, the symptoms of malignant melanoma are not obvious in the early stage. It is difficult to be diagnosed with human observation. Meanwhile, it is easy to spread due to missed diagnosis. In order to accurately diagnose melanoma, end-to-end skin lesion attribute segmentation framework is presented in this paper. It is applied to facilitate the digitalization process of attributes segmentation. The framework was improved on the U-Net construction that use the channel context feature fusion module between the encoder and decoder to further merge context information. A dual-domain attention module is proposed to get more effective information from the feature map. It shows that the proposed method effectively segments the lesion attributes and achieves good result in the ISIC2018 task2 dataset.

Keywords: Lesion attribute segmentation · Melanoma · Channel context feature fusion · Dual-domain attention

1 Introduction

Melanoma generally refers to malignant melanoma which is transformed by a tumor derived from the melanocytes. The malignancy degree of malignant melanoma is high. In death cases of skin cancer, the percent of malignant melanoma is about 75%. The symptoms of pre-melanoma are very similar to those of normal naevus on human skin. Because the location of melanoma is variable, it may be occurred on the surface of the skin, the soles of the feet or the eyes. The melanoma is not easy to be detected by patients. The above factors will lead to the diseases that found by patients are usually in the middle and late stages. The cancer cells will spread throughout the body, and the cure rate will reduce. Dermoscopy is a widely used technique for diagnosing melanoma. It is used to obtain a magnified image of the skin examination area, which can

© Springer Nature Singapore Pte Ltd. 2021
J. Zeng et al. (Eds.): ICPCSEE 2021, CCIS 1451, pp. 528–541, 2021.
https://doi.org/10.1007/978-981-16-5940-9_40

increase the clarity of skin lesions and provide more color and texture information. Therefore, compared with direct manual judgment, dermatologists can reduce the misdiagnosis rate through early observation and identification on dermoscopic images of melanoma. The lesion areas in dermoscopic images are accurately segmented that can help doctors and computer-aided system make better decisions. Early melanomas are very similar to benign naevus in the appearance of human skin. Although dermoscopy images can be used to depict more details, they may also be misdiagnosed. Early melanoma can be cured, so early detection and accurate diagnosis of melanoma become particularly important. The melanoma images obtained by dermoscopic imaging technology can clearly show the texture and shape features of the skin lesion area within melanoma. These features can better reflect the difference between benign naevus and malignant melanomas, and they can be used as a dermatology department. This is the foundation to correctly distinguish benign naevus and malignant melanomas for doctors. In order to accurately diagnose melanoma, it is particularly important to determine the specific lesion attributes (the texture and shape features of lesion areas) and perform attributes segmentation.

Dermoscopic attributes segmentation is a key step in the diagnosis and treatment of melanoma. At present, it mainly relies on the doctor's manual segmentation. The large amount of dermoscopic attributes image data brings huge pressure to the doctor. Manual segmentation spends a lot of time and labor. However, it mainly depends on the doctor's experience. Even experienced dermatologists are prone to misdiagnosis, so the research on automatic segmentation method for dermoscopic attributes images has great significance to assist doctors in diagnosis. With the development of computer-aided diagnosis and deep learning, segmentation methods based on deep learning have become a hot research topic. Fully Convolutional Neural Network (FCN) [13] is one of the typical end-to-end image segmentation deep network. In 2015, Olaf Ronneberger et al. proposed an U-Net network structure on the basis of FCN for biomedical image segmentation [18]. The proposed network structure reduced the spatial dimension by encoding and extracted high-level semantic features. It combined decoding with jump connection operations to restore the spatial dimension and the detail information of images. It has better performance on medical image segmentation. Eric Chen et al. used an architecture which was similar to U-Net to automatically segment the attributes of dermoscopic images [5]. In addition, two classification branches were added to the segmentation network to help the network detect empty attributes masks. However, it ignored the crucial context information during image segmentation.

In order to solve the above problems, This paper proposes Channel context and Dual-domain attention based U-Net (CDU-Net) for segmentation of skin lesion attributes. The Channel Context feature fusion Module (CCM) is used to further extract context information. It is not only increases the receptive field but also extracts higher-level semantic and spatial information. Then, Dual-domain Attention (DA) modules are further combined to improve the accuracy of segmentation result. The information interaction between spatial feature and channel feature are realized, so that the model has the ability to screen the feature learning. The anti-interference ability and the feature extraction efficiency of the network are improved. The experimental results confirm that CDU-Net has better segmentation on the ISIC2018 skin lesion attributes dataset.

2   Related Works

2.1   Skin Lesion Segmentation

Skin lesion segmentation is a major problem in clinical medical research, it can help doctors and computer-assisted systems determine specific lesion areas. Traditional skin lesion segmentation methods mainly include clustering, threshold-based, region growing and active contour models [2,7,17]. In recent years, deep learning methods have received extensive attention in solving medical image segmentation problems. Jonathan Long et al. Proposed the Fully Convolutional Network, it has achieved satisfactory results in segmentation tasks [13]. Yuan Y et al. used an end-to-end method to train a 19-layer deep convolutional neural network. they employed a new loss function based on Jaccard distance that contributed to the elimination of the imbalance between foreground and background pixel weights [24]. Yu et al. proposed a Fully convolutional Residual Network (FcRN) for end-to-end training and achieved good segmentation results [23]. Lei et al. used Generative Adversarial Networks (GANs) to enhance the segmentation of skin lesions [3]. The skin segmentation network proposed by Adegun et al. integrate encoder-decoder full convolutional network, dense block and conditional random field (CRF) modules. These modules were connected through cascading strategies and transition layers to decrease model complexity while enhancing performance [1]. Although the above methods have achieved good results in skin lesions segmentation, they did not consider the segmentation on the attributes level. The paper further studies the task of attributes segmentation.

2.2   Lesion Attributes Segmentation

Lesion attributes segmentation is a segmentation task of skin lesion patterns in dermoscopic images. The most common clinical attributes that used for diagnosing melanoma are pigment network, negative network, streaks, milia-like cysts and globules. In clinical practice, dermatologists use these attributes to better characterize the type and the malignant degree of lesions. Inspired by FCN, Olaf Ronneberger et al. proposed the U-Net network for biomedical image segmentation, the proposed network was divided into compression path and expansion path. Low-level semantic information and high-level semantic information can be obtained at the same time. The model can be trained with a small amount of data to obtain high segmentation accuracy and fast speed, which is often used in attribute segmentation tasks [18]. Kawahara et al. proposed a segmentation model based on a fully convolutional network, which combined a new multi-channel Dice loss to predict five attributes [12]. Bissoto et al. used a model which was similar to U-Net, its encoding path was pre-trained on the ImageNet dataset [4]. Sorokin et al. used the masked R-CNN model for lesion attribute segmentation tasks [19]. Eric Chen et al. also used an architecture which is similar to U-Net, they replaced the network's encoding path with a pre-trained VGG16 model. In addition, two secondary classification branches are additionally added to the segmentation network to help the network detect empty attribute masks [5]. Although the above methods are existed on the task of attribute segmentation, their accuracy and precision need to be improved. This paper combines the CCM and DA modules and proposes a new network (CDU-Net) to effectively improve the accuracy of segmentation.

3   Methods

In order to achieve improved performance in the task of skin lesion attribute segmentation, our method further integrates context information and attention mechanism on the basis of mutli-task U-Net [5]. CDU-Net includes four parts: Encoder module, Channel Context Feature Fusion module, Dual-domain attention module and Decoder module. The framework is shown in Fig. 1.

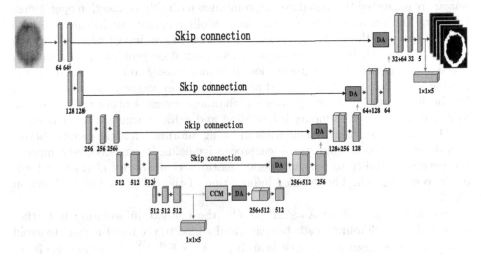

Fig. 1. The framework of channel context and dual-domain attention based U-Net (CDU-Net) for skin lesion attributes segmentation.

3.1   Encoder-Decoder Module

In the original U-Net framework, each encoder block contains two convolutional layers and a maxpooling layer. In this paper, the U-Net is replaced with pre-trained VGG16 [11] network weights in the encoder part. Existing experiments have proved that the training speed can be greatly accelerated if the existing weight model files are available to be loaded for deep learning model training. The encoder contains a total of 5 convolution blocks, namely conv1, conv2, conv3, conv4 and conv5. The first two convolution blocks contain two 3×3 convolutions, and the last three convolution blocks contain three 3×3 convolutions. The relu nonlinear activation function accompanies each convolution operation, and a 2×2 maxpooling operation is added after each convolution block which can reduce the amount of parameters and increase the receptive field. The direction of the decoder module is opposite to the encoder module. The decoder performs upsampling of the feature map from the channel context feature fusion module. These up-sampled features are sent to the dual-domain attention module to re-adjust the output features of the encoder, and connect with the upper-layer feature to feed to the next-layer network. It is observed that it is not all lesion attributes appear in the mask for each skin image. As in the literature [5], two classification branches are added to classify empty masks and non-empty masks.

3.2   Channel Context Feature Fusion Module

The spatial dimension is reduced and the high-level semantic features are extracted in the encoding. The spatial dimension and detail information of the image are recovered through decoding combined with skip connection operations. However, the continuous convolution and pooling operations in this structure also lose part of the spatial context information in the image while extracting high-level semantic features. In the actual segmentation problem, particularly in the skin lesion attributes segmentation, it is more easily to lost boundary information since the area of lesions and normal skin more vague boundary. Inspired by the literature [8,22], we proposes a channel context feature fusion module to reduce the loss of boundary information and achieve accurate segmentation. CCM enlarges the receptive field without losing information and excavates better deep-level features through multi-scale receptive fields. It can effectively improve the network's ability to recognize lesion features that are not obvious and difficult to distinguish. The detailed information of the CCM module is shown in Fig. 2.

For any scale feature $X \in R^{H \times W \times C}$, the context information is further extracted through four cascade branches and added to the input feature to avoid gradient disappearance. The first branch $f_1 : X \to E_1^{H' \times W' \times C'}$ kernel size is set to 3, the rate is 1. The second branch kernel size is set to 3, the rate is 3, expressed as $f_2 : X \to E_2^{H' \times W' \times C'}$. The third branch $f_3 : X \to E_3^{H' \times W' \times C'}$ firstly performs the convolution operation of kernel size which is set to 3, the rate is set to 1 and then performs the operation of kernel size is equal to 3, rate is equal to 3. The last branch performs the operations of kernel size which is set to 3, rate is equal to 1, 3, 5 respectively, which can be expressed as $f_4 : X \to E_4^{H' \times W' \times C'}$ At the same time, 1×1 convolution is added after each cascade branch to reduce the amount of parameters. After multi-scale features are extracted, the weight information between image channels are further extracted through channel attention. The output of CCM can be expressed as:

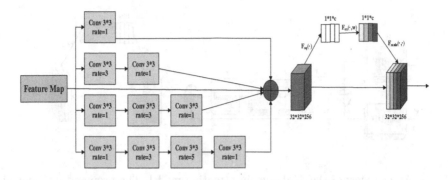

Fig. 2. Overview of channel context feature fusion module.

$$X' = att_{channel-wise}(X \oplus E_1 \oplus E_2 \oplus E_3 \oplus E_4) \qquad (1)$$

Where $\oplus$ represents element-wise addition, and $att_{channel-wise}$ represents inter-channel attention. E1 to E4 represent the output of each branch respectively.

3.3 Dual-Domain Attention Module

Attention is a mechanism in human perception. The attention mechanism in deep learning [16,21] enables to assign different weights to the input features to help the model extract more critical information. Therefore, introducing the attention mechanism into the medical image segmentation network can enhance the effect of skin lesion attributes segmentation. Usually tasks only focus on one attention domain of the feature, such as the channel domain or the spatial domain. In order to enable the network to pay attention to more effective information in the feature map, we combine the channel domain and the spatial domain to create a dual-domain attention module for the segmentation of skin lesion attributes, as shown in Fig. 3.

Attention Gate. The Attention Gate (AG) in the U-Net model was first proposed by Oktay et al. [15]. AG automatically learns the different shapes and sizes of the target structure in the focused medical image in the spatial position. AG can highlight salient features useful for specific tasks, while suppressing irrelevant areas in the input image. AG selects the spatial region by analyzing the context information provided by the gated signals collected from the coarser scale, and the input features are scaled according to the attention coefficient of AG. Attention factors $a_i \in [0,1]$ is used to identify important image areas and determine focus areas. AG takes in two inputs x^l and g, where a single scalar attention value is computed for each pixel vector $x_i^l \in R^{F_l}$, corresponds to the number of feature maps in layer l. The vector $g \in R^{F_g}$ is taken from the up-sampling layers in the network. The output of AG is the multiplication of the

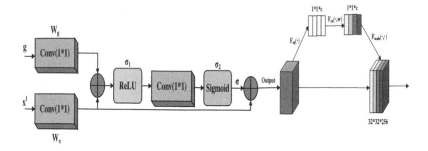

Fig. 3. An overview of the dual-domain attention model. The proposed dual-domain attention model weights features from the spatial domain and the channel domain, effectively improving the model's feature representation ability

elements of the input feature mapping and the attention factor $\hat{x}^l_{i,c} = x^l_{i,c} \cdot a^l_i$. Additive attention is formulated as follows:

$$q^l_{att} = \psi^T (\sigma_1(W^T_x x^l_i + W^T_g g_i + b_g)) + b_\psi \tag{2}$$

$$a^l_i = \sigma_2(q^l_{att}(x^l_i, g_i; \Theta_{att})) \tag{3}$$

Where $\sigma_2 = \frac{1}{1+exp(-x_i)}$ corresponds to the sigmoid activation function, σ_1 refers to the ReLU activation function. The linear transformation $W_x \in R^{F_l \times F_{int}}$, $W_g \in R^{F_g \times F_{int}}$, $\psi \in R^{F_{int} \times 1}$, and the bias term $b_\psi \in R$, $b_g \in R^{F_{int}}$, form a set of Θ_{att} parameters, which is used to characterize the AG. The linear transformation is calculated using a $1 \times 1 \times 1$ convolution in the channel direction of the input tensor. The concatenated features x^l and g linearly mapped to a $R^{F_{int}}$ dimensional intermediate space are called vector-based connected attention.

In order to eliminate noise and irrelevant responses from jumping connections, AG calculates the focus value for each pixel of the input feature and determines the importance to control different spatial location features. To pay attention to more effective information in the feature map, our method further integrates the channel attention module to learn the nonlinear interaction and non-repulsive relationship between channels.

Channel-Wise Attention. Spatial attention captures the spatial relationship between features, and channel attention can learn the importance of channels. The Squeeze and Excitation (SE) module is used to improve the feature expression ability of the model in combination with the relationship between the feature channels [10]. The output of AG first performs a squeeze operation. It is achieved by using global average pooling to generate channel-wise statistics. The formula is as follows:

$$z_c = F_{sq}(f^{H \times W \times C}) = \frac{1}{H \times W} \sum_{i=1}^{H} \sum_{j=1}^{W} f^{H \times W \times C} \tag{4}$$

where $F_{sq}(\cdot)$ is the global average pooling operation. $f \in R^{H \times W \times C}$ represents the output of the AG module and H×W correspond to the height and width dimensions, respectively. z_c is generated from each two-dimensional feature map by GAP compression. The second step of the channel attention module contains two full connection layers with a sigmoid activation:

$$s = F_{ex}(z, W) = \sigma(W_2 \delta(W_1 z)) \tag{5}$$

Where σ refers to sigmoid function and δ refers to the ReLU function, $W_1 \in R^{\frac{C}{r} \times C}$ and $W_2 = R^{C \times \frac{C}{r}}$ are the parameters for the first FC layer and the second FC layer, respectively. In addition, r is the reduction ratio. Finally, the weight information of each channel is fused into the feature image through $F_{scale}(,)$. The output of the SE module can be expressed as:

$$\hat{f}_c = F_{scale}(f_c, s_c) = f_c \times s_c \tag{6}$$

Such as ResNet [9], we connect the channel attention output feature with the input feature, and its purpose is to prevent the gradient from disappearing. The dual-domain attention module proposed in this paper realizes information interaction between spatial features and channel features, so that the model has the ability to filter feature learning, and improves the anti-interference ability of the network and the efficiency of network feature extraction.

4   Experiments

4.1   Dataset and Preprocessing

To validate the performance of the proposed model, a publicly available dataset, the ISIC2018 Task 2 dataset, is used in this paper. The dataset is from the International Skin Imaging Collaborative (ISIC) archives [6,20]. Five attributes are detected in this task, there are pigment network, negative network, streaks, milia-like cysts and globules. The 2,594 images and 12,970 ground truth masks in our ISIC2018 task2 dataset were used as training data. In a skin image, these five masks correspond to five lesion properties. The validation set includes 100 images, and the test set includes 1000 images. We resized all skin images with their corresponding masks to 512×512 in the pre-processing process. In addition, we used random flips, scaling, and rotation for skin images and masks, as well as random adjustments of saturation and brightness for data enhancement.

Table 1. Model parameters and corresponding output size of each block

Layer	Encoder (down-sampling)			Decoder (up-sampling)		
	Size of input	Size of output	Functions used	Size of input	Size of output	Functions used
Layer 1	512 × 512 × 3	512 × 512 × 64	conv(3 × 3), ReLU	32 × 32 × 768	64×64×256	deconv (4 × 4, s = 2), DA, ReLU
Layer 2	256 × 256 × 64	256 × 256 × 128	conv(3 × 3), ReLU	64 × 64 × 768	128 × 128 × 256	deconv (4 × 4, s = 2), DA, ReLU
Layer 3	128 × 128 × 128	128 × 128 × 256	conv(3 × 3), ReLU	128 × 128 × 384	256 × 256 × 128	deconv (4 × 4, s = 2), DA, ReLU
Layer 4	64 × 64 × 256	64 × 64 × 512	conv(3 × 3), ReLU	256 × 256 × 192	512 × 512 × 64	deconv (4 × 4, s = 2), DA,ReLU
Layer 5	32 × 32 × 512	32 × 32 × 512	conv(3 × 3), ReLU	512 × 512 × 96	512 × 512 × 32	deconv (4 × 4, s = 2), DA,ReLU
CCM	32 × 32 × 256	32 × 32 × 256	dilation rate (1,3,5)	512 × 512 × 32	512×512×5	Output

4.2 Implementation Details and Evaluation

In this paper, we use a segmented training strategy, first freezing all encoder layers and training only the decoder layer for 50 epochs. Then, we train the full network for 200 epochs. Figure 4 drew the distribution of train_loss, valid_loss, train_Jaccard and valid_Jaccard for the proposed network within 250 epochs. The network uses Adam optimizer with a learning rate of 0.01 and batchsize of 4. The model parameters and the output size of each convolution block are shown in Table 1.

To better evaluate the performance of the model proposed in this paper, Jaccard, sensitivity, specificity, and accuracy metrics are used in this paper. The calculation formula is as follows:

$$Jaccard = \frac{TP}{TP + FP + FN} \tag{7}$$

$$Sen = \frac{TP}{TP + FN} \tag{8}$$

$$Spe = \frac{TN}{TN + FP} \tag{9}$$

$$Acc = \frac{TN + TP}{TN + TP + FN + FP} \tag{10}$$

where TP (True Positive) indicates the proportion of correctly resolved lesioned pixels, TN (True Negative) represents the ratio of correctly distinguishing non-pathological pixels, FP (False Positive) indicates the labeling of non-lesioned pixels as lesioned pixels, and FN (False Negative) describes the proportion of incorrectly labeled lesioned pixels.

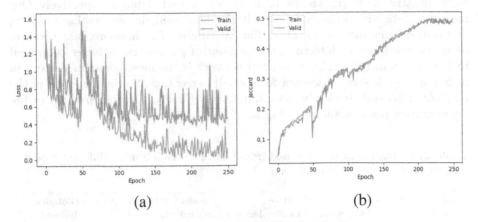

(a)   (b)

Fig. 4. Graph depicting the comparison of train_loss and valid_loss,train_Jaccard and valid_Jaccard for 250 epochs for the proposed approach.

Cross-entropy loss functions are often used in classification as well as segmentation tasks, but due to the inhomogeneous distribution of lesion attributes in skin images cross-entropy loss is not an optimal solution for this task. In this paper, the loss function is defined as:

$$loss = loss_1 + \lambda_1 loss_2 + \lambda_2 loss_3 \tag{11}$$

$$loss_1 = loss_{BCE} - log\frac{\sum y_{target}y_{output} + \alpha}{\sum y_{target} + \sum y_{output} - \sum y_{target}y_{output} + \beta} \tag{12}$$

Where $loss_1$ denotes pixel segmentation loss, $loss_2$ denotes middle classification branch loss, and $loss_3$ denotes final classification branch loss. $loss_2$ and $loss_3$ are calculated using cross-entropy loss. λ_1 and λ_2 are hyperparameters, and after experiments this paper takes $\lambda_1 = \lambda_2 = 0.5$.

4.3 Analysis of Experimental Results

Comparison with the State-of-the-Art. In order to verify the efficiency and performance of our proposed model, we compare our method with the multi-task U-net proposed by Eric Chen et al., the multi-scale attention U-net proposed by Duy Minh Ho Nguyen et al. [14] and one of the best ISIC 2018 structures LeHealth's method [25]. Table 2 shows the corresponding comparative predictions. Jaccard coefficients are calculated for five attributes, and the performance of each network is recorded. Our proposed architecture gains Jaccard of 0.525,

0.236, 0.124, 0.166 and 0.369 for five attributes, respectively. It can be seen that compared to mutli-task U-Net, our proposed model outperformed it with an enhancement of 1.4%, 1.7%, 0.3%, 2.4%, 2.9% on Jaccard from pigment network, negative network, streaks, milia-like cysts, and globules respectively. Our model only slightly compares poorly with LeHealth only in the streaks attribute, and outperforms him in the other four attributes. Furthermore, since we are based on the u-net architecture, the number of parameters is lower compared to LeHealth's method. The pigmented network is the most common pattern in melanoma, while streaks account for a small percentage. So the diagnosis of the pigmented network is clearly more important. Finally, the visualization of the segmentation result is shown in Fig. 5.

Table 2. Evaluation metrics resulted from ISIC 2018 Task 2 using different model in the proposed attributes segmentation.

Base networks	Pigment networks Jaccard	Negative networks Jaccard	Streaks Jaccard	Milia_like_Cysts Jaccard	Globules Jaccard
Multi-task U-Net	0.511	0.219	0.121	0.142	0.340
LeHealth	0.482	0.225	0.145	0.132	0.239
CDU-Net	**0.525**	**0.236**	**0.124**	**0.166**	**0.369**

Ablation Study. In order to prove the effectiveness of the dual-domain attention module and channel context extraction module in our proposed skin lesion attributes segmentation network, we use the ISIC2018 task2 data set as an example to perform the following ablation experiments:

Ablation Study for Adopting Channel Context Feature Fusion Module. The network that we proposed is based on mutli-task U-Net. Therefore mutli-task U-Net is the most fundamental baseline model. Table 3 shows the effect of the channel context feature fusion module, which improves the performance of the network. The baseline with CCM module is called baseline+CCM. Compared with the baseline, Jaccard, sen, spe, Acc increased by 2.4%, 0.8%, 0.3%, 1.1% respectively.

Ablation Study for Adopting Dual-Domain Attention Module. The dual-domain attention module considers the spatial domain and the channel domain at the same time, and more effective information can be focused in the feature map. The dual-domain attention module is composed of AG and channel attention. We first verified the effectiveness of AG in the task of attributes segmentation, and then added a channel attention module to further verify our points. The experimental results are shown in Table 3, which shows that our proposed dual-domain attention module can capture more effective information in the feature map and effectively improve the network segmentation performance.

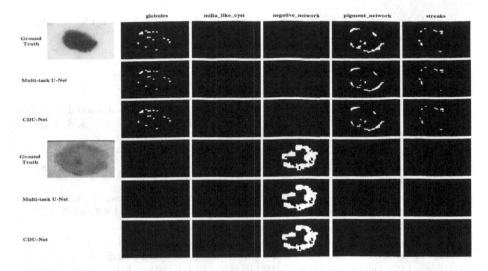

Fig. 5. Sample images of skin lesion attributes segmentation from ISIC2018 task2 dataset.

Table 3. Ablation study for each component on ISIC 2018 Task 2 datasets.

Method	Jaccard	Sen	Spe	Acc
Baseline	0.433	0.802	0.846	0.830
Baseline + CCM	0.457	0.810	0.849	0.841
Baseline + CCM + AG	0.469	0.813	0.852	0.844
Baseline + CCM + DA	**0.481**	**0.817**	**0.856**	**0.848**

5   Conclusion

In this paper, we propose CDU-Net for automatic segmentation of lesion attributes. The proposed model can help expert dermatologists to automatically detect and locate the lesions attributes accurately. CDU-Net can be divided into four parts: encoder module, channel context feature fusion module, dual-domain attention module and decoder module. We use the channel context feature fusion module to increase the receptive field in the lowest part of the down-sampling part. A dual-domain attention domain module for the segmentation of skin lesion attributes is proposed in this paper. Compared with other complex models that use integrated strategies, the model proposed in this paper is slightly simpler and has slightly worse performance. However, we believe that although model integration usually helps to improve prediction accuracy, in practical applications, especially in the field of biomedical domain, simple models are usually the first choice. In the future work, we can further try to add prior knowledge into the skin attributes segmentation network to effectively increase the value of Jaccard coefficient.

Acknowledgements. The paper is supported by the National Natural Science Foundation of China under Grant No. 62072135 and No. 61672181.

References

1. Adegun, A.A., Viriri, S.: FCN-based DenseNet framework for automated detection and classification of skin lesions in dermoscopy images. IEEE Access **8**, 150377–150396 (2020). https://doi.org/10.1109/ACCESS.2020.3016651
2. Ahn, E., et al.: Saliency-based lesion segmentation via background detection in dermoscopic images. IEEE J. Biomed. Health Inform. **21**(6), 1685–1693 (2017). https://doi.org/10.1109/JBHI.2017.2653179
3. Bi, L., Feng, D., Kim, J.: Improving automatic skin lesion segmentation using adversarial learning based data augmentation. CoRR abs/1807.08392 (2018). http://arxiv.org/abs/1807.08392
4. Bissoto, A., Perez, F., Ribeiro, V., Fornaciali, M., Avila, S., Valle, E.: Deeplearning ensembles for skin-lesion segmentation, analysis, classification: RECOD titans at ISIC challenge 2018. CoRR abs/1808.08480 (2018). http://arxiv.org/abs/1808.08480
5. Chen, E.Z., Dong, X., Li, X., Jiang, H., Rong, R., Wu, J.: Lesion attributes segmentation for melanoma detection with multi-task U-Net. In: 16th IEEE International Symposium on Biomedical Imaging, ISBI 2019, Venice, Italy, 8–11 April 2019, pp. 485–488. IEEE (2019). https://doi.org/10.1109/ISBI.2019.8759483
6. Codella, N.C.F., et al.: Skin lesion analysis toward melanoma detection 2018: a challenge hosted by the international skin imaging collaboration (ISIC). CoRR abs/1902.03368 (2019). http://arxiv.org/abs/1902.03368
7. Patiño, D., Avendaño, J., Branch, J.W.: Automatic skin lesion segmentation on dermoscopic images by the means of superpixel merging. In: Frangi, A.F., Schnabel, J.A., Davatzikos, C., Alberola-López, C., Fichtinger, G. (eds.) MICCAI 2018. LNCS, vol. 11073, pp. 728–736. Springer, Cham (2018). https://doi.org/10.1007/978-3-030-00937-3_83
8. Gu, Z., et al.: CE-Net: context encoder network for 2D medical image segmentation. IEEE Trans. Med. Imaging **38**(10), 2281–2292 (2019). https://doi.org/10.1109/TMI.2019.2903562
9. He, K., Zhang, X., Ren, S., Sun, J.: Deep residual learning for image recognition. In: 2016 IEEE Conference on Computer Vision and Pattern Recognition, CVPR 2016, Las Vegas, NV, USA, 27–30 June 2016, pp. 770–778. IEEE Computer Society (2016). https://doi.org/10.1109/CVPR.2016.90
10. Hu, J., Shen, L., Sun, G.: Squeeze-and-excitation networks. In: 2018 IEEE Conference on Computer Vision and Pattern Recognition, CVPR 2018, Salt Lake City, UT, USA, 18–22 June 2018, pp. 7132–7141. IEEE Computer Society (2018). https://doi.org/10.1109/CVPR.2018.00745. http://openaccess.thecvf.com/content_cvpr_2018/html/Hu_Squeeze-and-Excitation_Networks_CVPR_2018_paper.html
11. Iglovikov, V., Seferbekov, S., Buslaev, A., Shvets, A.: TernausNetV2: fully convolutional network for instance segmentation. In: Proceedings of the IEEE Conference on Computer Vision and Pattern Recognition (CVPR) Workshops, June 2018
12. Kawahara, J., Hamarneh, G.: Fully convolutional neural networks to detect clinical dermoscopic features. IEEE J. Biomed. Health Inform. **23**(2), 578–585 (2019). https://doi.org/10.1109/JBHI.2018.2831680

13. Long, J., Shelhamer, E., Darrell, T.: Fully convolutional networks for semantic segmentation. In: IEEE Conference on Computer Vision and Pattern Recognition, CVPR 2015, Boston, MA, USA, 7–12 June 2015, pp. 3431–3440. IEEE Computer Society (2015). https://doi.org/10.1109/CVPR.2015.7298965

14. Nguyen, D.M.H., Ezema, A., Nunnari, F., Sonntag, D.: A visually explainable learning system for skin lesion detection using multiscale input with attention U-Net. In: Schmid, U., Klügl, F., Wolter, D. (eds.) KI 2020. LNCS (LNAI), vol. 12325, pp. 313–319. Springer, Cham (2020). https://doi.org/10.1007/978-3-030-58285-2_28

15. Oktay, O., et al.: Attention U-Net: learning where to look for the pancreas. CoRR abs/1804.03999 (2018). http://arxiv.org/abs/1804.03999

16. Peng, C., Zhang, X., Yu, G., Luo, G., Sun, J.: Large kernel matters - improve semantic segmentation by global convolutional network. In: 2017 IEEE Conference on Computer Vision and Pattern Recognition, CVPR 2017, Honolulu, HI, USA, 21–26 July 2017, pp. 1743–1751. IEEE Computer Society (2017). https://doi.org/10.1109/CVPR.2017.189

17. Peruch, F., Bogo, F., Bonazza, M., Cappelleri, V., Peserico, E.: Simpler, faster, more accurate melanocytic lesion segmentation through MEDS. IEEE Trans. Biomed. Eng. 61(2), 557–565 (2014). https://doi.org/10.1109/TBME.2013.2283803

18. Ronneberger, O., Fischer, P., Brox, T.: U-Net: convolutional networks for biomedical image segmentation. In: Navab, N., Hornegger, J., Wells, W.M., Frangi, A.F. (eds.) MICCAI 2015. LNCS, vol. 9351, pp. 234–241. Springer, Cham (2015). https://doi.org/10.1007/978-3-319-24574-4_28

19. Sorokin, A.: Lesion analysis and diagnosis with mask-RCNN. arXiv preprint arXiv:1807.05979 (2018)

20. Tschandl, P., Rosendahl, C., Kittler, H.: The HAM10000 dataset: a large collection of multi-source dermatoscopic images of common pigmented skin lesions. CoRR abs/1803.10417 (2018). http://arxiv.org/abs/1803.10417

21. Woo, S., Park, J., Lee, J.-Y., Kweon, I.S.: CBAM: convolutional block attention module. In: Ferrari, V., Hebert, M., Sminchisescu, C., Weiss, Y. (eds.) ECCV 2018. LNCS, vol. 11211, pp. 3–19. Springer, Cham (2018). https://doi.org/10.1007/978-3-030-01234-2_1

22. Yu, F., Koltun, V.: Multi-scale context aggregation by dilated convolutions. In: Bengio, Y., LeCun, Y. (eds.) 4th International Conference on Learning Representations, ICLR 2016, San Juan, Puerto Rico, 2–4 May 2016, Conference Track Proceedings (2016). http://arxiv.org/abs/1511.07122

23. Yu, Z., et al.: Melanoma recognition in dermoscopy images via aggregated deep convolutional features. IEEE Trans. Biomed. Eng. 66(4), 1006–1016 (2019). https://doi.org/10.1109/TBME.2018.2866166

24. Yuan, Y., Lo, Y.: Improving dermoscopic image segmentation with enhanced convolutional-deconvolutional networks. IEEE J. Biomed. Health Inform. 23(2), 519–526 (2019). https://doi.org/10.1109/JBHI.2017.2787487

25. Zou, J., Ma, X., Zhong, C., Zhang, Y.: Dermoscopic image analysis for ISIC challenge 2018. CoRR abs/1807.08948 (2018). http://arxiv.org/abs/1807.08948

Correction to: ECG-Based Arrhythmia Detection Using Attention-Based Convolutional Neural Network

Renxing Zhao ⓘ and Runnan He ⓘ

Correction to:
Chapter "ECG-Based Arrhythmia Detection Using
Attention-Based Convolutional Neural Network"
in: J. Zeng et al. (Eds.): *Data Science*, CCIS 1451,
https://doi.org/10.1007/978-981-16-5940-9_37

In the originally published version, in chapter "ECG-Based Arrhythmia Detection Using Attention-Based Convolutional Neural Network" the author referred to a wrong table – Table 5. The citation at the bottom of page 498 was changed from Table 5 to Table 3.

The updated version of this chapter can be found at
https://doi.org/10.1007/978-981-16-5940-9_37

Correction to: Research and Simulation of Mass Random Data Association Rules Based on Fuzzy Cluster Analysis

Huaisheng Wu, Qin Li, and Xiuming Li

Correction to:
Chapter "Research and Simulation of Mass Random Data Association Rules Based on Fuzzy Cluster Analysis" in: J. Zeng et al. (Eds.): *Data Science*, CCIS 1451, https://doi.org/10.1007/978-981-16-5940-9_6

The originally published version of chapter 6 contained a few errors: the name of the third author was spelled wrong, the acknowledgment section was erroneously omitted. The name of the third author has been corrected as "Xiuming Li" and the acknowledgement section has been added.

The updated version of this chapter can be found at
https://doi.org/10.1007/978-981-16-5940-9_6

Author Index

Printed in the United States
by Baker & Taylor Publisher Services